WEAPON TESTS AND EVALUATIONS

PETER G. KOKALIS

WEAPON TESTS AND EVALUATIONS

THE BEST OF *Soldier Of Fortune*

Weapon Tests and Evaluations: The Best of Soldier Of Fortune
by Peter G. Kokalis

ISBN 1-58160-122-0
Printed in the United States of America

Published by Paladin Press, a division of
Paladin Enterprises, Inc.
Gunbarrel Tech Center
7077 Winchester Circle
Boulder, Colorado 80301 USA
+1.303.443.7250

Direct inquiries and/or orders to the above address.

Unless otherwise noted, all photography is by Peter G. Kokalis.

Visit our Web site at www.paladin-press.com

TABLE OF CONTENTS

Introduction . . . 1

Section 1
El Salvador . . . 5

ARMS AND THE ATLACATL: *SOF* Trains Salvadoran Immediate Reaction Battalion 7
WEIRD WEAPONS IN EL SALVADOR: *SOF* Uncovers What's Shooting Whom 13
ATLACATL ASSAULT:
SOF in Combat with El Salvador's Elite Immediate Reaction Battalion 21
RETURN OF LA PANTERA ROSA: *SOF*'s Tech Editor Trains Salvo SWAT 28

Section 2
Afghanistan . . . 37

RAIDERS OF THE LOST GRENADE LAUNCHER: Staff Tests USSR's AGS-17 in Darra 39
SOVIET AKS-74: Kokalis Debunks Russian Rifle 45

Section 3
China . . . 53

Guns Behind the Great Wall
Part 1: SOF SCOOPS THE CIA . . . AGAIN 55
Part 2: CHINA'S HIND-KILLER HMGS 64
Part 3: CHINA'S LOW-SIGNATURE SMGS 71
Part 4: CHINA'S HANDHELD TANK KILLER 78
Part 5: CHINA'S GENERAL PURPOSE MACHINE GUNS 83
Part 6: SETTING THE RECORD STRAIGHT ON THE TYPE 63 91
Part 7: PRC PISTOLS: FROM HOT TO PUNY 96
Part 8: PRC PINEAPPLES 106

Section 4
Africa . . . 111

DEATH FROM A DISTANCE: South Africa's Homegrown G5/G6 155s 113
GUNS OF OVAMBOLAND: *SOF* Tech Editor T&Es 101 Battalion 119

Section 5
Full Auto . . . 127

UZI REBORN: Vector Arms' New Model HR4332 129
M60: The Great GPMG SNAFU 138
MAC ATTACK: Poor Boy's SMG Has Indigestion 150
ALL-OUT SHOOT OUT: Knob Creek's Full-Auto Reunion 156
STONER'S SUPER 63: Legendary Weapon of Navy SEALs 160
RUSSIA'S BUFFALO BURP GUN:
New Bizon SMG Conceived by Kalashnikov, Designed by Dragunov 172
RUSSIAN KLIN/KEDR SUBGUNS 178

Section 6
Sniper Rifles . . . 185

COMBLOC SNIPER RIFLES: Crosshairs of the Warsaw Pact 187
STEYR SCOUT RIFLE: A Gun for All Seasons 195
GALIL'S NEW SNIPER RIFLE:
Israelis Take After Sovs and Design Mass-Produced Sniper System 202
THE DEFINITIVE SNIPING RIFLE: Accuracy International's Incredible AW/AWP 207
SHORT BARREL, LONG RANGE: Steyr's Compact Sniper 217
20-MIKE-MIKE MAYHEM: Aerotek's Shoulder-Fired Sniping Cannon 223
AS FAR AS YOU CAN SEE: The 1,000-Yard Chandler Rifle 228

Section 7
Rifles . . . 235

THE M16A2: The Final Verdict 237
HECKLER & KOCH'S REVOLUTIONARY G-11:
Caseless Gun Highlights ADPA's International Symposium on Small Arms 248
ISRAEL'S DEADLY DESERT FIGHTER: *SOF*'s Kokalis Evaluates Galil's AK 252
SMALL ARMS OF THE PERSIAN GULF WAR 258
SILENT SIX-SHOOTERS 265
RETURN OF THE BAR: Old Soldiers Never Die 271
BEST OF THE BULLPUPS?: *SOF* T&Es Vektor's CR21 278
KALASHNIKOV'S KOMBAT KLASSIC: *SOF* Celebrates the 50th Anniversary of the AK47 285
KALASHNIKLONES: A Consumer's Guide to AKs 288

Section 8
Handguns . . . 297

BERETTA'S 93R MACHINE PISTOL: A Burst Controlled Blaster . . . 299
CZ 83: Czech Pistol Checks Out . . . 304
TITANIUM BULLS: Taurus' Ultra-Lite Guns Lead the Herd . . . 310
PLASTIC PERFECTION IN .45 ACP: Glock's Model 21 . . . 319
BLACK AVENGER: Kokalis/Novak Custom Colt Commander . . . 325
RUGER'S NEW 22/45: Pop Gun in the Body of a Big Bore Boomer . . . 331
H&K'S UNIVERSAL SELF-LOADING PISTOL:
One Platform, 10 Variants Equals *Volks*handgun for the '90s . . . 337
BELLY GUN INS & OUTS: Half-Pint Wheelguns for Backup and Personal Defense . . . 345

Section 9
Shotguns . . . 355

BENELLI'S BLASTER: Modification Improves Revolutionary Shotgun . . . 357
ITHACA MAG-10: Big-Bore Blasts Birds But Not Bad Guys . . . 362
LOVABLE, YET LETHAL: Scattergun Technologies' "Politically Correct" Combat Shotgun . . . 368

Section 10
Training . . . 375

HITTS AND MYTHS: Awerbuck's Course on Cutting Edge of Sniping . . . 377
WHO'S AFRAID OF THE DARK?: Shooting in the Shadows at Thunder Ranch . . . 384

Sources . . . 393

About the Author . . . 397

Dedication

To my wife, Matina, faithful companion of 40 years,
who patiently waited at home with our children,
while I ricocheted from adventure to adventure

"They also serve who only stand and wait."
—John Milton

To Robert K. Brown, Editor/Publisher and founder,
Soldier Of Fortune Magazine,
who, of course, made all of this possible

And to El Salvador's Atlacatl Battalion,
especially those who didn't make it,
we were the best of the best

"Go tell the Spartans, thou that passeth by,
That here, obedient to their laws, we lie."
—Simonides of Ceos—Thermopylae

INTRODUCTION

This is my 20th year at *Soldier Of Fortune* magazine. It's been one hell of a roller coaster ride. Enough laughter and tears to fill several lifetimes. Laughter and camaraderie with soldiers on a number of battlefields. Tears over too many comrades fallen in a cause I dedicated my life to: the destruction of the Evil Empire.

I was a charter subscriber to *SOF*. Mine was one of the checks Robert K. Brown, *SOF's* Publisher, kept in a shoe box under the front seat of his car waiting to deposit until he was certain his new magazine was definitely a go. In 1981, seven years after the first issue of *Soldier Of Fortune* appeared on the newsstands, I was the informal marketing director for Mike Dillon's relatively new company, Dillon Precision, Inc. I had prepared press releases and photos on our mainstay, the powerful RL1000 progressive reloader and also the new, economical RL300. I sent this material to all the popular gun press magazines and to *SOF* as well.

When *SOF* printed the information, but transposed the photos, I wrote a personal letter to Bob Brown correcting this. I also commented that I had noticed that Chuck Taylor had been dropped from the masthead and was no longer writing for *SOF* (he had in fact taken a position at Jeff Cooper's American Pistol Institute). I stated that I was more than a little familiar with military small arms and also a professional advertising copywriter and that I felt I could contribute to the magazine.

I received a cagey letter in return, stating that *SOF* was looking for a "number" of writers in this field and I should submit a list of prospective topics. I did and the subject selected was the UZI submachine gun, my first article for *SOF*, which appeared in the September 1981 issue. Two months later, the article entitled "Doc Dater's Deadly Devices" was published. Phil Dater stated many times in subsequent years that this article provided the impetus for his business as a sound suppressor designer and manufacturer. Bob Brown flew up to Albuquerque, New Mexico, to pose for the pictures used in this piece and that's really where our close friendship began. My article on Jim Leatherwood's submachine gun appeared in the January 1982 issue. By that time I was on the masthead as a contributing editor. The inside back cover of the January 1982 issue contained a photo of me shooting my water-cooled Vickers at the first of the now well-known Firepower Demos I put on each year at the *SOF* Convention. In March 1982 I started up the Full Auto column again. I have now written about 150 of these columns alone and close to a thousand articles in total in *SOF, SURVIVE,* and *Fighting Firearms.*

Within the pages of those thousand pieces of print, I have brought more exclusives to *SOF's* readers than any other writer on small arms. Some of them included the first detailed exposure of the M249 SAW, the Spanish CETME MG82, Soviet body armor, the Russian AGS-17 30mm automatic grenade launcher, the Russian AKS-74 assault rifle and 5.45x39mm ammunition, the Soviet RPK-74 SAW (both this, the AKS-74 and the AGS-17 were tested and evaluated in Afghanistan), the Taiwanese T65 assault rifle, the first test and evaluation of the Glock pistol in the English language (October 1984), the M16A2, an eight-part series on the small arms of China, the South African SS-77 GPMG and Mini SS SAW, the Soviet 7.62x62.8mm low-signature assassination cartridge, the Combloc M26A2 shoulder-mounted 30mm grenade launcher, Reed Knight's suppressed revolver, South African CP1 and SP1/2 pistols, South African Neostead shotgun, Russian Bizon and KLIN/KEDR submachine guns, Glock Models 26, 27, 29 and 30, South African 20mm anti-materiel sniper rifle, Vektor 40mm automatic grenade launcher, South African CR21 bullpup assault rifle, and numerous others. I have written articles about everything from pen guns to 155mm self-propelled artillery, with a few armored fighting vehicles thrown in as well.

To obtain these stories I went into four war zones—Afghanistan, El Salvador, Southwest Africa/Angola and Bosnia Herzegovina—and about a dozen other countries throughout the world. I firmly believe that you cannot write about small arms with any degree of credibility unless you have actually made your bones and in fact seen the elephant. While there are a substantial number of competent writers in the fields of game hunting and sport shooting, the vast majority of those who write about combat arms and tactics for the popular gun press are little more than advanced hobbyists with no bona fides whatsoever. A few are even worse. Having no military experience but seeking law enforcement credentials, they have joined the reserve forces of small police departments and then proceeded to proclaim themselves as the genuine article. There is a minuscule number of writers out there with legitimate military and/or law enforcement backgrounds. A few that come to mind are Ken Hackathorn, Clint Smith, Chuck Karwan, Roy Huntington, Gary Paul Johnston, Jim Wilson, Walt Rauch and Louis Awerbuck. They are a good lot, but oh so small.

The others, knowing nothing about wound ballistics and having no formal training in trauma medicine or human physiology, tout magic bullets that will get a police officer killed in the real gunfight that these jaspers have never seen nor ever will. They discuss combat tactics, fighting at night, and weapon selection and yet have less time on active duty in any branch of the military than most of you reading this have had in the pay line. They have, in short, never fired a shot in anger, never been shot at or hit, and yet write about it profusely and continuously.

But what credentials can I lay upon the table to establish my credibility in the field of small arms and wound ballistics that would justify my criticism of these "gunzine wannabes?"

First of all, to truly understand the technology of small arms and their munitions, you must have a proper academic background. My graduate studies program, which led to an M.S. degree, emphasized advanced mathematics (i.e., calculus), chemistry, physics, and mechanical engineering. This has provided me with the ability to interface with small arms project engineers in arsenals throughout the world. Without a thorough understanding of Newtonian physics, you cannot really understand the mechanisms of complex automatic weapons systems. Without a solid foundation in chemistry, you cannot adequately cope with the technology of modern small arms ammunition. I have been reloading ammunition for more than four decades and that was certainly no substitute for my formal training in chemistry.

I enlisted in the "brown shoe" army four decades ago and had a regular army (RA) prefix before my serial number (not my Social Security number!). Serving in the area of technical intelligence, I went places and did things that still cannot be discussed. Suffice to say, I worked with captured enemy personnel, small arms and munitions.

In May 1982, a year after joining the *SOF* staff, I left for Afghanistan. It was there I learned that the mujahideen were neither born with a rifle in their hands nor could they shoot worth a damn, contrary to folklore filling the mainstream media at the time. After some adventures as serious as a heart attack, I left under some rather hilarious circumstances. On our last day in Peshawar, Pakistan (near the Khyber Pass), *SOF's* foreign correspondent, Jim Coyne, decided I had packed far too many inert Mills Bombs, 5.45x39mm cartridges, AKM bayonets, and other militaria in my luggage to ever get through Pakistan customs. I begrudgingly agreed and we spent the afternoon throwing Mills Bombs and other munitions onto the fairway of the British golf course adjacent to the hotel.

I never really felt committed to the mujahideen cause, as they were Muslims and we were Christian infidels. My own personal war was to come in El Salvador, where I first traveled in the fall of 1983. From that time until 1992, I made a total of 21 trips to El Salvador, training numerous Immediate Reaction Battalions, the Airborne Battalion, and the Special Reaction Team of the *Policia Nacional*. However, my most poignant memories are those associated with my long tour of duty with the famed Atlacatl Immediate Reaction Battalion. So feared were they by the FMLN Marxist terrorists that their demobilization was made part of the peace treaty. When that document was signed and the United Nations peacekeeping forces moved into El Salvador, "mercenaries" like me became persona non grata. I went back only once after that just to visit. A police helicopter pilot flew me over the former Atlacatl *cuartel* and while the full-size bronze statute of the Pipil Indian holding an M16 rifle above his head and the wall containing the bronze plaques with the list of our dead remained, there was no other reminder that the Atlacatl had ever existed. I wept bitterly when I thought of the battalion's dead as we flew back to San Salvador. However, in retrospect they were more likely selfish tears for a war I loved that was gone forever. I have never again felt as vibrant, as alive, as charged as I was in the bush with the Atlacatl. Life back in the world can never approach the twin emotions of excitement and dread, or match the senses of adventure and foreboding found on the battlefield. To be a warrior you must go to war.

Yet, war has its lighter moments. In the field I have always been partial to hand grenades. Although they account for more accidental injuries and deaths than any other weapons in the infantryman's arsenal, they can effectively complement the rifle and machine gun. I would much rather carry half dozen grenades than a pistol. During the early years of the war, when I was with the Atlacatl, we were one day preparing a relief column of several companies to assist the main body of the battalion, which was in heavy contact with the FMLN in Usulutan province. Approaching the map table where Lt. David Koch was briefing the platoon leaders, I bent over to examine the topographic map, and one of the six Argentine FM GME-FMK2-MO hand grenades I had hung on my web gear fell off to the ground and rolled between Koch's legs. Turning calmly and deliberately toward me, he said, "Well Peter, everyone in the battalion knows you are crazy about grenades and tries to stay away from you on operations, and now you are trying to blow my balls off."

On another occasion during the latter part of the war, when I trained the *Equipo de Reaccion Especial* (*ERE*—Special Reaction Team) *de La Policia Nacional,* we found ourselves on a range immediately adjacent to a police substation on the outskirts of San Salvador. The substation's phone lines ran above the range about

100 yards downrange from the firing line. A member of my training team had never fired an M203 40mm grenade launcher before, and during the lunch break one of the police officers volunteered his M16 to which was attached an M203 (this unit was equipped as light infantry and often so deployed). I told the instructor to aim for the top of some volcanic ash about 150 yards downrange. He had little luck in hitting the target as his strikes were all impacting far too low. The cop to whom the M16/M203 was issued asked for it back, stating that he would demonstrate how to hit the designated target. As he fired, I could see the HE round sailing slowly downrange with impossible ill luck. I stood frozen as I saw the grenade head straight for the phone line and said "Oh shit!" at about the same time the grenade exploded and cut the line in two. It took only a few seconds for the entire substation to empty, as all their phones went completely dead. Since I was OIC, they asked me what had happened. I told them I couldn't possibly know as we were just shooting the M203 and it was, of course, completely impossible to hit a target that small with a 40mm grenade round.

Subsequent trips to Southwest Africa/Angola and Bosnia Herzegovina in the early 1990s never brought the same rush as my years in El Salvador. From Bosnia Herzegovina I returned with only angry memories. I worked with Roman Catholic Croatians in their war against Eastern Orthodox Serbs. As a Greek Orthodox Christian, this was not an easy task. Furthermore, the HVO troops were arrogant, undisciplined, and knew far less about deploying machine guns in combat than they thought they did.

To train others in the art of war, you must both know war from the trenches and undergo constant training from others, both to keep the sharp edge and be exposed to the ever-evolving tactical concepts of combat at the down and dirty level. Several have asked why an "expert" (God, how I loathe that word) like me would *need* to participate in training at a firearms school. The answer is simple: for the same reason tennis and golf pros constantly train under other tennis and golf pros. You cannot observe yourself while shooting, but the professional firearms instructors under whom I train can instantly detect slight nuances of incorrect movement that need to be reprogrammed. To that end, I have eight graduation certificates from Louis Awerbuck's Yavapai Firearms Academy Ltd., six from Gunsite Training Center, Inc., and nine from Thunder Ranch. I attend a minimum of two firearms training courses per year, more if I can fit them into my schedule.

I have been a technical consultant to several arsenals, including 205 Arsenal in Taiwan and LIW in South Africa. I was one of two primary technical witnesses for Glock in their lawsuit against Smith & Wesson.

I am a member of the board of directors of the International Wound Ballistics Association and studied wound ballistics for many years under the skilled medical guidance of Dr. Martin L. Fackler, the generally acknowledged father of modern wound ballistics research. This has involved a mighty struggle against a group of gun writers championing proven counterfeit statistics. Unfortunately, these petty con artists have control of the popular gun press in a manner analogous to the liberal left's control of the mainstream media.

So what does all this make me? An "expert"? God, please don't use that noun to describe me. Maybe an "authority," but one who is continually expanding his database at all levels in his field of research.

Whatever amount of sweat, tears, and blood it has taken to write these articles, I hope you enjoy this potpourri that Paladin Press has assembled. And until we meet again, keep your heads down and watch "six."

Peter G. Kokalis
February 2000
Phoenix, Arizona

Section 1

El Salvador

ARMS AND THE ATLACATL

SOF Trains Salvadoran Immediate Reaction Battalion

Before Kokalis led arms repair instruction, Atlacatl Bn. could expect at least 20 percent of M60s out at any one time.

"Papa Kilo, Papa Kilo, this is Juliet Echo, over."

"Juliet Echo, this is Papa Kilo, over."

"Papa Kilo, move out, over."

"Juliet Echo, wilco, out."

Fifteen minutes after I hung up the handset on the PRC-77, every member of my 24-man patrol, except for myself and the point man, was dead. Killed in a classic "line" ambush in the dense jungle just seven klicks north of the Atlacatl Battalion *cuartel* (headquarters). Blown away by claymores and small-arms fire. On paper at least. For this was the final day of an intensive three-week-retraining cycle I had just conducted for the elite Atlacatl Immediate Reaction Battalion of the Salvadoran Army.

Since its inception on 1 March 1982, the Atlacatl Battalion has seen more combat than any unit in El Salvador. Composed entirely of battle-hardened volunteers, the Atlacatls have been commanded from their origin by Lt. Col. Domingo Monterrosa, a charismatic leader. The battalion has spent almost 80 percent of the past year on combat operations. This leaves little time for retraining, usually fitted in on a catch-as-catch-can basis that is quickly suspended when the battalion is called out.

Working closely with the battalion XO, Maj. Armando Azmitia, I had timed my arrival to correspond with the battalion's return from a two-month operation in guerrilla-infested Morazan Province. Hopefully, we'd have two to three weeks in the cuartel before having to move out on another operation.

The Atlacatl Battalion supply room (or *almacen* as it is called in Spanish) is a treasure-trove of undocumented technical intelligence.

After logging in the donated supplies, I was free to prowl with 1st Lt. David Koch, commander of the 7th Company, who was assigned to assist me and interpret during my stay with the battalion. Hours passed as we poked through crates and racks of captured materiel.

A wooden crate of surgical instruments captured from a guerrilla hospital contained a number of items marked "H.C.M.S.A. de C.V. (*Hospital Centro Medico Sociedad Anonima de Capital Variable*), which is the general hospital in San Salvador. Stolen? Not hardly. More likely a gift to the guerrillas from communist sympathizers on the hospital staff.

But it was the racks of captured weapons that produced the most intriguing information. How about an Erma M1 Carbine look-alike in .22LR? There were 10 Cuban Army FALs, all with the Cuban coat-of-arms removed. There was also an FN-produced heavy-barrel FAL with the Venezuelan crest on the right side of the upper receiver's magazine well, a gift to the Nicaraguan guerrillas from the Venezuelan government, no doubt. The American press corps was not the only early paramour of the Sandinista movement.

Resting next to a 1937-dated Czech Brno Mauser rifle were two BARs. One was a Model 1918 Winchester brought up to M1918A2 configuration and the other was a M1918A2 manufactured by the Royal Typewriter Company, Inc. As they both had been fitted with the late Korean War-vintage three-prong flash suppressor, it's a sure bet they were shipped to Nicaragua from stores sent to Vietnam in the early 1960s.

The six M16s I examined all carried very low serial numbers and represented a complete cross section of M16 evolution. Four of them carried the commercial Colt AR-15 logo on the lower receiver with "Property of U.S. Govt." marked below. Serial No. 142098 was marked XM16E1! Serial No. 234448 was marked M16. Serial Nos. 1251401 and 1644015 were marked M16A1. The final two M16A1s were manufactured by Hydra-Matic Div., General Motors Corp., and were both numbered in the low three-million-digit serial range. Without question, all of these rifles had been shipped from North Vietnam.

A Madsen LMG in caliber 7x57mm lay next to a ChiCom RPG-2 and a basket full of rockets and boosters. Stacks of Madsen M50 submachine guns (the infamous "banana-peel" gun—if the barrel nut loosens while you're firing the M50, the receiver can open up and drop its guts at your feet) collect dust next to old FN HiPowers. The Madsens and HiPowers all carry the Salvadoran crest.

But I digress.

One of my major objectives was to train some battalion armorers. To that end I had brought down a complete gunsmith's tool and supply kit (including aircraft safety wire pliers and stainless-steel wire for the M60). A sergeant and corporal were assigned to me for this pur-

Atlacatl Bn. patch.

pose, as well as a permanent room in the cuartel. Both men were enthusiastic and highly motivated. Before this, all defective small arms were sent to the *Maestranza* (central ordnance depot) for repair, which often took one to two months. In addition, once a month a truck from the Maestranza drove to the battalion cuartel to effect minor repairs.

Before I arrived, the typical chain of futility was as follows. There were no headspace gauges for the .50-cal. Browning M2 HB machine guns. Incorrect headspace adjustment frequently caused split cases. As there were no .50-cal. broken-case extractors, the barrel had to be sent to the Maestranza. The complete scenario could take up to two months.

Since there are no spare barrels, the gun was lost to the battalion during this entire time frame. So, what's one gun? Plenty. The small, but excellent, Salvadoran Air Force and the artillery regiment are spread too thin. They cannot be depended upon for consistent fire support. The battalion's supplies of ammunition for the 81mm and 60mm mortars, 90mm recoilless rifles and M79 40mm grenade launchers (no beehive rounds allowed, only HE) are running precariously low.

Therefore, the battalion's major fire-support weapons are its machine guns! During one live-fire training session I had five M60s down, out of 26 guns, for one reason or another, during the first three minutes of fire. That's a 20-percent loss in firepower.

To reduce down-time on the battalion's all-important machine guns, I initiated a three-echelon repair system for use during combat operations. The armorers and I would move several times daily, via the helicopter resupply runs, from company to company in the operational area. Weapons that could be repaired in the field would be put back into action immediately.

Those that required more extensive maintenance would be brought back to Ilopango Airport where a battalion armorer's shed would be set up. If the necessary parts and service were not available at this level, the weapon would be taken to the Maestranza and the parts obtained—or services effected on the spot. By the next morning, the gun would be returned to action on the first slick out.

The next several days were spent instructing the armorers in disassembly-assembly, troubleshooting, advanced maintenance and repair procedures on the Browning .50-cal. M2 HB, M60 GPMG, M16A1, M1D Garand sniper rifle, G3 and M79 grenade launcher. These men must eventually be trained to work on the 81mm and 60mm mortars and the 90mm recoilless rifle also.

No less true today than 60 years ago are Maj. Walter C. Short's words, in his classic *Employment of Machine Guns*:

"Of all of the developments of the World War none is more striking than the remarkable increase in the use of machine guns. Tactically nothing is of greater importance to the infantry and cavalry than the proper coordination and cooperation of the rifle companies and troops with the machine guns that support them."

Starting with this basic tactical canon, I scheduled a section weapons seminar for all the officers and NCOs of the Atlacatl Battalion. Topics included an in-depth examination of such timeless concepts as the cone of fire, beaten zone, plunging versus grazing fire, frontal, flanking, oblique and enfilade fire, techniques of fixed fire, traversing fire, searching fire, and the swinging traverse and free gun systems as applied to the M3 and M122 tripods. The battlefield rates of fire required of the section machine gun: single-shot (used with the .50-cal. M2 HB only, when engaging targets beyond 800 meters), sustained fire (called suppressive fire when used to hinder enemy movement), rapid fire and cyclic rate of fire (employed only to win the fire fight or beat off a determined enemy attack) were defined and specific applications given. Although only marginally useful in the modern battlefield of fire and movement, defilade fire and techniques were also described. The seminar ended with an analysis of the two roles of the machine gun during attack: fire support and assault fire.

The next several days were devoted to the M60 gun crews. During the first session, which stressed disassembly/assembly and maintenance, a specimen was selected at random and inspected. The entire weapon was filthy. The gas system was so badly fouled that the piston was frozen in place and had to be driven out with a punch and hammer. The gas-cylinder nut and extension and gas-port plug were likewise locked tight by fouling. The bolt-plug pin and cocking-lever guide screw were missing. The main spring had so many flat spots that it appeared square.

The black anodized finish was completely worn off the top cover, as it was on all the battalion's other M60s. I immediately started the armorers painting all the top covers with flat-black stove paint I had brought for this purpose.

Instruction in the correct prone, kneeling and standing assault position followed. They have only one M122 tripod, so all support fire with the M60 must be delivered off the bipod.

The M60 gunners were issued bore cleaner, LSA oil, patches, cleaning rods, brass bore brushes, broken-case extractors and combination tools for the very first time—with an explanation of their use. I discussed immediate action for clearing stoppages.

I also initiated a new battalion SOP for carrying the M60 on patrol movements with a 25-rd. teaser belt; 100-rd. belts will be carried and used in the issue cardboard box in conjunction with the cloth bandolier only. Carrying 100-rd. belts across the chest, exposed, like Pancho Villa, is very macho—and very stupid. It exposes the cartridges and M13 links to dirt, corrosion and damage, causing misalignment of the link tab and cartridge groove and frequent stoppages. It is no longer tolerated in the Atlacatl Battalion.

The M60s were zeroed as follows. At 25 meters a one-inch-square paster was used as a target. Flipping up the rear sight, the windage was centered by eye-balling. Firing single shots only, the elevation knob was adjusted up or down until the group impacted on the paster. The graduated range scale was disregarded during this adjustment. After the shot group impacted on the paster, without moving the elevation adjustment knob, the screw on the graduated range scale was loosened and the scale slid up or down until the rear sight was set for 500 meters. The graduated range scale's set screw was locked down since the M60 GPMG had then been zeroed for all ranges.

After this we were able to commence live-fire drills. The crews' range estimation was poor and work in this area needs to be expanded. They were taught fixed, traversing, searching and traversing/searching fire techniques. The assistant gunners were instructed to spot down-range impact and call corrections to their respective gunners. They had never done this before. Three guns went down with failures to extract from fouled gas systems, another had the firing pin installed backward and still another had a cracked chamber.

Our finale with the M60, several days later, was a GI party held in the *Plaza de Armas* (the cuartel's open pavilion). Hosting were the ever-present Lt. Koch, the armorers and myself. The somewhat reticent guests were the crews and their M60s. The men used their newly-issued tools for the first time as well as a vat of cleaning solvent, and—overall—did an excellent job. I walked away as the armorers safety-wired the gas systems, feeling confident most of our problems with this beast had been solved.

Working with the Browning .50-caliber M2 HB machine gun, after the M60, "'tis a consummation devoutly to be wish'd" (to quote Hamlet completely out of context). When headspaced and timed and fed a diet of properly linked ammunition, the Ma Deuce will rattle on forever. And, just try to reassemble any John Browning designed machine gun with parts in backward!

The Atlacatl Battalion's M2 HBs are all AC Spark Plugs manufactured during WWII. Arsenal-reworked and refinished in the late 1960s, they have been upgraded by addition of the bolt-latch assembly for single-shot fire and by minor improvements such as the newer barrel locking spring and substitution of the old sear stop and pin with the modified accelerator stop and lock.

Unfortunately, the dovetailed slot on top of the receiver and to the right of the rear sight has been removed. This prevents installation of a telescopic sight, essential when this machine gun is employed as a sniper weapon on the M3 tripod. With a proper scope (not the Telescope,

M1, manufactured by the Prefix Corp. in 1942 for this dovetail groove; although intended for precise long-range shooting, it was little more than a low-powered dial sight), single-shot sniper fire can be very accurate at ranges up to 2,000 meters.

The first meeting of the .50-cal. (*La Cincuenta*) gun crews was attended by the battalion's senior sergeants as well. I asked the sergeants if any were familiar with M2 HB. One who had been to Panama volunteered; he never got past headspacing because he did not know how to use the gauge. I explained headspace and timing adjustments, and sergeants and gun crews alike practiced on the five guns we had set up in the Plaza de Armas. The crews then worked on quick employment drills, learning to place the gun into and take it out of action with precision, speed, skill, and teamwork. Both two- and three-man carries were performed. The gunners were instructed to fire from the prone rather than sitting position.

During the live-fire exercises, range estimation was again stressed (even more important with the M2 HB with its very real potential for long-range killing). Use of the rear sights was explained. When these guns were refinished, many of the parts were sandblasted before Parkerizing and the rear sight markings are indistinct on several of the guns. These crews were also taught fixed, traversing, searching and traversing/searching fire techniques, as well as the swinging traverse and free gun methods as applied to the M3 tripod. Once set up properly, no down time was experienced with the Browning guns.

Placing five Ma Deuces together on a ledge 30 meters above an immense lava field, suitable targets were located at ranges of 1,200 to 1,500 meters. The guns were fired single-shot, but in rapid sequence, like an artillery battery. In this fashion, a sustained, full-auto effect is achieved with high hit probability at extremely long ranges.

Successful employment of squad weapons is to no small degree a consequence of the effective implementation of proven, modern concepts of small-unit tactics modified to fit a specific combat environment.

Ambush is an important counterguerrilla measure since it not only forces the guerrillas to engage in combat at places and times unfavorable to them, but it curtails the freedom of movement on which their success in El Salvador so greatly depends. The final portion of my training schedule covered ambush and counter-ambush techniques. In addition to general terminology, the fundamentals of successful ambush (surprise, coordinated fire and control) were discussed in detail, as well as three-point ambush formations: the "Line," "L," and "T." I was assisted by John Early, who conducted training in demolition ambush, convoy ambush and convoy immediate action counter-ambush drills for vehicles.

Participating were 126 NCOs. Many of these men have received training from U.S. Army personnel in Panama.

Las Cincuentas, *favored by both Kokalis and Salvadoran Atlacatl Battalion. Now steadily employed in El Salvador, thanks to* SOF *trainers.*

They are highly motivated combat veterans of many operations. I could fault them in only two areas. Sound discipline during ambush operations was poor, and on patrol far too many individuals used the carrying handle or sling on the M16A1 rifle. It should always be carried in the combat-ready position. British SAS troops remove not only the slings and sling swivels, but the carrying handle from their SLRs. It's just too damn bad that the carrying handle is an integral part of the M16's geometry.

Textbook tactics must often be altered by battlefield realities. The TMs state that in an "L" ambush the support element with the squad automatics should be positioned on the short leg. This will, of course, provide enfilade fire which makes the most effective use of the machine gun's beaten zone.

The Salvadorans, however, have not been trained to use sector limit stakes and do not have M122 tripods on which to mount the M60 GPMG. Some of the gunners are inexperienced, making it possible for them to take out some of their own people in the assault element on the long leg of the formation. I therefore instructed them to position the M60s on the long leg, where they can be controlled by the commander, so that each could deliver flanking fire along the entire killing zone (or when this is not possible, overlapping sectors of fire so that the entire killing zone is covered).

American journalists, while writing their stories in the bar of the El Camino Real Hotel, are quite fond of tacking a 9-to-5 label on the Salvadoran officer corps. The numbers are correct, but, in their drunken haze they have transposed them. First Lt. David Koch and I, along with all the other officers of the Atlacatl Battalion, most frequently worked from 0500 to 2100 while in the cuartel and not on alert. Out in the field on operations the working hours are, of course, extended even more.

Given good and proper equipment and the correct training the Salvadoran Army can defeat the communist guerrillas. They have the will: Let us see to it that they have the means. El Salvador may well be our last chance.

Originally appeared in *Soldier Of Fortune*
January 1984.

Argentine 9mm FMK 3 SMG: Heavy, badly designed, unreliable and, perhaps, the worst SMG ever marketed.

WEIRD WEAPONS IN EL SALVADOR

SOF Uncovers What's Shooting Whom

Take that, BATF! Shop-modified Browning HiPower becomes machine pistol.

Central America presents a fascinating opportunity for the student of military small arms. With an almost continuous history of warfare, these small countries have provided gun merchants with more than a century of lucrative transactions. During *SOF*'s recent expedition to El Salvador, I was able to test and evaluate a most interesting potpourri of weaponry.

The Argentine 9mm FMK 3 submachine gun, manufactured by Fabrica Militar de Armas Portatiles "D.M." Rosario, is one of the worst in both design and execution that I have ever fired. It is blowback-operated and fires from the open-bolt position. The model purchased in large quantity by the Salvadoran government has a retracting wire stock copied from the U.S. M3 submachine gun. Few features on the M3 "grease gun" deserve imitation and the feeble wire stock is certainly not one of them.

The FMK 3s upper and lower receivers are stamped sheet-metal pressings held together by two spring-loaded pins, which are removed for disassembly. The barrel is retained by a threaded cap, which one unscrews—for what seems to be an eternity—for field-stripping. The 11.4-inch barrel has six grooves with a right-hand twist of 1:9.85 inches. The wrap-around bolt encloses 7.1 inches of the barrel. The oversize, cumbersome forearm is made of plastic. The entire weapon has been finished in black baked enamel.

The front sight is an adjustable post and the rear a flip type with 50- and 100-meter apertures. (Because the magazine well is located in the grip assembly, some casual observers have thought they were looking at an UZI.) The magazine is a modern two-position-feed Beretta-type with a 40-rd. capacity. There is a useless grip-safety. The selector has three positions: "S" (safe), "R" (*repetition* = single shot) and "A" (automatic).

This black turd weighs 7.6 pounds empty, more than the M16A1 rifle! The weapon submitted to me was a brand-new, unfired specimen. During the test-fire sequences, I determined the cyclic rate to be approximately 650 rpm. When it can be made to function, the FMK 3 is quite stable in burst fire, as well it should be at that weight.

During the first 40-rd. magazine, the FMK 3 failed to feed twice and failed to extract three times. The weapon also doubled twice when in the semiauto mode. These stoppages continued and I was told they are typical.

I have an unconfirmed report that no less than 14 National Police officers have lost their lives in gun fights as a direct consequence of FMK 3 malfunctions. It seems the magazine well-to-chamber angle is incorrect. The extractor spring also seemed weak. The latter problem can be corrected, but the former is an inherent design failure.

Some of the extraction failures are a result of the erratic, low-powered Argentine ammunition. The best use of these submachine guns would be to distribute all of them with large quantities of Argentine ammunition along trails frequented by the communist guerrillas.

Unfortunately, the Salvadoran Army is replacing all of its FN-manufactured Browning HiPower pistols with those made under license in Argentina (also by Fabrica Militar de Armas). However, unlike the excellent specimens recently brought into the United States by Pacific International Merchandising Corp., the ones I examined in El Salvador had neither slide markings nor any semblance of quality control. The feed ramps and throats were rough and burred, the trigger pulls dreadful and many of the magazines defective. The Argentine HiPowers all have black baked-enamel finish over phosphate—superior to bluing in the humid heat of Central America.

While visiting the Cavalry Regiment, I fired an FN HiPower that had been locally altered

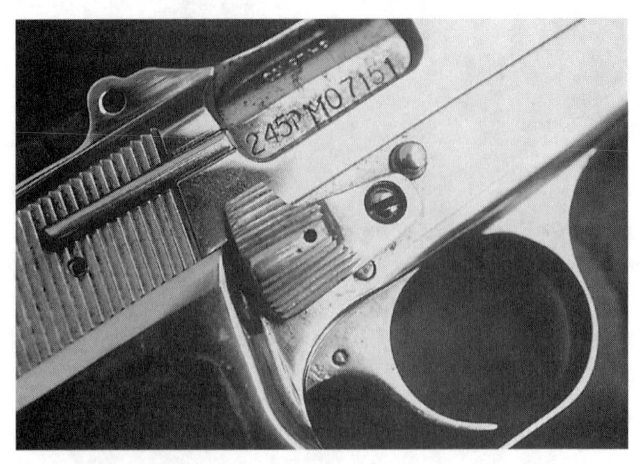

Fire selector switch above right side of trigger on home-grown Browning machine pistol.

7x57mm Madsen LMG used by Salvadoran Army.

Crest shows Salvadoran original issue of recaptured Madsen LMG.

into a machine pistol. A selector lever, fitted to the right side of the frame above the trigger, modified the function of the tripping lever and plunger to permit full-auto fire at a cyclic rate of 1,250 rpm. A 32-rd. magazine had also been fabricated for this pistol. This well-executed, interesting exercise in ingenuity has little practical application. Without either shoulder stock or front foregrip, with no muzzle compensator to reduce climb and without a burst-control mechanism, the weapon is difficult to control, except for the most experienced operators.

The Salvadoran Army's earliest true squad automatic was the Danish Madsen Model 1934 Light Machine Gun in caliber 7x57mm Mauser. It is still encountered in the hands of communist guerrillas and in battalion supply rooms. Introduced in 1902, the Madsen LMG has had a long history with minor users: Having been sold to 34 countries, it saw extensive service in wars, big and small, for more than 80 years.

The Madsen's method of operation is somewhat peculiar and merits close examination. The hinged bolt is similar to that of the lever-action Peabody/Martini rifle. The unusual recoil operation is part short and part long recoil. As the projectile leaves the muzzle, recoil forces move the barrel, barrel extension, and bolt to the rear. A pin on the right side of the bolt moves backward in grooves in an operating cam plate mounted to the side of the receiver. After 1/2-inch of travel the bolt is cammed upward, away from the breech (the "short" portion of the recoil system). The barrel extension and barrel continue to move rearward to a point slightly exceeding the combined overall length of the cartridge case and projectile (the long portion of the recoil system, responsible for the extremely low cyclic rate of about 400 rpm).

After the breech is exposed, an odd lever-type extractor/ejector, mounted under the barrel, is pivoted to the rear, extracting the empty case and ejecting it through the bottom of the receiver. The bolt's operating cam then forces the bolt face to pivot downward, aligning a cartridge feed groove in the left side of the bolt with the chamber. While the bolt and barrel are returning forward, a really strange cartridge-rammer lever, mounted on the barrel extension, is pivoted forward, loading a cartridge.

Just too many things take place during the Madsen's operating cycle and the principle is unsound in a machine gun. Unless the very highest quality ammunition is used, stoppages are frequent and difficult to clear rapidly. And most of the action must be removed to change barrels. Its complex mechanism required extensive machining and the Madsen was expensive to manufacture. The box magazines used on this gun are also prone to failure, as they feed from double column down to single column and the feed lips are in the receiver rather than on the magazine (also a feature of the Johnson M1941 LMG).

For all of its deficiencies, the Madsen was generally well-regarded over its lifespan. It was well-built and accurate, with high hit probability. With its extremely slow rate of fire and low recoil (especially in caliber 7x57mm) there is little to disturb the gunner's aim.

Lt. Col. Steben of the Cavalry Regiment, SOF's Kokalis, and Atlacatl 1st Lt. David Koch pose with Panhard AML 245 Armored Car.

With good ammunition and undamaged magazines, it will do. Although it cannot compare to the Bren or even the BAR, the Madsen's done its share of killing, and then some, from the Russo-Japanese War of 1904 to the present.

Prior to fielding the Heckler & Koch MP5 submachine gun, the Salvadoran military made use of another product from Dansk Industri Syndikat, the Madsen Model 1950 submachine gun (as well as a few of the earlier Model 1946). Innovative and designed to be produced at minimum cost, the M50 receiver is made of two mirror-image sheet-metal pressings, connected by a hinge at the rear and held together by the barrel nut. Unfortunately, this muzzle nut has been known to unscrew during firing sequences, opening the receiver shell and dropping the bolt group onto the ground. This embarrassing predicament has led to its nickname, the "Banana Peel Gun."

In addition, the M50 has a wretched bolt safety, located at the rear of the magazine well, which must be depressed when the weapon is fired. As if that were not enough, when the stock is folded it blocks access to this bolt safety. Otherwise, the M50 is quite unremarkable, but reliable, firing by straight blowback from the open bolt position.

The Salvadoran Cavalry Regiment is equipped with the Panhard AML (*Automittrailleuse Legere*) 245 Armored Car. It is one of the very best. Its 7.62mm NATO coaxial machine gun is one of the very worst. The main gun is the 90mm D921, firing HE and HEAT ammunition up to 1,500 meters in range. The coaxial machine gun for this three-man (commander/loader, gunner and driver) armored car is the AAT 7.62 NF1. This is nothing more than the woeful French AA 52 GPMG chambered in 7.62mm NATO instead of 7.5x54mm French. The vehicular configuration is without butt stock, has a heavy barrel and fires electronically.

The AA 52 (*Arme Automatique* Model 52 = GPMG model 1952) machine guns are delayed-blowback in operation. Designed for ease of manufacture, the receiver body is made of semi-cylindrical tubes welded together. The barrel

AAT NF1 mounted coaxially on gun turret.

Coaxial-configuration AAT 7.62 NF1 removed from Panhard Armoured Car.

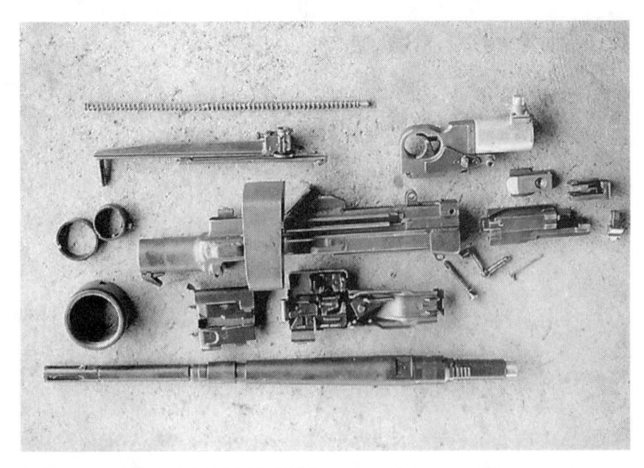

Disassembled AAT 7.62 NF1.

chambers are fluted to ease extraction, as are Heckler & Koch's retarded blowback weapons.

To prevent premature unlocking while chamber pressures are still too high, the AA 52 employs an unusual, troublesome, two-piece bolt. (The bolt's two parts are joined by a fragile T-shaped connecting pin about one inch long. If this little pin is broken or lost the weapon cannot be made to work.) The bolt head contains a lever, the short end of which rests in a receiver slot while the long end bears against the bolt body. Gas pressure on the bolt head forces this lever to rotate, accelerating the bolt body's rearward movement but restraining the bolt head until the lever clears the receiver slot. After extraction the empty case hits two ejectors on the bottom rear of the feed tray and is expelled down through the bottom of the gun.

Headspace is critical in this system and although the receiver bearing surface can easily be replaced, severe deformation is often evident in a bulge in which the case expands into the bullet guide. At the time of its adoption, French designers argued this defect would in fact prevent the cases from being reloaded by their opponents . . . as if anyone else in all the world except the French and their colonies uses, or would want to use, the 7.5x54mm cartridge.

The AA 52 series fires from the open bolt position. The floating firing pin can be forced forward only when the bolt body butts up against the bolt head. As this cannot occur until the lever has rotated into the receiver slot and lockup is complete, firing out of battery is impossible. To reduce wear on the sear surface, a sear trip mechanism similar to that first used on the MG 42 has been incorporated.

The AAT 7.62 NF1 has a cyclic rate of about 900 rpm, acceptable for vehicular hard-mounted machine guns. As expected, hit probability was high with the gun attached to a 5.5-ton tripod. However—as this poor grunt found out—even with the evacuator fan operating, one cannot fire for more than two minutes, due to the unbearable fumes within the Panhard's cramped quarters.

The AA 52 machine guns cannot make use of the usual roller in a reciprocating feedway to operate the feed pawl, since the two-piece bolt stays separated until just before ignition. A cam groove on the bolt holds a lever which operates the feed pawl. Unlike most modern GPMGs which move rounds over one half pitch for each stroke of the bolt, the AA 52 collects a round on the forward stroke of the bolt and positions it for firing as the bolt goes back. Thus the gun must remain cocked before a loaded bullet can be put in place. The AA 52 disintegrating belt, although based on the U.S. M13, is not interchangeable (on this gun). I found the Salvadorans using HK 21 non-disintegrating belts in the AAT 7.62 NF1 with moderate success.

Barrel changes are hazardous. A barrel release catch must be pressed and the hot barrel (no carrying handle on the vehicular version)

Theoretical perfection fails in practice as SOF's Kokalis testifies that the HK21 is the world's most punishing GPMG. ***(Photo: David Koch)***

rotated and pushed forward. The barrels are neither stellite-lined nor chrome-plated.

The excellent Panhard Armored Car deserves a much better coaxial gun than the AAT 7.62 NF1, which ranks in infamy with the U.S. M73/M219. The M240, coaxial version of the famed FN MAG would be just the ticket to permanently collect on some communist laundry.

Prior to the advent of U.S. aid and fielding of the M60, the HK21 was the GPMG of issue in the Salvadoran Army. It has not been the most successful branch on the fine Heckler & Koch tree of weaponry. The HK21 is a belt-fed GPMG in caliber 7.62mm NATO.

In theory the HK21 is the best of all possible worlds. It can feed either the German DM6, U.S. M13 or French AA 52 disintegrating belts, as well as the German DM1 nondisintegrating belt. However, the Salvadorans have had numerous failures they feel were caused by belts, and they reported to me that the Madsen-Saetter nondisintegrating belt yielded the best results. By installation of an adapter, 20-rd. and 30-rd. G3 magazines or 80-rd double-drum plastic magazine can be fitted.

Since the bolt passes over the belt, the belt must be placed links down in the feed tray. Unless a starter tab is used, loading is clumsy since the gun must be cocked and the first round placed in the feed sprocket, which is then rotated to the right until it locks. A curved cam slot on the bottom of the bolt engages an actuator on the feed mechanism. As the bolt moves rearward, the double sprockets are rotated and a new round positioned in the feedway.

Options abound. Changing the barrel, feed plate, and bolt will convert the HK21 to either 5.56 NATO or 7.62x39mm ComBloc. The robust bipod can be mounted either in front of the feed mechanism for greater traverse or at the muzzle for greater hit probability.

The black sheep of the family: HK21.

A tripod and three different vehicle mounts are available and scopes and passive night sights can be fitted. The gun is usually equipped with a 1,200-meter aperture drum rear sight, although those sold to El Salvador have only the G3 rear sight.

The HK21 can be field-stripped in one minute and reassembled in 1.5 minutes. The heavy barrel has an excellent quick-change mechanism. As the trigger mechanism is identical to the G3's, the gun can be fired either full-auto or semiautomatically. In fact, the HK21 has some 48 percent parts interchangeability with the G3.

All is for naught, however, since the HK21 weighs only 17.4 lbs., empty, with the bipod and has a cyclic rate of 900 rpm. Entirely unacceptable, in caliber 7.62 NATO, except for 300-lb. Nubian warriors. This is, without doubt, the most physically punishing GPMG I have ever fired. Counter-recoiling forces throw the gun about like a jackhammer. Before the brain is jarred into an amorphous mass of gelatin, its last thought is inevitably, "When will this nightmare end?"

In addition, it should be noted that, unlike gas-operated squad automatics with adjustable regulators, blowback-operated machine guns

HK21 feed ramp.

HK21 quick-change heavy barrel.

HK21 feed block with twin feed sprockets speckled with powder from dud round which lodged a bullet in the barrel. Fortunately, it was the last cartridge used in the author's test.

offer no power reserve to deal with the increased fouling associated with the sustained-fire role.

The HK21 has been replaced in the Heckler & Koch lineup by the HK21A1, which addresses some of these problems. The weight has been increased and a hooked butt stock with improved buffer mechanism has been added. The magazine feed option has been abandoned, but the feed mechanism is now hinged to allow easier insertion of the belts.

There is some good news from El Salvador. The Mauser Model 57 submachine gun I tested is innovative, reliable and well-executed. Now the bad news: It is the only one in El Salvador. Captured from the communist guerrillas, it's now the property of Carlos E. Cucalon. Carrying serial number 21, it is, in fact, one of only 25 ever made.

The Mauser Model 57 SMG was designed in France during the early 1950s by Louis B. Camillis. Early development of the weapon took place during 1955-56 at the relocated Erma plant near Munich. However, the gun's promoter eventually gave the contract to Mauser-Werke A.G. in Oberndorf/Neckar where it was extensively redesigned.

Pure blowback in operation and firing from the open bolt position, the Mauser 57 exemplifies many of the features we have come to expect in state-of-the-art machine guns. The magazine well is in the handgrip (which is a bit on the large side). The lower receiver assembly is made of black-anodized aluminum alloy. The black plastic grips display the famous Mauser banner. The other components are a combination of stamped sheet-metal pressings and milled forgings. All steel parts are well-finished and blued. A rear grip safety locks the bolt. The selector switch, located on the left side just above the trigger, has three positions: "D" (full-auto), "S" (safe) and "E" (single shot). Safe is the middle position and one pushes forward to go into full-auto or rather awkwardly pulls back for semiautomatic fire.

The telescoping two-piece bolt wraps around the greater part of the 10.25-inch barrel allowing an overall length of only 17 inches, stock folded. The recoil spring is wound around the barrel and rides inside the bolt assembly. A short, heavy buffer spring is positioned at the end of the bolt assembly. The weight, empty, is about 7 lbs.

The butt stock folds over the top of the gun and does not interfere with normal operation of the weapon when in this position. A folding foregrip is useful in controlling muzzle climb during burst fire when firing from either the shoulder or in the hip-assault position. The front sight is adjustable for windage zero by lateral movement in its dovetail. The rear sight is fixed with no protective hood.

The cyclic rate appeared to be 650 rpm. No stoppages were encountered, except as a function of the really bad Argentine ammunition. The 32-rd. magazines are tool-room modifications of single-position feed MP38/40 magazines. The weapon itself was robust and exhibited excellent human engineering for a 1950s design. It's a pity Mauser was unable to elicit any interest in its first serious attempt at a submachine gun after WWII, since the Model 57 showed great promise. But by this time the major contenders were already crowding the submachine gun's limited arena. What I'd like to know is how good old serial number 21 got into the hands of communist guerrillas.

El Salvador has become the elephants' graveyard of military small arms: It need not continue to be that way. The ultimate test and evaluation of small-arms technology—the flame and heat of actual combat—awaits those enterprising enough to offer their wares, on a limited basis, to the Salvadoran military establishment for this purpose.

Originally appeared in *Soldier Of Fortune*
March 1984.

Carlos Cucalon fires Mauser Model 57.

Fore grip, trigger housing, grip and modified MP38/40 magazine of Mauser Model 57.

Mauser Model 57 SMG, serial number 21, covered with camouflage tape for protection from wet Salvadoran climate.

ATLACATL ASSAULT

SOF in Combat with El Salvador's Elite Immediate Reaction Battalion

Pride and security are represented by the front gate of Atonal Bn. cuartel in Usulatan.

In Central America, "immediate reaction" often translates "walk faster." Hobbled by a lack of chopper transport, Atlacatl reserves advance to op area over remaining Rio Lempa railroad bridge.

Indian country—just east of the Rio Lempa in Usulutan Province, El Salvador. The rugged terrain is infested with communist guerillas, swarming gnats, tall grass and choking heat. I'm part of a 217-man relief column, commanded by 1st Lt. David Koch. Our objective is the Atlacatl Immediate Reaction Battalion TOC (tactical operations center) established the day before, after fierce and bloody fighting, on a hill called Hacienda El Carmen 4.5 klicks from the village of San Marcos Lempa where we have left the trucks which carried us from the battalion cuartel.

Waiting for a truck of rations to be brought up from the city of Usulutan, the troops lie in the meager shade of the village square. I sample strips of dried, candied papaya and sip Coca-Cola out of a plastic bag as I listen to 105 rounds screech overhead. The artillery battery is located on the west bank of the Rio Lempa next to the railroad trestle bridge we have crossed moments before. They're pounding guerrilla forces fleeing to the northeast in the sector on our right flank that is occupied by the Atonal Immediate Reaction Battalion. Like killer bees, swooping A37 Dragonflies repeatedly unload more presents for the Gs, which impact with thunderous cracks on the nearby hills, turning at least 16 of the bastards into meat for the vultures.

Sixteen hundred guerrillas have massed to destroy the remaining bridge over the Rio Lempa. Our job: Interdict the enemy advance on the bridge, engage and eliminate as many of them as possible, wipe out their nine base camps in our area of operations, capture a cache of weapons reported by intelligence sources, and begin the pacification of Usulutan Province.

My part in this effort began the day before the battalion left for the operation. All of the battalion's machine guns and crews were assembled in the Plaza de Armas for my inspection. All weapons had been superbly maintained. The training they had received from me in August '83 had paid life-saving dividends. Equipped now with the safety-wire pliers and stainless-steel wire I had provided them, the crews were no longer inhibited from disassembling the M60 gas systems.

I have never seen cleaner M60s! I needed only to swiss-file some operating rods and bolts, straighten a bent trigger spring, and replace the parts usually found missing or defective on this system: cocking lever guide screws, trigger-housing retaining leaf springs, actuator rollers, bolt plug pins, recoil springs and gas plugs. I discovered two bolt assemblies which had been incorrectly assembled.

The .50-cal. Browning M2 HB machine guns were also well maintained and, of course, no parts needed replacement in this finest of all heavy machine guns. A refresher course on use of the headspace and timing gauges satisfied me that these crews were also competent with La Cincuenta.

As the M60 crews safety-wired the barrel assemblies, I walked away, feeling confident neither they nor their weapons would fail us in the days ahead.

While still in the cuartel I was able to examine and fire an unusual captured FN FAL. One of the Cuban contract, this FAL's flash suppressor had been discarded and the barrel chopped back about six inches to the front sling swivel's retaining groove. Sling swivels and butt plate were missing and the wood buttstock had been shortened as well. The inner recoil spring had been removed, and the rifle fired without stoppages of any kind. As expected, the muzzle blast was ferocious and controllability in the full-auto mode diminished even further. But hit probability remained high and felt recoil was not appreciably increased. All in all, it was an interesting experiment in reducing the FAL's bulk and weight. The usual hole had been cut with an end mill to obliterate the Cuban coat-of-arms.

The Atlacatl Battalion moved out the next morning at 0300 under the able and experienced command of their new CO, Lt. Col. Armando Azmitia, their former XO. The operational area was a strip of highly dissected topography eight kilometers wide. Flanked by the Rio Lempa on the west and the Atonal Battalion to our immediate east, we would approach from the south. With guerrilla movement to the flanks thus inhibited, the situation was a textbook hammer-and-anvil set-piece begging for a blocking force to the north. We could smash the entire guerrilla concentration in one classic maneuver.

But this was not to be, since 80 helicopters would be required to execute such tactics. The entire inventory of the Salvadoran Air Force consists of only 18 well-worn UH-1H helicopters—only half or less of which may be flyable on any given day.

This one frustrating incident serves to spotlight the real dilemma of the Salvadoran Army. While many of the U.S. press corps sit in the Camino Real bar, gloating over each new guerrilla spectacular, sneering at the "incompetence" of the Salvadoran Army officers and playing the same tune they composed in Vietnam, the truth (as usual) can be found elsewhere. To force "immediate reaction" battalions such the Atlacatl, Belloso and Atonal to move their troops by truck and then hump 10 to 20 klicks through the bush to reach an operational area is more than just a macabre joke. (Meanwhile, the press insinuates the four to eight hours such response takes is a demonstration of "poor leadership" and "lack of the will to fight." That smacks of KGB disinformation.)

Give the Salvadoran Army the assistance and equipment they need to become airmobile and they will crush the communist rebel forces.

Contact with the enemy was not expected until the battalion moved eight kilometers north of San Marcos Lempa, but within two klicks

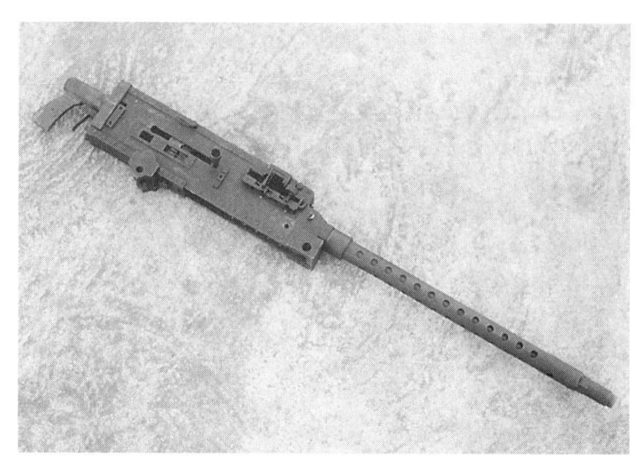

Modified WWII .30 Browning—made by Buffalo Arms Corp. of Buffalo, N.Y.—captured from communists by Atonal Bn.

Effective sniping is a recent Atlacatl development, with help from the Leatherwood ART.

Chopped Cuban contract FAL—captured by Atlacatl Bn.—handled surprisingly well, and was lighter and shorter than author's Para model.

guerrilla fire met our advance units. Moving slowly, the Atlacatls encountered sporadic resistance until they arrived at the base of Hacienda El Carmen, 300 feet from the summit.

Firing from defensive trenches surrounding the hilltop's perimeter, the guerrillas cut loose with all the firepower at their disposal. A withering hail from their M16s, Cuban FALs, M60s and grenades fell upon the assault units at the hill's base. While the main force pushed upward, one platoon was able to flank the enemy to the north and reach the summit unnoticed. The guerrillas counterattacked almost immediately. Engagement distances were less than 30 feet. The platoon was forced to withdraw, leaving three of their comrades holding the hilltop forever. One of those killed was the battalion's most popular sergeant. Men wept openly at the news of his death.

Meanwhile, the Atlacatl assault unit continued its drive up the south slope. Every tree on the summit was scarred by fire. The guerrillas fled northward, taking, as usual, their dead and wounded to prevent accurate body counts. Our snipers, delivering farewell wishes with deadly precision, watched bodies fall through scopes, only to see them dragged off by others. We suffered seven KIA and 18 WIA, but the guerrillas lost many more. They had not expected to fight the feared Atlacatl Battalion that day, and would not soon forget who had bloodied their noses.

Two of those killed were in Dave Koch's 7th Company. The battalion's best sniper, also a member of the 7th Company, had his kneecap blown off. He will be hard to replace.

The Atlacatls have the best sniper program in the Salvadoran Army. The snipers were chosen by a rigorous selection process that clearly defined the battalion's top marksmen. They were then run through an intensive two-week course by one of the world's most competent military small-arms instructors. Twenty-six snipers are now trained, equipped and employed with excellent results.

The weapon system stresses cost effectiveness and ease of training. Large quantities of G3 rifles were already inventoried. Since its 7.62 NATO cartridge is fine for this purpose and in general issue, the G3 was adopted—in spite of its trigger's inadequacy for precision shooting. Bases constructed in El Salvador were machined from industrial steel and spot-welded in four places to the G3 receiver. The superb Leatherwood MPC/ART scope was purchased. With its unique universal-ranging cam, the MPC/ART scope allows the user to "curve-fit" and fine-tune the optical system to whatever caliber or load he is shooting. Special lots of West German ammunition are supplied to the snipers.

The MPC/ART uses a set of standard crosshairs with a finer wire in position below the scope's horizontal crosshair. The distance suspended between these two horizontal lines is 18 inches at 200 meters. After initial zero has been accomplished, the sniper looks through the 3-9X scope at his target and proceeds to increase or decrease the power of the scope until 18 inches of the target is fitted between the two stadia wires. The scope has then been zeroed at this particular distance and the shooter may take dead-on aim without any holdover. Weeks of extensive training in range estimation, ballistics and hold-over techniques are no longer required.

I fired one of these G3s with an MPC/ART scope at the battalion range. Even with its 15-lb. trigger pull, the combination consistently produced 4 MOA. Some careful work on the trigger might cut this to 2-3 MOA.

I roll out with the relief column at 0500, just about the time the press corps has finished turning the FMLN's (Marti National Liberation Front) daily propaganda into the hot, smoking poop after a rough night downing Pilsners at Gloria's, the local cat house.

Intelligence reports indicate that some of the guerrillas have slipped past the Atonal Battalion and moved south to intercept us. Consequently, the convoy moves slowly once we pass Zacatecoluca and Koch stops the trucks several times to send out flankers at likely ambush points. We arrive at San Marcos Lempa without incident.

After the two-day rations have been distributed, the troops are formed up and, with the help of a local tracker, we begin to move north through the bush toward the Atlacatl TOC atop Hacienda El Carmen. A solid mat of vines underneath the tall grass awaits the unwary, and to avoid tripping constantly I'm forced to walk like a ballet dancer. Try this sometime with a "Fritz" helmet (the current issue U.S. ballistic helmet), flak jacket, chest pouches, canteens and FN FAL in the ready position. Within two klicks I'm reconsidering the wisdom of wearing the flak jacket. I become convinced it was designed for Marines guarding static positions in Lebanon and airmobile troops who are flown to the point of engagement, but definitely not for humping the hills of Usulutan.

After three klicks, the searing heat, stifling humidity, and cumbersome flak jacket are really getting to me. Koch trots back and informs me that I'm slowing them down. He puts on the flak jacket and over the next two days loses six pounds.

Shed of this millstone, I begin to match their pace. In this environment, gringos need at least a gallon of water a day. Crossing the Rio Roldan—actually nothing more than an intermittent stream—I replenish my canteens with its tepid, brackish water and pop two Halazone tablets into each quart. We continue grinding out the kilometers, ever conscious that hard contact may be only steps away.

Eventually, as in all COIN ops, you begin to pray the guerrillas will show themselves: first, to relieve the tension of not knowing

SOF's *Kokalis watches TOC and guerrilla camps fade as he leaves combat zone by chopper.*

where they are and second, you'll get to stop, at least momentarily. All the while, the 105 rounds continue to pass overhead and the A-37s work away. Now 60mm and 81mm mortar rounds can be heard plopping out of their tubes. Soon I hear the hard knocking sound of a Ma Deuce, fired in well-regulated three-shot bursts. I smile inwardly, as my .50-cal. gun crews have learned their lessons well and we're closing on the TOC.

We arrive at the command group by late afternoon and Azmitia debriefs me on the previous day's events. We discuss the need for faster reaction time from the A-37s on tactical air-support missions and better coordination between the artillery batteries and the infantry battalions.

After a dinner of meatballs in barbecue sauce MREs (Meals, Ready-To-Eat, the new U.S. Army field rations), dusk settles in and the Salvadoran gnats signal their fellows that a *Norte Americano* has arrived. They move in by the hundreds to drink gringo blood and the insect repellant proves to be mere frosting on their cake. The A-37s bomb the adjacent hills after sunset and Azmitia chortles as fiery shit rains down upon the *putas*. Koch checks his perimeter defenses and we retire to the tent to listen to Radio Venceremos from Nicaragua.

Four hours later we're up and on our way to join 7th Company, about one klick to the northwest. The terrain deteriorates with every step northward. As we move deeper into the foothills, the topography becomes more crenulated and steep hills loom before me every 100 meters.

Kokalis' abandoned flak jacket protects 1st Lt. Koch on UH-1H extraction from op area.

We arrive at an abandoned ranch house—the 7th Company's temporary CP—and Koch immediately holds a debriefing with his second lieutenants. I examine obscene graffiti the guerrillas have scrawled on the walls of the building, and then begin to assemble our M60 night sniping system. The 7th Company already has an AN/PVS-2 night vision scope provided by the U.S. government, but the scope lacks a mount. I attach the proper scope mount and an M122 tripod, complete with pintle and T&E mechanism.

Disgruntled, I am attached to the 4th (weapons) Platoon, along with the M60 night sniping system. We will march in reserve behind the three rifle platoons to our objective, the first guerrilla base camp 3.5 kilometers to the north.

The searing Salvadoran sun is overhead now and the humidity saps my remaining reserves as I trudge slowly up and down the seeming infinity of volcanic hills. Sporadic fire to the front and flanks constantly reminds me that there's only one way to hell—north.

As the day drags on, I reflect on the Para FAL I'm carrying. An impressive classic, it would warm the cockles of the heart of any armchair expert. But, to me at this moment, it has become a device of infernal torture. *Pesa* (heavy), the Atlacatls keep telling me. At 8.6 lbs., empty, it's all that—and much more, with 240 rounds of 7.62mm NATO ball strapped to my web gear. I curse the day I selected it—my final attempt to justify its existence on the modern battlefield.

In the last several days, men all around me have fallen—silent for eternity—to the M16, as they have for more than 20 years now. Yet, this does not satisfy those who unload their treasures from the trunk of a car and waddle 10 feet to the firing line to shoot at paper silhouettes. Their sarcasm, fed by the faulty memories of deeds decades past, casts doubt on the manhood of poor wretches unwilling to carry the iron burdens endorsed by the toy soldiers of the popular gun press. A pox on them.

As I top another hill and drop to its military crest, I come upon Koch who points ahead.

ARGIE GOLFBALL

While on the Usulutan combat operation, I carried an innovative Argentine grenade that is now in general service throughout the Salvadoran Army, although the U.S. M67 is still more commonly encountered.

Designed by Fabricaciones Militares of Argentina, this Doble Porposito Granada De Mano (Double Purpose Hand Grenade) is called the FMK2 Model 0. The grenade body, about two inches in diameter, is slightly larger than a golf ball. It follows the mini-grenade pattern of the Belgian MECAR 60mm anti-personnel grenade. The body is cast, but acquires its excellent fragmenting characteristics from a special after-casting heat treatment. This heat treatment makes the 5.8-oz. grenade body break regularly into three- to five-gram fragments.

The explosive charge consists of a 2.7-oz. mixture of recrystallized hexogene and flaked TNT. This combination has considerable brisance (shattering effect) and the chemical proportions can be varied to suit differing fragmentation characteristics.

The fuse assembly, a 2.1-inch aluminum tube, is permanently attached to the grenade body. It contains the detonator and firing mechanism. The 1.1-inch detonator is removable and can be stored upside down, in which position the grenade is inert. To arm: Unscrew and remove the fuse assembly cap along with the coil spring and plastic guide plug, drop out the detonator and replace it with the red end down (an arrow and the word *abajo*—down—on the side of the detonator indicate the correct position). The inert end of the detonator is painted black. When the spoon is released, a brass bar is pulled from the fuse tube and the spring-loaded detonator is driven downward onto a fixed firing pin. The time delay varies from 3.5 to 4.5 seconds.

The FMK2 has two additional safety devices. A red-painted spring-steel wire is wrapped around the grenade body and spoon and can be removed at any time by thumb pressure. A cotter pin on a safety pin ring also holds the spoon in place. Its removal is the final step taken in arming the grenade. A spare cotter pin is attached to the ring. Should the first be damaged in removal, this spare can be used to make the grenade safe again. I found this 3-safety system to be foolproof and comforting. Over the years, all too many grunts have been vaporized into a red mist through mishandling of their own grenades.

Because of its superior ergonomics, the FMK2 can easily be pitched 30 feet from the prone position and up to 100 feet while standing or kneeling. Although 100 percent lethality is assured within a 15-foot radius from the point of detonation, by 30 feet the casualty capacity has fallen almost to zero due to the extremely small size of the fragments. This fits the salient parameters required of a modern dual-purpose offensive/defensive grenade. In comparison, the U.S. M67 has a maximum casualty radius of 50 feet, which handicaps its versatility. Inside a contained area, the FMK2 will produce at least severe concussion within a 10-foot radius and within a four-foot radius, steel helmets will be penetrated.

A tail assembly is available for rifle launching with ballistite (blank) cartridges. With this device, the FMK2 can be fired to 400 meters. But this component is not in use by the Salvadorans.

Argentine FMK2 dual-purpose hand grenade.

There I see his M60 gunners raking the guerrilla base camp with reconnaisance by fire. Our objective is clearly in sight and my only bitterness now is that this time I have not counted coup over the enemy.

The summit above the guerrilla camp, which appears so close, is four grueling hills beyond. I climb the last incline to the cadence of a UH-1H rotor and by these means, I extract. With satisfaction and frustration, mixed, I leave an operation that has begun well. Hobbled by a lack of modern equipment that is available, but forbidden, the Atlacatls have taken the fight to rebels whose suppliers feel no such curbs of misguided conscience. Only the weeks to come will tell exactly how effective this campaign will be.

As we dust off, I watch the men huddled below shield their faces from the swirling clouds of debris and fade from view, but not from my heart. In the silence imposed by the engine's roar I reflect on the plight of the brave Atlacatls.

Scorned by the media and poorly equipped by the U.S. government, they will fight on.

The Reagan administration will determine what assistance the Salvadorans require to win. And a vacillating Congress will just as surely fail to provide the funds required for victory . . . only enough to continue the stalemate . . . only enough for the war and the killing and the destruction to go on, for another decade, or more, until the countryside lies charred, the people scarred and ravished in body and spirit, and all hope is obliterated.

We off-load in the city of Usulutan, home base of the Atonal Battalion, so the helicopter can fly a resupply sortie for them. While waiting in the cuartel, I examine a display of captured weapons. From the wall a sad-faced icon of Christ looks incongruously down upon the menacing assemblage. In my mind's eye I see a tear roll down His cheek—and I know why.

Originally appeared in *Soldier Of Fortune*
June 1984.

Atlacatls sight-in an M60 after Kokalis installs a tripod and mount, plus an AN/PVS-2 night sighting device provided by USG.

Return of La Pantera Rosa

SOF's Tech Editor Trains Salvo SWAT

Commander of the Equipo de Reaccion Especial (ERE) of El Salvador's Policia Nacional fires Heckler & Koch MP5 A3 submachine gun while Special Response Team member crouches next to him ready to fire a 40mm grenade from his M203-equipped M16A1. Note Steiner binoculars as recently adopted by the U.S. Army.

"Capture, kill and neutralize terrorists." That was the mission with which the *Equipo de Reaccion Especial* (Special Reaction Team, or ERE) of El Salvador's *Policia Nacional* (PN) was charged when it was established 25 May 1989, by Lieutenant Colonel Jose Antonio Almendariz Rivas, at that time head of the PN's intelligence section.

The ERE set to work almost immediately, when on 30 May 1989 they raided the house of Mario Gonzales, who belonged to the PRTC (*Partido Revolucionario de los Trabajadores Centro Americanos*—Revolutionary Party of Central American Workers) and was in charge of FMLN logistics. Uncovered during the raid was the first concrete evidence of massive ComBloc support to the FMLN terrorists. Included in the captured materiel were 283 Kalashnikovs, 88 Hungarian PA-63 pistols, 10 RPG-7s, 30 RPG-18s, more than one million rounds of 7.62x39mm ammunition and a large quantity of explosives.

The ERE more than proved its mettle in combat during the November 1989 offensive. During an intense three-hour contact in Colonia Santa Marta four men of the unit were KIA—one of whom ran out of shells for his shotgun—when they were struck down 25 meters in front of a terrorist base of fire. Three groups from the Policia Hacienda, Bracamonte Battalion and Signal Corps Security unit attempted without success to retrieve the four fallen ERE members.

Realizing they would have to take care of their own, 16 members of ERE broke through the FMLN lines to recover the bodies after four days of unrelenting contact without resupply, water or food and equipped with no more than their basic load. No amount of training can duplicate the fire discipline instilled in troops forced to fight for extended periods with only the ammunition they bring into contact. During this firefight, the ERE accounted for 20 confirmed terrorist KIA, for a kill ratio of 5:1.

Small arms fielded by the ERE throughout the November offensive were no more than M16A1s (some equipped with M203 40mm grenade launchers), 12 gauge shotguns and Heckler & Koch MP5 A3 submachine guns. Forty millimeter grenade ammunition employed were the HEDP Dual Purpose and M576 Multi-Purpose (containing 21 No. 4 buckshot pellets) rounds. The ERE also carried the golfball-sized Dutch V40 mini-grenade. Weighing only 4.2 ounces, the V40 provides 100 percent casualties at 3 meters from point of detonation.

Shortly after this, on 26 November, the ERE raided the house of Jennifer Jean Casolo—the

ERE shoulder tab above the emblem of El Salvador's Policia Nacional signifies "Equipo de Reaccion Especial" (Special Reaction Team). This unit has experienced a great deal of action during its short time in existence.

American Left's apparent replacement for the aging Jane Fonda—and captured, among other things, 103 60mm mortar rounds, 213 blocks of TNT, 405 blasting caps, 12,510 7.62x39mm rounds, 9,110 5.56x45mm NATO rounds and 325 7.62x51mm NATO rounds. Oddly enough, no Bibles were found among the effects of this innocent little church worker.

Based upon interrogation reports and information from undercover agents, the ERE continues to raid FMLN safehouses on an almost daily basis, capturing terrorists and large caches of weapons. Their most recent raid uncovered 800 blocks of TNT and a substantial quantity of Flex-X, an RDX-based plasticized explosive manufactured in Canada by C.I.L. In April 1990, a cache of ammunition was captured that points directly to FMLN sympathizers in the United States (see sidebar).

For all of their accomplishments, the ERE is a relatively small unit. The group presently consists of 45 enlisted personnel and one officer. First Lieutenant Aristides Merlos Flores has been commander of the unit since its inception. It began with only seven men. A call for volunteers throughout the Policia Nacional resulted in 40 additional recruits. Qualifications for enlistment in the ERE are high by Salvadoran standards. A minimum 9th grade education is required. A minimum of two years prior military or PN service is mandatory. Successful applicants must pass a complete public and private investigation, a polygraph test, and a physical test that includes a 2-mile run, push-ups, sit-ups, chin-ups, squat jumps and swimming.

Training for the ERE had consisted almost entirely of material obtained from the U.S. Army MP School Field Circular 19-152, "Special Reaction Teams, Operational Concepts/Training." While it is a valuable conceptual guide, this Field Circular provides no useful information concerning the specifics of small arms employment.

Their lack of weapons training was displayed to me in February 1990 when I was introduced to the unit through a live-fire demonstration presented for my benefit. Most of the firing techniques were obsolete and the hit

CAPTURED MATERIEL

We were on full alert. Intelligence sources indicated the FMLN was planning to send its greetings via an 81mm mortar attack on the PN HQ. After the usual dinner of eggs, refried beans, crema (sour cream) and thick Salvadoran tortillas at *el comidor de los oficiales* (officers' mess) and with little desire to go to bed with my boots on, I wandered through the dark, pseudo-Gothic hallways of the *Direccion General De Policia Nacional* (built in 1935) and down into the basement where the abandoned dungeons are now used as enlisted barracks, a tailor shop and storage. It is here that all materiel captured by the ERE from FMLN safehouses, with the exception of explosives, is kept. In February I had requested that all materiel captured after I left be segregated so that upon my return in June I might examine it with an eye to a possible change in the FMLN's logistical pipeline.

In one of these former cells, illuminated by only a single bare incandescent bulb, was the cache uncovered by the ERE in April 1990.

Stacked in one corner were six East German MPiKMS-72 Kalashnikovs, two Hungarian AKMs and one Soviet AKM in almost new condition along with about three dozen magazines—mostly steel and of unknown origin, although four were red plastic types of Russian manufacture. All of this was probably residual material left over from the FMLN buildup prior to the November 1989 offensive. Incongruously, lying next to the Soviet AKM was a 19th-century, hinged-frame, pinfire revolver, of undetermined caliber or manufacture, with 12 chambers.

Next to these weapons were what I estimated to be between 50,000 and 100,000 rounds of caliber 7.62x39mm and 5.56x45mm NATO ball ammunition. The 7.62x39mm cartridges were in grain sacks containing three to four thousand rounds each—the standard container by which the FMLN transports ammunition of this caliber across El Salvador The 5.56x45mm NATO cartridges were tightly and neatly stacked in vertical columns on plastic pallets containing approximately 1,000 rounds each and wrapped in clear plastic.

The ammunition had been removed from the original 20-round boxes in an apparent attempt to disguise its origin. It never ceases to amaze me that terrorist organizations and even some foreign governments seem to be oblivious to the information that can be obtained from a cartridge's headstamp.

The 7.62x39mm ammunition was mostly of two origins. Much of it was headstamped "WINCHESTER 7.62x39" with Boxer-primed, brass cases. The remainder—with either copper-washed or lacquered, Berdan-primed steel cases—was manufactured in the People's Republic of China. The following headstamps were found on the latter cartridges: "31 73" (in SKS stripper clips), "71 88," "946 79,"

Cartridges captured from FMLN safe house in April 1990. Top row: 5.56x45mm NATO headstamps include "FC 223 REM" (Federal Cartridge Co.), "85 IVI" (Canadian), "FNM 86-2" (Portuguese) and "C J 8" (People's Republic of China). Bottom row: 7.62x39mm headstamps include "WINCHESTER 7.62x39" and PRC lots "31 73," "71 88," "946 80" and "6201 72." All of this ammunition was smuggled from the United States by FMLN sympathizers.

"946 80," "6201 71" and "6201 72." All of these are lots exported to the United States for commercial consumption.

I found five different headstamps on the 5.56x45mm NATO cartridges: "WCC 89" (Western Cartridge Co.; i.e. Winchester, 1989), "FC 223 REM" (Federal Cartridge Co.), "IVI 85" with NATO cross in circle (rejected Canadian government SS 109 type manufactured by Industry Valcartier, Inc., 1985, and imported by Pacific International Merchandising Corporation, Sacramento, California, for commercial sale in the United States), "FNM 86-2" with NATO cross in circle and green tip (SS-109 type manufactured by *Fabrica Nacional Municoes e Armas Legeiras*, Moscavide, Portugal, and exported to the United States for commercial sale) and "C J 8" (manufactured in the PRC and exported to the United States for commercial sale).

It appears that, at least temporarily, the supply route from Nicaragua has been interdicted. All of this ammunition was obtained in the United States. The FMLN fields Kalashnikovs and M16s (obtained from Vietnam via Cuba and Nicaragua). Thus, 7.62x39mm and 5.56x45mm NATO ammunition are their two most important small arms calibers.

The great mix of headstamps indicates to me they are purchasing this ammunition in small lots (probably no more than 1,000 rounds with each buy) at normal retail outlets to avoid arousing suspicion. With hundreds of dupes and sympathizers from New York to Los Angeles, this should present no problem. After accumulating a substantial quantity, the shipment is probably smuggled across the border to Mexico and then down through Guatemala into El Salvador.

While the FMLN failed to drop any mortar bombs on us that evening, this alarming information more than made my day. Need I remind you that the leftist mindset in the United States that would condone this outrage would most often also support efforts to relieve you of your Second Amendment rights? "Hypocrisy" is a synonym for "left liberal."

probability was poor. When Lieutenant Colonel Almendariz asked for my opinion of the unit's performance, I gave him an honest appraisal and suggested that *SOF* might be able to provide the weapons training required to enhance the ERE's combat effectiveness. He gave his enthusiastic approval and upon my return to the United States, I contacted *SOF*'s Editor/Publisher, Lt. Col. Robert K. Brown, who signed off on the mission without hesitation.

An intensive five-day course was designed which would provide training in the handgun, MP5 submachine gun, combat shotgun, and M16A1. As it is the most difficult weapon to master, two days were allocated to the handgun, leaving only one day for each of the other weapons systems. This short time frame would require 10- to 12-hour days with no more than 20 minutes for lunch and only infrequent five-minute breaks, during which time the students would be required to load magazines. Mark Evan, an experienced small arms employment authority, was selected as my co-instructor. Six different men would be trained in each of the four course segments. This would permit the two of us to monitor each of the students carefully throughout the training cycle. It was anticipated that the 24 men exposed to the courses would cross-train the other members of ERE after our departure.

To be successful, a tight rein on logistics would be necessary. We arrived in El Salvador several days early in June 1990 to inspect the range facility—located at the PN substation in San Marcos, about six klicks south of the PN HQ in San Salvador—and made certain that the target frames, swinging steel gongs and heavy steel plates had been constructed to our specifications. These were supplemented by Dualatron, Know Your Enemy, and Yavapai Firearms Academy camouflage pattern paper targets that we brought with us.

On the first day of the training cycle we were presented with the option of either riding back and forth to the range facility in an unmarked car with our translator, Carlos Salinas, or in the truck with the troops. A 2-ton, stake-bed truck loaded with armed, uniformed soldiers is an inviting target for FMLN terrorists.

And, the danger of ambush in the urban environment of San Salvador is now every bit as high as it always has been in the bush. Throughout our stay, we heard sporadic machine gun fire and explosions within the city, both night and day. To establish immediate rapport with the ERE and indicate we were willing to share their exposure to contact, we opted for the truck. Fortunately, we never had a contact with the Gs and the truck's driver proved to be more dangerous than the terrorists.

Immediately after our arrival we were each issued a Kalashnikov with five magazines—Mark Evan a Hungarian AKM with rigid stock and I an East German MPiKMS-72 with a side-folding stock. We carried these weapons everywhere during our entire stay, except on a few occasions when we ate in a restaurant, where I packed a Wayne Novak modified Browning High Power in a DeSantis shoulder rig with an S&W Model 640 Centennial .38 Special as backup in one of Bruce Nelson's inside-the-pants Summer Specials. Mark toted a Colt Officer's Model .45 ACP in Nelson's No. 1 Professional and an S&W Model 649 Bodyguard .38 Special, in a Nelson ankle holster.

The handgun course was scheduled for the first two days. At 0700 on the first day, we were pleased and encouraged to see all 24 men selected for training, not just the six individuals assigned to the handgun course. All of them enthusiastically attended every day of the training cycle to observe, learn, and assist those actually taking the course. In addition, their commander, Lt. Merlos, participated as a seventh student in all five days of training. With the exception of the Atlacatl Battalion, all too often the commanders of units I have trained in El Salvador have never exhibited enough interest to even visit the training site, let alone participate in the training with their men.

Except for Lt. Merlos, who was armed with the M9 (Beretta 92F), recently adopted by the Salvadoran armed forces in emulation of the United States, the ERE has been issued Browning High Powers. It is, without doubt, still one of the finest 9mm combat pistols ever fielded. Unfortunately, however, the ones with which the ERE is equipped—a mixture of Belgian and Argentine manufacture—are all in woeful condition. To further compound the problem, 9mm Parabellum ammunition is difficult to obtain in quantity in El Salvador. We had specified a total of 10,800 rounds for both the handgun and MP5 courses. As Merlos explained, this had required an appeal to divine authority. As a result, our inventory included no less than a dozen different headstamps from eight different countries (Argentina, Canada, Czechoslovakia, Germany, Great Britain, Israel, Sweden, and the United States). Some of these cartridges were more than a half-century old. Most prevalent and of lowest quality was the Argentine ball, a few of which had been manufactured with the primers set into the pocket sideways.

As a consequence of both of the above, we had no difficulty illustrating malfunction procedures. During the two days of the course, the students encountered every stoppage conceivable, including: light hits (weak firing pin spring),

ERE members who participated in SOF's training cycle pose with their instructors, Peter Kokalis and Mark Evan, and Carlos Salinas, the translator.

failure to extract, failure of the slide to go into battery, failure to retain or seat the magazine (worn magazine catch/release), no firing pin protrusion (firing pin hole plugged with brass shavings), stove pipes, dead rounds, failure of the hammer to fall, slide locked open before last round has been fired (the so-called "Commander Syndrome"), double feed, and low stripping pressure (weak magazine follower spring).

In addition to the time devoted to clearing these stoppages, the students spent considerable time with the Weaver stance; the draw stroke; Weaver low ready position; trigger control; follow through and front sight reacquisition; standing, kneeling, prone and barricade firing positions; shooting on the move; speed and tactical reloading; weak hand firing and reloading; mindset and tactical procedures.

With the exception of Lt. Merlos, who had prior advanced training, all of the students started very poorly. However, by the end of the two days, all were performing considerably better than most students I have observed in similar courses in the United States. Why? In the United States most of those participating in courses of this type are playing games at Fantasy Island, albeit learning something that someday might conceivably save their lives. The ERE has contact with terrorists on the streets of San Salvador on an almost daily basis. Their motivation is high. What they learn today, they will most likely employ tomorrow in a deadly confrontation with the FMLN. We observed the same progress in the other three courses as well.

The MP5 course was scheduled for day three. All of the ERE's MP5 A3s (retractable stock version) are well used and were probably manufactured in the 1970s, but are in generally excellent condition overall. However, the rear sight elevation drums were all frozen with rust and could not be adjusted. Fortunately, the weapons were within tolerable limits for elevation zero and only the horizontal point of impact needed adjustment. The ERE MP5s are equipped with older style forearms and trigger packs that permit semiautomatic and full-auto fire only. In my opinion, the three-shot burst mechanism is a superfluous feature on this weapon as its 700 rpm cyclic rate permits experienced operators to fire two-shot bursts with ease. Our students stepped up to the firing line with slings ranging from the superb Heckler & Koch multi-purpose carrying sling to boot laces.

During the course of the day, only one stoppage was encountered. It is a one-in-a-million MP5 malfunction that I have experienced only once before. It occurs when an empty case is extracted from the chamber and instead of being propelled out the ejection port, spins under the bolt carrier to wedge itself between the carrier and trigger group. Clearing this stoppage requires removal of the buttstock and swinging the trigger pack forward to expose the case. If this should occur during a firefight, your best alternative would be to discard the weapon and immediately transition to your handgun. Many of those familiar with the MP5 have heard of this stoppage, but few have ever seen it. Since it is rare and cannot be induced, no one

SOF's Technical Editor Peter Kokalis observes ERE student in the submachine gun course as he assumes the standing position required for accurate semiautomatic fire with the MP5. *(Photo: Mark Evan)*

knows for sure what initiates it, although most authorities surmise it might be the result of a weak extractor spring.

The MP5 course covered mindset, ready positions, sight alignment, trigger control, follow through, general manipulation, weapon retention techniques, malfunction procedures, tactical procedures, standing, kneeling, squatting (developed many years ago by the USMC and sometimes referred to as the "rice paddy prone"), barricade, prone and low prone firing positions, shooting on the move, reloading and semiauto versus full-auto firing stances.

The students were taught to load only 29 rounds in the magazine because of the weapon's closed-bolt method of operation. They were instructed to observe (whenever the tactical situation permitted) whether the top round was on the left or right of the staggered column prior to insertion in the magazine-well and then to remove the magazine to establish that the top round had moved over (indicating that a round had been chambered).

Also stressed was the ability to employ the MP5 at longer distances than possible with an open-bolt submachine gun. During full-auto firing drills only two-shot bursts were permitted. It was also emphasized that when engaging precision targets (such as head shots in hostage situations) at close ranges, the operator must compensate for the MP5's high line of sight.

No other weapon offers the close-range effectiveness and ammunition versatility of the combat shotgun. It should be selected over the handgun whenever concealment is not a required element. Its potential for self-defense is limited only by the proficiency of its operator. It provides massive stopping power while minimizing the potential for over-penetration—an important consideration in urban environments.

On the day of the shotgun course we were dismayed to see a mix of only No. 4 buckshot and No. 7 1/2 birdshot (some of which was ancient paper-cased Canadian Eley) in the combat vests of those selected for this training. Except at point-blank ranges, birdshot serves no function in a combat shotgun,

For many years, law enforcement agencies have depended on No. 4 buckshot, which has 27 .24-inch diameter pellets in each standard 2 3/4-inch shell. Penetration problems through car bodies and even clothing have resulted in a return to 00 buckshot with its superior lethality, albeit somewhat lower hit probability. You can depend on 00 buckshot to put an immediate stop to any confrontation up to 25 yards.

In my experience, Federal Premium Buckshot (Federal Cartridge Co.) will significantly improve 00 buckshot performance if you select their 9 pellet load (the 12-pellet 2 3/4-inch Magnum load just plain kicks too much). These tower-dropped pellets are 97.5 percent pure lead with 2.5 percent antimony for added hardness. Two polishings guarantee sphericity. Copper-plating further increases resistance to deformation during firing. Shot is arranged in a spiral configuration within a long-range shot cup with granulated buffer added to fill the gaps. The granulated buffer will eventually leak into the gun's action and chamber and the mouth of these shells should be sealed with nail polish.

Shooting on the move is required training for the effective combat employment of most military small arms, including the MP5.

Shotgun ammunition of any type is extremely difficult to obtain in large quantities in El Salvador. Through its in-country sources, *SOF* was able to provide 1,000 rounds of 12 gauge ammunition for the course. We collected all the ERE's No. 4 buckshot and No. 7 1/2 birdshot for use in the training cycle. At the end of the day we issued them more than twice as many Federal Premium 00 buckshot shells as the number of rounds they had turned in to us.

Except for one Remington Model 870 with an extended magazine, the ERE is armed with the Winchester Model 1200 slide-action 12 gauge shotgun. First introduced in 1964, the Model 1200 has a four-round magazine and a 20-inch barrel in the military/police version.

As every shotgun, even those of the same make, model and sequential serial numbers, will throw a pattern different than any other, the students first patterned their weapons on paper at varying distances. No matter how crude it may appear, the importance of the shotgun's front bead sight was stressed throughout the course. Also emphasized was the fact that pump guns must be racked smartly and completely, both rearward and then forward.

This segment of the training cycle also covered mindset once again, ready positions, the critical importance of constantly reloading the shotgun's limited-capacity magazine with the weapon on target, tactical procedures, trigger control, follow through and firing positions to include standing, kneeling, barricade, squatting and prone.

On the final day of training we worked with the M16A1 rifle. All of those issued to the ERE are well-worn. Most exhibit excessive play between the upper and lower receivers. Little remains of the original finish. Some are equipped with M203 40mm grenade launchers. Lieutenant Merlos and several members of the unit carry M16A2 Commandos while several others have M16A1 Carbines. All of the unit's M16A1 rifles are equipped with Vortex flash suppressors donated by *SOF*. This muzzle device almost completely eliminates the weapon's flash signature.

Proper and periodic maintenance of both the weapon and its magazines (and never loading more than 29 rounds in a 30-round magazine) are the key to reliable operation of the entire M16 series. We retired several magazines with bulged bodies and floorplates before the course commenced. Approximately 6,000 rounds were fired and yet we experienced only one "bolt-over-base" stoppage. Most of the unit's 5.56x45mm NATO ball ammunition is drawn from FMLN stockpiles captured during the November 1989 offensive and is headstamped "LC 73," indicating an origin from material abandoned by the U.S. in Vietnam.

"Geronimo" fires the MP5 from the "rice paddy prone" position.

To be effective, the M16 must be zeroed by the individual to whom it is issued and this was accomplished before proceeding further. With the proper technique, two-shot bursts with the M16 will deliver reliable hits out to 50 meters. However, the principal employment of the infantry rifle should be in the semiautomatic mode and this was heavily stressed throughout the day. Another key concept is balancing speed against accuracy, which requires selection of the firing position that will insure the highest hit probability within the time frame permitted by the specific combat scenario. With that in mind, the standing, braced standing, squatting, kneeling, standard prone and low prone positions were covered. As with the MP5, the M16's high line of sight requires the operator to aim at the top of the forehead to strike between the eyes at extremely close distances.

Other topics covered were the high ready and low ready (muzzle depressed) positions, trigger control, follow through, shooting on the move, malfunction procedures, reloading techniques and tactical procedures.

Feelings of camaraderie are usually intense in small, elite units, especially under the tension of an almost constant combat environment. As the men of the ERE soaked up our training, they began to draw us ever more closely into their midst. We became, at least while we were with them, members of the unit. There were some amusing aspects to this.

Everyone in the ERE has a nickname. Since I could remember neither these nor their given names, I began to call them by names of my own choosing, which usually described their demeanor or physical attributes. There was "Mafia," "Flaco" (Thin), "Feo," (Ugly), "Cabo Indio" (Indian Corporal), "Mudo" (Mute), "Peter Pan," "Geronimo," "Cosa Seria" (Serious Matter), "Shorty," "Antiguo" (Ancient One), "Pequena Lulu" (Little Lulu), "Sargento Muela" (literally "Sergeant Molar Tooth," however, in El Salvador "muela" is an idiom for "stupid"), "Dopey," "Jimmy," "Viejo Tigre" (Old Tiger), "Cristo de Lata" (Tinny Jesus), "Atlacatl," "Bebe" (Baby), "Guapo" (Handsome) and "Sargento del Almacen" (Sergeant of the Supply Room). All of this was received with a great deal of laughter by the comrades of the individual whose nickname was being evoked. There is, however, always a payback.

During the lunchbreak of the last day of training, I noticed all of the men gathered under the shade of a mango tree laughing and shouting loudly. Whenever I passed by, they lapsed into silence. When I questioned Carlos Salinas, he informed me that the men were conducting an important meeting. After I asked what it was about, he informed me its purpose was to decide upon what my name was to be in the unit. When I walked over to the tree with my Kalashnikov in hand, I announced that I knew what they were about and assumed the dispute was over a name such as "Rattlesnake" or "Scorpion." Sorry, they said, but they had already selected my name. Because I always walked quickly and hunched over, I would always be known to the ERE as "Pantera Rosa" (Pink Panther).

On Saturday the ERE hosted a party in our honor at the PN's private beach in La Libertad, about 30 klicks south of San Salvador. Upon our return, Geronimo volunteered to ride in our car as a bodyguard since terrorists sometimes infest the highway along this route. It was only after we departed that we noticed that he was almost comatose from the beer he had ingested at the party. He had commandeered Mark's Kalashnikov and kept sliding the selector lever off safe while fingering the trigger and pointing the muzzle at the back of my head. Carlos was driving and armed with only a rusted Colt Detective Special and one speed strip. Having spent the morning pretending I was once again 24 years old, I was fast asleep in the front seat. Mark, surrounded by a drunken bodyguard, sleeping comrade and poorly armed driver, spent the trip back with his Colt OM in the Weaver ready position. The Gs failed to make a play and Mark suffered no more than a case of high anxiety.

On Monday we were escorted to the airport by eight of the boys riding shotgun in a pickup truck in front of our car. As I watched them roll forward at point down the road to Comalapa airport, I began to conjure reasons for my return to the ERE. They need training in dynamic entries and help with the three sniper systems they employ (M21, Steyr SSG, and Dragunov).

When we arrived at the TACA ticket counter, their last words to me were, "When will you return?"

"Soon," I said. And I meant it.

Originally appeared in *Soldier Of Fortune*
November 1990.

Section 2

Afghanistan

RAIDERS OF THE LOST GRENADE LAUNCHER

Russian AGS-17 complete with tripod mount, loaded and ready to fire.* SOF *Military Small Arms Editor, Peter G. Kokalis, was the first American to test this Soviet full-auto grenade launcher.

Staff Tests USSR's AGS-17 in Darra

Blowback operation, heavy bolt and light tripod gave high recoil and low control, so Coyne impersonates a sandbag by sitting on the rear tripod legs for firing session. The villagers are not visible, since they hid behind a rock berm, thinking the launcher unreliable.

After reflecting upon the numerous obstacles encountered in tracking down and finally locating a new Soviet weapon being used against Afghanistan freedom fighters, SOF *Editor/Publisher Robert K. Brown, one of the bloodhounds, called in this report on the lengthy search from vacation overseas:*

*"*SOF *has been tracking down the elusive Russian AGS-17 automatic 30mm grenade launcher for almost 18 months. We had a team in Pakistan in 1980, seeking out the AK-74, additional AK-74 5.45mm ammo and the AGS-17. We found the "74" and its ammo, but not the AGS. There were numerous live-sighting reports and promises to show one to us, which gradually started vying for the spot of one of the three greatest lies in the world. As the months and trips to Afghanistan continued, the AGS started to become our own personal Big-Foot or Loch Ness monster. In early 1981 we obtained the first AGS-17 30mm round. In February '82 Jim Coyne got a hurried picture of the weapon in Pakistan.*

"We took another giant step when Coyne and I went into Afghanistan in late May 1982. En route to our departure point, we stopped for lunch in Darra, the weapons mecca of mid-Asia, in the Northwest Frontier Province (NWFP) of Pakistan. Our host, a wily gun dealer from whom we had obtained the AK-74 for testing and who therefore had some credibility, said, "Yes, I have an AGS-17, but it's in my village a few kilometers from here. Yes, you may photograph and shoot it."

"No time. Our presence was required on the border. However, we

arranged to contact him on our way out. When we returned, we met and scheduled a short session for the following morning. We were to be transported by our Afghan driver (as Westerners/Journalists were banned from going in the NWFP without a Pakistani official as an escort) but the driver did not show up, or so we thought. We subsequently learned he had been ejected from the hotel lobby and did not call us. By the time we found this out, we were more interested in negotiating for an alleged engine manual from an Mi-24 Russian gunship. We eventually found the manual was a fake—and we were out of time. Still no AGS-17.

"We gave it another shot last September. Patience paid. And over to you, Peter . . ."

•••••

Darra Adam Khel. Forty klicks from Peshawar and well within the forbidding Tribal Lands on the border between Pakistan and Afghanistan. In this untamed, violent land, *Pukhtunwali* (tribal law) prevails once you step off the road.

Darra Adam Khel. Closed to foreigners for more than a year now, ever since a large drug bust at the local heroin factory.

Darra Adam Khel. A small, yet overwhelming, village as its narrow, congested main street literally lunges at all the senses. The ear is attacked most brutally of all. Incessant gun fire, as customers step out into the street and test-fire weapons they are interested in, aiming directly into the burning sun above. The constant honking of horns as vehicles of all descriptions swerve in and out on direct collision courses with sheep, people, cattle, vendors' stands and trucks painted in the most garish style imaginable. The sense of taste is overloaded and finally dulled by raw onions, harsh cigarettes, half-ripe tomatoes, Coca-Cola, heavily spiced lamb kabob and mint tea. The smells promote turbulence in the gut: the odor of human and animal wastes, vegetables, fruits, spices, hashish, sweat, the gasoline engine's poison vapors, the primitive gunsmith's hearth and above all the acrid smell of gun powder. This melange assails the eye as well, reinforcing the frenzied messages being sent without interruption to my brain.

But, I am getting used to it. Twice before *SOF*'s foreign correspondent, Jim Coyne, and I have been wedged into the back of the mujahideen jeep and smuggled past the Pakistan Army checkpoint. The first time a 100-rupee bribe got us across. Subsequently, we sailed past the checkpoint hidden in the rear of the vehicle. On each of our two previous trips we were told the Russian AGS-17 (*Avtomaticeski Granatomojot Stankovi*) grenade machine gun, which I had come 12,000 miles to test, was still not available, that we should return in a few days and *Inshallah* (if God is willing) it would be there. Due to my schedule, this would be the last attempt. If it is not here today, I will never see the AGS-17.

As Coyne and I wait in the jeep parked outside the shop of gun dealer Hakim Gaz, I find myself hot, sweaty, cramped and bummed out, since I now feel there is almost no chance we will ever examine the AGS-17. Hakim Gaz's glassy-eyed vagueness and lack of enthusiasm kindle little expectation that my hands would ever press the trigger on this new Russian weapon, about which there has been so much speculation and so little information.

The mujahid beckons us and we enter again the candy store of Hakim Gaz. Gaz and Darra's other dealers in guns and intrigue sit amid an exhaustive inventory of military small arms. My pulse quickens each time I step inside his dingy shop. The walls and ceiling are covered with the most amazing assemblage of weaponry one could possibly conjure up into a single 15x40-foot room. Proud of their wares, Gaz and his brother allow me to examine everything.

On the back wall hang the pistols. Browning Hi-Powers, Walther P-38s, old commercial .30-caliber Lugers, Spanish-made Star and Llama .32 ACP pocket pistols, TT-33 Tokarevs, a long-barreled, stocked Mauser broom-handle sporting carbine, a large assortment of flare guns and row after row of Darra-made .38-caliber Webley revolvers. On shelves behind the pistols are boxes of ammo, fountain-pen pistols in .25 ACP, Russian AKM bayonets,

Khyber knives, and Kalashnikov and submachine-gun magazines of all types.

From the ceiling hang M1 carbines, Russian PPSh-41 submachine guns and dozens of locally made Mk III and Mk V Sten guns. Suspended along the walls are hundreds of handmade .303 Enfields in every configuration imaginable: P14s, No. 1 Mk.IIIs, No. 4 Mk.1s, and No.5 Mk.1s (the "Jungle Carbine"). Even the British proof marks have been duplicated. In some cases only the recent date of manufacture and the presence of file marks where they should not be betray the rifle's origin. A few puzzle the eye as they combine Mauser 98k muzzles, front sights and barrel bands with Enfield-type actions in 7.92mm caliber. There is even a Mauser-type bolt-action rifle in 7.62x39mm ComBloc using Kalashnikov magazines and fitted with a pistol-grip wooden stock. Several dozens of the Enfields are actually British-made. A steel vault contains the most highly prized items: FN FALs, G-3s, Chinese and Russian AK-47s, Chinese and Egyptian AKMs, and Russian AK-74s and RPK-74s. The floor is littered with cans of .303 ammo in Vickers cloth belts, and 7.62x39mm and 5.45mm ComBloc ammunition is strewn everywhere.

Dealer Gaz' gunsmith fiddles with recalcitrant damaged links while trying to load the AGS-17.

Since all of this presumably goes better with a Coke, we are seated and offered one. A half hour later a man appears silhouetted in the doorway, struggling to bring the AGS-17 (also known as the *Plymya*, Russian for "flame") inside. Its presumed part-owner follows and we are immediately told that we cannot fire the weapon until we return with the cash to purchase it—$45,000. My hopes dim again as we have no intention of buying the AGS-17 and could not in any event produce $45,000 cash. The corpulent, red-faced, orange-haired individual leaves eventually and Hakim Gaz, who has said nothing during this interval, asks me how many rounds I wish to fire through the AGS-17. I tell him 40, a not inconsiderable number as they are $20 per round. That settled we pile into the jeep and follow Gaz to a small rural village several miles deeper in the tribal area. It is nestled against low-lying hills and steep bedrock escarpments that will prove to be our targets. The AGS-17 is set up—by a gunsmith Gaz has brought with him—facing the sheer cliffs about 400 meters away.

The 30mm rounds are loaded into the non-disintegrating belt. The individual links are held together by cotter pins. A drum-type belt carrier, which will hold 29 rounds, is available, but we did not use it during the firing test.

The links are defective and a great deal of time is expended in attempting to load the weapon. Soon the entire male population of the village, about a hundred, has assembled—to watch Jim and me more than the AGS-17.

The AGS-17 feeds from the right side. Two slightly curved, parallel prongs on top of the sheet-metal feed block are inserted between the cartridge and the link, holding the round in alignment for loading. The belt is placed onto these prongs, links up.

The sequence of operation is as follows. With a belt properly positioned, the rubber-covered cocking handle, located at the rear of the receiver, is pulled rearward. It is attached to a steel cable, which in turn is connected to the rear of the bolt. A pull of 16.5 inches is required to bring the bolt back against the end of the receiver and this represents a considerable amount of over-travel. During this movement the trigger mechanism is cocked. After the cable is released smartly, the recoil spring forces the massive bolt forward. As the bolt approaches the feed block, a semicircular claw rises vertically, grabs a round, pushes it from the link and falls back down in alignment with the bolt face. The bolt continues its forward travel until the cartridge is fully chambered. The AGS-17 fires from the closed-bolt position.

There are no locking lugs and this weapon operates on pure blowback. In contrast, the U.S. Mk. 19 40mm machine gun has a fixed firing pin and fires with advanced primer ignition. The AGS-17 has a spring-loaded floating firing pin. As the weapon operates on blowback, the bolt is stationary when the chambered round is fired and recoil accelerates the bolt to the rear. This requires an extremely heavy bolt to keep counter recoil velocity within safe limits. In addition, the Soviets have incorporated a hydraulic recoil damper on the left side of the receiver to further reduce counter recoiling energy (the damper was marked "MAKC").

Firing is initiated by first holding the rubber-coated horizontal finger grips at the rear of the receiver and then pressing the thumbs against a sheet-metal, spring-loaded panel in the center, somewhat reminiscent of the Schwarzlose. The AGS-17 will continue to fire as long as the trigger panel is depressed and there is ammo supplied to the system. The cyclic rate is quite low, probably around 300 rpm, and it is easy to fire single rounds.

As the bolt moves rearward, a standard configuration extractor removes the case from the chamber. The empty case (only 1-1/8 inches long) then strikes a nub on the feed block and is ejected downward out the bottom of the open receiver.

The single large feed pawl inside the top cover is positioned over the cartridge's center of gravity. It reciprocates by virtue of an arm which travels in a lateral channel on top of the bolt. Two spring-loaded holding pawls on either side of the feed pawl assure the cartridge's proper retention while the feed pawl is moving over to pick up the next round.

The 16-groove, right-hand twist barrel is 11.5 inches long. A series of 11 large cooling fins toward the rear of the barrel are carry-overs from the Hotchkiss-type machine guns. This feature adds considerably to the cost of fabrication. While of dubious value in small-caliber arms, such as the Thompson submachine gun, it is of demonstrable usefulness in air-cooled heavy machine guns. A version with cooling fins all the way out to the muzzle has been reported. It is most likely the helicopter model, since the higher rate of fire required by an aircraft-borne weapon would require greater cooling capacity. The AGS-17 I fired had been hit several times at the muzzle end of the barrel by mujahideen bullets. The deformation produced caused consternation on my part during the initial firing sequence of the high explosive rounds.

Overall length of the AGS-17 is about 33 inches. Of all-steel construction, its weight with the tripod and belt carrier with 29 rounds is close to 90 pounds. This is greater than our Mk. 19 40mm machine gun.

The feed cover was marked BA 465, 1974r, obviously indicating the serial number and date of manufacture.

The AGS-17 incorporates a telescopic sight mounted on the left side of the receiver. This compact angle-prism device was missing and I was forced to use line of sight over the top of the barrel. The lack of permanently mounted auxiliary iron sights is a serious design defect. Range tables for the telescopic sight are stamped on a plate attached to the top of the receiver. The tables cover direct-fire sight settings from 50 to 1,730 meters (the apparent maximum effective range of this weapon) and settings for indirect-fire from 1,000 to 1,730 meters. The cartridge's maximum-time-of-flight mechanism, in essence a self-destruct device, is used to achieve airbursts with indirect fire.

The tripod itself deserves description. It is finished, as is the gun, with a semi-gloss black enamel. The legs are constructed of rectangular sheet-metal tubing crudely welded together by hand. The elliptical traversing bar is similar to that of the Bren-gun tripod. Elevation is accomplished by operating a folding crank on a simple turn gear which is meshed with a curve-shaped rack-and-pinion gear. The leg adjustments are notched teeth similar in design to those of the M1917A1 Browning and British Vickers tripods. They have large locking levers that are easy to use, but sure to be damaged during field use. The tripod's legs have notched paws to better grip the ground surface. However, I found the recoil severe and the tripod should be sandbagged to exploit the AGS-17's inherent accuracy. Unsupported, the AGS-17 climbs excessively in full automatic fire. This would not be the case when it is vehicle- or helicopter-mounted.

The majority of all automatic weapon malfunctions occur in the vicinity of the barrel's chamber in the form of extraction failures, ejection hangups or stubbing of the incoming round. The stubbing of an impact detonated high-explosive projectile against the barrel's breech is a serious consideration. Because of this, point-detonating munitions (except incendiary types) almost invariably have bore-safe fuses which spin-arm themselves at least 10 meters from the muzzle. Stubbing of such a round will usually produce only a low-order detonation.

To alleviate this problem completely the designers of the U.S. Mk. 19 made use of the so-called curved stationary rail. The curved stationary rail not only eliminated the extractor and ejector, but made it impossible to stub a round since the cartridge is positioned with the center line of the chamber on the recoil movement of the bolt.

Not so with the AGS-17. The possibility of a stubbed round is very real and my hands sweated every time I fired. The heavy 30mm round is held only at its rim by the semicircular claw. The chamber's small feed ramp fails to diminish this unstable situation. Everytime you press the trigger on the AGS-17, you are literally throwing the dice. Although you will usually survive a low-order detonation, 40 rounds were more than enough for me. The Soviets, with their low regard for the individual soldier, are apparently willing to take the calculated risk. I'm not. Neither are the mujahideen, who scattered behind a six-foot rock berm each time I fired the AGS-17.

We used two different rounds. One was a high-explosive fragmentation type and the other an anti-personnel beehive (darts or shot pellets) cartridge. The fragmentation round is of the wire-wrapped variety, but a few earlier examples contained flat washers. Illumination and smoke rounds may also be available. A new HEAT round is also being issued in limited quantities.

The AGS-17 case is belted, which means there is a pronounced raised rim around the cartridge-case body ahead of the extractor groove. This combines the feeding advantages of a rimless case with the solid support of the rimmed case. In addition, the safety factor is increased over the level provided by rimless cases. The belt's forward shoulder abuts against a corresponding shoulder in the chamber.

A copper driving band, to increase the round's rotational thrust, is sweated onto the rear of each projectile. The AGS-17 case mouth is heavily crimped on the projectile. This indicates use of an extremely slow-burning propellant that necessitates a heavy projectile pull. The powder charge is not much more than that found in a 12-gauge shotgun shell. The primer appears to be similar to a conventional shotgun battery cup type.

The Soviet 30mm round offers a much smaller explosive envelope than the powerful 40mm high-velocity M384 grenade fired in the U.S. Mk. 19. The M384 grenade yields limited artillery capabilities, with a maximum effective range of 2,200 meters, and resulted in a ballistic breakthrough that the AGS-17 grenade falls far short of. Still, the Soviet grenade is not to be taken lightly, since the HE round is filled with A-IX-1 explosive. This is an extremely fast explosive, made of 95-percent RDX and six percent wax.

An automatic, belt-fed, grenade machine gun with feeding, loading, firing, extraction and ejection operated solely by the energy generated by the explosion of the propellant charge, with vehicular and tripod mounting, light weight, and capable of delivering a high rate of suppressive fire at comparatively short ranges is almost a necessity to the modern, highly mobile infantry unit.

Both the Mk. 19 and the AGS-17 meet these criteria. In the areas of design superiority, safety and destructiveness of the explosive package, the Mk. 19 wins hands down. The AGS-17's only edge can be in the realm of economy—a consistent strong point of Soviet small arms technology.

Firepower of all USMC divisions is to be greatly enhanced by the addition of 180 Mk. 19 40mm machine guns each. We can only hope the U.S. Army follows suit.

Further information concerning the Soviet AGS-17 grenade machine gun can be obtained by traveling 12,000 miles to Peshawar, Pakistan, smuggling yourself into Darra Adam Khel and laying $45,000 on Hakim Gaz. But, even then, you'd still have to figure some way of getting the damned thing back!

Originally appeared in *Soldier Of Fortune*
February 1983.

Soviet AKS-74

Kokalis Debunks Russian Rifle

AK-74 Specifications

CALIBER: 5.45x39mm

MUZZEL VELOCITY: . . . . 2,950 fps

WEIGHT, empty: 7.9lbs

LENGTH (AKS variant):
. . . . .overall: 36.5 inches
. . . . with stock folded: 27.25 inches

BARREL: . . . . Length: 18.5 inches, chamber and bore chrome-plated 4-groove, right-hand twist, 1 turn in 7.7 inches

FEED: . . . . 30 round plastic magazine, will also accept 40-round RPK-74 SAW magazine

SIGHTS: . . . . Front: adjustable post Rear: tangent, U-notch, elevation to 1,000 meters

OPERATION: . . . . Gas, no regulator, selective fire

CYCLIC RATE: . . . . 600-650 rpm

METHOD OF LOCKING:
. . . . two-lug rotary bolt

MANUFACTURER:
. . . . unidentified Soviet arsenals

We are on the road to a mujahideen training camp near the the Afghanistan border. The hot, humid fall afternoon is made all the more oppressive by our cramped quarters in the red jeep. The three mujahideen who ride with us are armed with an incongruous melange of weaponry: a Tokarev pistol (marked M20), a rare DWM marked 1906 Commercial Luger (in caliber .30 Luger), an H&K 91 rifle (the U.S. import version) and the AKS-74 assault rifle (folding-stock version) which I have come to test and evaluate.

We are met at the camp's entrance by guards carrying Egyptian-made AKMs and a P14 Enfield (in caliber .303 British). A long row of open-sided tents extend the length of the camp's firing range. A table has been set up for me and I race toward it, clutching the AKS-74 in my sweating hands. I begin immediately to disassemble this fascinating new Soviet battle rifle.

With an overall length of just 36.5 inches (only 27.25 inches with the stock folded) and weighing in at 7.9 pounds (empty), the AKS-74 feels right even before you fire it. The barrel is 18.5 inches long. The AK-74 series are basically AKMs chambered for the new 5.45x39mm ComBloc round, incorporating a number of important improvements. In fact, reports from British intelligence sources indicate at least one AK-74 has been examined which had a barrel resleeved from 7.62x39mm to 5.45x39mm. The receiver remains the same. It is a one-millimeter-thick "U" section of stamped sheet metal extensively supported by pins and rivets. The rails on which the bolt reciprocates are stamped and spotwelded to the inside of the receiver.

All Kalashnikovs are gas-operated, but have no gas regulator. I have never seen a Kalashnikov malfunction as a result of fouling. But, without a regulator, the gas port cannot be cut off to utilize blank cartridges to fire rifle grenades. (The only exception: the Polish PMK-DGN-60 assault rifle, whose gas cylinder has been modified to accept a gas cut-off valve for this purpose.) Of course, grenades fitted with a bullet trap, now being used ever more widely, could be adapted, although, because of their small size, they deliver a substantially reduced explosive package. However, except for the little-used RKG-3 rifle grenade, the Soviets prefer portable rocket launchers, such as the familiar RPG-7.

The AK-74 operates as follows: After ignition of the primer and propellant, gases are diverted into the gas cylinder on top of the barrel. The piston is driven rearward and the bolt carrier, attached to the piston extension, goes through the necessary amount of free-play travel until the gas pressure drops to a safe level. A cam-slot milled into the bolt carrier engages the bolt's cam stud and rotates the bolt about 35 degrees to unlock it from the receiver. Unlike the BAR (Browning Automatic Rifle) and many other designs, the Kalashnikov offers no primary extraction during bolt rotation. Thus, a large extractor claw is required on the AK-74 and is even more massive than its AKM predecessor.

As the bolt travels back it rolls the hammer over and compresses the recoil spring. The bolt ceases its rearward travel by slamming into the rear end of the receiver. The recoil spring then drives the bolt forward, another round is stripped from the magazine and chambered, and the bolt then comes to rest. The bolt carrier continues onward for about 5.5mm after the two-lug rotary-bolt locking has been completed.

In the case of the older AK-47, the heavy forged receiver alone is sufficient to prevent the bolt's unlocking when the carrier hits the receiver stop. The danger of the weapon's firing out of battery is thus eliminated. It is also during this last forward movement of the bolt carrier

Mohammad Kareem, mujahideen camp commander and former brigadier in the Afghan Army, demonstrates the Soviet AKS-74 at a secret base just inside Afghanistan.

PRC's New Assault Rifle

A new variant of the Kalashnikov assault rifle has surfaced in Afghanistan. Supplied to the mujahideen by the People's Republic of China, this AKM-type spin-off differs considerably from its Type 56-1 predecessor. The Type 56-1 was essentially a folding stock version of the AK-47. Its forged receiver was manufactured by conventional milling operations. It is distinguished from its Russian counterpart by the selector markings ("L" for full-auto and "D" for semiautomatic) and the conspicuous extra rivets on the buttstock struts.

The new Chinese assault rifle has the stamped sheet-metal receiver of the AKM. It can be instantly identified by the prominent magazine-guide dimples on each side of the receiver, located directly over the magazine. There is no bayonet lug. The barrel has a plain end. There is no muzzle compensator (AKM) or muzzle nut (AK-47). The selector markings remain as before, "L" and "D." The extra rivets are gone from the buttstock struts. The receiver cover does not have the transverse ribs found on the Russian AKM.

Missing also is the so-called "rate reducer" of the Russian AKM/AK-74 series. The Chinese must feel this hammer-delaying fail-safe is not necessary.

The most distinctive external feature of the Chinese AKM is its furniture. The upper and lower handguards are composed of a bright reddish-brown fiber-reinforced phenolic plastic. There are no finger swells on the lower handguard. The pistol grip is likewise fabricated from the same material. Its unique shape somewhat resembles the pistol grip found on the Colt Browning 1917A1 and 1919A4/A6 light machine guns, which were in turn patterned after the famous Colt Single Action Army revolver.

Very interesting.

that the secondary, or safety, sear is released and control of the hammer's rotation goes back to the primary, or trigger, sear.

The pinned, sheet-metal receivers of the AKM and AK-74 are much lighter than that of the AK-47. They undoubtedly induce more severe bounce characteristics on the bolt carrier. This brings us to the so-called "rate reducer" found on the AKM and AK-74. Western small-arms technologists unanimously observe there is little difference in cyclic rate between the AK-74 and the AKM. Why then add a complex five-component device to the weapon? Because, it is actually not a "rate reducer," but a final fail-safe, delaying hammer drop until the complete cessation of all bolt-carrier bounce.

It's a shame but Ivan just won't tell us a damn thing about his weapons systems. We always have to figure them out for ourselves.

In typical Kalashnikov fashion, the AK-74 hammer and trigger spring is made of three-strand cable, patterned after the Germans' multiple-strand cables in the recoil spring of the MG-42. It is a common misconception that this was done so that if one strand broke the weapon could continue to function. In truth, it is simply less expensive to manufacture springs in this manner. The AK-74 trigger and firing mechanisms are borrowed from the U.S. M1 Garand rifle.

The selector lever on the AK-74 also remains the same. It is a stamped, sheet-metal bar on the right side of the receiver which is manipulated by the thumb. Noisy, stiff and difficult to operate with gloves, it remains a decided defect in the Kalashnikov design. The top position is safe, locking the trigger so the bolt cannot be retracted. The center position, marked "AB," is for full-auto as the spring-loaded safety sear is depressed and deactivated entirely from control on the hammer. The lower position, marked "OA," is for semiautomatic fire and places no pressure on the safety sear, which is free to catch the hammer. In the "OA" position, when the trigger is released, control of the hammer is restored to the primary sear and pulling the trigger will fire another round.

The original Russian AK-47 and AKM folding stock is a double-strut type controlled by a large press-button release located above the pistol grip on the left side. It folds under the receiver and the magazine passes through it. Patterned directly after the German MP-40 submachine-gun stock, it is adequate for firing pistol-caliber ammunition, but makes a flimsy firing platform for larger caliber weapons. The new AKS-74 folding stock is a serious attempt to correct this deficiency, rather than a consequence of the older stock's inability to clear the new magazine (as others have stated); the AK-74 30-round magazine is slightly shorter than the 30-round AK-47/AKM magazine.

The AKS-74 folding buttstock is fabricated from stamped sheet-metal struts, bent into a U-shape and assembled by punch fit and welding. It folds to the left, and is held open by a spring-loaded button latch located at the rear of the receiver on the left side. The stock is held closed by a spring-loaded hook on the left forward end of the receiver. This new stock is robust and rigid—an excellent, much needed improvement.

The pistol grip, forearm (which has AKM-type finger swells) and buttstock, on the less-common AK-74 version, are the usual laminated wood.

The gas tube on the AK-47, AKM and AK-74 contains longitudinal depressions around the entire circumference to provide a reduced bearing surface for the piston and permit excess gas blow-by. The gas tube on the Finnish M-62 and the Israeli Galil have smooth surfaces, and thus the piston itself is notched to create this effect. The AK-74 gas tube has a spring washer attached to its rear end that is not found on other Kalashnikovs. Designed to retain the gas tube more securely, it requires a sharp slap when reassembling.

The AK-74 rear sight is a tangent type with an open U-shaped notch. It is adjustable for elevation only out to 1,000 meters—a complete fantasy since this is well beyond the accuracy potential of both weapon and operator. The front sight is a threaded post type, adjustable for elevation zero with the standard Kalashnikov combination tool. Windage adjustment can be done only by sliding the front sight in its dovetail.

The AK-74 is finished in a high-gloss black enamel. The issue web sling is of the spring-hook variety common to Soviet small arms.

Does the AK-74 have a bayonet? I don't know; I never saw one and neither had any of

Field-stripped AKS-74.

the mujahideen I queried. The AK-74 muzzle device would certainly prevent use of the AKM bayonet, yet the weapon has a bayonet stud below and to the rear of the front sight.

The specimen I fired was manufactured in 1976 (serial number 597780). Seven years after the manufacture of almost 600,000 units, it seems certain we would have heard of a bayonet had one ever been fielded.

The AK-74's most interesting feature is certainly the 5.45x39mm cartridge for which it is chambered. First reports hinted at a ballistic breakthrough and rumors were widely circulated that the muzzle velocity exceeded 4,000 fps. Stories of massive tissue damage filtered out of Afghanistan. The first large quantities of this ammunition were provided by *SOF* Editor/Publisher Robert K. Brown, and turned over to the U.S. Army and the National Rifle Association for testing. Test results indicate there is nothing particularly astounding about the 5.45x39mm round.

New AK-47 Magazine

The Russians have recently fielded an entirely new 30-round magazine for the AK-74. Uncovered during *SOF*'s most recent incursion into Afghanistan, it has until now remained unreported by Western intelligence sources.

The original AK-74 magazine, except for changes in geometry required by the 5.45x39mm cartridge, was identical to that last produced for the AKM. It was fabricated from a glass-reinforced, rust-colored, polyethylene plastic. Molded in two parts, the magazine body was assembled using a viscous two-part epoxy resin adhesive. The adhesive residue was removed by hand-grinding.

The four prominent notches on the upper front of the magazine come from clamp marks used in gluing the two segments together. While not phenolic, the AKM/AK-74 magazine was noted for its great strength and durability.

The new AK-74 magazine is composed of a dark-brown buterate plastic (also called ABS) of the type commonly used in appliance manufacture in the United States. The two body components, probably vacuum-formed, have been heat-molded together rather than glued.

The four clamping notches remain and the entire assembly has been finish-ground also. The magazine follower, spring and floor plate are identical to the original. All AK-74 magazines have a raised. horizontal rib on each side of the rear lug to prevent their insertion into a Kalashnikov of 7.62x39mm caliber.

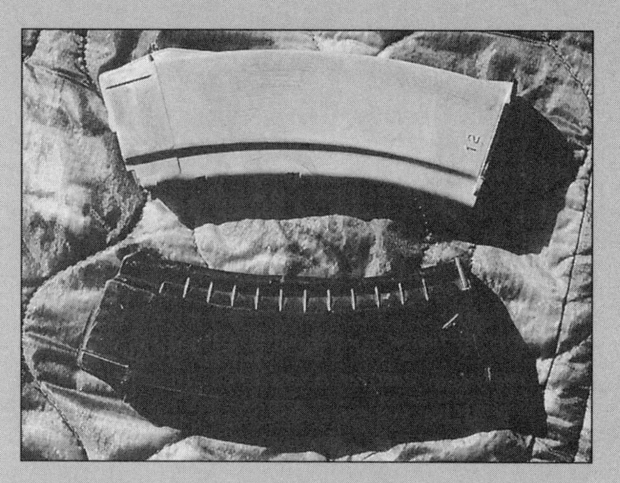

Original rust-colored AK-74 magazine (top) was identical to that last produced for AKM. New AK-74 magazine (bottom) is dark brown ABS plastic. Change to darker shade was dictated by the realization that the original color compromised camouflage.

The change to dark brown was obviously dictated by realization that the bright rust color seriously compromised positions by visual disclosure. Wall strength of the new magazine is considerably less than that achieved by use of glass-reinforced polyethylene. However, the new AK-74 magazine must be far less expensive to produce. It undoubtedly represents an attempt to design a really cost-effective, totally disposable assault-rifle magazine. It also represents the Soviets' tendency to push improved technology into service more rapidly than the United States, illustrating their preference for field-testing over the laboratory.

The brown-lacquered steel case is Berdan-primed. Its 39.37mm length makes it slightly longer than the 7.62x39mm ComBloc case which measures 38.60mm. The primer has a copper cup and is sealed with a heavy red lacquer.

The propellant charge is an unremarkable ball powder—almost primitive by U.S. standards—and similar in burning characteristics to the WC 844 powder used in our 5.56mm NATO military ball ammunition. The average charge weight is 23.0 grains, but plus or minus 1.5 grains.

The 56-grain boattail projectile has a gilding-metal-clad steel jacket. The unhardened steel core is covered by a thin lead coating which does not fill the entire point end, leaving a hollow cavity inside the nose—the cause of most of the speculation concerning its wounding potential. Others have commented that the bullet design is complex and expensive to fabricate, but in reality it is not as difficult to produce as our WWII M2 armor-piercing bullet.

The 50-grain tracer projectile has a shorter ogival profile and is green-tipped—although the igniter mix appears to be red. The amount of igniter material is only slightly less than our new L110 tracer projectile (which has a 900 meter burn) and so a burn of up to 800 meters can be expected.

The ballistic performance of the 5.45x39mm cartridge is no better than the .222 Remington. The muzzle velocity is only 2,950 fps, compared to 3,270 fps for the U.S. M193 ball ammunition. Steel-plate penetration is inferior to the M193 projectile and not even in the same ballpark with our new SS109 bullet, which will penetrate the U.S. steel helmet at 1,300 meters. Tests indicate that on soft targets, even with its hollow cavity, the 5.45x39mm round has no greater wounding capacity than the M193 bullet.

The 5.45mm bullet, largely because of its greater length, has outstanding aerodynamic characteristics and exhibits far less drag than the M193 projectile. The Soviet bullet's overall length of .990 inch requires an extremely fast twist for adequate stabilization. The four-groove AK-74 barrel has a right-hand twist of 1:7.7 inches. This is fast (compared with the M193 bullet, which needs only 1:12 inches). So fast, in fact that the barrel lands on the leading edge are beveled to minimize distortion of the bullet's steel jacket.

In conclusion, there is nothing spectacular about the 5.45x39mm cartridge. It barely approaches the performance level of our Vietnam-era M193 ammunition. It is an entire generation behind our new SS109 ball ammunition.

The most immediately distinctive feature of the AK-74 is its muzzle device. The brake is threaded onto the barrel's muzzle and held in place by a spring-loaded button. It is easily

Darra Adam Khel:
Everything in Weaponry and Then Some

So you have always wanted a bolt-action carbine in caliber 7.62x39mm ComBloc that used Kalashnikov 30-round magazines? Furthermore, you wanted it to be fitted with the most exotic looking—but probably useless—muzzle compensator that anyone has ever seen?

As *SOF*'s Foreign Correspondent Mekong Jim Coyne and I found out, all that and much more is available from the gun dealers of Darra Adam Khel in Pakistan. Using foot-pedal lathes, primitive forging methods and lots of files, Darra's gunsmiths can make or duplicate almost anything that does not involve stamped sheet-metal processes.

The rifles made in Darra are mostly bolt-action. Exhibiting many years of British influence, Enfields and their local hybrids predominate. Strange Mauser/Enfield combinations are also common. Darra-made submachine guns are always variants of the Sten, usually Mk Vs with the vertical foregrip. In pistols, the Webley prevails, as do small pocket autos that are Spanish Star spin offs.

Modern assault rifles, which depend so much on spot welded, die-stamped sheet metal are either smuggled in from other areas (such as Pakistan-produced G3s) or more often, in the case of Kalashnikovs, taken off very dead Ivans.

removed. The major portion of the brake consists of a large, two-inch-long expansion chamber. Three vent holes have been drilled into the rear end of the chamber, one to the right and two toward the top. This positioning of the holes prevents muzzle climb upward and to the right when fired by a right-handed shooter, and deflects gases sideways, preventing backblast from reaching the firer. Two large vertical cuts have been made at the forward end of the brake on each side, and are offset toward upper dead center to further drive the muzzle downward. Finally, a flat plate (open in the center for passage of the projectile) deflects the gases and produces a forward thrust.

The shooter's most lasting impression of the AK-74 is the remarkable effectiveness of its muzzle brake. Felt recoil is almost nonexistent. During full-auto bursts my line of sight never moved from the target. In the kneeling position, using a hasty sling, two-round full-auto bursts were consistently held to only six-to eight-inch groups on 200-meter targets. This is no worse than the accuracy potential of the rifle and ammunition, which after several hundred rounds I judged to be no better than about six true minutes of angle (MOA). The Soviets consider this combat-acceptable since they do not build small-arms systems with the Camp Perry matches in mind.

The stability of the AK-74 muzzle brake for the shooter is somewhat counterbalanced by its effect on those poor souls located on his immediate flanks: The side blast is simply horrendous, as I learned when I photographed Mohammad Kareem, the mujahideen camp commander (a former brigadier in the Afghan Army) during our firing sequences. As they say, everything involves some trade-offs.

The AK-74's trigger pull is typical Kalashnikov. It's not two-stage, just one long scratchy creep with a sudden let-off at the end. It takes some getting used to. Mohammad Kareem was never able to achieve anything shorter than a three-round burst. Too many years with an Enfield, I guess.

As expected, the AKS-74 operated without malfunctions of any kind. However, I noted that the lack of a hold-open was often disconcerting even to the mujahideen. Balance characteristics of the AKS-74 are superb, even when the 40-round RPK-74 magazine was inserted. Cyclic rate of the AKS-74 appeared to be in the area of 600 to 650 rpm.

The AK-74, and more commonly the AKS-74, are now ubiquitous among Soviet troops throughout Afghanistan. The mujahideen are impressed with it and it sells for more than $4,500 in the gun shops of Darra Adam Khel. It is extremely robust and reliable, following in the Kalashnikov tradition. With the exception of its muzzle brake, there is nothing revolutionary about the AK-74 series. It is a most logical and natural evolution of the Kalashnikov system.

The AK-74 carries the Soviets into the current mainstream of military small-arms technology. The overwhelming adoption by NATO and ComBloc armies of the 5.56mm and 5.45mm cartridges, respectively, is correct, as I've said before. It will continue, despite the anachronistic carping of those on the sidelines.

Originally appeared in *Soldier Of Fortune*
May 1983.

Section 3
China

SOF *Editor/Publisher Robert K. Brown fires PRC Type 81 Squad Automatic Weapon (SAW) with 75-round drum.*

Guns Behind the Great Wall

SOF Scoops the CIA . . . Again

PRC Type 81 rifle, caliber 7.62x39mm, fieldstripped with sling and 30-round magazine.

Publisher's note: SOF *in communist China? You've got to be kidding. We've been to a lot of weird places during the last 12 years, but we never thought we'd be jogging the Great Wall, firing PRC small arms and eating bizarre, "I don't want to know" parts of plants, fish and fowl at the invitation of the Chinese Communist government. But then that's how it is with* SOF.

SOF *was approached by contacts from the PRC about a year and a half ago to test and evaluate a number of small arms never before seen outside the Bamboo Curtain. Now, we've never been moderate about our anti-communist position in general, and have never been enthusiastic about the PRC's form of government in particular. We fought them in Korea. They backed the North Vietnamese and Robert Mugabe's terrorists in Rhodesia. No love lost there.*

But times and governments change. SOF *decided to adopt "the enemy of my enemy is my friend" concept. It would also appear that so have the Chinese. After all, for all its warts and past sins, the PRC is fighting the Vietnamese on its borders, aiding the anti-Communists in Cambodia and Laos, and providing large quantities of Russian-killing items to the Afghan freedom fighters. (In April 1985, when we were training the contras, we saw a half-million PRC manufactured 7.62x39mm rounds in a contra base camp. How they got there, we don't know.) And, conspiracy theorists notwithstanding, the Chinese are no friends of the Soviets. Neither, of course, are we.*

But why SOF? *Why not* International Defense Review, Jane's, Armed Forces Journal *or any of a number of prestigious military trade maga-*

zines? Or, they could have invited a number of journalists from various publications if in fact the PRCs main objective was to gain maximum exposure for its line of small arms for military sales. We wouldn't get an answer to that mystery until we got to China.

My curiosity as to why we were chosen prompted me to call a number of foreign affairs experts. One source speculated it was some "Byzantine Chinese plot" with unknown objectives. General Jack Singlaub quoted an old Chinese proverb: "It is better to sit down across a table from an enemy you know than a friend you don't know." A well-known international defense consultant and military author agreed with Singlaub, stating, "The PRC would rather sit down with a known, hard-core anti-Communist than a wishy-washy liberal. If Carter had been president, no rapprochement would have been effected with the PRC by a Jimmy Carter-type president. The Chinese knew where Nixon stood and therefore felt comfortable in dealing with him."

There were questions to be answered, mysteries to solve and weapons to fire. It was time to go.

•••••

With a strength of some 3,625,000 regulars, the People's Liberation Army (PLA) of China is the largest in the world. This is a most misleading statistic.

Undermechanized and largely equipped with outdated weaponry at all levels, the PLA is primarily a foot-mobile army that would find itself at a severe dis-

TYPE 81 SPECIFICATIONS

CALIBER 7.62x39mm

OPERATION . . . . .Gas operated with adjustable three-position regulator; piston not attached to bolt carrier. Locking by means of rotary two-lug bolt. Fire from the closed-bolt position.

CYCLIC RATE . . . . .Varies with gas regulator setting: 650 rpm with regulator set to position "1," 740 rpm at position "2."

FEED . . . . .Detachable 75-round spring-wound drum or 30-round staggered box-type magazine. Not interchangeable with Kalashnikov magazines.

WEIGHT, ... empty 7.5 pounds (rifle);
. . . . .7.7 pounds (Type 81-1, stock folded);
. . . . .11.4 pounds (SAW).

LENGTH, ...overall 38.2 inches (rifle);
. . . . .29.2 inches (Type 81-1, stock folded);
. . . . .40.2 inches (SAW).

BARREL . . . . .Four-groove with a right-hand twist of one turn in 9.6 inches. Chrome-plated chamber and bore.

BARREL LENGTH . . . . .17.6 inches (rifle);
. . . . .20.8 inches (SAW).

SIGHTS . . . . .Hooded, round front post; adjustable for windage and elevation zero. Open, square-notch rear; adjustable for elevation from 100 to 500 meters (SAW: 600 meters) in 100-meter increments by means of knurled range drums on either side which lift or lower sight bar on their eccentric axis pin. SAW uses notch in top bar of rear sight assembly for engagement at 700-800 meters.

FINISH . . . . .Metal components salt-blued, except for phosphated bolt and carrier and hard-chromed gas regulator and piston head.

FURNITURE . . . . .Wood, either mahogany- (rifle) or walnut- (SAW) stained with varnish finish.

ACCESSORIES . . . . .Bayonet (rifle only), rifle grenades (rifle only), sling, cleaning kit and spare parts.

STATUS . . . . .Currently in production; adopted for service in the People's Liberation Army in 1984.

MANUFACTURER . . . . .PRC government arsenals.

EXPORTER . . . . .Poly Technologies, Inc., 5/F, CiticBuilding, 19, Jian Guo Men Wai Street, Beijing, People's Republic of China.

T&E SUMMARY . . . . .With 80 percent parts interchangeability, an excellent example of a modern infantry squad weapons system; based upon highly reliable Kalashnikov-type locking; adequate accuracy potential; low cyclic rate permits short-burst fire; low perceived recoil with high hit probability; cost effective; SAW needs quick-change barrel and improved heat shield.

advantage were it to engage the Soviet Union in a major military confrontation at this time.

However, the recent move from a previously Leninist society to "market socialism" has provided the PLA with an important opportunity to upgrade its military potential through importation of Western technology and the development of indigenous designs. Western authorities still regard the Chinese defense industry as geared to the production of Soviet copies dating back to the 1950s. As we shall see, this assessment no longer holds true, at least in the area of military small arms.

Adopted in 1984 by the PLA, the new Type 81 assault rifle and Squad Automatic Weapon (SAW) represent convincing proof that PRC designers are acutely tuned to the combat user's requirements and are fully capable of executing designs that incorporate time-proven concepts with numerous innovative features.

Let's first take a close look at the PRC's new assault rifle, which up to now has only been described, inaccurately, by *Jane's Defence Weekly*.

TYPE 81 ASSAULT RIFLE

Chambered for the ubiquitous 7.62x39mm cartridge, the Type 81 series (which refers to the year development commenced, in this case 1981) is gas operated and fires from the closed-bolt position. Equipped with either a fixed or folding stock (designated Type 81-1), the weights, empty, are 7.5 pounds and 7.7 pounds respectively. This is somewhat heavier than the AKM. Overall length is 38.2 inches, regardless of stock. With the stock folded, this is reduced to 29.2 inches. Except for the bolt group, which is phosphate finished, and the hard-chromed gas components, all metal surfaces have been salt blued.

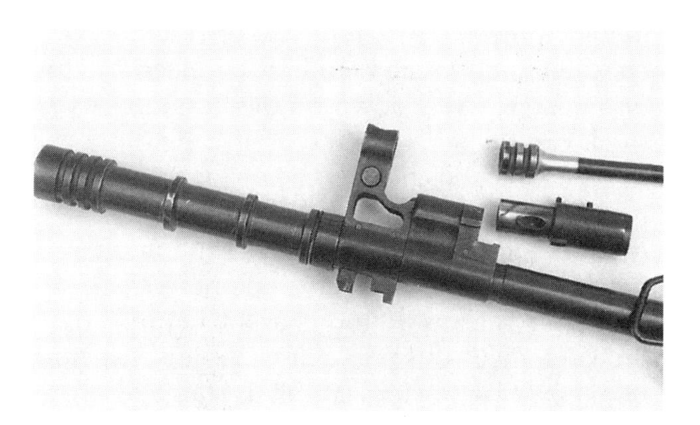

Close-up of PRC Type 81 rifle, showing integral grenade launcher, front sight assembly, gas block, gas regulator and piston.

Both weapons in the Type 81 series have adjustable gas regulators marked "1," "0" and "2." The "0" position is used for launching rifle grenades with ballistite (blank) ammunition. The "1" position is used for normal conditions and "2" is the adverse setting. The base of a cartridge case is used to rotate the regulator. Setting "1" provides a cyclic rate of 650 rpm. This is increased by approximately 100 rpm at the "2" position, unless the weapon is badly fouled.

The retracting handle, shaped like that of the SKS, is permanently attached to the right side of the bolt carrier and reciprocates with it.

The receiver has been fabricated from a pinned and riveted, sheet-metal U-section. Missing are the small depressions over the magazine on each side of the receiver, which serve as magazine guides on the AKM. The sheet-metal receiver cover has four transverse ribs in the manner of the Soviet AKM. A projection on the end of the guide rod, quite different in appearance from that of the AKM, locks the receiver cover in place.

Type 81 series barrels have four grooves with a right-hand twist of one turn in 9.6 inches. Bores and chambers are chrome plated. The rifle barrels are 17.6 inches long and have been fitted with a muzzle brake combined with gas rings for grenade launching. This appears to be integral with the bayonet lug, front sight assembly and gas block.

Front sight units are standard Kalashnikov. Adjustable for elevation and windage zero, the post-type front sight rests within a protective hood with a hole in the top for insertion of an adjustment tool. An open square-notch rear sight has been attached to the front end of the receiver, above the chamber. Rotation of a knurled knob on either side of the sight base raises or lowers the sight leaf in 100-meter increments from 100 to 500 meters (far more realistic than the 800- and 1,000-meter settings found on the AK-47 and AKM).

The selector switch has been moved to the left side of the receiver by placing the Kalashnikov's sheet-metal bar selector inside the receiver on the left side. This thumb-operated manual lever mimics that of the M16, except that the three positions (marked "1," "2" and "0") are the mirror image of the M16's and their location is thus ergonomically incorrect. When set to safe ("0"), the lever must be pulled rearward, first to full auto ("2") and then to semiautomatic ("1"). These positions should be transposed so that pushing forward will move the lever off safe into the semiauto mode and then into full auto.

The magazine-well, equipped with a flapper-type magazine release taken from the Kalashnikov, will accept either a steel 30-round box magazine or 75-round drum. Unless altered, neither can be inserted into a Kalashnikov-series weapon. A raised edge on the box magazine's follower activates a spring-loaded hold-open inside the receiver. This is a useful feature, omitted on the Kalashnikov series. The hold-open is inoperative when the drum is used. Both magazine and drum must be rolled rearward into the locked position in the magazine-well.

Primarily intended for use with the Type 81 SAW, the drum outwardly resembles the Soviet 75-round RPK drum, but its internal mechanism is partly derived from the Soviet 71-round PPSh-41 submachine-gun drum.

To charge the PRC Type 81 drum, first open the hinged back cover. Depress the spring-tension release button at the center of the drum and rotate the spindle clockwise until the cartridge carrier stops at the end of the spiral track. Load five rounds, bullets down, into their slots on the outer rack, immediately to the right of the magazine's feed slot. Hold the spindle and rotate the spring-tension knob one complete revolution. Release the spindle slowly and allow the five cartridges to move up into the feed slot. Drop in the remaining 70 rounds. Close the rear cover and secure the latches.

This drum can be stored loaded, but unwound, for an indefinite period. Only six more complete revolutions of the key winder on the outside of the drum are required to ready the drum for firing. Modified to function in all Kalashnikov derivatives, the Type 81 drum is available from Keng's Firearms Specialty, Inc.

Consisting of the buttstock, pistol grip and handguards, the wood furniture on the Type 81 series is either walnut- (SAW) or mahogany- (rifle) stained and with a varnish finish. The pistol grip has eight longitudinal grooves, rather crudely cut, on each side . The upper handguard contains an open, sheet-metal piston housing that also serves as a heat shield. Sling swivels are attached to the front of the handguard and buttstock, both on the left side. A green web sling with leather end tabs and aluminum mounting studs is standard issue. A cleaning rod rides under the barrel, partially housed in the lower handguard.

The side-folding stock has been taken from one of PRC's AKS variants. This stock's skeletonized frame has been fabricated from a single piece of heavy-gauge sheet metal with five horizontal ribs stamped into the buttplate. A 4.75-inch long, bright reddish-brown plastic panel on each side of the stock hides the cleaning kit (depress a small, spring-loaded pin protruding through the top stock strut to remove the kit containing a bore brush, jag tip and combo tool). These panels have ribbed gripping surfaces and are retained by a threaded screw and pin.

This entire assembly is securely attached to the latch mechanism, which is operated by a checkered, spring-loaded button-release on top of the latch. Press down on the button and the stock can be swung to the right, where it locks in the closed position by means of the same latch mechanism. No locking hook is thus required on the receiver body. The button must be depressed again to open the stock. While the release button is easily depressed, the latch mechanism is quite substantial, helping make this stock every bit as stable as the rigid wooden buttstock.

The Type 81 bayonet (available from Keng's Firearms Specialty, Inc. for $50) represents a somewhat startling departure from the AKM style. Although the pommel, grips and cross-piece resemble those of the 2nd-pattern AKM bayonet, they are thinner; this bayonet cannot be fitted to a Kalashnikov. Even more unusual is the plated, spear-point, unsharpened blade which has double fullers (so-called "blood grooves") on each side and appears to be little more than a flattened and shortened version of the PRC cruciform spike bayonet. Dubious features, such as sawteeth and wire cutters, have been omitted. The blade is 6.7 inches long and the overall length is 11.8 inches. The green plastic scabbard has an aluminum mouthpiece and plastic bell hanger. There is no provision for attaching this bayonet to the Type 81 SAW.

Type 81 Method of Operation

A two-lug rotary bolt of the Kalashnikov type has been employed on both the Type 81 assault rifle and Squad Automatic Weapon (SAW). Operation is as follows: After ignition of the primer and propellant, gases are diverted through the gas block on top of the barrel and into the three-position regulator. The piston, in this case not connected to the bolt carrier and with a short, multiple-strand, helical spring wrapped around its shaft, is driven rearward to strike the bolt carrier. The bolt carrier goes through the necessary amount of free-play travel until the gas pressure drops to a safe level. A cam slot milled into the bolt carrier engages the bolt's cam stud and rotates the bolt about 35 degrees to unlock it from the barrel extension's locking recesses. As there is no primary extraction, a large extractor claw is required.

When the bolt carrier travels back, it rolls the hammer over and compresses the recoil spring, which together with its guide rod rides in a hollow in the rear of the carrier. The bolt ceases its rearward travel by slamming into the rear end of the receiver. The recoil spring then drives the bolt forward, another round is stripped from the magazine (or drum) and chambered, and the bolt then comes to rest. The bolt carrier continues onward for about 5.5mm after locking has been completed. The piston returns to the regulator's cup under the force of its own spring.

Except for the selector system, a trigger mechanism of the Kalashnikov-type has been utilized, although the components are not interchangeable. Based on the firing mechanism of the M1 Garand, the hammer has two contact surfaces to engage a primary and auxiliary sear.

When cocked, the hammer is held back by the primary sear, which is part of the trigger lever. In semiautomatic fire, when the hammer is rolled back by the recoiling bolt carrier, it is caught by the auxiliary sear (spring-loaded by a single-strand coil). When the trigger is released, the primary sear moves back to catch the hammer as the auxiliary sear is rotated clear.

In full-auto fire, the auxiliary sear is rotated out of contact with the hammer. As long as the trigger is depressed, firing is controlled by the auto safety sear. The auto safety sear, a spring-loaded (single-strand coil) lever, is operated by the counter-recoiling bolt carrier and permits the hammer to fall on the inertia (non-spring-loaded) firing pin only after the carrier is fully forward. There is no anti-bounce device of the type encountered on the Soviet AKM. The hammer spring is a multiple-strand coil. Trigger pull weight on the specimen we tested was a scratchy 5.5 pounds.

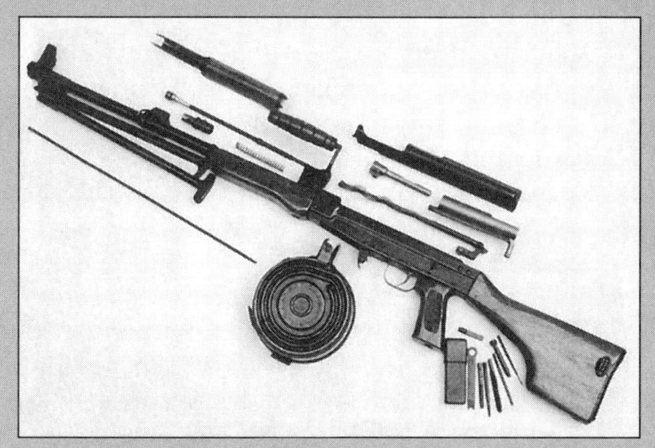

Type 81 Squad Automatic Weapon (SAW), fieldstripped with 75-round drum and cleaning kit.

TYPE 81 SAW

While more than 80 percent of the Type 81 assault rifle's reciprocating components are interchangeable with those of the Type 81 squad machine gun, there remain some significant differences. Although unfortunately not of the quick-change type, the Type 81 SAW's barrel is longer (20.8 inches), giving the weapon an overall length of 40.2 inches, and heavier, bringing the weight, empty, up to 11.4 pounds. The added barrel length only increases the 7.62x39mm cartridge's muzzle velocity by 50 fps (to a total of 2,411 fps). Cyclic rates remain essentially the same.

Potential for grenade launching has been removed and an RPD-type muzzle nut attached in the AK-47 manner. An RPD-type, non-adjustable bipod has been mounted at the muzzle end of the barrel. The front sight remains the same. A 600-meter elevation increment has been added to the rear sight, and a square notch cut into the rear sight assembly's top crossbar is used to engage targets at 700 to 800 meters. A wooden carrying handle with three spring-locked positions is fixed to the rear sight base on the left side. Buttstock and buttplate configurations are also derived from the RPD. A plastic box within the butt trap contains a combo wrench, two disassembly punches, broken case extractor, front sight adjustment tool, bore brush and jag tip for the cleaning rod.

Disassembly procedures of both Type 81 weapons are identical to the Kalashnikov series, with the following exception. To disassemble the gas system, first rotate the rear sight elevation knobs past the highest elevation setting. Lift up the upper handguard and heat shield from the rear end and separate them from the gas regulator. Rotate the gas regulator either to the right or left, past the numbered positions, and withdraw it from the gas block. Draw the piston and spring forward and remove.

SOF's test team fired hundreds of rounds through both the Type 81 rifle and SAW. There were no stoppages of any kind. Both weapons have excellent handling characteristics. Perceived recoil and muzzle jump were minimal. At either cyclic rate, the test team had no trouble in consistently firing two- to three-shot bursts. Burst-fire hit probability is quite high. Accuracy testing was conducted using targets electronically monitored by equipment developed and manufactured in the PRC. Both weapons are capable of firing semiautomatic groups of 3 to 4 MOA (minutes of angle), about all you can expect from a sheet-metal receiver. Ejection is typically violent and most of the empty cases sail at least 20 feet to the right front.

During sustained-fire tests, the Type 81 SAW's upper handguard became hot and started to smoke. An improved heat shield and quick-change barrel would eliminate this fire hazard. Altering the selector's sequence of positions is the only other modification I can suggest. Overall, this appears to be an excellent and highly integrated squad weapon system. None of the *SOF* test team would hesitate to carry these weapons into a combat environment.

TYPE 74 LMG

Hidden behind the bamboo curtain was another surprise for *SOF*'s test team—the Type 74 Squad Light Machine Gun. Once more a clever blend of combat-proven designs and indigenous adaptations, this interesting squad automatic has also been chambered for the 7.62x39mm cartridge. It would, however, be a serious mistake to suppose that the PRC is forever committed to this caliber. Although *20 billion rounds* of 7.62x39mm remain in the Chinese inventory, *at least 50 different small arms cartridges*, ranging in caliber from 5.2mm to 6.2mm, are currently under development and experimentation. Hottest contender at this time is a unique 5.8x42mm cartridge. This latter project is classified and no additional info is presently available.

But, back to the Type 74. It weighs only 14.1 pounds, empty. Overall length is 43.6 inches. The nine-groove barrel is 20.8 inches long and has the standard right-hand twist with one turn in 9.6 inches. Chamber and bore are, of course, chrome lined. All components, with the exception of the lightly phosphated bolt and

Type 74 Specifications

CALIBER 7.62x39mm

OPERATION . . . . .Gas operated with adjustable four-position regulator. Goryunov-type (SG-43) propped breech locking: After bolt reaches battery position, continued advance of the slide cams rear end of bolt in front of locking shoulder on left receiver wall. Fires from the open-bolt position.

CYCLIC RATE . . . . .750 rpm.

FEED . . . . .Detachable 100-round spring-wound drum or 30-round staggered box-type magazine (Kalashnikov—Type 56).

WEIGHT, empty . . . . .14.1 pounds.

LENGTH, overall . . . . .43.6 inches.

BARREL . . . . .Four-groove with a right-hand twist of one turn in 9.6 inches. Chrome-plated chamber and bore.

BARREL LENGTH . . . . .20.8 inches

SIGHTS . . . . .Hooded, round front post in RPD-type housing; adjustable for windage and elevation zero. Open, square-notch rear; sliding tangent-type; adjustable for elevation from 100 to 800 meters with battle-sight setting.

FINISH . . . . .Metal surfaces salt-blued, except for phosphated bolt and slide and hard-chromed piston and gas regulator.

FURNITURE . . . . .Wood, RPD-style buttstock; checkered, black plastic pistol grip panels; red, fiber-reinforced phenolic handguards.

ACCESSORIES . . . . .Sling, spare drums and magazines, cleaning kit and spare parts.

STATUS . . . . .Currently in production; in service with People's Liberation Army.

MANUFACTURER . . . . . .PRC government arsenals.

EXPORTER . . . . .Poly Technologies, Inc., 5/F, Citic Building, 19, Jian Guo Men Wai Street, Beijing, People's Republic of China.

T&E SUMMARY . . .Lightweight squad machine gun; innovative combination of combat-proven design features; cost-effective; sturdy and reliable; low perceived recoil; excellent accuracy potential and high hit probability; lacks quick-change barrel.

slide and the hard-chromed piston and gas regulator, are salt blued.

There is a spring-loaded dust cover on the ejection port that flies open when the bolt group moves to the rear. The retracting handle is non-reciprocating. Both the recoil spring and its guide rod ride within the slide's piston extension. The end of the guide rod is used to retain the sheet-metal receiver cover in the Kalashnikov manner.

The gas system has been taken from the PRC Type 63 rifle (incorrectly referred to as the Type "68" by Western sources). To adjust the four-position regulator, rotate the regulator pin's spring clip from its notch on the gas block and withdraw the pin. Turn the regulator in either direction until the desired hole is aligned with the two holes in the gas cylinder. Reinsert the retaining pin and pivot its spring clip back into the locked position. The normal operating position is "2." A small, single-strand coil spring between the regulator and piston head prevents the piston from impinging directly upon the regulator's face.

The trigger mechanism is an adaptation from the RPD (PRC Type 56-1). A hook on the spring-loaded trigger enters an opening in the sear. When the trigger is pulled, the hook draws the sear downward, out of engagement with the slide's bent (notch). Rotating the safety lever (located on the right side, above the trigger guard) forward locks the sear in the upward

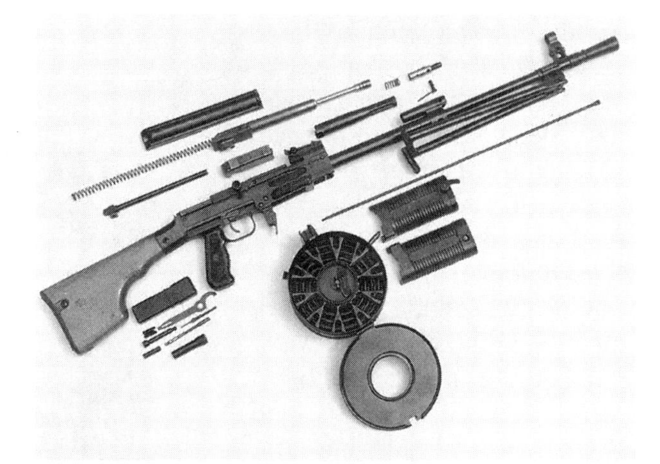

PRC Type 74 squad machine gun, caliber 7.62x39mm, fieldstripped with cleaning kit and 100-round "music box" drum.

position. There is no provision for semiautomatic fire. The cyclic rate is 750 rpm.

A ratchet-locked, conical flash hider can be removed from the muzzle by means of a spanner on the end of the combination tool included in the cleaning kit. A raised ring at the end of the flash hider holds the cleaning rod in place under the barrel.

Both the front sight assembly and the bipod are right off the RPD. Covered by a protective hood, the conventional front post is adjustable for both elevation and windage zero. A sliding, tangent-type rear sight with open square notch accommodates elevation adjustments in 100-meter increments from 100 to 800 meters along with a battle-sight setting.

No attempt has been made to "color coordinate" the Type 74's furniture. Fabricated from reddish, fiber-reinforced, phenolic plastic, the two-piece handguards are retained by a single captive pin that is easily removed. The two-piece, checkered, black plastic pistol-grip panels have two finger grooves and are held in place by one screw. Reminiscent of the RPD, the orange-stained buttstock has a single swell for gripping with the support hand. It is attached to the receiver by a crossbolt, again in the RPD manner. Its butt trap contains a plastic box with slotted cleaning rod tip, broken case extractor, sheet-metal reamer for the gas tube, gas port reamer, combo tool, bore brush, punch and oiler. Another compartment in the buttstock carries a spare gas regulator spring. Sling swivels have been attached to the gas block and rear of the buttstock, both on the left side.

Either standard 30-round Kalashnikov magazines or a unique 100-round drum can be inserted in the magazine-well. Dubbed the "music box" by *SOF*'s test team, this 100-round drum makes a loud whirring sound after the last round has been fired as its mainspring unwinds, immediately notifying friend and foe alike of your predicament. Loading procedures generally duplicate those of the Type 81 75-round drum.

Disassembly of the Type 74 is every bit as simple as its design. After removal of the receiver cover, in the usual manner, withdraw the recoil spring and guide rod. With the safety lever in the firing mode, pull back on the charging handle and draw the slide and bolt group fully rearward. Lift them out of the receiver and separate the two components. Rotate the handguard retaining pin upward 90 degrees and pull it out to the right. Remove the handguards. Pull the gas tube away from the gas block and lift it up. Rotate the gas regulator's pin out of its notch in the gas block and pull it to the left and out of the gas block. Withdraw the gas regulator and its spring from the front end of the gas block. No further disassembly is required to service the Type 74.

Uncomplicated, but robust and reliable, the Type 74 needs only a quick-change barrel to move it up to the superlative. At 14.1 pounds, perceived recoil is negligible. Resting directly under the weapon's center of mass, the 100-round drum neither disturbs the handling characteristics nor interferes with firing from the

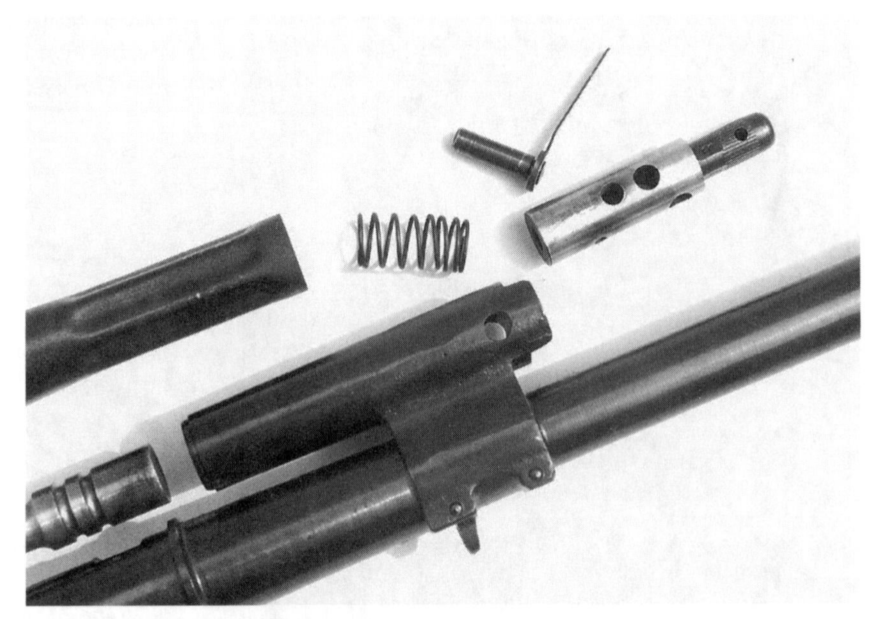

Type 74 gas system, disassembled.

Type 74 Method of Operation

The receiver on the Type 74 Squad Light Machine Gun has been fabricated as a substantial milled forging which acts as a heat reservoir to prevent "cook-offs" during sustained fire. Gas operated and firing from the open-bolt position, the system of operation is essentially that of the Russian SG-43 (Stankovaya Goryunov 1943) Medium Machine Gun (piston actuated with propped breech locking), except that it has been turned upside down.

Thus, the gas system rests on top of the barrel, as the ammunition is fed from the bottom of the receiver, by magazines instead of belts. After the bolt (breech) goes into battery, the slide (attached to the piston) continues forward to cam the rear end of the bolt over in front of its locking shoulder on the left side of the receiver.

This principle was patented by John M. Browning in 1895 but never used, as dangerous side loads are supposedly transferred to the weaker walls of the receiver. However, this theoretical consideration never seemed to affect the SG-43's endurance or reliability.

The cam projection on the bottom of the slide also serves as the hammer. After locking, the slide continues forward and this stud strikes the inertia firing pin. At the initiation of the recoil momentum, the piston and slide move rearward in free travel until the chamber pressure drops to a safe level. Then the cam projection on the slide engages the unlocking slot in the bolt, pulling the bolt away from its locking shoulder and driving the entire assembly rearward.

Ejection occurs during the recoil stroke. The ejector pin rides in a tunnel on the left side of the bolt body. When the bolt is thrust away from its locking shoulder, the ejector pin protrudes through the bolt face after its wedge-shaped end hits the locking shoulder, propelling the empty case out the ejection port on the right side of the receiver.

prone position. If firing is kept to short bursts, the beaten zone remains quite small at all practical ranges, as the bipod's location close to the muzzle maximizes accuracy potential and hit probability at the expense of some lateral mobility. After several drums of sustained fire, the handguards were only slightly warm to the touch. When used in the assault role, the Type 74 can be fired with effective accuracy from the standing, kneeling and hip positions.

No doubt about it, the rugged Type 74 squad machine gun represents a significant improvement over the RPK. In addition to a quick-change barrel, I can recommend only a bipod with adjustable command heights, an "unmusical" drum and possibly a slightly lower cyclic rate.

No longer the Soviet Union's clone, the People's Liberation Army has surged forward without hesitation into the rapidly moving mainstream of modern military small aims technology. Its products should find great favor with military forces seeking simple and reliable small arms that complement the current concepts of fire and movement in a cost-effective manner.

Originally appeared in *Soldier Of Fortune*
September 1987.

GUNS BEHIND THE GREAT WALL, PART 2

SOF Editor/Publisher Robert K. Brown firing Type 77 HMG in People's Republic of China.

China's Hind-Killer HMGs

SOF Technical Editor Peter G. Kokalis fires PRC Type 54, caliber 12.7x108mm HMG. Note huge fireball which invariably throws burning powder embers back into the operator's face.

No matter what the cyclic rate, the sharp, painful pounding of a heavy machine gun can be instantly distinguished from its more feeble cousins. But do 80-pound machine guns firing ammunition designed in World War I really have any legitimate applications in modern counterinsurgency warfare? You bet they do.

Russia's Model 38/46, caliber 12.7x108mm heavy machine gun remains a principal weapon in the inventory of the Afghan freedom fighters. While originally designed to penetrate light vehicular armor and now overshadowed by Stingers and Blowpipes, the "Dashika" has popped its share of Soviet aircraft. In El Salvador, Immediate Reaction Battalions have found John Browning's .50-caliber M2 HB (so-called "Ma Deuce") to be an effective long-range antipersonnel weapon.

Military planners in the PRC are equally convinced of the HMG's utility on the modern battlefield. During *SOF*'s recent trip behind the Bamboo Curtain, we had the opportunity to thoroughly test and evaluate two PRC HMGs, one a direct copy of the Russian Model 38/46, the other a new and unique lightweight 12.7x108mm gun never previously exposed to Western small arms authorities.

PRC TYPE 54 HMG

All PRC small arms carry type designations indicating the last two digits of the year of adoption or commencement of development. Those with type designations in the 50s range were invariably Soviet clones. The Type 54 (1954) HMG is an exact duplicate of the Degtyarev Pekhotnyy (DP) Model 38/46 machine gun. This weapon was itself a modification of the DShK 38, which featured a rotary feed system designed by G.S. Shpagin. In 1946 this rather complicated feed mechanism was replaced by the more conventional shuttle system used on the RP46 (a belt-fed version of the pan-fed DP).

While usually encountered mounted on armored vehicles, pity the poor grunts who have to tote the ground version of this horrendous weapon. Sans the barrel, the receiver group alone weighs 78.5 pounds, empty—18.5 pounds more than Ma Deuce. Add another 28 pounds for the barrel and you can see why it's commonly dragged along the ground on a wheeled mount. Let's not forget the ammo load either. Each 70-round belt comes close to 35 pounds. Overall length of the Type 54 is 62.5 inches.

Barrels on the DShK 38 were difficult to change and this problem was also addressed when the Model 38/46 (PRC Type 54) was developed. Using the combination tool usually supplied with the weapon, you need only unscrew the barrel lock's retaining nut, push the barrel lock to the side and pull the barrel forward and out of the receiver. Since the barrel has no carrying handle, removal when it's overheated (precisely the time a barrel change is required) is still an unwieldy process.

Type 54 barrels have radial cooling fins over almost their entire length, supposedly to increase the barrel's surface area and improve the rate of heat loss to the atmosphere. In practice, however, the improvement in heat loss is minimal unless there is a constant air flow over the barrel. The majority of post-World War II weapons designs have abandoned cooling fins for this reason. Also, the labor-intensive machining process required to make the fins is not a cost-effective production technique.

These barrels are 42.1 inches in length. The bores have eight grooves with a right-hand twist of one turn in 15.2 inches. The distinctive bulbous muzzle device is moderately effective in controlling muzzle jump at the expense of considerable side blast and muzzle flash. Great balls of fire spew out of the muzzle and burning embers of propellant all too frequently fly back into the operator's face.

Both the ground and antiaircraft sights on this weapon leave a great deal to be desired. Little thought was expended on the design of the ground sights. A protected, round, post-type front sight can be adjusted for windage and elevation zero. An open, U-notch rear sight of the sliding tangent-type can be adjusted in elevation from zero to 3,300 meters (markings: 0 to 33) in

BACKGROUND OF 12.7X108MM HMG ROUND

As a result of General Pershing's demand for a large-caliber machine gun, John Browning commenced development of an up-scaled version of his .30-caliber water-cooled machine gun in July 1917. It took Browning little more than a year to complete his efforts, using a cartridge based upon the U.S. .30-06 cartridge.

Soviet work in this area did not start until the early 1930s and eventually resulted in the DShK 38 (Degtyarev Shpagin Heavy Machine Gun—1938). Based upon the 13mm German round, the 12.7x108mm cartridge's case is 9mm longer than the U.S. .50-caliber case (12.7x99mm). At approximately 765 grains, depending upon the bullet type, its projectiles are about 50 to 60 grains heavier than the U.S. equivalents, and performance is marginally superior. Muzzle velocity, usually about 2,750 fps, is almost 100 fps slower than the U.S. round. Both are exceptionally accurate and reliable.

Most commonly encountered 12.7x108mm rounds are Armor Piercing (AP), Armor Piercing Incendiary (API), Armor Piercing Incendiary Tracer (API-T) and sometimes a High Explosive Incendiary (HEI Type ZP) with a flat-tipped, hollow-point projectile containing an air-gap fuze and PETN explosive charge. Handle this latter cartridge with caution.

The metallic "push-through" links are non-disintegrating and difficult to charge by hand. I have observed the mujahideen beating fresh cartridges into these belts with rocks.

100-meter increments. There is a windage knob on the rear sight's base. Attempting to engage a target at 3,300 meters with an open U-notch is an example of Marxist optimism, especially so since the maximum effective range of this weapon is no more than 2,000 meters.

The antiaircraft sight is, if anything, worse and requires the coordinated efforts of two individuals, as it consists of two sights coupled together. By laboriously turning a hand crank, the assistant gunner rotates his and the gunner's sights to align them with the target (certainly moving much faster than he can turn the crank handle) and thus indicate its angle of approach. Estimating the target's speed, the gunner aligns his sight with an appropriate amount of lead using one of the markers on the main ring. In most instances, before all of this has been completed, the gun crew has been zapped by the enemy plane.

Usually fitted with an armored shield, Russian two-wheeled mounts for this weapon would be more appropriate on an artillery piece, as these atavistic devices weigh 259 pounds. While the mount permits coarse free-traverse and elevation movement, there is provision for the fine adjustment of elevation only. The improved PRC Type 54-1 ground mount, with a telescoping rear leg for antiaircraft applications, weighs a more modest 64.3 pounds and is equipped with adjustable elbow rests. Adjustable padded shoulder braces can be attached to the receiver.

Within the limitations imposed by its salient features, the Type 54 HMG is a robust and reliable weapon. It needs only a reduction in overall weight, a less complex antiaircraft sight, a more cost-effective aircooled barrel and an improved method of barrel-changing to effectively meet the needs of modern, highly mobile infantry units. PRC designers have addressed these problems and presented the People's Liberation Army with an intriguing new lightweight HMG.

TYPE 54 SPECIFICATIONS

CALIBER 12.7x108mm

OPERATION . . . . .Gas operated with adjustable three-position regulator. Flapper-type locking lugs. Fire from the open-bolt position. Employs conventional slide and piston.

CYCLIC RATE . . . . .550 rpm.

FEED MECHANISM . . . . . .Conventional shuttle-type with feed lever pivoted by operating stud on slide. Seventy-round non-disintegrating metallic belts with "push through" links. Feeds from right or left.

WEIGHT, empty, without tripod . . . . . .106.5 pounds

WEIGHT, tripod (Type 54-1) . . . . .64.3 pounds

LENGTH, overall .62.5 inches

BARREL . . . . .Eight-groove with a right-hand twist of one turn in 15.2 inches. Chrome-plated chamber and bore. Radial cooling fins.

BARREL LENGTH . . . . .42.1 inches.

SIGHTS (ground) . . . . .Hooded, post-type front; adjustable for windage and elevation zero. Open, U-notch rear, sliding tangent-type; adjustable for elevation in 100-meter increments from zero to 3,300 meters; windage adjustment knob on base.

SIGHTS (Antiaircraft) . . . . .Double rings for two-man operation; assistant gunner aligns sight by means of hand crank; lead gauge on gunner's ring.

FINISH . . . . .Black oxide and phosphate; hard-chromed gas components.

ACCESSORIES . . . . .Spare belts, ammo chests and padded shoulder braces.

STATUS . .Currently in production; in service with the People's Liberation Army.

MANUFACTURER . . . . . . .PRC government arsenals.

EXPORTER . . . . .Poly Technologies, Inc., 5/F, Citic Building, 19, Jian Guo Men Wai Street, Beijing, People's Republic of China.

T&E SUMMARY: Sturdy and reliable; excellent accuracy potential and high hit probability; heavy; barrel has no carrying handle and its radial cooling fins are of dubious value; excessive muzzle flash; antiaircraft sight requires two operators and is slow to place in action.

PRC TYPE 77 HMG

As the Type 77's receiver group weighs but 47 pounds, PRC designers have managed to whittle 31.5 pounds off the weight of the Type 54's receiver group. This is also 13 pounds lighter than the Browning M2 HB's receiver group. At 18 pounds, the Type 77's barrel is 10 pounds less than the Dashika barrel and 6 pounds lighter than Ma Deuce's. While the 62-pound tripod weighs 18 pounds more than the U.S. M3 ground mount for the M2 HB, it also serves in the antiaircraft role and so its legs are of necessity considerably longer. Removal or attachment of the weapon to the tripod is rapidly accomplished by means of a hinged, U-shaped clamp with a single locking screw.

Overall length of the Type 77 HMG is 86 inches (21 inches longer than the M2 HB).

To both reduce weight and improve heat dispersion, the Type 77's 40-inch barrel has been fluted from the gas block to the receiver. Bore and chamber are chrome lined and the rifling remains eight grooves with a right-hand twist of one turn in 15.2 inches. Barrels can be changed rapidly by turning the carrying handle on top of the receiver, which also acts as a wrench to loosen the barrel's locking ring. The carrying handle is then placed in a slot on the front sight base and the barrel is pulled forward

Type 54 Method of Operation

Gas operated, with an adjustable three-position gas regulator (marked "3," "4" and "5"), the Type 54 fires from the open-bolt position. There is no provision for semiautomatic fire. No matter, as the cyclic rate is a mere 550 rpm and experienced operators can easily tap off single shots. For the first 1,000 rounds, the gas regulator should be set to the middle position. After that, the cyclic rate will increase, causing excessive wear on the reciprocating parts, and the regulator must be rotated to the smallest port (position "3"). As firing continues, increased fouling will mandate readjustment to a larger port.

After the primer has been ignited and the bullet passes down the bore, gases enter the cylinder, mounted under the barrel, and drive the piston rearward. After about 5/8-inch of free travel, to allow pressures to drop to a safe level, the slide, attached to the piston, carries the firing pin to the rear. This permits the flapper-type locks to pivot back into the bolt body. Primary extraction occurs during this initial unlocking movement. As the empty case is pulled rearward by the extractor claw, it strikes the ejector and is propelled out an ejection port in the bottom of the receiver. The reciprocating parts continue rearward until they strike the buffer in the spade-grip assembly and begin their counter-recoil momentum.

If the trigger is still depressed, the bolt group moves forward under the force from the compressed recoil spring and the feed rib pushes the next round forward and into the chamber. When the bolt starts into battery, the two flapper-type locks reach a point beyond the locking shoulders milled in the receiver. As the firing pin, attached to the slide, travels forward, it cams the locks into their recesses.

When the slide recoils rearward, a so-called "operating stud" projecting out the right side of the receiver enters the open hook of the feed lever and pivots it back. This operates the feed slide under the top cover. As the feed slide moves inward on its cam path, the feed pawl transports the next round to the cartridge guide and up over the stop pawl. When the operating stud moves forward in counter-recoil, it rotates the feed lever forward which, in turn, draws the feed slide back outward while the feed pawl moves out to engage the next cartridge in the belt. Feeding is normally from the left out of a 70-round ammo can attached to the tripod's cradle, although the Type 54 can be altered to feed from the right. The bolt remains closed on an empty chamber when the last round in the belt has been fired.

An uncomplicated and robust trigger mechanism controls this problem firing sequence. When the trigger is pressed, it rotates a trigger release lever upward. This lever pivots and depresses the sear to release the slide. A safety lever on the right side of the receiver blocks the trigger release lever when rotated forward.

and out of the receiver. Headspace is fixed. A large brake on the end of the barrel effectively reduces muzzle climb and suppresses the flash signature but does nothing to moderate the horrendous muzzle blast, which generates target-obscuring clouds of dust when the Type 77 is fired from the prone position.

While the front sight remains a hooded, round post-type, adjustable for both elevation and windage zero, the rear sight, an open U-notch of the sliding tangent type, has more realistic elevation gradations from zero to 2,400 meters in 100-meter increments. A knurled ring at the bottom right adjusts windage and the entire unit can be folded forward when not in use.

A decided improvement over the Type 54's clumsy antiaircraft sight, the Type 77's optical unit can be attached or removed instantly from its dovetail in back of the rear ground sight. A single locking screw with a sliding handle holds the unit firmly in place. With 2X magnification, the reticle pattern is a typical series of antiaircraft rings. There is a built-in sliding, neutral-density light diffuser. A lead gauge is mounted to the right of the optical housing and the entire system is operated by the gunner alone. A separate battery pack provides illumination for use at night.

A simple, uncomplicated design, disassembly procedures for the Type 77 are easily mastered by even the most inexperienced operators. First, remove the belt and clear the weapon. Allow the bolt group to move forward under control. Remove the antiaircraft sight. Remove the barrel in the manner already described,

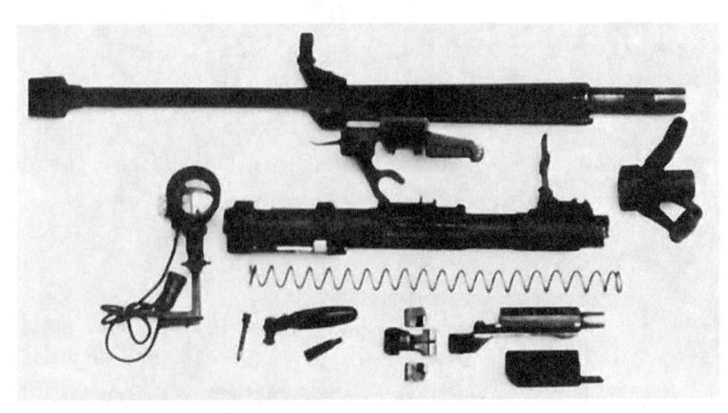

Type 77 HMG, fieldstripped.

Type 77 Specifications

CALIBER 12.7x108mm

OPERATION . . . . .Gas operated with adjustable three-position regulator. Flapper-type locking lugs. Fire from the open-bolt position. Uses M16-type direct gas impingement on bolt carrier.

CYCLIC RATE .760-780 rpm.

FEED MECHANISM . . . . . . .Conventional shuttle-type with feed lever pivoted by operating stud on bolt carrier's extension. Seventy-round non-disintegrating metallic belts with "push through" links. feeds from left only.

WEIGHT, empty
without tripod . . . . .65 pounds
tripod . . . . .62 pounds

LENGTH, overall . . . . .86 inches

BARREL . . . . .Eight-groove with a right-hand twist of one turn in 15.2 inches. Chrome-plated chamber and bore. Fluted cooling fins.

BARREL LENGTH .40 inches.

SIGHTS (ground) . . . . .Hooded, post-type front; adjustable for windage and elevation zero. Open, U-notch rear, sliding tangent-type; adjustable for elevation in 100-meter increments from zero to 2,400 meters; windage adjustment knob on base.

SIGHTS (Antiaircraft) . .Optical with 2X magnification and multiple-ring reticle pattern; built-in sliding, neutral-density light diffuser; exterior lead gauge; auxiliary battery pack for night illumination.

FINISH . . . . .Black oxide and phosphate; hard-chromed gas components.

ACCESSORIES . . . . .Spare belts, ammo chests and padded shoulder braces.

STATUS . . . . .Currently in production; in service with the People's Liberation Army.

MANUFACTURER . . . . . . .PRC government arsenals.

EXPORTER . . . . .Poly Technologies, Inc., 5/F, Citic Building, 19, Jian Guo Men Wai Street, Beijing, People's Republic of China.

T&E SUMMARY: Acceptable reliability; excellent accuracy potential and high hit probability; lightweight; carrying handle permits barrel to be easily changed when overheated; excessive muzzle blast; superb optical antiaircraft sight; outstanding dual-purpose tripod.

while the weapon remains on the tripod. Depress the spring-loaded locking latch at the aft end of the rear sight base and rotate the spade-grips assembly in either direction until this unit is free of the receiver body. Unscrew the mounting clamp and remove the weapon from the tripod. Withdraw the sear housing's retaining pin and separate this component from the receiver. Tilt the receiver upward and withdraw the recoil spring and bolt assembly. Pull the bolt head away from the carrier and remove the flapper locks. Pull the feed cover's axis pin out to the right and lift the feed cover off the receiver. Both segments of the gas tube can be removed for cleaning. No further disassembly is required. Since all of the gas action takes place outside of the receiver group, the Type 77 is considerably easier to clean and maintain than conventional gas-operated systems. Reassemble in the reverse manner.

Mounted on its sturdy tripod, the Type 77 HMG has excellent accuracy potential. Although the cyclic rate is somewhat high, most operators can be readily trained to fire short, three- to four-round bursts with high hit probability. At the tail end of *SOF*'s test and evaluation there were a few failures to extract, due only to inadequate lubrication of the reciprocating compo-

Type 77 Method of Operation

Patterned directly after that of the M16 series the method of gas operation is unique for a weapon of this type. There is no traveling piston A long gas tube attached to the gas block (which contains a three-position adjustable regulator) on the underside of the barrel, connects to a shorter segment on the receiver with a piston-shaped head. This hard-chromed, double-ringed head mates with a hollow extension on the bottom of the bolt carrier that projects below the receiver. Gas traveling down the tube impinges directly onto the bolt carrier to blow it rearward during the recoil cycle. Empty cases are ejected downward through a port in the bottom of the receiver. After the bolt group has been driven rearward to compress the multiple-strand recoil spring, it strikes the sealed buffer (a series of saucer-shaped Belleville washers) in the spade-grip assembly and commences forward in counter-recoil.

Locking is by means of two "flappers" resting in curved notches on each side of the bolt head. As the bolt group moves into battery, the carrier pivots these flapper-shaped locking lugs into recesses on the sides of the tubular-shaped receiver body (of both increased diameter and fluted in the area below the feed mechanism). The bolt carrier continues forward in free-travel to strike the firing pin and drive it against the cartridge's primer. During counter-recoil the locking lugs pivot back against the bolt head.

Firing is from the open-bolt position. Due to the decreased mass of the reciprocating parts, the cyclic rate has been driven upward to 760-780 rpm. There is no provision for semiautomatic fire.

When the curved sheet-metal trigger bars adjacent to each spade grip are pulled rearward, they drive a steel rod forward to rotate the sear downward, away from its bent (notch) on the bolt carrier. Releasing the trigger bar permits the spring-loaded sear to pivot upward and catch the bolt carrier before it travels forward. A safety lever on the right side of the sear housing can be rotated to block the sear in its upward position.

A padded shoulder brace can be inserted into the end of the buffer housing when the weapon is employed in the antiaircraft role.

The shuttle-type feed mechanism, duplicating that of the Type 54, permits feeding from the left side only. In this instance the open hook on the feed lever engages a round knob on the left side of the bolt carrier's bottom extension. Also reminiscent of the Type 54 is the steel rod, attached to a bracket on the tripod's pintle and connected to the bolt carrier by means of a removable pin, which is used to retract the bolt group for charging or clearing the weapon. The end of this retracting rod has been bent to the right and fitted with a finger-grooved plastic handle.

nents but, overall, reliability was more than acceptable. While the muzzle brake needs modification to reduce the possibility of position disclosure when the weapon is fired close to the ground, the Type 77 HMG represents an ingenious, and long overdue, attempt to eliminate the Dashika's shortcomings.

Nevertheless, 12.7mm-caliber machine guns need dramatically improved ammunition to overtake the strides made in the armor plating of their potential hard targets. *SOF* has learned that Soviet Mi-24 Hind-D helicopters carry titanium armor plate, varying in thickness from 7 to 18mm, under the nose and cockpit.

An example of the type of response required to counter this threat is the Norwegian Raufoss Multipurpose NM-140 .50-caliber BMG cartridge adopted by the Norwegian army in 1979. Developed for attacking helicopters of the Soviet genre, it resembles a normal AP projectile but contains a tungsten carbide core and a small charge of RDX high explosive/incendiary mixture. Detonation is delayed somewhat after impact so the bullet can first penetrate the target's armor plate before exploding to deliver its fragments and incendiary material inside the aircraft. Capable of penetrating 11mm of armor plate at ranges in excess of 1,000 meters, performance is supposedly comparable to 20mm cannons. Another alternative appears to be the U.S. SLAP (Saboted Light Armor Penetrator) round with either fin-stabilized or spin-stabilized projectiles.

You can be sure that PLA designers are experimenting with similar 12.7x108mm cartridges. With armor-piercing ammunition of the Raufoss Arsenal type, PRC's Type 77 HMG would provide the Afghan mujahideen with a cost-effective and highly portable Hind killer.

Originally appeared in *Soldier Of Fortune*
October 1987

GUNS BEHIND THE GREAT WALL, PART 3

China's Low Signature SMGs

PRC Type 85 (top) and Type 64 (bottom) sound-suppressed submachine guns were designed specifically for clandestine operations.

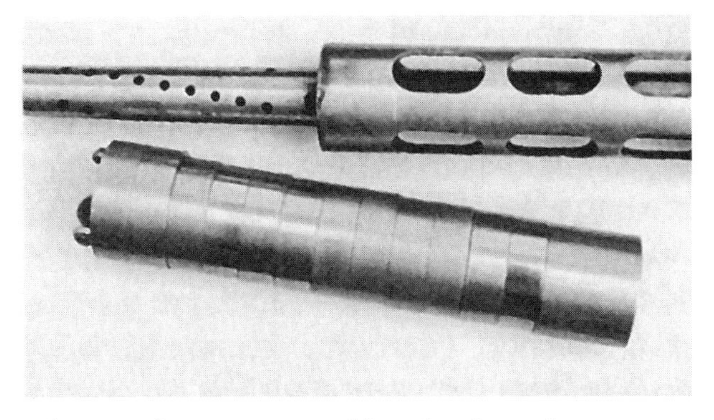

Close-up of Type 64's ported barrel and sound suppressor's baffle stack and expansion chamber.

Submachine guns—by definition selective-fire, shoulder-held weapons chambered for pistol ammunition—are moribund. The development of lightweight, short-barreled assault rifles has signaled their death knell. While 20 million submachine guns of one sort or another were fielded during World War II, their short effective range of rarely more than 100 meters, limited accuracy due to their open-bolt operation, and relatively low power lend serious doubt about their military future. Millions still reside in dead storage and no major military service has adopted an SMG in more than a quarter century. There remains but one highly specialized application for these relics of the past. When their barrels are shrouded by a sound suppressor, submachine guns can be effectively deployed by elite units engaged in clandestine operations including, but not limited to, ambush, assassination, prisoner recovery and reconnaissance.

If the submachine gun will live on only in this form, why not design one specifically for this role? The People's Liberation Army did just that 23 years ago. Two years ago the People's Republic of China introduced another unique suppressed SMG that, until *SOF*'s team tested and evaluated the entire series of weapons developed at the Small Arms Research Institute outside Beijing, had never been exposed to anyone beyond the Great Wall of China. Both the Type 64 and new Type 85 suppressed submachine guns will be described here in detail hitherto unpublished.

TYPE 64 SMG

There are few accounts of the Type 64 SMG in Western sources and all contain errors of one degree or another. Firing from the open-bolt position, the Type 64 SMG is operated by means of unlocked, pure blowback. It weighs 7.5 pounds, empty. The overall length is 33.8 inches with the stock extended and 25.5 inches after the stock has been folded. The barrel length is 9.8 inches. Except for the bolt and suppressor assemblies, all components have been salt blued.

Type 64 receivers are milled forgings, very similar in configuration to that of the AK-47, even to the extent of the characteristic lightening grooves above the magazine well. The receiver cover is a half-cylinder, sheet-metal pressing with slots cut on the right side for the ejection port and bolt's retracting handle.

Type 64 bolts are cylindrical in shape and unfinished. Extractors consist of two components, the claw and a flat spring. Firing pins are press-fit into the bolt face. A small-diameter single coil wrapped around a two-piece guide rod serves as the driving spring. There is a red fiber buffer at the end of the receiver, similar to the one found on the Soviet PPSh 41 SMG.

A folding stock, taken directly from the PRC Type 56-1 assault rifle, has been attached to the receiver in the Kalashnikov manner with a spring-loaded catch button on the left side. The rear sling swivel has been fitted to this catch button. The front sling swivel is mounted halfway up the suppressor tube on the left side.

The trigger mechanism is patterned after that of the Czech ZB 26 and British Bren series of light machine guns. A tripping lever projects through a "window" in the sear. In semiautomatic fire, the head of the tripping lever is depressed by the moving bolt to release the sear, which then engages the bolt's bent (notch) and holds the bolt to the rear. Releasing the trigger pivots the tripping lever upward. Pulling the trigger will repeat

TYPE 64 SPECIFICATIONS

CALIBER . . . . .7.62x25mm; subsonic ammunition or standard ball.

OPERATION . . . . .Unlocked blowback. Fire from the open-bolt position by means of advance primer ignition. Selector provides semiautomatic and full auto modes.

CYCLIC RATE . . . . .1,000 rpm.

FEED . . . . .Detachable, 30-round, two-position-feed, staggered-column box magazine. Type 64 and 85 magazines are not interchangeable.

WEIGHT, empty . . . . .7.5 pounds.

LENGTH, overall,
stock extended . . . . .33.8 inches.
stock folded . . . . .25.5 inches.

BARREL . . . . .Four-groove with a right-hand twist. Chrome-plated chamber, bore and exterior. Four rows of nine holes, each 3mm in diameter, spiral with the rifling grooves around the front end of the barrel. Barrel cannot be removed.

BARREL LENGTH . . . . .9.8 inches.

SOUND SUPPRESSOR . . . . .12 dished baffles stacked in front of muzzle with expansion chamber at rear end of barrel. No screens or porous filler employed. Suppressor tube enclosing system and entire barrel is threaded to front end of receiver. User maintainable.

SIGHTS . . . . .Hooded, round front post; adjustable for windage and evation zero. Protected, flip-type rear with 100- and 200-meter apertures.

FINISH . . . . .Metal surfaces salt blued, except for unfinished bolt; chrome-plated barrel and sound suppressor internal parts.

FURNITURE . . . . .Pistol grip; checkered wood or longitudinally grooved plastic.

ACCESSORIES . . . . .Sling, spare magazines and parts.

STATUS . . . . . .In service with the People's Liberation Army, but no longer in production.

MANUFACTURER . . . . .PRC government arsenals.

EXPORTER . . . . .Poly Technologies, Inc., 5/F, Citic Building, 19, Jian Guo Men Wai Street, Beijing, People's Republic of China.

T&E SUMMARY: Effective, user-maintainable sound suppression. Needs asbestos cover over suppressor tube. Cost-effective and reliable. Hit probability and accuracy potential are acceptable for intended applications.

the process. In the full-auto mode, both the tripping lever and sear are pulled down and clear of the bolt.

A lever on the left side of the receiver controls the firing modes. When rotated to the forward notch, the weapon fires semiautomatically. Full-auto fire is obtained when the lever is rotated to the rear notch. A manual safety on the right side, in back of the trigger, can be pivoted downward to block the trigger mechanism. A pivoting, sheet-metal dust cover for the retracting handle slot acts as a transport safety. When the bolt is forward on an empty chamber, this cover, which looks very much like the selector bar on a Kalashnikov, can be rotated upward to block the bolt's rearward movement. Pistol grips, interchangeable with those of the Kalashnikov, are either checkered wood (early) or plastic with longitudinal grooves (late). A "flapper" magazine catch/release assembly of the Kalashnikov type has been riveted to the receiver body in front of the trigger guard.

All very interesting, but the heart of this weapon is its integral suppressor system. The four-groove barrel has a right-hand twist with three flutes in the chamber, each .1mm wide, .075mm deep and 10mm in length. These flutes ease extraction and theoretically assist in velocity reduction. Four rows of nine holes, each 3mm in diameter, follow the spirals of the rifling grooves. As these 36 ports have been milled at the muzzle end of the barrel, they can have no effect on the bullet's velocity, which has almost reached maximum by the time it has traveled that far down the barrel. Barrels must be ported at least 2 inches from the chamber mouth for significant velocity reduction to occur. These ports are an obvious attempt to dump gas and nothing more. The escaping gas is supposed to bleed into an expansion chamber covering the rear half of the barrel. Four rows of three large oval ports surround this expansion chamber in an attempt to create turbulence and dissipate thermal energy. Between the muzzle and the suppressor tube's front cap is a stack of 12 dished baffles held together by two rods with a pivoting handle. All of the suppressor components and both the exterior and interior surfaces of the barrel have been hard-chrome plated. The forward end of the suppressor tube has four longitudinal depressions designed to secure the baffle stack. The unit is entirely user maintainable and no screens or porous fillers are used in the expansion chamber. Increased back pressure from the suppressor has driven the weapon's cyclic rate to 1,000 rpm.

Both sights are mounted to the suppressor tube. The hooded front post is adjustable for windage and elevation zero in the Kalashnikov manner. The flip-type rear sight has protective ears and peep apertures for 100 and 200 meters.

Curved, 30-round, two-position-feed, staggered-column box magazines were designed for this weapon and they can be used in no other. They are, of course, slightly wider than usual for this caliber to accept the longer Type 64 subsonic cartridge. They are well-built, substantial and reliable.

Disassembly procedures parallel the simplicity encountered with most ComBloc weaponry. Remove the magazine and clear the weapon, returning the bolt to its forward position under control. Rotate the suppressor tube's

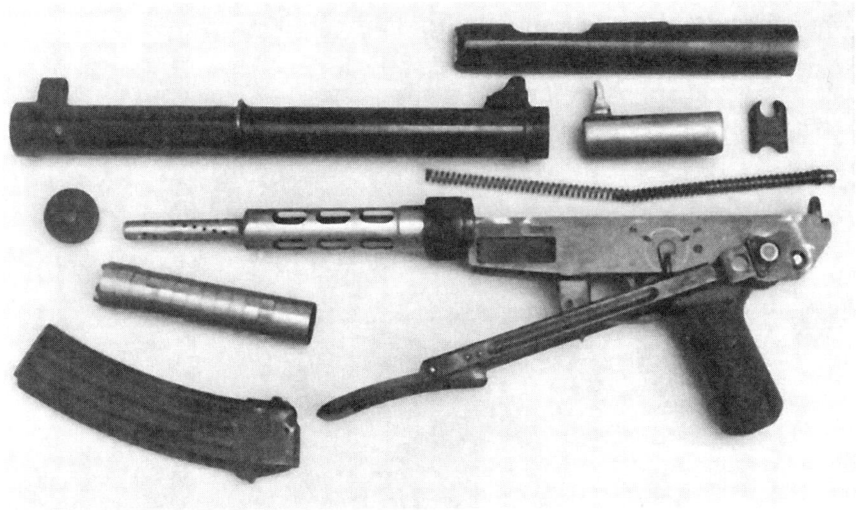

PRC Type 64 submachine gun, fieldstripped.

front cap and remove it. Grasp the suppressor stack's handle and withdraw this assembly from the tube. Depress the spring-loaded locking lever in back of the rear sight and twist the suppressor tube in either direction to separate its interrupted threads from the end cap. Pull the tube forward and off the barrel assembly. Lift the end of the recoil spring's guide rod out of its retaining notch at the rear of the receiver. Push it forward and lift off the receiver cover. Withdraw the recoil spring, guide rod, buffer and bolt. No further disassembly is required of the operator. Reassemble in the reverse order.

Approximately 500 rounds were fired through the Type 64 presented to *SOF* for test and evaluation. Although the cyclic rate is quite high, there were no failures to feed, extract or eject, no double feeds or failures of the bolt to close into battery. The ammunition appears to be well-balanced with the suppressor and there were no runaways. It was difficult to fire less than four-shot bursts in the full-auto mode. Both the accuracy potential and hit probability were adequate for the intended applications. There was no muzzle flash and the sound suppression was effective, but inferior to the Type 85. Bolt clatter was only moderate and there was, of course, no downrange supersonic crack. The suppressor tube became quite hot as the test proceeded and a protective soft cover would be well advised. There was no gas leakage into the operator's face. Altogether, this is a cost-effective, easily maintained and reasonably effective low-signature device. Although still in service with the PLA, it is no longer in production, as the Chinese were anxious to move forward with superior technology.

TYPE 85 SMG

The juxtaposition is clear-cut. Twenty years of advanced technology have resulted in a noticeable improvement. While the Type 85's overall length with the stock unfolded is, at 34.8 inches, an inch longer than the Type 64, when the stock is folded, the length (25.2 inches) is slightly shorter. Most dramatic is a weight reduction of 2 pounds, as the Type 85 tips the scales at only 5.5 pounds, empty.

The method of operation remains the same and thus the Type 85 suppressed submachine gun fires from the open-bolt position by means of unlocked blowback. As the firing pin is fixed, the cartridge is detonated by advanced primer ignition. Except for the bolt and barrel group, the entire weapon has been fabricated from seamless steel tubing with stamped sheet-metal components riveted, welded or pinned to the main body. All the steel parts, with the exception of the hard-chromed barrel/suppressor group and the phosphated bolt, have been salt blued.

Cylindrical seamless steel tubing, to which has been welded a threaded front collar/socket to accept both the suppressor tube and barrel and a rear collar for the buttstock and trigger assembly, has been used for the receiver body. Slots for the ejection port, cocking handle and magazine-well are cut in the appropriate places.

A massive bolt has been employed to reduce the cyclic rate to a more sensible 800 rpm. A large-diameter, single-coil driving spring rides partially over a rear projection on the bolt in a manner reminiscent of the British Sten, eliminating the need for a guide rod. The spring-loaded retracting handle must be withdrawn from the bolt body before the bolt can be removed from the tubular receiver.

PRC Type 85 submachine gun, fieldstripped.

Type 85 Specifications

CALIBER 7.62x25mm; subsonic ammunition or standard ball.

OPERATION . . . . .Unlocked blowback. Fire from the open-bolt position by means of advanced primer ignition. Selector provides semiautomatic and full auto modes.

CYCLIC RATE . . . . .800 rpm.

FEED . . . . .Detachable, 30-round, two-position-feed, staggered-column box magazine. Type 64 and 85 magazines are not interchangeable.

WEIGHT, empty . . . . .5.5 pounds.

LENGTH, overall,
stock extended . . . . .34.8 inches.
stock folded . . . . .25.2 inches.

BARREL . . . . .Four-groove with a right-hand twist. Chrome-plated chamber, bore and exterior. Four rows of nine holes, each 3mm in diameter, spiral with the rifling grooves around the front end of the barrel. Barrel can be removed.

BARREL LENGTH . . . . .9.8 inches.

SOUND SUPPRESSOR . . . . .11 dished baffles stacked in front of muzzle with expansion chamber at rear end of barrel. No screens or porous filler employed. Suppressor tube enclosing system and entire barrel is threaded to front end of receiver. User maintainable.

SIGHTS . . . . .Hooded, round front post; adjustable for windage and elevation zero. Protected, flip-type rear with 100- and 200-meter apertures.

FINISH . . . . .Metal surfaces salt blued, except for phosphated bolt; chrome-plated barrel and sound suppressor internal parts.

FURNITURE . . . . .Pistol grip; reddish-brown phenolic.

ACCESSORIES . . . . .Sling, spare magazines and parts.

STATUS . . . . . .In service with the People's Liberation Army.

MANUFACTURER . . . . .PRC government arsenals.

EXPORTER . . . . .Poly Technologies, Inc., 5/F, Citic Building, 19, Jian Guo Men Wai Street, Beijing, People's Republic of China.

T&E SUMMARY: Effective, user-maintainable sound suppression. Needs asbestos cover over suppressor tube. Cost-effective and reliable. Hit probability and accuracy potential are acceptable for intended applications. By virtue of its lighter weight and lower cyclic rate than the Type 64, Type 85 must receive superior rating.

The stock is formed from a single oval-shaped steel tube riveted to the folding latch mechanism, and to which has been welded a simple sheet-metal buttplate. The stock folds to the right. Its latch mechanism is attached to the receiver's end cap, which is held in place by a single pin. The rear sling swivel is fitted to the end of this pin and also serves as a handle to grasp the pin during disassembly. The front sling swivel has been mounted halfway up the suppressor tube on the left side.

The trigger mechanism has been greatly simplified and now rests in a separate subassembly. Located on the left side of the sheet-metal trigger housing, the three-position selector lever provides for semiautomatic fire at the upper position (marked "1"). Rotating the lever to the center position (marked "2") places the weapon into the full-auto mode. The lowest, or "safe," position (marked "0") blocks all sear and trigger movement. This arrangement needs to be reversed, as the more natural sequence is to move downward from safe to semiautomatic and finally to full auto. A reddish brown phenolic pistol grip has been attached to the trigger housing at the proper grip-to-frame angle.

A stamped sheet-metal magazine-well with flapper-type catch/release is riveted to the receiver tube. The curved, 30-round, two-position-feed magazines resemble those of the Type 64 but cannot be interchanged.

The sound suppressor on the new Type 85 SMG has been altered extensively. Although the four-groove barrel with its right-hand twist remains 9.8 inches in length, it can now be removed from the receiver by the operator. The expansion chamber over the rear half of the Type 64's barrel has been eliminated, but the four spiral rows of nine ports each at the front end of the barrel remain. A ported collar has been attached to the barrel at midpoint. It serves as the front end of an expansion chamber formed by the suppressor tube and

MYSTERIOUS SUBSONIC AMMO

The ammunition developed for the PRC Type 64 and Type 85 submachine guns is at least as intriguing as the weapons themselves.

Both of these submachine guns are chambered for the ComBloc 7.62x25mm rimless bottleneck cartridge and can, in fact, fire standard ball ammunition (with its 85-grain Full Metal Jacket round-nose bullet) in this caliber. Unfortunately, for the purposes of sound suppression, this is one hot little number. Soviet ammunition steps out of my PPS 43 SMG with an average velocity of 1,750 fps and Czech ammo will average 1,900 fps (which duplicates the velocity of the locked-breech .30 M1 Carbine's 112-grain projectile!).

PRC designers could have milled enough holes into the barrel to drop the velocity below the speed of sound (1,087 fps at 32 degrees Fahrenheit at sea level) and eliminate the projectile's down-range supersonic "crack," but the tradeoff would have been unacceptable. Dropping the small 85-grain bullet's velocity to subsonic levels would lower the stopping power (already marginal) and effective range to objectionable levels and most likely have degraded functional reliability.

A far superior alternative is to employ subsonic ammunition with heavy bullets that will yield the same stopping power as standard ball ammunition at velocities below the speed of sound, and port the barrel only enough for effective sound suppression. Experiments to this end commenced in World War II when the Germans tested 9mm Parabellum ammunition with heavy bullets. The concept was picked up by the British during the 1950s and further expanded during the Vietnam War when U.S. Navy SEAL teams were provided with subsonic 9mm ammunition manufactured by Super Vel for Smith & Wesson Model 39 pistols fitted with sound suppressors known as "Hush Puppies."

This was also the course of action chosen by PRC designers in 1964. Type 64 7.62x25mm subsonic ammunition is loaded into the standard copper-washed steel case (Berdan primed) common to most PRC ammunition and in this caliber actually only 24.7mm in length. But the bullet itself is quite unorthodox for a pistol cartridge. This 121-grain FMJ projectile has the boat-tail configuration usually associated with high-power rifle ammunition. The jacket material is brass-washed steel with a red case-mouth sealant. The bullet is 25.9mm in length and 7.85mm (.314 inches) in diameter. With regard to this latter dimension, keep in mind that, while the maximum bore diameter (groove-to-groove) of so-called "7.62mm" caliber weapons in NATO countries is nominally .308 inches, it is usually .311 inches in ComBloc firearms. A bullet only .003 inches larger than .311 inches will have negligible effect on the chamber pressure. The propellant charge is 3.5 grains of a cut-sheet flake, similar in appearance to Hercules Bullseye. This is about 1.5 grains less than provided in standard ball ammunition for this caliber. Overall length of the cartridge is 36.8mm (2.1mm longer than standard 7.62x25mm ball). Cited specifications indicate this subsonic 7.62x25mm ammunition produces a maximum chamber pressure of 2,400kg/cm^2 with muzzle velocities from 935 to 1,000 fps and a claimed effective range of 400 meters.

Left to right: 7.62x25mm standard ball, 7.62x25mm Type 64 subsonic ammunition, and 7.62x39mm standard ball (the latter used in the Kalashnikov, not Type 64 or Type 85 SMGs).

All of the subsonic ammunition fired in *SOF*'s test of the Type 64 and Type 85 submachine guns was headstamped either "11 67" or "11 82." Factory 11 is located at Mudanjiang, Heilongjiang Province in northern China. It is interesting to note that the only specimen observed previously was manufactured in 1966. That this ammunition was still being produced in 1982 indicates a steady and consistent demand.

a nonported collar at the chamber end of the barrel. This front barrel collar also prevents the barrel from sliding forward once the suppressor tube is locked in place. There are now only 11 dished baffles in the stack, still held together by two rods. A knurled surface surrounding the suppressor tube in back of the front sling swivel acts as a gripping point to unscrew the tube from the receiver. As before, both the exterior and interior surfaces of the barrel and the baffle stack have been hard-chrome plated. Both sights remain on the suppressor tube and are identical to those of the Type 64.

As a consequence of the new design, disassembly procedures have been further simplified. Withdraw the magazine, clear the weapon and move the bolt forward under control. Unscrew the suppressor tube's front cap and withdraw the baffle stack. Depress the spring-loaded locking stud on top of the receiver above the barrel socket and unscrew the suppressor tube. Pull it forward and away from the barrel. Pull the barrel away from the receiver body. Push the trigger housing's retaining pin from the right and remove it. Swing the trigger housing down and away from its front locking pin. Pull out the end cap's retaining pin and remove the stock group. Withdraw the recoil spring from the receiver's rear opening. Draw the bolt to the rear and pull out the retracting handle and its spring. Tip up the front of the receiver tube and the bolt will slide out the rear. Nothing more is necessary. Reassemble in the reverse order.

There were no stoppages of any kind during *SOF*'s 500-round test and evaluation of the Type 85 SMG. Because of its lower cyclic rate, experienced operators will have little difficulty firing consistent two- or three-shot bursts. Although the Type 85's buttstock looks insubstantial, it proved to be an adequate firing platform, and both the accuracy potential and hit probability paralleled the Type 64's performance. Bolt clatter remained the same, but the sound suppression appeared to be qualitatively superior. I cannot explain this, as the suppressor seems to have been redesigned with an eye toward increased simplicity rather than enhanced efficiency. Once more, during burst-fire sequences the suppressor tube overheated. I strongly recommend adoption of an asbestos cover.

Both these weapons indicate the PLA's awareness of the submachine gun's limited role on the battlefield, both currently and in the future. SMGs will continue to exist only with their barrels surrounded by baffles and cans. They will continue to find favor only with elite units trained for highly specialized operations. Stressing design simplicity, long life and operator maintainability, PRC's Type 85 SMG demonstrates sufficient ingenuity to warrant consideration by any group whose mission essential need statement includes low-signature devices.

Originally appeared in *Soldier Of Fortune*
November 1987.

GUNS BEHIND THE GREAT WALL, PART 4

China's Handheld Tank Killer

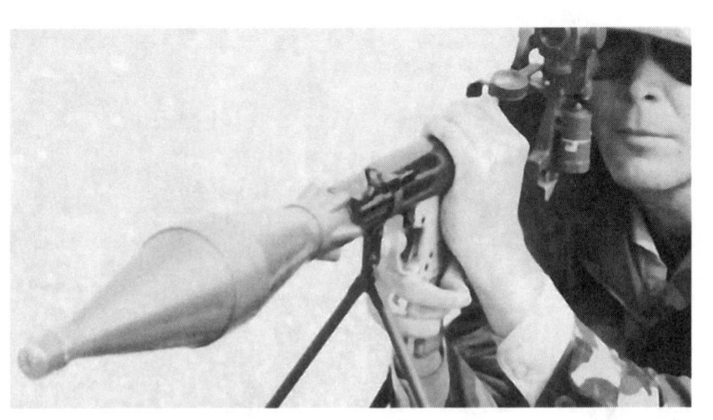

SOF contributor Bob Jordan prepares to fire Type 69 rocket launcher from kneeling position.

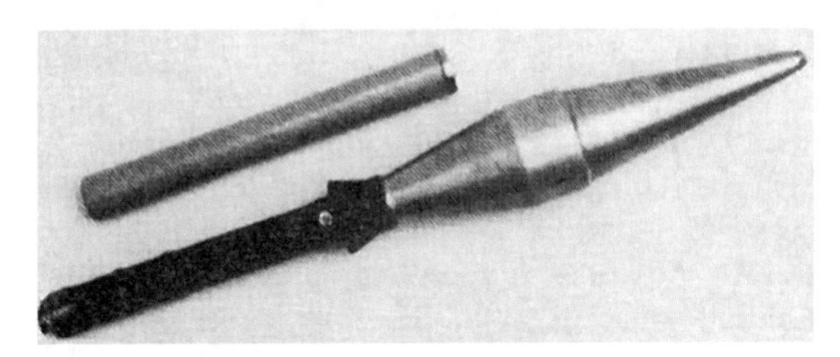

Type 69 dummy HEAT round for training purposes.

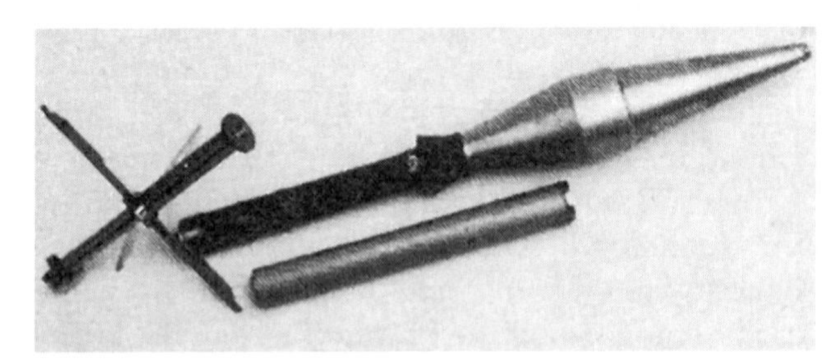

Type 69 dummy HEAT round with stabilizing fins removed and extended.

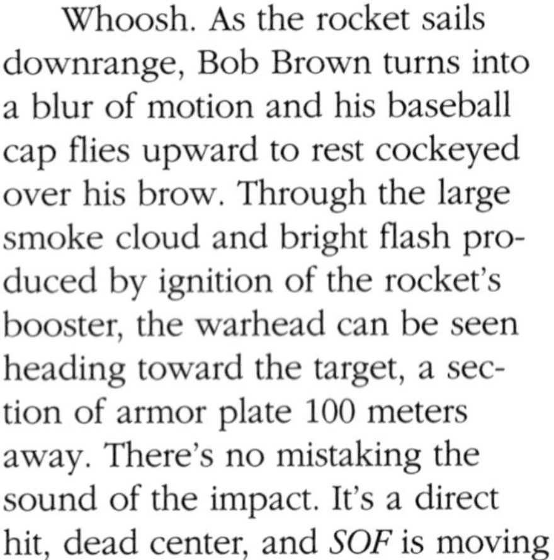

Whoosh. As the rocket sails downrange, Bob Brown turns into a blur of motion and his baseball cap flies upward to rest cockeyed over his brow. Through the large smoke cloud and bright flash produced by ignition of the rocket's booster, the warhead can be seen heading toward the target, a section of armor plate 100 meters away. There's no mistaking the sound of the impact. It's a direct hit, dead center, and *SOF* is moving forward with its test and evaluation of the PRC Type 69 40mm rocket launcher at the People's Liberation Army Small Arms Research Institute just outside Beijing.

The RPG-7 (*Reaktivniy Protivotankovyi Granatomet*: rocket antitank grenade launcher) was introduced in 1962. Modified shortly thereafter to the RPG-7V, it remains in use, in one form or another, throughout the world. By every standard of current technology, the RPG-7V has been outmoded by weapons like the U.S.-made TOW (Tube-launched, Optically tracked, Wire command-link guided missile) or French MILAN (Missile d'Infanterie Léger Antichar). Yet it is still blowing the hell out of buildings in Beirut, knocking out its makers' BMDs in Angola, and almost every infantry squad of the Israel Defense Forces still has an "RPGist."

Why? Because the RPG-7V weighs only 19.6 pounds, and gunners with more guts than brains can expect a second-round hit probability on stationary targets of more than 50 percent at close-up ranges of

300 meters and less, with penetration through 330mm of armor (at normal). Designed principally for use against armored vehicles, it can be employed as well against fortified bunkers and buildings.

It is most readily distinguished from its predecessor, the RPG-2, by a large, conical-shaped blast-shield at the rear end. The method of operation is essentially that of the World War II German *Panzerfaust*, consisting of a tube, open at both ends, with a pistol grip and trigger mechanism. It was copied in 1969 by the PRC and it was their version, the Type 69, that we tested. There is no Chinese equivalent of the Soviet RPG-7D, a two-piece collapsible version for airborne troops.

Anyone can master the procedures for firing an RPG-7V or Type 69 in just a few minutes, and you'd be well advised to do so if you're planning a vacation to either Lebanon or Afghanistan in the near future. Proficiency with the optical sight requires more extensive training. Here's how it's done, step by step:

- Remove a grenade from the packing crate (Soviet PG-7 rockets are packed six to a box, PRC only four to a box) and inspect the fuze, nose and nozzle tube for external damage.
- Remove the shipping cap from the end of the rocket and attach the booster element by screwing it clockwise until it's hand tight.
- Make sure the launcher's hammer is up. If not, pull the trigger. Push the crossbolt safety in back of the trigger to the right into the safe position. Raise both the front and rear iron sights. PRC rear sights can be adjusted for windage, Soviet rear sights cannot. Leave the front and rear sights down if you're going to use the optical sight.
- Insert the assembled grenade into the launcher muzzle and rotate the grenade until the slotted-head indicator stem, located to the rear of the pocket nozzle, moves into the notch in the top edge of

PG-7 Grenade: Method of Operation

Although the launcher tube's diameter is 40mm, the PG-7 grenade has a maximum diameter of 85mm. The 5-pound rocket-assisted fin-stabilized, shaped-charge HEAT (High Explosive Anti-Tank) round is muzzle loaded and percussion fired A piezo-electric fuze produces sufficient voltage to activate an electric detonator at the rear of the round when the grenade's nose is crushed against an inner skin. After ignition of the explosive charge, the explosion is focused by the coned shape of the outer shell into a super-hot gas jet. This jet burns through armor plate with enough residual energy to expand inside the vehicle after penetration and initiate a secondary explosion. Subsequent to this, the copper cone collapses, forming a thumb-sized, teardrop-shaped, momentarily molten slug which passes through the hole in the armor plate as a solidified projectile.

Four large, knife-like fins snap out when the projectile emerges from the tube. At the rear end of the missile are some small offset fins designed to improve stability by causing a very slow rate of roll. The warhead arms after 5 meters. Initial velocity is 580 fps, but the booster motor ignites after the rocket has passed 10 meters from the muzzle and the velocity is increased to 965 fps. The effective range is 300 meters against moving targets and 500 meters against stationary objects.

Soviet PG-7 HEAT rockets will self-destruct at 920 meters if not detonated prior to this range. PRC HEAT rounds have no self-destruct mechanism. In flight; Russian PG-7 rounds can be distinguished by a bright red tracer element that burns as brightly as a highway flare. In the early 1970s, the Soviets introduced the OG-7 antipersonnel rocket, which employs the 0-4M impact fuze used in their 82mm mortar series. OG-7 rockets have been encountered in Afghanistan.

The PRC Type 69 launcher will accept all Russian munitions designed for the RPG-7V. China also produces a yellow illumination rocket of 600,000 candlepower for the Type 69 launcher with an illuminating radius of not less than 250 meters and an illumination time of more than 40 seconds.

the muzzle. This is absolutely essential to insure alignment of the firing pin and primer—failure to do so is a primary cause of misfires.

- Remove the safety cap retaining pin from the grenade's nose by pulling the tape attached to it and removing the safety cap. Retain them in case you decide not to fire the round. The safety cap should not be removed during heavy rain or hail, as the piezo-electric fuze may short-circuit.
- Place the launcher on the right shoulder and grasp the insulator on the tube with the left hand. Grasp the pistol grip and trigger mechanism with the right hand. Point the launcher toward the target.
- Cock the hammer by applying downward pressure with the thumb.
- Push the cross-bolt safety to the left.
- Press the right cheek against the insulator and align the sights on target.
- Hold your breath and squeeze the trigger slowly. Felt recoil is almost imperceptible. Stay on the target (if you're not under fire) and watch as Ivan Ivanovitch goes straight to hell by means of an instrument of his own creation.

Inspect the launcher tube after each shot, as bits and pieces of the booster's cardboard wall often remain to block the firing pin.

There are three basic firing positions and in each instance you need to pay close heed to the backblast, as this weapon is almost as dangerous to the rear as it is to the front. The danger area stretches 20 meters to the rear and 8 meters in width.

Firing from the prone position provides the most protection from enemy fire for the gunner and his assistant when there is little cover or concealment. It is the most stable position, especially when firing the PRC Type 69 launcher since it is equipped with a bipod. It's also the least comfortable, as the elbows must be placed on the ground and the body pivoted 45 degrees away from the launcher's axis to avoid the backblast. If you are firing an RPG-7V without a bipod, make certain the muzzle is at least 6 inches above the ground, otherwise the stabilizing fins will strike the ground after they extend and deflect the rocket. Exercise caution in arid regions also, as the backblast can set dry grass on fire.

When there is sufficient cover, the kneeling position should be employed, as it is both stable and permits the gunner to move quickly after firing a round. When assuming this position, prop the left elbow against the thigh of the left leg and hold the right elbow as close to the body as possible to stabilize the weapon. When firing from the kneeling position over a wall or window ledge, there must again be at least 6 inches of clearance for the fins, to avoid deflection of the rocket. If you are firing from inside a build-

Type 69/RPG-7 Optical Sight

The PGO-7V optical sight is a 2.5X prismatic telescope with a 13-degree field of view. In my opinion, it's inappropriate for the frenzy of combat.

There is a temperature compensator under objective lens. Turn it to "+" or "–," depending on whether the temperature is above or below freezing. The reticle pattern has a stadia scale for estimating ranges from 200 to 1,000 meters. But the procedures for its use are too complex.

To determine the range, you must first set the bottom of the target on the baseline, then note the single digit on the stepped scale that aligns with the top of the target. It the actual height of the target is more or less than the scale's given target height, you must take this difference (in tenths of a meter) and multiply it by a constant (four) and the digit on top of the target. This product is then added to, or subtracted from, the range indicated by the stadia scale. Imagine performing this mathematical feat while tracers are slowly arcing your way! You're not through, however, as you still have to estimate the amount of crosswind deflection.

Forget it. Leave the optical sight in the barracks and depend upon the open sights and your ability to estimate the range.

Type 69 40mm Rocket Launcher Specifications

CALIBER
launcher .... .40mm
munition .... .85mm

MUNITIONS AND METHOD OF OPERATION .... .Rocket-assisted, fin-stabilized, shaped-charge HEAT round with piezo-electric fuze. Muzzle velocity, 580 fps; velocity after booster ignition, 965 fps. Muzzle loaded and percussion fired. 600,000 candlepower illumination rocket also available.

OVERALL LENGTH
launcher .... .37.4 inches.
rocket with booster .... .36.4 inches.

WEIGHT
launcher .... .19.6 pounds.
munition .... .5 pounds.

RANGE .... .300 meters (moving target); 500 meters (stationary target).

ARMOR PENETRATION .... .330mm at normal; 110mm at 65 degrees angle of incidence.

SIGHTS .... .Square post, hooded front sight for temperatures below freezing with secondary post for temperatures above freezing. Rear sight adjustable for both elevation and windage deflection.

OPTICAL SIGHT .... .2.5X prismatic telescope with field of view of 13 degrees. Reticle has stadia scale for estimating ranges from 200 to 1,000 meters and range scale graduated from 200 to 500 meters with deflection scale. Temperature compensation knob and battery-operated reticle illumination.

ACCESSORIES .... .Spare parts and tool kit, night vision scope, canvas rocket/accessory pouches and subcaliber (7.62x39mm) training device.

STATUS .... .In service with People's Liberation Army, Afghan mujahideen and numerous other countries.

MANUFACTURER .... .PRC government arsenals.

EXPORTER .... .Poly Technologies, Inc., 5/F, Citic Building, 19, Jian Guo Men Wai Street, Beijing, People's Republic of China.

T&E SUMMARY: Improved version of Soviet RPG-7V with windage-adjustable rear sight, folding bipod, heavier launcher tube insulation and folding carrying handle, HEAT round has no self-destruct mechanism. Effective and battle-proven anti-armor weapon.

ing, you need at least 6 feet of clearance to the rear to prevent injury from the backblast.

Although this is a "fire and forget" weapon, the standing position subjects the gunner to the most exposure from enemy fire and is the least stable. Keep both elbows pressed close against the body and lean forward with the left foot. If you attempt to engage overhead aircraft from this position, make certain you're standing on at least a 6-foot boulder with the rear of the launcher projecting over the ledge. If not, the backblast will reflect off the ground in back of you and incinerate the lower half of your body.

Strong head winds will drop the RPG-7V/Type 69 rocket's point of impact, while tail winds will raise the point of impact. Compensate accordingly by aiming either higher or lower. However, crosswinds are even more detrimental to the RPG-7's performance. Crosswinds affect the stabilizer fins more than the nose so that the head of the rocket, and thus the flight pattern, are invariably turned into the wind.

Stay with the iron sights, as the optical sight is too complicated. Use the primary front sight when temperatures are below freezing. Rotate the higher secondary sight into position when temperatures are above freezing. Dial the range into the rear sight. Adjust the Type 69's rear sight for crosswind deflection, if necessary, and hope you get a second shot for correction if there are high crosswinds.

In addition to a folding bipod and windage-adjustable rear sight, there are some other differences between the Soviet RPG-7V and the PRC Type 69. The Type 69's launcher-tube insulator is grooved and quite a bit more substantial. It provides noticeably improved protection from heat transmitted through the launcher tube by the backblast. There is also a folding carrying handle in back of the rear sight. A version called the Type 69-1 differs only by the addition of a rear, folding pistol grip attached to the bottom of the insulator by a wing nut.

A spare parts and tool kit is provided with each Type 69 launcher. Disassembly procedures at the operator's level are quite simple. Using a brass or rubber mallet, tap out the pistol grip's front retaining pin. Swing the pistol grip down to the rear to separate it from the launcher tube. With the open-end wrench on the combo tool, remove the firing pin cap by turning it counterclockwise, then drop out the firing pin, return spring and guide. No further disassembly should be attempted. A shotgun-type rod with expandable jag tip, carried in the assistant gunner's bag, should be used to clean and lubricate the launcher tube's smooth bore. Clean the hammer mechanism and remove all cardboard residue left from the booster. Reassemble in the reverse order.

COUNTERMEASURES

What if someone points an RPG-7V or Type 69 in your direction? Israeli armored vehicles can purportedly detonate the PG-7 warhead in flight by electronic means. Some have suggested protection by sandbags. It's not likely you'll be near an Israeli AFV when a PG-7 rocket is heading your way, and at least six layers of sandbags are required to stop the rocket. To date, the only practical countermeasure is the use of a chain-link fence set up 12 feet from an armored vehicle and somewhat further if you're protecting a bunker or building.

When the PG-7 rocket impacts on the chain-link mesh, any one of the following will occur. At least 50 percent of the time, the wire mesh will short-circuit the piezo-electric fuze and the warhead will fail to detonate. If the rocket does detonate, the gas jet dissipates into the air and you must contend with the copper slug only. If the rocket manages to penetrate the chain-link, the stabilizing fins will be ripped off and the warhead will usually veer off on an erratic flight path. Finally, the Soviet HEAT round's self-destruct mechanism might be triggered and the rocket will flop harmlessly about after making a popping sound. Remember, for full effect the shaped-charge PG-7 warhead must make hard contact. Rounds impacting into soft ground are no more effective than the explosion of an aluminum beer can.

Chain-link fence material is not always available, cannot be issued with every vehicle and is useless when a vehicle is on the move. It is difficult to encircle large buildings and bunkers with mesh. Expect to see the RPG-7V and Type 69 for many more years in many more wars. With its modest improvements, PRC's Type 69 is the preferred version of an effective, battle-proven, man-portable antitank weapon.

Originally appeared in *Soldier Of Fortune*
January 1988.

GUNS BEHIND THE GREAT WALL, PART 5

China's General Purpose Machine Guns

Front to rear: PRC Type 67-2, Type 67-1 and Type 80 machine guns.

SOF *Technical Editor Peter G. Kokalis fires Type 67-1 medium machine gun on heavy tripod mount.*

General Purpose Machine Gun (GPMG). What is it? It's difficult to define precisely. We owe the concept to the Germans and their employment of the MG 34 and MG 42 during World War II. General Purpose Machine Guns are supposed to be capable of serving as both medium machine guns and light squad automatics. Thus, they should be able to deliver sustained fire over fixed lines in support of advancing infantry and in perimeter defense and at the same time be light enough for carry by the assaulting troops themselves. They are invariably belt fed and feature quick-change barrels.

Like all things that are supposed to do everything, in practice the GPMG has never fully satisfied either requirement. While some are stable enough when mounted to a tripod, they are air-cooled and so frequent barrel changes are required when they are fired for sustained periods of time. Ranging in weight from 22 to 24 pounds, they are too heavy, when combined with an adequate load of ammunition, for employment at the squad level.

Although the GPMG concept appears to be fading in popularity, machine guns like the FN MAG 58, M60 and MG1/3 will continue in service well into the 21st century. Originally equipped with duplicates of the Soviet DPM (PRC Type 53) and belt-fed RP-46 (PRC Type 58), in 1967 the People's Liberation Army adopted an indigenous design that cleverly mixes some of the very best features of no less than five earlier machine guns.

PRC TYPE 67-1/2 MGs

Is the Type 67 a true GPMG? Yes and no, because there are actually two machine guns in this series, both chambered for the 7.62x54R cartridge: the Type 67-1 medium machine gun and the somewhat lighter Type 67-2, the GPMG version. Mounted on its tripod, the Type 67-1 weighs 55.2 pounds with the weight distributed as follows: 29.8 pounds for the tripod with 25.4 pounds for the weapon. Type 67-1 barrels weigh 9.5 pounds each. The Type 67-2 GPMG weighs 34.3 pounds, complete with tripod. Equipped with an 8.1-pound barrel fabricated from lightweight steel alloy, the gun's weight has been reduced to 22.1 pounds. A 12.2-pound tripod of the PKMS type is issued with the Type 67-2. Both weapons are 50 inches in overall length. Receiver bodies in each case are fabricated from substantial milled forgings. All the steel components, with the exception of the hard-chromed piston, have a black oxide or phosphate finish.

Type 67-1 and 67-2 barrels are approximately 24 inches in length with four grooves and a right-hand twist of one turn in 9 inches. Chambers and bores are chrome-plated and a conical flash suppressor is attached to the muzzle. Early Type 67 barrels were fluted in back of the gas cylinder. Barrels can be changed in less than 6 seconds in a manner reminiscent of the SGM Goryunov. Lift the top cover, press the barrel lock to the left, grab the carrying handle and pull the barrel forward out of the receiver. Insert the spare barrel, align the gas cylinder with the gas tube and press the barrel lock to the right. A screw on the barrel lock permits adjustment of the headspace (by qualified armorers only) in a manner identical to the SGM. This concept was also employed, somewhat differently, on the Japanese Type 99 LMG.

Fitted with a protective hood, the post-type front sight is adjustable for both elevation and windage zero. The open-notch leaf-type rear sight, attached to the receiver body, must be lifted to the vertical position for use. Rotation of the left knob adjusts elevation. Up to 1,000 meters, each click changes the point of impact by approximately 25 meters. At ranges beyond 1,000 meters, each click represents a change of about 20 meters. Each click of the right-hand windage knob will change the lateral point of impact about 1 inch for each 100 meters of range. An optional anti-aircraft ring sight can be attached to the left side of the receiver and the heavy tripod opened for this purpose.

Non-adjustable bipods were fitted to the gas tube on early versions of the Type 67. They are now mounted to a fitting under the front sight base on both the Type 67-1 and 67-2. The Type 67-1 is not normally equipped with a bipod, as it is most often employed on its heavy tripod.

Early buttstocks were plastic. They are now fabricated from wood. The steel buttplate has an RPD-type trap cover which retains a buttstock cleaning kit holding the standard assortment of tools and spare parts. The ochre-colored plastic two-piece pistol grip panels are retained by a single screw and nut.

PRC Type 67-1, fieldstripped.

PRC Type 67-1/2 Specifications

CALIBER 7.62x54R

OPERATION ....Gas operated with adjustable three-position regulator. Zb26/30-Bren type locking. Fire from the open-bolt position.

CYCLIC RATE ....650 rpm.

FEED . . .Belt fed. Non-disintegrating metallic belts with "push through" links.

WEIGHT
empty 67-1 ....25.4 pounds;
empty 67-2 ....22.1 pounds;
tripod 67-1 ....29.8 pounds;
tripod 67-2 ....12.2 pounds.

LENGTH, overall ....50 inches.

BARREL ....Four-groove with right-hand twist of one turn in 9 inches. Chrome-plated chamber and bore. Quick-change system based upon SG43 (Goryunov). Conical flash suppressor.

BARREL LENGTH ....24 inches.

SIGHTS ....Hooded, post-type front; adjustable for windage and elevation zero. Folding leaf-type rear with open square-notch; adjustable for elevation and windage. Optional antiaircraft ring sight.

FINISH ....Black oxide and phosphate; hard-chromed piston.

FURNITURE Wood buttstock; plastic pistol grip panels.

ACCESSORIES ....Spare belts, spare barrels, cleaning kit, tripods, ammo chests and 50-round drum-type belt carrier.

STATUS ....Currently in production; in service with the People's Liberation Army.

MANUFACTURER ....PRC government arsenals.

EXPORTER ....Poly Technologies, Inc., 5/F, Citic Building, 19, Jian Guo Men Wai Street, Beijing, People's Republic of China.

T&E SUMMARY: Sturdy and reliable in both configurations; 67-1 offers excellent accuracy potential and low group dispersion when fired from its heavy tripod. Sufficient energy to operate feed mechanism with any anticipated belt load. Clever adaptation of rimmed cartridge to "push through" belt. Heavy-barreled version is superior choice for battalion and company level issue.

Except for the gas regulators, all Type 67-1 and 67-2 components can be interchanged. Disassembly procedures are similar to those of the Bren. Lift the top cover and remove the belt. Rotate the safety lever rearward to the "fire" position. Lift the retracting handle from its vertical position upward 45 degrees to the cocking position. Clear the weapon and allow the bolt and piston/slide assembly to move forward into battery under control by holding the retracting handle. Remove the barrel in the manner described. Press the takedown pin at the end of the receiver out to the right and pull the buttstock and trigger group straight to the rear until separated from the receiver. Remove the recoil spring and guide rod. Pull the retracting handle to the rear until you can grasp the end of the slide. Withdraw the piston/slide assembly and bolt from the receiver. Separate the bolt and slide. Pull the retracting handle to the rear until it separates from the receiver. Remove the gas regulator in the manner already described and separate the bipod from the barrel. Drive out the feed tray's axis pin and lift it off the receiver. No further disassembly is required. Re-assemble in the reverse order. After placing the bolt over its post on the slide make sure it's fully forward before attempting to insert this group into the receiver. Be careful not to kink the recoil spring when reassembling the buttstock and trigger group to the receiver.

Both the Type 67-1 and 67-2 are exceptionally robust and reliable. There were no stoppages of any kind during *SOF*'s test and evaluation. When fired from its heavy tripod, the Type 67-1 is capable of almost incredible accuracy, with an effective range of 1,000 meters. To maximize hit probability, both of these weapons should be fired from their tripods whenever possible and the bipods only when necessary. Save the hip-assault position for movie actors. Perceptions of felt recoil are of no consequence with machine guns in this weight category. There appears to be more than adequate energy to drive the feed mechanism with any anticipated belt load.

The Type 67 system was somewhat eclipsed in 1980 with the introduction of the PRC's dupli-

Type 67-1/2 Method of Operation

Gas operated and firing from the open-bolt position to reduce the possibility of a "cook-off," the method of operation has been taken from the Czech Zb26/30-British Bren series, the finest magazine-fed Light Machine Guns ever fielded and abandoned in large quantity by the Chinese Nationalist Forces when they fled the mainland.

When the bolt flies forward, a cartridge is pushed downward and into the chamber. After the bolt ceases moving, the piston/slide assembly continues forward and the bolt is tilted upward to butt against a locking shoulder in the top of the receiver. The piston/slide assembly continues its forward travel after locking is completed and the flat front surface of the hammer claw/piston post drives the firing pin into the primer.

Gas bleeding through the barrel vent passes into the gas cylinder to drive the piston rearward. After a short period of free travel to permit gas pressures to drop to a safe level, the ramp on the rear face of the hammer claw/piston post pulls the rear end of the bolt down and out of its locking recess. As the piston/slide assembly carries the bolt rearward, the empty case is extracted and thrown out the ejection port in the bottom of the receiver. The recoil spring is compressed on its guide rod. When the piston/slide assembly stops its rearward travel the recoil spring drives it forward in counter-recoil and, as long as the trigger is pressed and ammunition remains in the belt, the process will be repeated.

The gas block and cylinder are pinned to the barrel. There is a three-position gas regulator attached to the gas block that has been taken directly from the RPD (PRC Type 56-1). The settings are marked "l," "2" and "3," any of which can be aligned with an index pin on the gas block. Number "1" is the normal setting. To adjust the regulator you must loosen and remove its retaining nut on the left side with the combination tool in the cleaning kit. Then press the regulator to the right to disengage it from the index pin. This can be a difficult procedure if the weapon is hot and/or the regulator is fouled.

The feed mechanism appears to be patterned after that of the belt-fed version of the Czech Vz52, which in turn had its origin with the Maxim/Vickers machine guns. All feed from the right, and this system is of proven reliability with a minimum of friction. Although the 7.62x54R cartridge is rimmed, a non-disintegrating belt, unique to the Type 67, has been designed for more efficient "push-through" operation. The links resemble those of the RPD, with a turned-down tab on the end of the link. When loaded correctly the tab must be behind the cartridge base. A drum-type belt carrier holding one 50-round belt can be attached to the receiver. Feeding is normally from an ammo can containing 250 linked rounds.

Operating by means of a lever system, the feed mechanism is driven by a cam groove on the top front of the reciprocating slide. As the slide travels rearward, a roller on the lower feed arm enters the cam groove and is moved sideways by the cam. A vertical shaft then transmits this movement to an upper feed arm. A slot in the upper feed arm engages a roller on the feed slide, and the upper arm's movement causes the feed slide to move outward to engage a cartridge.

As the bolt and piston/slide assembly move forward in counter-recoil, the bolt strips a round from the belts and pushes it forward and down into the chamber. When the two feed arms move back to their original position, forcing the feed slide inward, the cartridge held by the feed slide moves into the feed tray's slot. A holding pawl on the feed tray stops the belt from falling out of the tray, and a pair of cartridge guides in the top cover push downward on the cartridge holding it for the bolt's forward run.

The trigger mechanism has been lifted off the RPD and is equally simple and robust. There is no provision for semiautomatic fire as the cyclic rate is but 650 rpm in both versions of the Type 67. A hook on the spring-loaded trigger enters an opening in the sear. When the trigger is pulled, the hook draws the sear downward out of engagement with the slide. Rotating the safety lever (located on the right side of the trigger housing) forward locks the sear in the upward position. Do not retract the slide group while the sear is locked or the two components will bind and render the weapon inoperative until it has been disassembled.

cation of what is quite possibly the finest GPMG ever fielded.

PRC TYPE 80 GPMG

First introduced to the Soviet army in 1961, the PK GPMG was eventually product-improved and lightened into the PKM (*Pulemet Kalashnikova Modernizirovanniy*) series. A quarter century of fighting from arid deserts to tropical jungles has demonstrated it to be flawless, with the possible exception of an overly complex feed mechanism required to accommodate an almost 100-year-old rimmed cartridge (7.62x54R).

PRC's Type 80 GPMG is a further improved and lightened version of this highly regarded machine gun. Chambered for a full-size rifle cartridge, the Type 80 weighs only 17.5 pounds, empty, due in no small measure to its pinned and riveted sheet-metal receiver and the extensive use of sheet-metal stampings in other areas, especially the top cover and feed mechanism. This is 5.5 pounds less than the M60. Its excellent lightweight tripod (with aluminum legs) tips the scales at only 10.4 pounds. Overall length of this weapon is 47.7 inches. The four-groove, non-fluted barrels have a right-hand twist with one turn in 9 inches, chrome-plated chambers and bores and are 27 inches in length. There is a flash suppressor of the M14 type with five longitudinal slots.

Barrels are changed in a manner similar to the Type 67 series and SG43 Goryunov. Lift the top cover, remove the belt and lift the feed tray. Push the barrel lock to the left as far as possible. Pull up on the carrying handle and then pull the barrel forward and away from the receiver. Insert the new barrel, push the barrel lock to the right, drop the feed tray, place a loaded belt on the feed tray, with the bolt group forward make certain the first round is in the cartridge gripper, close the top cover, retract the cocking handle. Commence firing.

The front sight base resembles that of the Kalashnikov, with protective ears on each side of the round post which is adjustable for both elevation and windage zero. A sliding, tangent-type rear sight with open square notch is adjustable for both elevation and windage. Elevation graduations are in 100-meter increments from 100 to 1,500 ("1" to "15") meters. There is a 300-meter battle sight setting. A knurled knob on the right side corrects for wind deflection. Sheet-metal ears riveted to the top cover protect the sight blade.

7.62x54R Ammunition

The Russian 7.62x54R cartridge has outlived some glorious military contemporaries, such as the 8mm Lebel, .303 British and .30-06. Adopted in 1891 for the Mosin-Nagant bolt action rifle, it's still going strong as a machine gun and sniping round. First bullets for this cartridge were flat-based 150-grain projectiles called Type "L." In 1930 a boat-tailed 182-grain projectile with lead alloy core was introduced as the Type "D" (heavy ball—yellow tip). The current bullet called the Type "LPS" (light ball—white tip) has a mild steel core and weighs 150 grains. This latter bullet leaves the muzzle at 2,700 fps. Performance of this cartridge is equivalent to the .30-06 round.

Except for commercial ammunition manufactured by Norma, the brass, copper or brass-washed steel or lacquered-steel cases are Berdan primed. Bullets measure .311 inches in diameter. In addition to either heavy or light ball rounds this caliber will be encountered with tracer, API, APIT and ranging incendiary projectiles. In the PRC this cartridge is referred to as the Type 53.

Since 1945, 7.62x54R ammunition has been manufactured by PRC the Soviet Union Czechoslovakia, Egypt, Finland, France, Hungary, Poland, Spain, Syria and Yugoslavia. In recent years it has been difficult to locate in the United States except for occasional lots of Egyptian (captured by the Israelis) and Russian (1930s vintage) manufacture. Fresh lots of recent PRC manufacture are now available from Keng's Firearms Specialty, Inc.

A non-adjustable sheet-metal bipod is attached to the gas tube directly in back of the gas regulator. It can be folded to the rear or forward (when the weapon is mounted on the tripod). Its location on the weapon is excellent, as it permits the gunner to quickly engage flanking targets without seriously compromising accuracy potential. The tripod with its aluminum legs appears insubstantial, but is more than adequate for work at the squad level.

The most distinctive characteristic of any ground version of the PK machine gun is its skeletonized buttstock, which in 1967 also appeared on the Dragunov sniper rifle. Russian PK buttstocks are fabricated from wood laminate material, while the Type 80 stock is of solid wood construction, stained dark mahogany with a clear lacquer finish. There is a folding butt strap and butt trap compartment containing a cleaning kit that includes a bore brush, drift, broken case extractor and combo tool with a screwdriver on one end and gas cylinder reamer on the other. A three-piece cleaning rod and handle are attached to one of the tripod legs. Because the tripod is not always fielded at the squad level, this equipment is sometimes carried in a canvas pouch together with an oil bottle. The molded pistol grip panels are made of high-impact plastic.

Although disassembly procedures for the Type 80 approximate those of the Kalashnikov rifle, there are some important

PRC Type 80 GPMG mounted to its 10.4-pound aluminum tripod.

PRC Type 80 Specifications

CALIBER . . . . .7.62x54R.

OPERATIONGas operated with adjustable three-position regulator. Kalashnikov-type rotary-bolt locking. Fire from the open-bolt position.

CYCLIC RATE . . . . .650 rpm.

FEEDBelt fed. SG43 (Goryunov) non-disintegrating metallic belts with "pull out" links.

WEIGHT, empty .17.5 pounds.

WEIGHT, tripod .10.4 pounds.

LENGTH, overall .47.7 inches.

BARRELFour-groove with right-hand twist of one turn in 9 inches. Chrome-plated chamber and bore. Quick-change system based upon SG43 (Goryunov). M14-type flash suppressor with five longitudinal slots.

BARREL LENGTH .27 inches.

SIGHTS .Hooded, post-type front; adjustable for windage and elevation zero. Tangent-type rear with open square-notch; adjustable for elevation and windage. Elevation adjustments in 100-meter increments from 100 to 1,500 meters with 300-meter battle sight setting.

FINISHBlack oxide and phosphate; hard-chromed piston.

FURNITURE .Skeletonized wood buttstock; mahogany-stained with clear lacquer finish. Plastic pistol grip panels.

ACCESSORIES . .Spare belts, spare barrels, cleaning kit, ammo cans and lightweight aluminum tripod.

STATUS . .Uncertain with PLA, apparently produced for export purposes only.

MANUFACTURER . . . . .PRC government arsenals.

EXPORTER . . . . .Poly Technologies, Inc., 5/F, Citic Building, 19, Jian Guo Men Wai Street, Beijing, People's Republic of China.

T&E SUMMARY: Battle-proven in over a quarter century of fighting. Extremely reliable in spite of complex feed mechanism required by use of "pull out" links. Lightweight and no buffer system, yet low perceived recoil. Imperceptible muzzle climb during short-burst sequences. Most likely the world's finest GPMG.

differences. Raise the top cover by depressing the Kalashnikov-type spring-loaded cover latch. Remove the belt, clear the weapon and send the piston/slide group forward under control. Raise the feed tray. Grasp the recoil spring and guide rod. Push them forward and lift them out of the receiver. Grab the cartridge gripper and pull it to the rear until the slide is aligned with the notches on the receiver. Remove the piston/slide and bolt from the receiver. Pull the bolt forward and twist it free of the cam. Push the firing pin to the rear and extract it from the bolt body. Remove the barrel in the manner described. No further disassembly is required. Reassemble in the reverse order. The bolt must be in the forward position before seating the slide into the receiver. Pull the trigger and push the piston/slide group forward into battery.

This machine gun needs no further test and evaluation to demonstrate its already

Type 80 Method of Operation

Gas operated and firing from the open-bolt position, the Type 80 has a rotary-locking bolt of the Kalashnikov type. With the weapon charged, pulling the trigger releases the sear and permits the compressed recoil spring to drive the piston/slide (bolt carrier) assembly forward. The bolt picks up the cartridge resting on the feed tray's lips and pushes it into the chamber. The extractor grabs the rim, and the bolt's forward movement ceases. As the piston/slide continues forward, its cam rotates the bolt 35 degrees to its locked position. Locked to the piston/slide assembly, the firing pin strikes the primer. After the bullet passes the barrel's gas vent, some of the gases are tapped off to strike against the piston face, driving it to the rear. After enough free travel to allow gas pressures to drop to a safe level, the bolt is rotated and unlocked from its recesses in the receiver. As the bolt and piston/slide assembly move rearward, the firing pin is retracted, the recoil spring compressed and the empty case extracted. After striking the fixed ejector, the case is expelled out the ejection port on the left side of the receiver.

Meanwhile, up under the top cover, fresh cartridges are moving about in a reliable but rather complicated manner. To simplify matters, let's examine the belt's movement first. Feeding is from the right side. As the piston/slide assembly moves rearward in recoil, a cam along its side forces a roller on the belt feed lever outward. This causes the feed lever to pivot on its axis pin and the upper end, to which is attached the feed pawl, travels inward and the cartridge engaged by the pawl moves into position for pickup by the cartridge gripper mounted at the rear of the piston/slide assembly. When the piston/slide assembly moves forward in counter-recoil, a second cam strikes the feed lever, driving the feed pawl outward to grab another cartridge. Holding pawls in the top cover prevent the belt from slipping back.

When the piston/slide assembly is fully forward, two spring-loaded claws on the cartridge gripper slip over the rim of the cartridge. During the recoil stroke the cartridge is pulled out of the link. A spring-loaded depressor, moving in conjunction with the feed cam, drives the cartridge down, out of the gripper's claws, onto the feed tray's lips. None of this latter motion would be necessary were it not for the rimmed cartridge and SG43 Goryunov "pull out" links.

The adjustable gas regulator has three positions marked "1," "2" and "3." To adjust the regulator, slip the rim of a 7.62x54R cartridge in the regulator jaws and, using it as a handle, rotate the regulator until the detent tab covers the number you wish to use. This sleeve-type regulator bleeds off excessive gas into the atmosphere and can affect the cyclic rate. When the weapon is clean and operated at the "1" setting, the cyclic rate is 650 rpm. If the "2" or "3" settings are used before serious fouling occurs, the cyclic rate can increase by as much as 150 rpm and decrease the life span of the reciprocating parts accordingly.

The RPD's simple and reliable trigger mechanism has been incorporated in toto except for a reconfiguration of the safety lever.

proven reliability. At 17.5 pounds and with no buffer system of any kind, the lack of perceived recoil is nothing short of amazing. If bursts are kept to three or four shots, the muzzle climb is negligible. The accuracy potential when fired from the tripod is more than adequate out to its effective range of approximately 800 meters. The handling characteristics are excellent, with a consequence of exceptionally high hit probability in the hands of experienced operators. This is truly an outstanding machine gun.

Type 67 machine guns are a mainstay of the People's Liberation Army. Some were given to North Vietnam, back in the good old days when these nations were on friendly terms. The status of the Type 80 GPMG in the PLA is uncertain, and it may be produced for export only. Its belt is incompatible with the Type 67 series but has been in service for some time with the Type 57 (SGM) machine gun.

As for me, I'll take as many Type 67-1s as I can lay my hands on for issue at battalion and company levels. Until it is chambered for a modern rimless cartridge of smaller caliber and its feed mechanism simplified, I take the Type 80 for my squad just as it is.

Originally appeared in *Soldier Of Fortune*
February 1988.

SOF *Editor/Publisher Robert K. Brown fires PRC Type 80 GPMG, an improved version of the famed Soviet PKM.*

GUNS BEHIND THE GREAT WALL, PART 6

Setting the Record Straight on the Type 63

Firing the Type 63. **SOF's** ***test and evaluation of this unusual rifle was the first ever granted to Westerners.***

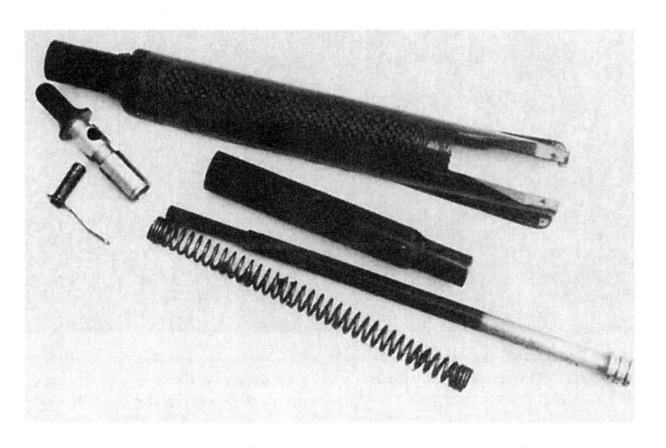

Type 63 gas system, showing gas regulator, regulator retaining pin, piston spring, piston, gas cylinder and upper handguard.

Production of the U.S. M14 rifle ceased in 1963. Only 1,376,031 were delivered by the three contractors and Springfield Armory. In that same year the People's Republic of China adopted its first modern assault rifle of more or less indigenous design. Close to 6,000,000 were eventually produced. Every Western authority refers to this rifle as the Type 68, yet the Chinese call it the Type 63. They should know, it's their rifle. At one time more than 100 factories were involved in the production of the Type 63 and SKS (PRC Type 56) rifles. Most are closed today, as are many of the PRC's small arms munitions factories. Series production of the Type 63 rifle commenced in 1969. Today it is found only in the hands of the People's Militia, which serves as the manpower reserve. It will soon be replaced by the Type 81 rifle. To my knowledge, the Type 63 rifle was never exported, at least not in any significant quantity. Errors and misconceptions about the Type 63 abound in Western small arms literature. *SOF*'s test and evaluation of this unusual rifle were the first ever granted to Westerners.

I found the Type 63 to be an odd blend of old and new, combining features of the Kalashnikov and SKS with some native innovations. It incorporates the two most salient characteristics of the modern assault rifle, as it is chambered for an intermediate-sized cartridge, the 7.62x39mm, and is capable of selective fire.

Overall length is 41.2 inches. The four-groove barrel has a right-hand twist with one turn in 9.6 inches. Total length of the barrel is 20.8 inches. Chambers and bores are chrome-plated. There is no muzzle device on the barrel and no grenade-launching capability. Somewhat heavy by today's standards, the weight, empty, is 8.6 pounds. Gas operated and firing from the closed-bolt position, the cyclic rate in full-auto fire is either 680 rpm or 725 rpm, depending upon the gas regulator position chosen.

Three major variants can be encountered. Some specimens have receivers fabricated from milled forgings. Others have pinned and riveted, stamped sheet-metal receivers but with two distinctly different bolt and bolt carrier configurations. Type 63 receivers consist of little more than a set of rails for the reciprocating parts, locking recesses for the bolt and a magazine-well. There are also three different types of recoil springs, guide rods, sheet-metal receiver covers and locking pins. The receiver covers protect only that portion of the receiver to the rear of the bolt carrier and have a sliding dust shield for the retracting-handle slot. Two different selector mechanisms were used at one time or another. And finally, both plastic and wood stocks were manufactured for the Type 63. Sounds confusing, and it is, although understandable with 100 arsenals cranking out the same rifle.

With the exception of the bolt and bolt carrier (left "in the white"), piston head and gas regulator (hard-chromed), the steel components have been salt blued.

Both a cleaning rod and folding, cruciform-shaped bayonet of the PRC Type 56 (AK-47 and SKS) configuration are mounted under the barrel.

A front sight of the round post-type is attached to the barrel's muzzle end, directly above the bayonet's locking stud. Its protective hood necessitates a tool for adjustment of elevation zero. Drifting the sight base to the right or left, after loosening its retaining screw, permits alteration of the windage zero. An open, U-notch rear sight of the conventional sliding tangent-type has been attached over the chamber to the barrel and receiver. Adjustable for elevation only, in 100-meter increments from 100 to 800 meters, a battlesight setting marked "III" is used for all ranges at 300 meters and below. The sight radius is 19 inches.

There is some confusion among Western authorities concerning the Type 63's magazine. Of all-steel construction, these detachable, staggered column, box magazines hold 20 rounds,

TYPE 63 SPECIFICATIONS

CALIBER . . . . .7.62x39mm.

OPERATION . . . . .Gas operated with adjustable two-position regulator; piston not attached to the bolt carrier. Locking by means of rotary two-lug bolt. Fire from the closed-bolt position.

CYCLIC RATE . . . . .Varies with gas regulator setting: 680 rpm with regulator set to normal position, 725 rpm at adverse position.

FEED . . . . .Detachable 20-round staggered box-type magazine with hold-open.

WEIGHT, empty . . . . .8.6 pounds.

LENGTH, overall . . . . .41.2 inches.

BARREL . . . . .Four-groove with a right-hand twist of one turn in 9.6 inches. Chrome-plated chamber and bore.

BARREL LENGTH . . . . .20.8 inches.

SIGHTS . . . . .Hooded, round front post; adjustable for windage and elevation zero. Open, U-notch rear, sliding tangent type; adjustable for elevation from 100 to 800 meters in 100-meter increments; 200-meter battlesight setting marked "III."

FINISH . . . . .Metal components salt-blued, except for unfinished bolt and carrier and hard-chromed gas regulator and piston head.

FURNITURE . . . . .Manchurian Chu wood or plastic buttstock; plastic upper handguard.

ACCESSORIES . . . . .Integral cruciform-type folding bayonet, sling, cleaning kit, plastic oil bottle, cleaning rod and spare parts and magazines.

STATUS . . . . .Adopted for service in the People's Liberation Army in 1963. Series production commenced in 1969. No longer in production. Currently in service with People's Militia only.

MANUFACTURER . . . . .PRC government arsenals.

EXPORTER . . . . .Poly Technologies, Inc., 5/F, Citic Building, 19, Jian Guo Men Wai Street, Beijing, People's Republic of China.

T&E SUMMARY: Sturdy and reliable, with features of an indigenous nature combined with those of the Kalashnikov and SKS. Somewhat heavy by today's standards. Acceptable accuracy potential with high hit probability in the semiauto mode only. Lack of a muzzle brake and pistol grip degrades capability in the full-auto mode.

Fieldstripping the Type 63 Rifle

Remove the magazine, clear the weapon and allow the bolt group to move forward into battery, under control. Rotate the disassembly lever, located on the left side of the trigger housing at the rear, downward 90 degrees. Withdraw the trigger housing. Separate the stock from the barreled action. Depress the spring-loaded button on the left side of the receiver cover and pull the cover rearward and off the receiver. Push the recoil spring and guide rod forward and lift them out of the receiver. Retract the bolt carrier to the rear, tilt it to the right and separate it from the receiver. Rotate the bolt and pull it away from the carrier. Remove the gas regulator in the manner described elsewhere. Pull the upper handguard away from the gas block. Remove the gas cylinder, piston and piston spring. Re-assemble in the reverse order.

not 15 as previously reported. The magazine follower is raised on the right side to operate a spring-loaded hold-open device in the magazine-well after the last round has been fired. When the magazine is removed or a loaded magazine inserted, retracting the bolt carrier to the rear slightly and then releasing it smartly will permit the bolt group to travel forward and chamber a round. Type 63 magazines can be loaded in the conventional manner after removal from the rifle or, when left in the rifle, with either single rounds or from 10-round SKS stripper clips using the clip guide on the bolt carrier. Magazines are inserted with the same rocking motion employed in the Kalashnikov series. A flapper-type magazine catch is attached to the front of the trigger housing.

The trigger mechanism is another variant of the much-copied system designed for the M1 Garand. There are three sears. The auto safety sear, which prevents premature ignition, is located in front of the hammer on the left side of the receiver and is actuated by the bolt carrier only after locking is completed. When the selector

PRC Type 63 rifle, fieldstripped.

lever (located on the right side of the trigger mechanism) is rotated downward to "1" (semiautomatic) and the trigger is pulled, the primary sear releases the hammer. A shot is fired and the recoiling bolt carrier rolls the hammer rearward. The auxiliary sear catches the hammer and prevents it from rotating forward until the trigger is released (at which time the auxiliary sear moves back and the hammer starts to fall forward until it is caught once more by the primary sear) and then pulled again. When the selector is rotated forward to "2" (full auto), the auxiliary sear is held back and firing will continue until the trigger is released and the hammer is engaged by the primary sear. If the selector is rotated fully rearward to "0" (safe), the selector lever's shaft prohibits any movement of the primary sear.

Our test specimens had stocks fabricated from Manchurian Chu wood, although I examined a plastic-stock version at the PLA's Small

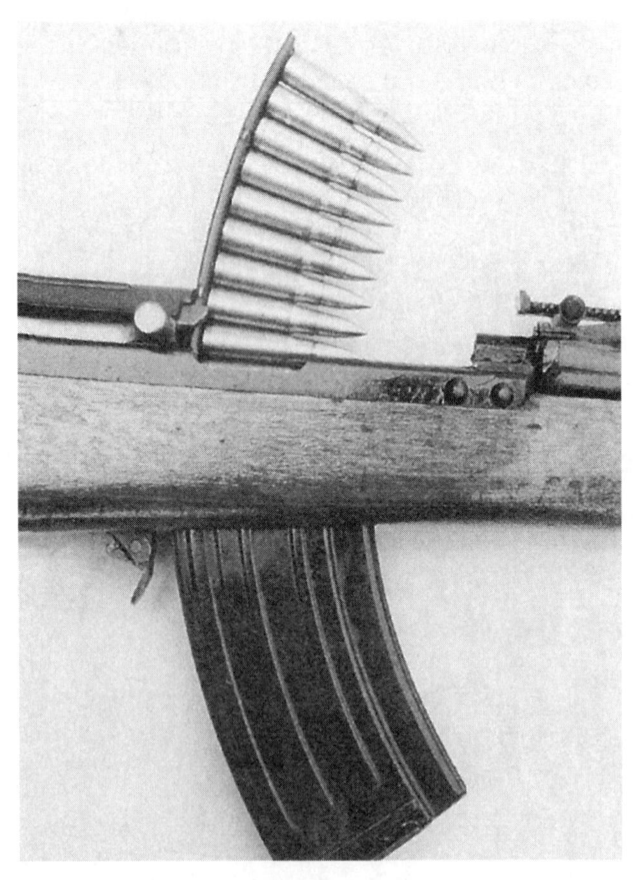

Type 63's 20-round magazine can be charged by means of clip guide in the bolt carrier using 10-round SKS stripper clips.

Type 63 Rifle: Method of Operation

When the trigger is pulled and the hammer strikes the firing pin (which is not spring-loaded) to ignite the primer, gas passes through a port in the gas block and into the gas cylinder housed in the upper handguard (protected by a sheet-metal heat shield). The piston, which has its own return spring and is not attached to the bolt carrier, is forced rearward to impinge upon the bolt carrier (in appearance similar to that of the SKS). The rotary bolt duplicates that of the Kalashnikov. After free travel of about 1/4 inch, a cam cut in the bolt carrier contacts the bolt's operating lug. Rotation of the bolt provides primary extraction. As the bolt moves rearward, the empty case is held to the bolt face by the spring-loaded extractor until it strikes the fixed ejector to be expelled out to the right. Ejection is violent and cases are frequently propelled as far as 40 feet. After the bolt carrier hits the rear wall of the receiver it is driven forward by the compressed recoil spring to strip and chamber another round from the magazine. After the bolt comes to rest, the carrier continues forward and its cam rotates the bolt into the locked position in the barrel extension. When the bolt is completely locked, the carrier trips the auto safety sear, permitting the hammer to roll forward once more.

The gas regulator functions as follows. For normal operation the small hole in the regulator should face the gas block's port. To position the larger hole over the barrel's gas vent and inject more gas into the system when adverse conditions prevail, press in the retaining pin's flat spring and pivot it downward away from its slot in the upper handguard. Withdraw the retaining pin. Using its serrated tip, rotate the gas regulator 180 degrees (be careful, it gets quite hot) and replace the retaining pin, rotating it until the flat spring enters the slot in the handguard. This gas regulator has been used without modification on the Type 74 squad machine gun.

Weapons Research Institute. The stock's configuration is essentially that of the SKS. The brown plastic upper handguard was molded with stippling on its top surface and vertical ribs on each side. The rear sling swivel is attached to the left side of the buttstock. The front sling swivel is attached to the gas block on the left side, just forward of the stock. Swinging aside the ribbed buttplate reveals two spring-loaded compartments in the buttcap and stock, which contain an oil bottle and the standard AK/SKS cleaning kit.

For our test and evaluation of the Type 63, *SOF* staffer Bob Jordan and I were transported to a People's Militia compound approximately 50 klicks north of Beijing. Three brand-new specimens, dripping with cosmoline, were selected at random, removed from their packing crates and degreased by the PLA troops assisting us. Serial numbers were in the 1.5-million range. After photographing the components, we broke for lunch and shared a meal of the saltiest noodles I've ever tasted with the PLA soldiers. At the range we were observed by an assemblage of unidentified local party cadres who remained inscrutable throughout the entire proceedings.

Handling characteristics of the Type 63 rifle approximate those of the SKS, as long as it is fired in the semiautomatic mode only. When firing three-shot bursts at the slow rate (680 rpm) and engaging targets at 100 meters, you can expect the second and third rounds to impact 10 meters high and to the right. Why? There is no muzzle brake and no pistol grip, two essential ingredients of controllable full-auto fire with a rifle weighing less than 10 pounds. In the semiautomatic mode, the Type 63's hit probability is quite high, as the recoil impulse is low and the accuracy potential is within 3 to 4 MOA (good enough to match its effective range of no more than 400 meters). Full-auto fire with this weapon should be reserved for emergency use at 50 meters or less.

We fired hundreds of rounds and there were no stoppages of any kind. The two-position regulator is a useful feature, as is the ability to charge an empty magazine from stripper clips while it remains in the rifle. This is a simple and reliable rifle—an admirable design effort for the time frame during which it was developed. But the People's Liberation Army has surged forward into the mainstream of current small arms technology with systems like the Type 81. Anyone wishing to equip a 6 million-man army with infantry rifles at extremely reasonable cost should contact Poly Technologies, Inc. for further information.

Originally appeared in *Soldier Of Fortune*
March 1988.

SOF *contributing author Bob Jordan fires the PRC Type 63 rifle, which proved to be uncontrollable in full auto.*

GUNS BEHIND THE GREAT WALL, PART 7

PRC Pistols: From Hot to Puny

Model 54 Tokarev pistol, caliber 7.62x25mm, with manual safety as required by BATF.

Type 77 pistol, caliber 7.62x17mm, a reincarnation of the 1920s German **Lignose Einhand**, *uses the trigger guard to cock and load.*

Pistols are a matter of small consequence to the military. Everywhere, that is, except in the United States. As big-bore blasters rant and rave over replacement of the .45 ACP cartridge with the 9mm Parabellum, and political intrigue over adoption of the M9 service pistol accelerates into a confused crescendo, the rest of the world observes from the sidelines in total bewilderment. Little more than insignia of rank and placebos, pistols and the caliber for which they are chambered are of little relevance on the field of battle to anyone, with the possible exception of tunnel rats.

PRC TYPE 54 PISTOL

The first pistol fielded by the People's Liberation Army, and still standard issue, was the Russian Tokarev TT-33. First Chinese copies of this design included a slot on the backstrap to accept a detachable shoulder stock/holster. Few were produced in this configuration. Production was established in Mukden, Manchuria, in 1951 at Arsenal Number 66. Both Russian and Chinese components were utilized, and these pistols were marked "Type 51." In 1954, after approximately 250,000 were manufactured, the nomenclature was changed to Type 54. By that time all parts were of Chinese origin. A sanitized version of the Type 54, devoid of all Chinese markings and the five-pointed star on the grip panels, was encountered in Vietnam with an enigmatic "M20" marked on the slide.

Known in this country as little more than a war trophy, the Tokarev pistol has several interesting features. Chambered for the 7.62x25mm cartridge, the Type 51/54 is essentially a much modified Colt-Browning design. With the exception of the magazine, springs and grip panels, all the components are milled forgings. All the steel components have been salt blued. With an empty weight of 29 ounces, the overall length is 7.77 inches, almost an inch less than the M1911A1 The method of operation is locked-breech, short recoil. After the cartridge has been fired, the barrel and slide travel rearward, locked together by two ribs on the barrel which engage two corresponding recesses inside the slide. After chamber pressures have dropped to a safe level, the swinging link on the bottom of the barrel, which is fastened to the slide stop pin, draws the barrel away from the slide. The barrel stops when it strikes a stop machined into the frame.

As the slide continues rearward, it compresses the recoil spring over its guide rod. The extractor, pinned to the right side of the slide, pulls the empty case out of the chamber and carries it back to hit against the ejector, mounted to the unique hammer and sear subassembly, propelling the case out the slide's ejection port. The compressed recoil spring drives the slide forward to strip another round from the magazine and chamber it.

All modifications to John Browning's original design were intended to simplify manufacture and enhance reliability. Type 51/54 barrels are 4.57 inches in length with chrome-plated bores and chambers. Retention of a 7.62mm bore permitted the Soviets and Chinese to use existing machinery for boring, reaming and rifling. These four-groove barrels have a right-hand twist of one turn in 9.45 inches, which duplicates that of all other Soviet 7.62mm caliber small arms. In addition, the barrel's two locking ribs are machined around the entire circumference of the barrel, which permits turning these ribs on a lathe and much simplifies this operation.

The Browning-type slide is machined to accept a heavy, forged bushing through which the barrel protrudes and to which is seated the recoil spring's plug. Type 54 slides have the 24 closely spaced serrations found on postwar Russian Tokarevs. The fixed, high-profile front and rear sights attached to the slide are surprisingly excellent for a military sidearm. The large, open U-notch rear sight can be aligned quickly with the front blade.

Most innovative of all is the hammer and sear subassembly, which drops into the frame. Machined and drilled to accommodate the hammer, sear and disconnector, this single block of steel has two arms of unequal length that rest on top of the frame's guide rails. The lower surfaces of these arms serve as cartridge guides to facilitate feeding into the chamber. Machining feed guides into a substantial steel surface instead of depending upon the lips of an easily deformed sheet-metal magazine eliminates the majority of the stoppages associated with semiautomatic pistols. The longer left arm of the subassembly acts also as the ejector. A hole in the hammer's body holds the coil hammer spring. The "burr"-type hammer is somewhat difficult to cock.

Type 51/54 trigger mechanisms are also of the Colt-Browning type, although simpler. Trigger pull weights are heavy and all that I have measured range from 7 to 9 pounds. A stirrup-shaped trigger controls the sear, to which is attached the disconnector. When the slide recoils rearward, it rides over the hammer and rolls it back so the tip of the sear can catch the hammer's full-cock notch and hold it as the slide goes forward. The slide also forces the lower end of the disconnector below the lever of the trigger's stirrup and thus another shot cannot be fired until pressure on the trigger is released. There is no manual safety or inertia-type firing pin. Although the hammer has a half-cock position, I would not carry this pistol with a chambered round.

To meet BATF regulations, a version of the Type 54 called the Model 54, which incorporates a frame-mounted manual safety, is imported by Pacific International

Merchandising Corp. By drilling a hole in the frame and machining a cut-out on the left side of the hammer subassembly, the safety lever's axis pin reaches and blocks the disconnector when the lever is rotated forward. A retaining screw on the frame and to the rear of the safety lever prevents the lever from rotating past the "fire" position and must be removed during disassembly. The lever's sense of direction is ergonomically incorrect. Pushing forward on the lever should place the mechanism in the "fire" position, not "safe."

Tokarev magazines are excellent. These eight-round single-column box magazines have a removable floorplate and no feed lips. The sides of the magazine are rolled slightly inward at the top to retain the follower and cartridges. Configured in the M1911A1 manner, the lower arm of the follower pushes up the slide-stop after the last cartridge has been fired, to hold the slide rearward. The magazine catch-release is positioned to the rear of the trigger on the left side. That's the correct location, but the button is too small and difficult to depress.

Disassembly presents no problems. Remove the magazine and clear the weapon. With a plastic-tipped punch or the point of a bullet, drive the retaining clip to the rear and off the end of the slide-stop. Remove the slide-stop. Push the slide forward and separate it from the frame. Withdraw the recoil spring, plug and guide rod. Rotate the barrel bushing 180 degrees and separate it from the slide. Rotate the link downward and pull the barrel forward and out of the slide. The extractor and spring-loaded firing pin are pinned to the slide and should be removed only by an armorer. Withdraw the hammer and sear subassembly. Reach inside the magazine-well with the point of a bullet and rotate the retaining clip on the left plastic grip panel to

Model 54 Tokarev pistol, fieldstripped

Type 54 Specifications

CALIBER 7.62x25mm

OPERATION . . . . .Locked-breech, short-recoil. Semiautomatic. Single-action.

FEED . . . . .Eight-round detachable, single-column box-type magazine; no feed lips.

WEIGHT, empty . . . . .29 ounces.

LENGTH, overall . . . . .7.77 inches.

BARREL . . . . .Four-groove with a right-hand twist of one turn in 9.45 inches; chrome-plated bore and chamber.

BARREL LENGTH . . . . .4.57 inches.

SIGHTS . . . . .Fixed, front and rear; blade-type front and open U-notch rear; high profile.

FINISH . . . . .Blued.

FURNITURE . . . . .Black plastic grip panels.

MANUFACTURER . . . . .Arsenal Number 66 in Mukden, Manchuria.

EXPORTER . . . . .NORINCO (China North Industries Corp.), 7A Yue Tan Nan Jie, P.O. Box 2137, Beijing, People's Republic of China.

IMPORTER . . . . .Pacific International Merchandising Corp., 2215 J Street, Sacramento, CA 95816.

T&E SUMMARY: Excellent and cost effective modification of Browning design. Robust and reliable, with feed lips built into the hammer and sear mechanism. Manual safety found only on imported version. Without an inertia-type firing pin; should be carried without a chambered round.

the rear until it's free of the frame. After removing the left grip panel, rotate the right panel's retaining clip until it can be lifted off the frame. Although the trigger and magazine catch-release can be removed and the hammer and sear assembly completely disassembled, this is not normally required. Reassemble in the reverse order.

TYPE 59 SPECIFICATIONS

CALIBER . . . . .9x18mm (Makarov)

OPERATION . . . . .Unlocked blowback. Semiautomatic. Double-action.

FEED . . . . .Eight-round detachable, single-column box-type magazine.

WEIGHT, empty . . . . .25.8 ounces.

LENGTH, overall . . . . .6.4 inches

HEIGHT . . . . .4.7 inches.

THICKNESS . . . . .1.125 inches.

BARREL . . . . .Four-groove with a right-hand twist of one turn in 9.45 inches; chrome-plated bore and chamber.

BARREL LENGTH . . . . .3.74 inches.

SIGHTS . . . . .Fixed, front and rear; blade-type front and square-notch rear.

FINISH . . . . .Blued.

FURNITURE . . . . .One-piece red plastic grips.

MANUFACTURER . . . . .Arsenal Number 66 in Mukden, Manchuria.

EXPORTER . . . . .Poly Technologies, Inc., 5/F, Citic Building, 19, Jian Guo Men Wai Street, Beijing, People's Republic of China.

IMPORTER . . . . .Keng's Firearms Specialty, Inc., Suite 222, 6030 Georgia Highway 85, Riverdale, GA 30274.

T&E SUMMARY: Robust and reliable; extensive modification of Walther PP. Midway between the .380 ACP and 9mm Parabellum, the 9x18mm Makarov cartridge is at the power limit for unlocked blowback operation.

In general, the handling characteristics of this pistol are more than adequate. Perceived recoil is sharp, but not at all unbearable. Muzzle whip is moderate and has little effect on recovery time for those who fire from the Weaver position. Accuracy potential is no better or worse than any other unmodified military pistol of this type. Hit probability parallels the shooter's skill level. Robust and reliable if fed quality ammunition, stoppages will be few and far between. Trigger pull weights can be lightened by a competent pistolsmith; if you can depress the magazine catch-release, the magazines fall cleanly away and the fixed lanyard ring, attached to the left side of the frame at the heel, does not interfere with rapid magazine changes like that of the M1911.

PRC TYPE 59 PISTOL

Apparently deciding that even the 7.62x25mm cartridge was too high up on the power scale for the pistol's limited applications in warfare, by 1951 the Soviets abandoned the Tokarev and adopted a pistol of lesser power. Once more, a previously successful design, this time the Walther PP, was extensively modified. The PM (*Pistolet Makarova*), or Makarov, was adopted by the People's Liberation Army in 1959 as the Type 59. Chinese versions of the Makarov differ only in minor cosmetics (i.e., the width of the slide's sight rail and configuration of the safety

Model 59 pistol, fieldstripped

PRC Pistol Ammunition

PRC pistols are chambered for any one of three cartridges, two of Soviet origin and one of more or less indigenous design. All PRC pistol ammunition is two-holed, Berdan primed.

7.62x25MM

The Russians, as well as the Chinese, were very much enamored of the Mauser Model 1896 (C96) "Broomhandle" pistol. In fact, most experts believe that the small-framed version of the Broomhandle is referred to as a "Bolo" because of its popularity with the Bolsheviks. Although some Broomhandles were chambered for the 9mm Parabellum cartridge, the majority will be encountered in caliber 7.63mm Mauser (.30 Mauser), a more powerful adaptation of the 7.65mm Borchardt round for which original Model 1896 prototypes were chambered. Until the advent of the .357 Magnum, the 7.63mm Mauser cartridge was the world's highest velocity pistol round, stepping out of the barrel at 1,410 fps. When the Soviets adopted the Tokarev pistol, they chambered it for the 7.62x25mm cartridge, which is dimensionally similar to the 7.63mm Mauser round and, in some weapons, interchangeable.

PRC production of this rimless, bottle-necked cartridge features a copper-plated steel case filled with 8.5 grains of an IMR-type extruded tubular kernal powder. Weighing 86 grains, the Round Nose (RN), Full Metal (steel) Jacket (FMJ) bullet is also copper-plated with a lead core. Both case-mouth sealant and primer annulus are light red in color.

Ammunition in this caliber has been difficult to obtain in the United States. Those with Tokarev pistols or PPSh41 SMGs have been forced to consume either Czech (headstamped "bxn") or Soviet (headstamped "539 48") ammo. Both are aging, hot and erratic—a recipe for disaster, as I can testify after shearing the receiver hinge, pin on my PPS43 SMG with this garbage. Keng's Firearms Specialty, Inc. currently imports PRC 7.62x25mm ammunition of recent vintage and excellent quality, Headstamped "11 84," it was manufactured in 1984 at factory 11, which is located at Mudanjiang, Heilongjiang Province in northern China. It zips out of the Tokarev's 4.46-inch barrel with an average velocity of 1,455 fps and an astounding standard deviation of only 8 fps.

Don't sell the 7.62x25mm cartridge short. While it will never satisfy the big boomers, it has counted final coup on tens of thousands of the revolution's enemies.

9MM MAKAROV

In 1959 the People's Liberation Army adopted another Soviet pistol, the Makarov, and its somewhat enigmatic 9mm cartridge. Still the standard pistol cartridge of the Warsaw Pact armies, the 9mm Makarov cartridge is thought to be derived from a 1936 German project, the 9mm Ultra. Neither should be confused with the post-WWII German 9mm Ultra. There are subtle dimensional differences among all three and they are not interchangeable.

Another pipsqueak by U.S. standards, the 9mm Makarov lies midway between the .380 ACP and 9mm Parabellum in both size and power. At 17,000 psi, chamber pressures are low enough to permit unlocked blowback operation (note: a few unlocked blowback-operated pistols, such as the Spanish Astra 400/600 series, have been chambered for more powerful cartridges—in this instance the 9mm Bayard Long and 9mm Parabellum, respectively).

The rimless, straight-sided case is 18.05mm in overall length. PRC production of the Type 59 cartridge features a copper-plated steel case loaded with 3.5 grains of a short, extruded tubular kernal powder. The squat, RN Full Metal (steel) Jacket bullet has a lead core and weighs 95 grains with a diameter of 9.27mm and a length of only 11.4mm. Muzzle velocities range from 951 to 1,033 fps.

The Polish Wz63 and Russian Stechkin machine pistols are also chambered for this cartridge. Both the Type 59 (Makarov) pistol and its ammunition are imported by Keng's Firearms Specialty, Inc. Keng's ammunition (headstamped "81 83") was manufactured at the unidentified factory 81 in 1983.

TYPE 64: 7.62x17MM

Type 64 and Type 77 pistols and both the Type 64 and Type 67 suppressed pistols fire the unusual Type 64 7.62x17mm cartridge, about which Western authorities knew little until *SOF* jumped over the Great Wall.

It appears to be derived from the 7.65mm Browning (.32 ACP) cartridge. However, although the case lengths are identical (17mm), the .32 ACP case is semi-rimmed, while the Type 64 case is rimless. Bullet weights are similar. Type 64 RN FMJ projectiles weigh 74 grains. Maximum weight for. 32 ACP bullets is 75 grains. Muzzle velocities in 7.62x17mm vary from 1,000 to 1,050 fps. With chamber pressures hovering between 17,000 and 20,000 psi, locked breech designs are not required.

All of the ammunition fired in this caliber at the PLA Small Arms Research Institute had brass cases (somewhat unusual for PRC small arms ammo) and were headstamped "301 83" (indicating manufacture at unidentified factory 301 in 1983). By any standards this is an underpowered cartridge with marginal performance at best, yet it has been estimated that more than 65 percent of all semiautomatic pistols manufactured since 1900 have been chambered for its performance equivalent, the .32 ACP.

PRC pistol cartridges. Left to right, the rimless 7.62x17mm is somewhat more powerful than the .32 ACP; the 9x18mm Makarov round lies between the .380 ACP and 9mm Parabellum; and the high velocity 7.62x25mm has also seen extensive use in submachine guns.

lever) and markings. PRC Type 59 pistols are marked on the frame with the serial number, arsenal ("66" in triangle for the factory at Mukden, Manchuria) and "59SHI" (Shi = Type). Soviet Makarovs carry a single star in a circle on the grip panels; PRC versions have five smaller stars within a shield. The only other countries known to manufacture this pistol are East Germany and Bulgaria. In addition to the frame, Type 59 pistols carry serial numbers on the slide, magazine, safety lever and sear. Most of the components are milled forgings and all the metal parts have been salt blued. Type 59 pistols are imported by Keng's Firearms Specialty, Inc.

An unusual variant of the Type 59 has recently surfaced. With a "ZZ" prefix, the serial numbers only appear on the left side of the frame and slide (with the last three digits on the sear). There are no arsenal or type markings of any kind. Even more peculiar are the grip panels, which are those of the Russian Makarov (a single star in a circle). Could this be the clandestine equivalent of the M20?

Although externally a scaled-up Walther PP, the Type 59 is but 6.4 inches in overall length, with a total height of 4.7 inches. Weight, empty, is 25.8 ounces.

Its relatively low-power 9x18mm cartridge permits operation by unlocked blowback. Like the Walther series, it has a double-action trigger mechanism.

The four-groove, chrome-plated 3.74-inch barrel, with a

right-hand twist of one turn in 9.45 inches, is pinned to the frame—conventional for blowback-operated handguns.

The Makarov's firing mechanism is unusual and differs considerably from that of the Walther PP/PPK series. When fired double-action, a sustained pull on the trigger pushes the sheet-metal trigger bar, to which it is attached, forward. Forward movement of the trigger bar causes the cocking lever, connected to its rear end, to pivot upward, engage a notch in the hammer and rotate it back. When the cocking lever slips out of this notch, it lifts the sear up, freeing the rebounding-type hammer to roll forward and strike the firing pin.

A cam slot in the recoiling slide forces the cocking lever to the right, clear of the sear. This permits the spring-loaded sear to engage its notch in the hammer and hold it back in the cocked position as the slide moves forward to strip and chamber another round. Subsequent rounds are fired single-action. When the trigger is released, the cocking lever moves back under the sear, ready to pull it away from the hammer after the trigger has been pulled once more. Trigger pull weights are typically on the heavy side—about 7 pounds in single-action and usually 14 pounds or more in double-action.

There are numerous other differences between the Makarov and Walther PP/PPK pistols. The Makarov has no loaded-chamber indicator. Walther pistols have a coil hammer spring, while the Makarov uses a leaf-type hammer spring. Makarov slide-stops have an external button on the left side of the frame which is pushed downward to release the slide. Walther slides must be retracted slightly by hand and then released. The rear end of the Makarov's sheet-metal slide-stop serves as the ejector. The two-piece, spring-loaded extractor is mounted on the slide to the rear of the ejection port in the Walther manner. The Makarov's slide-mounted manual safety must be pushed up for "safe" and pulled down for "fire." These positions are reversed on the Walther pistols. Engaging the Makarov safety lever drops the hammer if it's cocked, moves a bar between the hammer and firing pin and locks both the slide and hammer.

Walther magazine catch-release buttons are located on the left side of the frame to the rear of the trigger. Makarov magazine catch-releases are found at the heel of the frame in the European manner. Type 59 magazines are of the single-column type and hold eight rounds. The floorplate can be removed for disassembly and cleaning. Large cuts on each side of the magazine body clearly indicate the number of loaded cartridges remaining.

Disassembly procedures follow those of the Walther PP/PPK. Remove the magazine and clear the pistol. Set the safety lever to "fire." Pull down on the front of the trigger guard and press to the side. Pull the slide fully rearward and lift the rear end off the frame. Ease the slide forward and off the barrel. Separate the recoil spring from the barrel. To remove the one-piece red plastic grip assembly, just unscrew the single retaining screw on the backstrap and slide the grip rearward. No further disassembly is usually required. Reassemble in the reverse order. Install the recoil spring with the smaller diameter end toward the chamber.

Handling characteristics are about what one would expect for a large "pocket" pistol. As pistols go down in size, so, most often and unfortunately, do their sights. A fixed blade front sight and open square-notch rear (adjustable for windage zero only by drifting in its dovetail slot on the slide) are no more than adequate for the close-range capabilities of this pistol. A checkered and slightly raised rib between the sights serves no practical purpose.

Grip-to-frame angle is ergonomically correct and does not impede target acquisition. The frame is large enough to accommodate a normal-sized hand. Perceived recoil is hardly a consideration when we drop to these energy levels. At 7 meters or less, both hit probability and accuracy potential match those of other pistols in this class. In stopping power this cartridge should fall somewhere between the .380 ACP and the 9mm Parabellum, although performance is diminished by its Full Metal Jacket projectile. Reliability is high. If cleaned and maintained properly, there will be no stoppages attributable to the pistol.

TYPE 64 SPECIFICATIONS

CALIBER 7.62x17mm

OPERATION . . . . .Unlocked blowback. Semiautomatic. Double-action.

FEED . . . . .Seven-round detachable, single-column box-type magazine.

WEIGHT, empty . . . . .19.68 ounces.

HEIGHT . . . . .4.12 inches.

THICKNESS . . . . .1 inch.

LENGTH, overall .6.2 inches.

BARREL . . . . .Four-groove with a right-hand twist of one turn in 9.6 inches; chrome-plated chamber and bore.

BARREL LENGTH . . . . .3.46 inches.

SIGHTS .Fixed, front and rear; blade-type front and open square-notch rear.

FINISH . . . . .Blued.

FURNITURE . . . . . .Two-piece red plastic grip panels.

MANUFACTURER . . .Arsenal Number 66 in Mukden, Manchuria.

EXPORTER . . . . .Poly Technologies, Inc., 5/F, Citic Building, 19, Jian Guo Men Wai Street, Beijing, People's Republic of China.

T&E SUMMARY: Reliable clone of the Walther PPK chambered for a marginal cartridge.

Type 64 pistol, fieldstripped

PRC TYPE 64 PISTOL

Even in an army without recognized rank, someone must lead the charge up the hill. As cavalry sabers are no longer in vogue, pistols will suffice to identify those in control. If they serve no other function, then the smaller they are, the better. Could there be any other reason for the PLA's adoption in 1964 of the Type 64 pistol and its truly pipsqueak 7.62x17mm cartridge? I can't think of any.

The Type 64 pistol is a plain and simple copy of the Walther PPK. It should not be confused with the more well-known Type 64 suppressed pistol, which, although chambered for the same cartridge, can be fired single shot from a locked breech or semiautomatic by unlocked blowback. Manufactured at arsenal Number 66 in Mukden, Manchuria, with the exception of the magazine, springs, plastic grip panels and some internal parts, the Type 64's components are all fabricated from milled forgings. All the steel parts have been salt blued.

Blowback-operated with a double-action trigger mechanism, the Type 64 pistol is 6.2 inches in overall length, with a total height of 4.12 inches and a maximum thickness of only 1 inch. The 3.46-inch pinned barrel has a chrome-plated bore and chamber with four grooves and a right-hand twist of one turn in 9.6 inches. Total weight, empty, is 19.68 ounces.

In addition to the caliber and dimensions, there are a few differences between the Type 64 and the Walther PPK. The slide-stop/ejector, although identical in operation, has a different configuration. The Type 64's hammer is not ringed. The trigger mechanism has been somewhat simplified. The Type 64's red plastic grip panels are more rounded at the heel and embellished with a Chinese character (signifying 1 August 1927—birthdate of the PLA) inside a five-pointed star and a scroll pattern at the bottom. The slide-mounted manual safety's "fire" and "safe" positions are those of the Type 59 (Makarov). There is no loaded-chamber indicator.

The fixed-blade front sight and open square-notch rear sight (adjustable for windage zero only by drifting in its slotted dovetail on the slide) are of the conventional type. The sight radius is 4.7 inches.

Trigger pull weights range from 3 to 6 pounds in single-action to about 14 pounds in double-action. Trigger movement prior to hammer release is 0.24 inches in single-action and twice that in double-action.

There is nothing imaginative about the Type 64 pistol, with one exception. The chamber is fluted with four helical grooves of the same twist as the bore. Why? Certainly not to ease extraction, as is the intent of the longitudinal flutes in the chambers of the Heckler & Koch series of weapons. Operating within a modest chamber pressure range of 17,000 to 20,000 pounds per square inch (psi), the 7.62x17mm cartridge is nevertheless almost 150 fps faster out of the muzzle than the .32 ACP round. Furthermore, the Type 64's slide has slightly less mass than that of the Walther PPK. Together, this is just enough to cause premature blowback. When the case expands into the chamber's helical flutes, the increase in the coefficient of friction is enough to retard extraction until pressures have dropped to a safe level. A clever solution which requires no increase in the pistol's weight.

By anyone's standards this cartridge has to be marginal for self-defense. Unless placed between the eyes, it may only serve to infuriate your opponent. Hit probability and accuracy potential with the Type 64 are both acceptable at the close ranges for which this pistol was designed. Perceived recoil and muzzle whip are, again, of little consequence. Rugged and dependable, there were no stoppages of any kind throughout the course of *SOF*'s test and evaluation. All in all, rather humdrum, but the PRC's most recent handgun proved to be a more intriguing affair.

PRC TYPE 77 PISTOL

In 1977 the PLA adopted another Lilliputian sidearm in caliber 7.62x17mm, although the design was not finalized until 1978 and the final tooling study did not end until 1981. Currently

Type 77 Specifications

CALIBER 7.62x17mm

OPERATION . . . . .Unlocked blowback. Hammerless. Trigger guard used to cock and load.

FEED . . . . .Seven-round detachable, single-column box-type magazine.

WEIGHT, empty . . . . .17.6 ounces.

LENGTH, overall .5.9 inches.

BARREL . . . . .Four-groove with a right-hand twist of one turn in 9.6 inches; chrome-plated chamber and bore.

BARREL LENGTH . . . . .3.46 inches.

SIGHTS . . . . .Fixed, front and rear; blade-type front with forward taper and open square-notch rear.

FINISH . . . . .Blued.

FURNITURE . . . . .One-piece black plastic grips.

MANUFACTURER . . .Arsenal Number 316, unknown location.

EXPORTER . . . . .Poly Technologies, Inc., 5/F, Citic Building, 19, Jian Guo Men Wai Street, Beijing, People's Republic of China.

T&E SUMMARY: Reincarnation of 1920s German *Lignose Einhand*. Reliable, lightweight and concealable. Chambered for a marginal cartridge.

Type 77 pistol, fieldstripped

in series production, the Type 77 pistol weighs only 17.6 ounces, empty, and has a total length of 5.9 inches. At seven rounds, the magazine capacity is identical to the Type 64. Type 77 magazines appear to be scaled-down versions of the Czech Vz 52 magazine. Method of operation remains unlocked blowback. The majority of the components are salt blued, milled steel forgings. The one-piece wraparound black plastic grips are marked with a single five-pointed star in a circle on each side. There is decorative scrollwork molded into the bottom of the grips, which are retained by a single screw. A small lanyard ring has been attached to the heel of the frame.

The pinned barrel is also a duplicate of the Type 64's: 3.46 inches in length, chrome-plated chamber and bore, four grooves with a right-hand twist of one turn in 9.6 inches. The four helical chamber flutes remain as well.

The magazine catch-release is located on the left side of the frame behind the trigger. Although the fixed-blade front sight has a forward taper, the open square-notch rear sight is that of the Type 64.

However, at this point the similarities end. Type 77 pistols are hammerless. A projection on the firing pin engages the sear directly. Pulling the trigger drops the sear, which releases the spring-loaded firing pin to strike the primer. Trigger pull weight on the test specimen was 6 pounds. The manual safety, mounted on the left side of the frame, blocks the sear and also serves as a disassembly lever (after the lever has been rotated downward, the disassembly procedures follow those of the Types 59/64). The ejector is permanently attached to the left side of the frame. There is no slide stop or hold-open device.

The Type 77's most unique feature is the two-piece trigger guard, the upper portion of which can be used to retract the slide and charge the pistol. When the trigger finger is placed on the outside of the trigger guard and pressed rearward, the trigger guard engages a spring-loaded hook on the slide. This pushes the slide back to open the breech and cock the firing mechanism. As the slide reaches the end of its rearward travel, the hook hits a lug on the frame (directly under the barrel), separating the slide from the trigger guard, freeing it to fly forward, propelled by its recoil spring, and chamber a round.

We must go back more than 80 years to find the origins of this concept. One-hand cocking and loading was first used in the prototype White-Merril pistol entered in the U.S. trials of 1907. It is practical, however, only if the caliber is small and the recoil spring is thus easily compressed. It was adapted in 1913 by a Swiss, Chylewski, for a caliber 6.35mm (.25 ACP) pocket pistol. Marketed initially by Bergmann, it was placed into series production in Germany during the 1920s as the *Lignose Einhand*.

Very little remains to be invented when it comes to firearms. Modern small arms, from the Galil rifle to the FN Minimi (M249 SAW), are usually the more or less successful combination of previous designs.

There were no stoppages of any kind during *SOF*'s test and evaluation of the diminutive Type 77 pistol. Accurate and reliable, it's every bit as well executed as the other weapons *SOF* tested at the PLA Small Arms Research Institute. Lightweight, compact and very concealable, the Type 77 should find great favor with the high-ranking officers and Public Security personnel for whom it was obviously designed. As for me, I'll continue to pack a .45.

Originally appeared in *Soldier Of Fortune*
April 1988.

Guns Behind the Great Wall, Part 8

PRC Pineapples

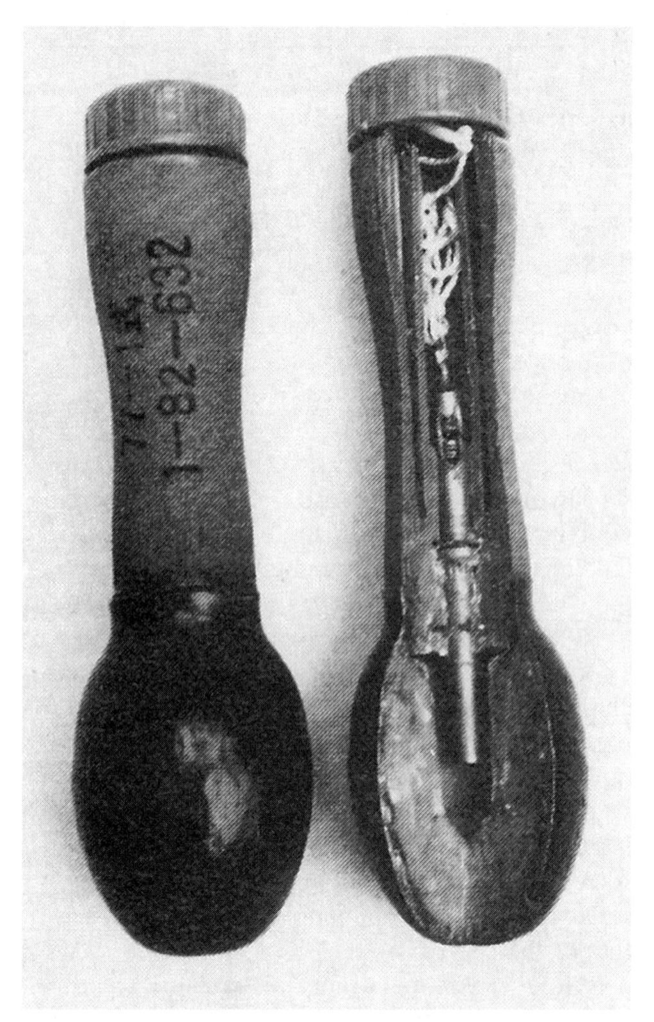

PRC Type 77-1 is the final word on stick grenades, as it features a synthetic handle with a plastic screw cap. Hungary is the only other nation still thought to be fielding stick grenades.

Hand grenades. Handle with extreme care. They are potentially the most dangerous item in the soldier's arsenal of personal weaponry. Throughout history they have produced more self-inflicted injuries than any other device available to the infantry. Yet they still remain an important ingredient in the grunt's ability to meet with, and destroy, the enemy. While they'll never replace a rifle, I'll take a sack of grenades over a pistol anytime.

Their first recorded use was at the siege of Arles in 1536. During that era grenades (from the Spanish word *grenada*, meaning pomegranate) consisted of an earthenware pot filled with blackpowder and stones and ignited by a piece of slow match. Used principally for the defense of fortifications, as siege warfare declined so did the grenade.

Grenades were revived during the Russo-Japanese War of 1904 at the siege of Port Arthur. World War I, with its static trench scenario, saw the universal adoption and massive use of hand grenades. By 1916 the British were producing more than a million grenades per week.

Modern grenades have evolved into two principal types, defensive and offensive. Defensive grenades are designed to produce a large number of fragments which will incapacitate at distances up to 40 meters. Earlier types, such as the U.S. Mk2, Soviet F1 and British Mills Bomb, have externally serrated cast-iron bodies which break unevenly into either large chunks or powder-sized particles. Unfortunately, the fuze assemblies and large fragments are often propelled as far as 200 meters from the point of impact. Defensive grenades such as these must be employed from adequate cover.

Offensive grenades are supposedly designed for use by assaulting troops fighting in the open. Featuring heavy explosive charges with thin walls and little potential to produce large fragments, they depend upon their blast effect to stun or kill the enemy.

In recent years, these distinctions have blurred. If incapacitation can be guaranteed within a 10-meter burst radius and the danger area limited to a maximum of 30 meters, the grenade can be used for both defensive and offensive purposes. For these parameters to be met, the fragments must be extremely small, with high initial velocity (up to 6,000 fps) and almost no residual kinetic energy at 30 meters.

Three methods have been used to successfully achieve these results: 1) a pre-notched wire coil inside a thin sheet-metal body (U.S. M26 series); 2) a large number (2,600 to 3,500) of steel balls within a plastic matrix (Austrian Arges Types 72 and 73); and 3) a pre-notched liner or sleeve within a thin sheet-metal body (PRC Type 82-2). This discounts the fuze assembly which, depending upon its construction, can still fly back and embed itself in your forehead.

Hand grenades are encountered in two broad configurations. Most common today is a hand-sized fragmentation body filled with the explosive charge and to which is directly attached the fuze assembly. This grenade takes several shapes: spherical ("baseball" or "golf ball"), oval ("egg" or "pineapple"), cylindrical or combinations of these geometries.

Most of these grenades use one variation or another of the Bouchon (from the World War I French inventor), or "mousetrap," igniter system. The fuze unit, which usually screws into the grenade body, consists of a detonator tube and a casting with a percussion cap in the center. Hinged to the casting is a spring-powered flap holding a firing pin that aligns with the percussion cap. A safety lever (sometimes called the "spoon") clips on to a lip on the opposite side of the casting with the firing pin flap forced back against its coil spring. The spoon is held in place by a cotter pin, with an attached pull ring, that passes through a hole in the fuze head casting.

While the cotter pin is being removed, the spoon is held by hand against the grenade body. When the grenade is thrown, the coil spring pivots the flap, flipping the safety lever away from the grenade and driving the firing pin into the percussion cap to ignite the fuze. Most hand grenade fuzes are designed with a 3- to 5-second delay—time enough to reach the target without permitting your opponent to pitch it back at you.

Never hook grenades of this type to your LBE (Load-Bearing Equipment) by either the pull ring or safety lever. And never tape the spoons. By the time you unwind or tear away the tape, you may already be on your way to the pearly gates. Grenades should always be carried in pouches attached to your web gear.

The other basic configuration is that of the so-called "stick grenade," which is any grenade attached to a handle, usually wood, supposedly for ease in throwing. Introduced by British inventor Martin Hale in 1908, the first stick grenade featured an impact fuze and a cane handle to which were attached canvas streamers to insure nose-first impact. In 1915 the Germans added a time-delay fuze and removed the streamers, as nose-first delivery was no longer required. Called the *Steilhandgranate*, it was better known to the Allies as the "potato masher."

Delay fuzes on stick grenades usually incorporate pull-friction igniters that work on the same principle as a sulfur match. An abrasive-coated wire is embedded in a sensitive composition and attached to a cord passing through the hollow handle. When the cord is pulled, friction causes the primer material to flash and ignite the fuze, which burns for 3 to 5 seconds until it sets off the detonator.

No country has ever fielded as many stick grenades in such a wide variety of forms as the People's Republic of China. With plain or scored, cylindrical or serrated oval bodies, the wooden handles have varied in length from 2 to 5.5 inches. Early on, PRC stick grenades were frequently weatherproofed with a coating of beeswax.

Explosive fillers ranged from picric acid to mixtures of TNT (TriNitroToluene) or nitroglycerin with ammonium nitrate, potassium nitrate or sawdust and schneiderite. Picric acid has almost universally been abandoned as a filler because, in contact with iron, highly unstable salts will form and grenade body interiors must be heavily varnished.

A study of PRC hand grenades reveals the same evolutionary trends we have seen in other areas of Chinese military small arms development. By 1967 the PRC stick grenade had reached a single, standardized form as the Type 67. Overall length is 8.16 inches with a weight of 21 ounces. A 1.33-ounce charge of TNT is used as the explosive filler in the plain cylindrical body and the usual pull-friction fuze has a time delay of 2.8 to 4 seconds. The effective casualty radius (that distance from the point of detonation at which 50 percent of all exposed personnel will be incapacitated) is 7 meters, with a danger radius of 32 meters.

Type 67 grenades are still in service with the People's Liberation Army. An inert practice version of this grenade called the Type 58 is available from Keng's Firearms Specialty, Inc. for only $11.95 plus $3.50 for shipping. The record throwing distance for this grenade is an incredible 90 meters, established by a PLA soldier.

The Chinese have also produced a number of grenades based on Soviet designs and differing from them only slightly. These include the Type 1 (Soviet F1), Type 42 (Soviet RG-42), Type 59 (Soviet RGD-5) and the RKG-3T (Soviet RKG-3 antitank grenade series). Although used with some success in the Arab-Israeli War of October 1973, the RKG-3 series has never been popular with the troops. Weighing almost 2.5 pounds, this shaped-charge, parachute-delivered grenade can be hurled but a short distance and requires a suicidal attitude on the part of its user.

In 1977 the PRC introduced what must surely rank as the final word on stick grenades. Known as the Type 77-1, it features a synthetic handle with a plastic screw cap. Overall length has been reduced to 6.84 inches and the weight dropped to 12.6 ounces, although the TNT explosive filler has been almost doubled (2.45 ounces). The egg-shaped body breaks into an average of 280 pieces (four times as many as the Type 67). A pull-friction fuze with a 2.8- to 4- second delay has been retained. The effective casualty radius and danger radius have been increased to 8.4 and 37.2 meters, respectively.

While this represents a significant improvement over previous types, Hungary is the only nation still thought to be fielding stick grenades. By the early 1970s, the PRC had begun its development of more compact, Western-style grenades.

First to roll down the Great Wall was the relatively diminutive Type 73. Weighing but 6.65 ounces and with an overall length of only 3.6 inches, it contains 580 steel balls within its 1.7-inch diameter. The effective casualty radius is 7.2 meters and it can be pitched 80 meters by almost any weak sister. A standard Bouchon-type igniter system is employed with a delay time of 3 to 4 seconds.

This was followed in 1982 by a series of three oval-shaped grenades. All three employ an inner liner with hexagonal serrations. Type 82-1 grenades are 4 inches in overall length and weigh 9.1 ounces with 2.6 ounces of TNT explosive filler. Producing 280 fragments, the effective casualty radius is 6.25 meters, with a danger radius of 30 meters. Surprisingly enough,

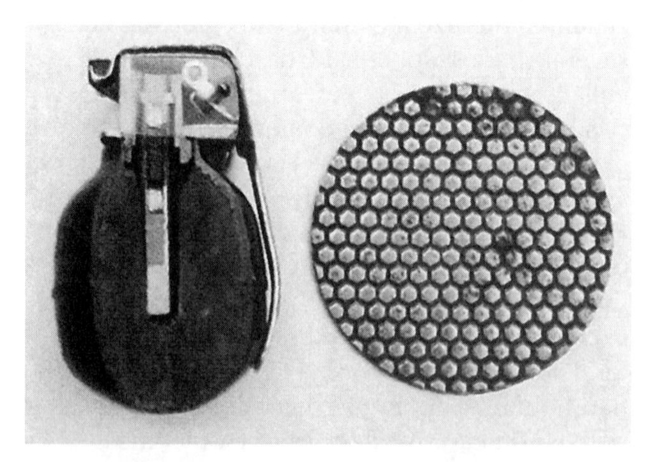

Type 82-2 grenade has a Bouchon-type igniter system with 2.2 ounces of TNT explosive filler. It breaks into 280 fragments.

a by-now atavistic pull-friction igniter system with a 3- to 4-second delay has been stuffed into the Type 82-1's short neck.

Type 82-2 grenades incorporate a Bouchon-type igniter system with a delay time of 2.8 to 3.8 seconds. (How the hell do you count 2.8 seconds when pitching a grenade?) Overall length is only 3.4 inches, with a total weight of 9.1 ounces and 2.2 ounces of TNT filler. Breaking into 280 fragments, the effective casualty radius and danger radius are the same as those of the Type 82-1.

Even smaller is the Type 82-3 grenade. Although it produces 366 fragments, the overall length is but 3 inches due to its more compact mousetrap igniter assembly. Weighing 8.75 ounces with 2.1 ounces of TNT, the effective casualty radius is 6 meters, with a danger radius of 27 meters.

Grenades launched by hand will remain a useful combat asset into the foreseeable future. As it has with assault rifles, automatic grenade launchers and machine guns, the PLA has launched itself into the developmental forefront of these short-range, but deadly, missiles. Legitimate governmental entities desiring further information concerning the entire line of PRC grenades should contact Poly Technologies, Inc., Dept. SOF, 5/F Citic Building, 19, Jian Guo Men Wai Street, Beijing, People's Republic of China.

Originally appeared in *Soldier Of Fortune* June 1988.

Section 4

Africa

DEATH FROM A DISTANCE

South Africa's Homegrown G5/G6 155s

G5 is equipped with a dial sight and direct-fire system mounted just to the left of the breech, which is shown in the open position immediately after a round has been fired. *(Photo: Armscor)*

Browning .50 caliber M2 HB machine gun, mounted to the G6 Rhino's left cupola, is loaded with the standard ratio of four ball rounds to one tracer. *(Photo: Armscor)*

Groundpounders like me develop a deep-seated distrust of artillery early on. A master sergeant once told me that the only thing more accurate than enemy incoming artillery fire was friendly incoming artillery fire.

Five years ago in El Salvador, as part of a 217-man relief column humping toward the Atlacatl Immediate Reaction Battalion tactical operations center, established the day before after fierce and bloody fighting on a hill called Hacienda El Carmen just east of the Rio Lempa in Usulutan Province, I found myself trudging across a railroad trestle bridge to reach our objective. Unbeknownst to me, a battery of 105mm howitzers were in place directly under the bridge on the west bank of the Rio Lempa. Without prior announcement, they cut loose with a salvo, and I clearly remember levitating what seemed like 3 feet off the bridge. Coming down with one foot between the railroad ties, my previous bad attitude about "cannon cockers" was intensified severalfold.

Be that as it may, cannon have their uses. The first certain record of the use of ordnance, as they are sometimes called, on battlefields in the western hemisphere was the employment of brass cannon firing iron balls during the siege of Metz in 1324. They've come a long way since then. Most of the significant advances in artillery technology took place during the 19th century. In that time frame the transition from smooth bore to rifled artillery occurred, and this resulted in important increases in accuracy and range. By 1867 Alfred Nobel had perfected dynamite, which increased the explosive power of the projectiles. At about the same time, Alfred Krupp designed and produced successful breech-loading artillery pieces that were used with great effect during the Franco-Prussian War of 1870-71. Finally, near the end of the 19th century, Colonel Hippolyte Langlois, a Frenchman, developed a 75mm cannon with a recoilless carriage and quick-firing semiautomatic action (propellant gases were used to open the breech and eject the shell casing).

Since then, enhancement of artillery effectiveness has largely consisted of sophisticated refinement of these basic elements developed more than a century ago. Today, no country fields artillery superior to that of the Republic of South Africa (RSA).

When the South African Defence Force (SADF) marched across the Angolan border in 1975 to counter Cuban troops supporting the communist MPLA forces, they were equipped with the same artillery they had used against the Germans in Italy during World War II. They were clearly outgunned by the more modern Soviet equipment provided to the Cubans. But within less than a decade, the South Africans reversed this imbalance in their favor.

Development of both a new artillery piece and ammunition system commenced in 1976. By 1983, the 155mm G5 towed howitzer had become operational with the SADF. Its origin, and that of its unique ammunition, can be traced to the now-defunct Space Research Corporation of Canada, although series production was established exclusively at Lyttelton Engineering Works (LEW), an Armscor affiliate, in Verwoerdburg outside of Pretoria.

Heart of the G5 is its 155mm/45 caliber barrel (i.e., the barrel length—6.975 meters—is 45 times the 155mm bore diameter) designed to complement Extended Range Full Bore Base Bleed (ERFB BB) ammunition. The monobloc autofrettaged (frettaging is the process of reinforcing the barrel, originally by placing heated hoops of wrought iron or steel around it that contracted as they cooled to strengthen it) barrel comes to LEW as a forging manufactured from high-grade steel billets at the Union Steel Corporation in Vereeniging. Its starting weight of 3.5 metric tons is reduced by machining to 1.6 metric tons. The bore has 48 grooves, with a twist of one turn for every 20 caliber lengths.

The G5 barrel has an open-type, single-baffle muzzle brake. The breech mechanism is very similar to that of the U.S. M198 155mm howitzer and is semiautomatic with an electrical or mechanical firing mechanism. Behind the breech is a manually operated pneumatic rammer, on whose tray the projectile is loaded and slides into the chamber before it is rammed. While the length of its barrel, almost 23 feet, is partially responsible for the G5's exceptional accuracy and long range, its superior performance is largely due to the base bleed ammunition developed for this weapon.

The range of an artillery shell can be increased by reducing "base drag." During its

South African G6 Rhino 155mm 45 caliber SP gun in action. (Photo: Armscor)

flight to the target, an area of low pressure behind the projectile, caused by its passage through the air, produces drag and reduces its velocity and hence its range. A cavity in the base of ERFB BB projectiles contains a small quantity of propellant, which when ignited bleeds gas into the base region to fill up the low pressure area and eliminate the drag. This system can add as much as 30 percent to the maximum range of the shell. With this ammunition, the G5 can reach out and touch someone more than 25 miles from the muzzle. This matches the range of the Mark 7 Mod O 16-inch/50 caliber gun (with barrel length of more than 66 feet) mounted in the turrets of USS Iowa-class battleships—albeit with a much smaller projectile. Others, such as Austria's Hirtenberger, Belgium's PRB, the Hellenic Arms Industry (EBO) of Greece, the People's Republic of China, Spain's SITECSA and Talley Defense Systems of the United States have all developed ERFB BB ammunition in recent years. However, to date, only South Africa has combat tested base bleed munitions.

South African extended-range projectiles include high explosive (HE), screening smoke (SCR SMK), illuminating (ILUM), red phosphorus (RP), propaganda (PROP) and submunition—anti-personnel/antiarmor (CLSTR). The hollow-charge subprojectiles contain 56 individual bomblets which cover an area of 30x100 meters in size and will self-destruct on the ground after three seconds. M57 HE ammunition is of the high-fragmentation type with welded splines. A TNT-filled version fragments into 3,032 pieces and another with Composition B filling breaks into 4,756 pieces. The projectiles can be standard (18.6-mile maximum range) or base bleed. A six-zone charge system is used to optimize the projectile/gun combination and give a range overlap for the total distance. Use of the newly developed M43A3 combustible cartridge increases the life expectancy of the barrel from 600 to more than 1,200 rounds with the maximum charge. Available fuzes comprise direct action (superquick or delay), radio proximity (with superquick selectable and back-up) and electronic time (with direct action back-up).

The G5 carriage has a slightly angled trunnion (the axis shaft on which the barrel pivots) for balancing the barrel and the recoil forces produced on firing. During firing the carriage is raised hydraulically on a firing platform by an auxiliary power unit which also supplies hydraulic power to raise and then lower the trail wheels and to open and close the trails. The hydraulic power is also used for the limited self-propelled mode, which enables the G5 to reach speeds exceeding 10 miles per hour and climb up gradients of 40 percent. The recoil system has a buffer, recuperator and replenisher. When traveling, the barrel rests over the trails in a clamp. Only two men are required to get the G5 into action. Elevation and traverse controls are manual.

To the left of the breech is a dial sight for indirect firing and direct-fire sight mounted on a mechanism that compensates for trunnion cant. The fire control computer has processing capability to compensate in elevation for range, projectile mass, zone of charge and temperature, all of which is communicated to the gun crew by means of the gun display unit. It also calculates the equivalent full charge (EFC), recoil length, number of rounds fired and pressure of each round. There are also alarms on the gun monitor unit for incomplete recoil run-up, low battery voltage, limits of firing arc, and a warning device that indicates high chamber temperature. The application range of the direct sight is 1,200 meters, but it can operate up to 5,000 meters. Additional optional equipment includes a fire-control computer, data-entry terminal, muzzle velocity analyzer, meterological ground station and special helmet radios for the gun crews.

Weighing 13.5 metric tons, the G5 is air-transportable in the Lockheed C130. The G5 is towed by a special gun tractor which also accommodates the full crew of eight, charges, fuses, primers, and 15 projectile pallets. A hydraulic crane is used to handle the ammunition.

Soviet bloc countries have traditionally focused on towed artillery for indirect fire with some recent interest on auxiliary-propelled towed guns. In general, they still seem to consider the self-propelled (SP) gun as no more than a second-class tank that moves forward with the infantry and armor to destroy targets at close range. Few, if any, of their SP weapons exhibit the stabilizing devices required for consistent indirect fire. These ComBloc assault guns are almost exclusively "offensive" in employment.

Western views on SP artillery have, on the other hand, stressed their ability to "keep up with the armor" (some authorities feel this is a dubious capability) and "shoot and scoot" to avoid retaliatory fire. Heavy, complicated and expensive, their indirect-fire capabilities emphasize the "defensive" mindset of the NATO alliance.

Recognizing that while the towed G5 could adequately support their tanks and highly mechanized infantry it lacked protection against enemy fire, the South Africans commenced development of an SP gun in 1978. The major design requirements were 1) fire power, with particular attention to a high rate of fire, long range, increased lethality and quick engagement of alternative targets; 2) mobility, with emphasis on high speed, long range, cross-country capability and quick in/out of action time; 3) protection against direct enemy fire and counter-bombardment; and 4) standardization with existing equipment.

The 155mm/45 caliber gun (somewhat modified) and most of its optional equipment were selected because they would be standardized with the proven G5 system and its ERFB BB ammunition and, as the 155mm caliber is worldwide in operation, other 155mm ammunition could be used interchangeably. Its fume extractor, fitted two-thirds of the way along the barrel, is constructed of carbon reinforced epoxy which is lightweight and reduces out-of-balance

Projectiles are placed by hand on the G5's semiautomatic, hydraulically operated, electronically controlled flick rammer. (Photo: Armscor)

moments. Directly behind the breech is a semiautomatic, hydraulically operated, electronically controlled flick rammer for ramming the projectile after the projectile has been placed on the rammer tray by hand.

Whether offensive or defensive, most SP artillery, with the exception of the Czech Vzor 77 152mm SP howitzer DANA, rolls about the countryside on tracks. However, after intensive investigations, the South Africans determined that, at least for the terrain in which they operated, wheeled vehicles provided overwhelming advantages over tracked equipment. In comparison with tracks, wheeled operation offers exceptional strategic mobility (albeit lower tactical mobility), lower cost of acquisition and maintenance, simplicity of design, less noise and reduced chance of detection, less crew exhaustion due to better ride characteristics, superior buoyancy for amphibious applications, ability to move out of battle range on run-flat tires, much simpler logistic support, less training, lower fuel consumption, higher maximum speed and a longer range without refueling.

The G6 Rhino SP six-wheeled gun is now in service with the SADF. It is used for offensive and defensive fire support with direct firing capability and used in indirect fire service as a gun, howitzer or mortar. The G6 has a crew of six, with a driver and five other crew members who travel inside the turret. During firing, four crew members operate the gun while the driver and one other member of the crew resupply the ammunition at the rear. The G6 carries 47 projectiles and 50 charges together with the fuzes and primers. Nineteen of these rounds are directly accessible from inside the turret. For continuous bombardments, ammunition can be fed from outside via a transfer chute. Three crew members can replenish all the main weapon ammunition in 15 minutes. A firing rate of four rounds per minute can be achieved for a period of 15 minutes. Experienced personnel can fire a burst of three rounds in 21 seconds.

The G6 is also equipped with a gun display unit to communicate firing orders to the crew. A gun-monitor system can also be provided, with all the alarm functions and a chamber temperature warning device. An optional direct link between the gun display unit and gun control system permits automatic laying of the gun.

To the left of the breech is a panoramic sight and direct-fire system mounted on a compensating mechanism. Range capabilities of the direct and indirect sight is the same as for the G5. To enhance its performance, the G6 can be equipped with a navigation system and inertial fire directing system while still retaining the direct fire capabilities.

In and out of action times are an astounding 60 and 30 seconds, respectively, including lowering and raising of the four hydraulically operated stablizer legs. From one position, the G6 can cover an area of 415 square miles. It can elevate from minus five degrees to 75 degrees in five seconds and traverse through 80 degrees in seven seconds.

Another superb feature of the G6 is its exceptional mobility. It can reach a speed of almost 55 mph on the open road, and it has a strategic range of 375 miles without refueling. Maximum speed over desert terrain is 20 mph with a turning circle of 105 feet in sand and 82 feet on the road.

The G6's mobility is, in no small measure, due to the run-flat tires used on the G5/G6 artillery series (as well as the Ratel AFV). South Africa originally employed Michelin tires of the type found on heavy-duty grading equipment, but quickly found the sidewall construction too thin. In addition, the tread design was incorrect for arid-region terrain, so the South Africans designed their own instead. Called "Sand Trails," they have a 28-ply rating, and with their special run-flat insert a vehicle equipped with them can be driven for about 45 miles at 25 mph with one or more tires flat. Air pressure in these 21.00x25 tires is controlled by the driver to provide maximum flotation.

The vehicle hull (manufactured by Sandock Austral) and turret are fabricated from high strength armor steel and offer all around protection from 7.62x51mm NATO armor piercing (AP) ammunition and artillery fragments, and protection from 20mm AP from a 60-degree frontal angle. The double-armored floor offers increased protection against mines. The driver's side and front windows provide the same degree of protection as the hull and excellent visibility through the front 180-degree arc. The center window can be covered by an armored shutter, in which case the driver views the terrain ahead through a periscope.

Entrance to the turret is by means of a main door on the right-hand side or two roof hatches. There are four firing ports on the turret, two on either side. I doubt that they are ever used in combat. They reduce the rifleman's hit probability and increase the possibility of an accidental discharge within the vehicle.

Some degree of NBC protection is provided, and the design permits additional armor to be fitted as a shield against gamma and neutron radiation. Because of its overpressure system, personnel inside the vehicle do not have to wear protective clothing. Blow-off doors at the rear cover a storage compartment for emergency ammunition. In the event of a pressure build-up, one of these doors will blow off. A fast reacting, automatic fire extinguisher for both engine and crew compartments (with manual backup) has a 12-millisecond flame detector and uses halon gas for fire suppression. An air-conditioning unit with NBC filters, directly to the rear of the turret, feeds air into the turret from the outside.

A wedge-shaped box in front of the driver's compartment holds 16 projectiles and serves as brush-clearing device capable of cutting down shrubs and small trees.

Overall length of the G6 is 33.8 feet with a total combat weight of 101,660 pounds (46 metric tons). Powered by an air-cooled diesel engine developing 525 hp, the G6 has an automatic transmission with six forward and two reverse gears and both automatic and manual gear selection. Steering is power-assisted and the drive system is a permanent 6x6. The engine and gearbox are mounted on a subframe and can be changed within two hours. Independent suspension with torsion bars and a hydropneumatic damping system helps reduce crew fatigue during extended combat operations,

Mounted on each side of the turret is a bank of four electronically operated 81mm grenade launchers that fire smoke grenades. Optional auxiliary weapons include the Browning .50 caliber M2 HB heavy machine gun (mounted to the left cupola on SADF G6s), Browning caliber 7.62x51mm NATO machine gun or the GA1 20mm automatic cannon (a derivative of the World War II Mauser MG-151).

In the SADF, G6 Rhinos are deployed in batteries of eight, with three batteries to a regiment. In Angola, the batteries were often split into two troops of four each while shooting and scooting about in the operational area.

Battle-proven in Angola, the G5 and G6 Rhino currently stand supreme among heavy ordnance. The G5 has also demonstrated its ability on the Iraqi side of the front in their recent war with the Ayatollah. It will soon be manufactured under license by Chile. Other nations will be hard-pressed to match the performance level of the G5/G6 series anytime in the foreseeable future. Military procurement agencies should contact Lyttelton Engineering Works (Pty) Limited for further information.

Originally appeared in *Soldier Of Fortune*
August 1989.

GUNS OF OVAMBOLAND

SOF Tech Editor T&Es 101 Battalion

SWATF sergeant fires Y2 40mm Multi-shot Grenade Launcher (MGL), one of which is in every Casspir MPV. Although too heavy for humping on foot, it provides mechanized troops with 18 rounds per minute and maximum effective range of 400 meters.

Two Casspirs in every team are equipped with the RPG-7V; Arabic markings on PG-7 warhead indicate Syrian manufacture. Note trooper with South African antipersonnel rifle grenade constructed by using body of M26 hand grenade and attaching it to base fuze assembly and finned tail unit.

A burst cracks from the Hispano-Suiza's muzzle, stunning the senses and vaporizing the bush in front of us. Shock waves from the 20mm gun bounce off the Casspir's turret and hammer the ear drums. Oil and burnt powder splatter my face and shirt. The exploding incendiary rounds painfully sear my eyes, like the blinding flash from an arc welder.

Nothing sharpens the nerves or whets the appetite for excitement as does the anticipation of contact. The desert pan that surrounds us stands in stark contrast to the jungles of El Salvador I've humped off and on for the last five years. I'm above the giant Etosha Pan in South West Africa (Namibia), wherein lies an immense, flat sand plain which stretches northward 80 miles to the border with Angola. Called Ovamboland, it's the hunting ground for 101 Battalion, a Reaction Force unit of the South West Africa Territory Force (SWATF). Their prey? SWAPO terrorists (South West Africa People's Organization), whose infiltration from Angola increases when the annual rains swell the muddy oshanas (seasonally flooded pans) from the beginning of December through the end of April.

This region is but one part of the great Kalahari Basin, the largest continuous cover of deep sand on the earth's surface. In South West Africa alone, Kalahari sands overlie more than 78,000 square miles. Although it comprises only eight percent of South West Africa, Ovamboland is the most densely populated part of SWA, containing nearly half the total inhabitants. Overgrazing, mostly by ubiquitous but useless herds of donkeys, has turned once-luxuriant grasslands into barren dust bowls. Fertile ground for discontent and recruiting by the minions of Marxism's evil empire.

And they're here, make no mistake about it. In December of 1986, 1,100 SWAPO terrs crossed the border. A total of 846 never lived to see their Russian masters in Luanda again. A similar group trotted down in December of 1987. During the first three months of 1988, 203 had been sent to hell.

Sector 10 of the SWATF, with its HQ in Oshakati—about 30 miles south of the Angolan border—comprises both Kaokoland to the west and Ovamboland. As 90 percent of all SWAPO terrorists are Bantu-speaking Ovambos, there is little SWAPO activity in the mountainous 19,000 square miles of Kaokoland with its sparse population of only 17,000. Ovamboland's 20,000 square miles are the AOR (area of responsibility) of 51, 53 and 54 Infantry Battalions, who run foot patrols throughout the area. Their mobile force, or Romeo Mike (Reaksie Mag, which is Afrikaans for Reaction Force) as it's called, is 101 Battalion.

Before deploying with a Romeo Mike team, I was coerced into taking a standard military test conducted by the battalion's RSM (Regimental Sergeant Major). Tests of this type are always held in the officer's club and involve the testee attempting to drink more than the testor before sliding into a comatose state, while trading insults of an increasingly obscene and personal nature. Although the explicit details will forever remain obscure to me, I apparently passed Parts I and II of the test, as I was eventually led away to spend the night with the RSM and his family after the lieutenants participating in this fracas had collapsed in the BOQ. Part III of the test consisted of firing a Hispano-Suiza 20mm automatic cannon the next day, without ear protection, and attempting to synchronize the pulses of the hangover with the rhythm of the gun. War is hell.

The 101 Battalion, with its HQ in Ondangwa (about 18 miles southeast of Oshakati), was established in January, 1976 as 1 Ovambo Battalion. Its current commander is Colonel Leon Marias. Originating out of the home guard concept, it was little more than company-size in strength. Formal training was first conducted in 1977 at 21 Battalion in Johannesburg. In January of 1978, the unit's name was changed to 35 Battalion and its mission was primarily relegated to a training function. By October of 1978 only one company was actively engaged in combat operations. While the soldiers were all Ovambos during this time frame, all other ranks, from lance corporal up, were whites. Today, 80 percent of the platoon sergeants and one of the company commanders are Ovambos. Ovambos move upward into the battalion's command structure as fast as they are trained and qualified to do so. Because of its Ovambo ethnicity, almost all training activities take place within the battalion itself. Some of the Ovambo troopers are ex-SWAPO, most of whom have proven to be reliable and dedicated to the struggle against communist incursion into their homeland. One I spoke to was a former SWAPO political commissar, who now has four years of service in 101 Battalion.

Some SWAPO cadre are less amenable to conversion, however. After capturing a SWAPO terr, one of the battalion's teams returned to their company base in the bush and placed him in a prisoner compound—nothing more than a sandbagged trench. Communication with HQ in Ondangwa revealed that he was a high-ranking SWAPO political commissar. The Casspir team was instructed to preserve this valuable intelligence asset at all cost and escort him immediately to the battalion. Charging back to the trench, they found the commissar dead—with his fist clenched and one finger outstretched in the universal symbol of defiance. Using an undetected knife, he had slit his own throat.

Candidates for enlistment in 101 Battalion must be Ovambos. Prior to enlistment, they must pass a series of physical fitness tests, including a 10-kilometer forced march. They must also possess a basic knowledge of tracking and bush survival. As a consequence, urbanized Ovambos, who have usually lost most of these tribal skills, are rarely selected.

After enlistment, recruits commence a one-year probationary period, during which time they participate in an intensive infantry basic training course. Subjects covered include counterinsurgency tactics, platoon weapons and some fairly elementary training in driving, communications and first aid. After successful completion of the one-year course, advanced specialized training is offered on such topics as 20mm automatic cannon, mortars, antitank warfare and advanced driving techniques. Everyone attends an advanced counterinsurgency course, which is unique to this unit as it is oriented around its specific AOR. Combat medical skills are stressed throughout the training cycles.

Incentives for enlistment are twofold. Unemployment is a chronic problem in Ovamboland, and a career in the battalion provides a stable source of income. Depending upon length of service and time in grade, soldiers earn an average of 600 to 700 rand per month (at .41 U.S. dollars per rand, that's $246 to $287). Finally, service in 101 Battalion is a positive means of retaliation against the SWAPO terrorists who have ravaged and murdered throughout all the Ovambo villages.

SOF's *Technical Editor holds folding-stock Soviet AK-47 captured from SWAPO terrorists by 101 Battalion's Team 18. Note early slab-sided magazine.*

In September of 1980, the battalion's mission was restructured to emphasize combat operations, and the unit's name was changed to 101 Battalion. In 1982, the Romeo Mike concept was initiated using the Buffel Mine Protected Vehicle (MPV). However, suppressive fire is limited to no more than twin-mounted L3s with this vehicle.

In July 1982, 101 Battalion received twelve Casspir Mk 2 MPVs, a vast improvement over the Buffel, and 901 Special Service Company (the designation in this battalion for a company-sized mobile strike force) was established. The 903 Special Service Company followed in September of that year, and 902 and 904 Special Service Companies were added to 101 Battalion rolls in April 1984.

Casspir Mk 2 MPVs are powered by a six-cylinder, 170-hp diesel engine mated to a five-speed manual transmission and two-speed transfer case. They have a maximum road speed of 60 mph and an operational range of 500 miles. The engine, transmission and fuel tank of these four-wheeled vehicles are contained within the all-welded steel hull. The unitary armored hull will withstand penetration of small arms fire up to 7.62 x 51mm NATO ball, as well as rifle grenade and mortar fragments. A 200-liter potable water tank in the hull's bottom serves only the troops, and does nothing to enhance protection against mines.

Mines are an everyday occurrence in Ovamboland, and the Casspir's hull can handle a Soviet TM-57 and be on the road again after no more than two hours of repair time, since only the axles and/or wheels need to be replaced. This vehicle was designed for quick axle changes.

The driver sits on the right, with the commander/gunner to his left. Access to the turret is by means of a central hatch over their positions. Their side and front windows, 2.2 inches thick, provide the same degree of protection as the hull. An L3 in a ball-joint mount is attached to the commander/gunner's front window. The troop compartment extends to the rear of the vehicle, and up to 10 troops can sit, five on each side facing each other, on individual seats with four-point safety harnesses. The troops enter and leave the vehicle via two power-assisted, remote-controlled rear doors.

There are six firing ports and three rectangular bulletproof windows on each side of the hull. Useful only for suppressive fire, the firing ports are never used. They reduce the rifleman's hit probability and increase the possibility of an accidental discharge within the vehicle. Casspirs have open roofs, and the troops fire their South African-manufactured R4 Galil-type assault rifles from the roof by standing on their seats. This not only increases hit probability, but also protects the upper torso from PG-7 shaped-charge grenade fragments—which can penetrate the Casspir's hull and ricochet about within the vehicle's interior.

In September of 1984, the first Hispano-Suiza HS 820 20mm automatic cannons were mounted on the turrets of 101 Battalion Casspirs. The HS 820 is one of the most successful 20mm guns ever produced. It has been manufactured by a number of countries throughout the world and used for a large variety of tasks. Operating by locked-breech blowback and firing from the open-bolt position, the cyclic rate is 1,000 rpm, and its ground-to-ground suppressive-fire capability is excellent.

I fired both 20x139mm HEI (High Explosive Incendiary) and HEI-T (High Explosive Incendiary Tracer) ammunition through a turret-mounted HS 820 during my brief deployment with 101 Battalion. Both are filled with Hexal P30, which delivers both a strong blast and powerful incendiary effect, along with excellent fragmentation of the projectile body. The impact-detonated fuse has a self-destruct time of four to 12 seconds.

HEI or HEI-T 20x139mm ammunition is superior to .50-caliber ball for antipersonnel applications, because it disperses numerous fragments into an area pattern often referred to as the "Effective Casualty Radius." This is defined as the distance from the point of detonation at which a certain percentage of exposed personnel will be casualties. Unfortunately, there is no consensus as to what constitutes the proper percentage of casualties to be used in the definition. Some authorities and governments use 20 percent and others up to 50 percent. Furthermore, the orientation of the projectile's long axis at the moment of detonation will affect its wounding capability. Finally, in most instances this information, if it has even been determined, is classified. Nevertheless, 20mm guns can wreak awesome destruction upon unprotected troops.

HS 820s are usually mounted on the Casspir's turret, coaxial with either an L3 or Browning .50-caliber M2 HB machine gun. Other configurations I encountered were twin L3s, twin .50-caliber M2 HBs and an L3 coaxial with Ma Deuce.

Everyone in the Casspir is issued an R4 assault rifle. Each Casspir has a MAG GPMG and a Y2 40mm Multi-shot Grenade Launcher (MGL). The MGL, a six-cylinder, revolver-type weapon, fires M406-type, high-low propulsion 40mm grenades. Its double-action firing mechanism is not linked to the cylinder. Cylinder advance is controlled by a gas-operated plunger, which is actuated when a round is fired. It's equipped with a single-point gunsight, and its maximum effective range is 400 meters. At almost 12 pounds, empty, it's a bit too heavy for the infantry, but mechanized troops can put its practical rate of fire of 18 rounds per minute to good use in the bush.

Each team in 101 Battalion's Special Service Companies consists of four Casspirs and one Kwevoel 50 logistic vehicle, which carries supplies and spare tires (in supplement of the two spares on each Casspir Mk 2). Casspirs do not use "run-flat" tires (used on the Ratel AFV and G5/G6 artillery, these tires feature inserts which allow the vehicle to be driven for 30 kilometers at a speed of 25 km/h on a hard surface with one or more tires flat) as they cannot be repaired in the field, and punctures from small tree stubs are a constant menace. Nor are tires any longer filled with water, as experience demonstrated this was of no benefit for mine protection. There are three teams in each Special Service Company, with 32 to 40 members in each team (approximately 10 men per Casspir).

The firepower capabilities of these teams is outstanding. Each team is equipped with two

Hispano-Suiza HS 820s, four window-mounted L3s, two twin-L3s or .50-caliber M2 HBs (single or twin) and L3s or Ma Deuces for coaxial mounting with the HS 820s, four MAGs, four 40mm MGLs, two RPG-7s and two 60mm M4 Mk1 patrol mortars, in addition to individual weaponry (the R4 rifle and 9mm Parabellum pistols for vehicle commanders).

Manufactured by Armscor, the 60mm M4 Mk1 patrol mortar weighs only 15.5 pounds. With a range of 100 to 2,000 meters, it will fire any NATO-standard ammunition. Its barrel hinges in one plane only on the interchangeable breechpiece for quick and accurate aiming, with remarkable stability for so light a mortar. The M4 breechpiece has a trigger mechanism which enables the operator to walk with a bomb in the barrel. If the fixed firing pin is installed or the M4 Mk1 breechpiece substituted, the mortar is fired by conventional muzzle-end loading. A clamp-on handgrip incorporates a special sight with all data needed (range, elevation and charge) engraved on the face plate. This is, in my opinion, the ultimate lightweight mortar, with a wide range of applications for special ops units.

But no matter how superb the equipment, its *proper tactical application* is the cutting edge between success and failure in a counterinsurgency. Area domination is 101 Batallion's primary mission. Incessant patrolling operations are its *raison d'etre*. Deploying with Team 18 permitted me to observe a typical patrol scenario.

Tracking is an all-important skill for successfully patrolling the desert pans of Ovamboland. Although the Himba tribesmen may be superior in mountainous terrain, no one can best an Ovambo tracker on sand and hard ground. Every trooper in the team is a tracker. Ovambos begin their study of tracking as children in the *kraal* (village). Military courses in this subject sharpen their ability to distinguish between individual tracks with the same certainty as fingerprint identification.

With the Casspirs moving forward at five to six mph and several trackers out front on foot, the other troopers scrutinize the ground from the windows and roof of the vehicles. Suspicious spoors are tracked to the kraal, and the local pops (populations) are questioned. If terrs are identified, the team will surge forward, moving back and forth across the spoor. Teams to the rear are notified and will begin to leapfrog with the lead team, as much as 10 clicks a jump. When the spoor is no more than 60 minutes old, the Alouette helicopter gun ships are called into the operation. If the terrs are sighted, the Alouette's 20mm automatic cannon will shred the bush, hoping to force the SWAPO group into a fatal confrontation with the Casspirs.

After contact, which usually occurs at a distance of between 25 to 100 meters, the Casspirs will circle the enemy at top speed to avoid hits from an RPG-7. To snare any of the quarry attempting to slip out of the net, the circumference of the circling movement is constantly enlarged. Romeo Mike troops fight on wheels, effectively utilizing the protection and mobility of their MPVs. They do not disembark and deploy from the ground as do U.S. Army mechanized infantry.

In the bush, each team sets up an NDP (Night Defensive Position) by placing their vehicles in a five-pointed star pattern. Team members are not supposed to leave the inner perimeter formed by the rear end of their vehicles at any time during the night. The consequences of moving outside the NDP can be severe. In one instance a troopie who awoke to relieve himself, groggily walked away from the vehicles instead of toward the center. Upon his return, he was blown away by the other members of his team.

How successful are the Romeo Mike's tactics of tracking and mobile encirclement? In 1987, 101 Battalion killed 257 SWAPO terrorists, captured 38 equipment caches and lifted 206 antivehicle mines and 193 antipersonnel mines. With a frequency of contact of up to three times a day, the battalion has maintained a casualty ratio heavily in their favor—ranging from 5:1 to 23:1 and in some instances as many as 100 terrorists snuffed to one member of the battalion.

An examination of the cache material and mines points to a conspicuous trail leading directly back to the Soviet Union and its Eastern

European cohorts—with only two somewhat surprising exceptions.

In small arms, the Kalashnikov, of course, predominates. Both AK-47s and AKMs with the following origins have been captured: USSR, Hungary (including the short-barreled AMD-65), Bulgaria, East Germany, Rumania, a very few from Yugoslavia (M70B1 and M70AB2 with grenade launching potential), North Korea, the People's Republic of China, and the Vz 58 series from Czechoslovakia.

Although only a few SKS rifles from the Soviet Union and PRC have been captured, the Yugoslav M59/66A1, with its spigot-type grenade launcher permanently attached to the muzzle, is a SWAPO favorite. The RPG-7 can be faulted by virtue of its large visual and audible muzzle signature, compelling the operator to move quickly immediately after firing in order to avoid detection. On the other hand, rifle grenades can be launched from the deep bush with little chance their point of origin will be located. To launch M60 antitank or antipersonnel grenades from the M59/66A1 rifle, the operator must cut off the piston's gas supply by pressing the gas valve and rotating it to the top of the gas cylinder. A ballistite (blank) cartridge must be used for grenade launching, and one comes packed in the Yugoslav grenade's tail boom. If ball ammunition is used, the grenade will explode on the launcher. The average SWAPO terrorist is close to brain dead when it comes to even the most rudimentary technical competence, and no doubt several have vaporized themselves in this manner.

A small number of both the Soviet bolt-action Mosin-Nagant and Dragunov (SVD) sniper rifles have been captured from SWAPO terrorists. In some instances the Dragunov's PSO-1 scope and mount have been missing. South African technical intelligence personnel who have evaluated the Dragunov are not impressed with its accuracy potential (no better than three MOA). Mosin-Nagant M1944 carbines were occasionally encountered early on in the conflict.

Submachine guns are also uncovered in caches, apparently stashed long ago. They include the entire Czech Vz 23/24/25/26 series, the M61(j)—a Yugoslav version of the Czech Skorpion, Soviet PPSh-41 and PPS-43 and the Polish Model 1943-52 (PPS-43 with wooden stock and longer receiver).

Pistols taken from SWAPO political commissars include Soviet, Hungarian and PRC versions of the Tokarev TT-33, Soviet and East German Makarovs and the Czech Vz 24 (predecessor of the blowback Vz 27, but in caliber .380 ACP and with a rotary locking barrel).

Squad-type machine guns employed by SWAPO are all of ComBloc origin. Originally they used DP, DPM, and RP46 LMGs, plus the occasional SGM. The following now prevail: RPDs, both Soviet and Chinese; RPKs from the Soviet Union and Yugoslavia (M72B1 and folding-stock M72AB1); M53 (Yugoslav version of the German World War II caliber 7.92x57mm MG42 GPMG); and both the Soviet PKM and PKT (solenoid-fired, coaxial variant) GPMGs.

In portable rocket launchers, the Soviet-made RPG-7V is dominant. Although issued with the PGO-7 optical sight and a complete maintenance and spare parts kit, these components are usually discarded by the SWAPO terrorists before they cross into South West Africa. No RPG-7Ds (folding version) have been encountered. PG-7 and PG-7M HEAT grenades are supplied by the Soviet Union, Bulgaria and Syria. OG-7 antipersonnel grenades are also fielded by SWAPO, but to a lesser extent. At present, RPG-2s are no longer found with any frequency. During a recent contact, 101 Battalion troops engaged 20 terrs who were carrying 18 Czech RPG-75s (a variant of the Soviet RPG-18, which in turn was copied from the U.S. M72).

Hand grenades that I examined included the Soviet F1 (and PRC Type 1), Soviet RGD-5, Soviet RG-42 (and PRC Type 42), Yugoslav M69 (egg-shaped with 4.5-second delay fuze), PRC stick grenades in various configurations, the Hungarian M42 stick grenade and the so-called Soviet M75.

The M75 is supposed to be a *copy* of the Austrian Arges Type HdGt 73 fragmentation grenade. It contains 2,600 steel-balls, varying in size from .086 to .109-inch, set in a plastic matrix. Upon detonation, the plastic matrix disintegrates and the pellets are driven outward at almost 6,000 fps. Because of their spherical shape, the balls lose velocity rapidly. Although 100-percent lethality is assured within five meters, by 20 meters velocity has fallen so low that the casualty capability is virtually zero. The outer skin of this grenade is a sprayed thermoplastic synthetic with a waffle pattern.

The explosive filler weighs 1.3 ounces and consists of granular PETN (pentaerythritol tetranitrate) blended with sufficient light grease to make it putty-like and moldable. PETN is a high-velocity explosive of considerable brisance (shattering effect) and has a detonation rate on the order of 20,000 fps. Stable under typical ambient conditions, PETN is more easily detonated than TNT.

Both the PETN explosive filler and the Bouchon-type igniter system (with a three- to four-second time delay) appear to be of Bulgarian origin. As a result of X-ray analysis, South African technical intelligence sources are now convinced that the body itself is manufactured by ARGES (*Armaturen-Ges.m.b.H.*) in neutral Austria through the subterfuge of merely changing the two molded identification markings on the surface of the outer skin.

Soviet-manufactured RKG-3 and RKG-3M antitank grenades are occasionally employed by SWAPO terrorists. These large stick grenades have a shaped charge which will penetrate 125mm and 165mm of armor plate, respectively, each with an instantaneous impact fuze. They are stabilized in flight by a small, four-panel parachute which pops out of the handle when the grenade is thrown. Since they're hand-propelled, they must be thrown from an almost suicidal distance to the target vehicle, and their velocity is so slow that hit probability on a moving target is extremely low.

SWAPO terrs rarely trudge across the border with equipment that is not readily man-portable, but sometimes larger weapons systems will make their way into South West Africa. Examples include the Soviet 82mm B-10 recoilless gun with both HEAT and fragmentation warheads, GRAD-P (DKZ-B) 122mm rockets fired from a light single tube, PRC 75mm Type 52 recoilless rifles (a copy of the U.S. M20) with U.S. and PRC ammunition, Soviet SA-7 surface-to-air missiles, Soviet ZPU-1 14.5mm KPV heavy machine guns, Soviet 30mm AGS-17 automatic grenade launchers, Yugoslav M-57 and PRC Type 63 60mm mortars, a new version of the Soviet 82mm M-37 mortars and on very rare occasions, the 12.7x107mm Degtyarev Model 38/46 heavy machine gun.

Mines are the manna upon which terrorism feeds. No weapon I have encountered in more than a quarter century on the world's battlefields is more repulsive and indiscriminate. Every soldier and civilian within a combat zone lives in horror of their potential to maim and destroy without warning. With the world's trendy-left media solidly entrenched against the Republic of South Africa, there has been no attempt to disguise the communist origin of these cruel devices. No need in South Africa, as in El Salvador, to feign an indigenous spontaneity to the revolution. With one disgusting exception, all of the mines I examined were gifts from comrade Gorbachev and his communist cronies.

Just about every ComBloc mine that has ever been manufactured has been encountered at one time or another in South West Africa. Antipersonnel mines play particular havoc with the civilian population in a counterinsurgency conflict, and those walking across the oshanas may detonate any of the following: Soviet POMZ-2 and POMZ-2M stake mines (or their Czech PP-Mi-Sk and Yugoslav PMR-1/2 equivalents), any of the numerous Soviet PMD-6/7 and PMD-57 series wooden box mines, the duroplastic Soviet PMN with a rubber cover over its pressure plate, the neoprene-covered East German PPM-2 and the plastic Yugoslav PMA-3.

Worst of all is a small, non-metallic antipersonnel mine with electronic circuit-board ignition and a mercury switch that serves as an anti-lift device. Once activated, the slightest movement will roll a drop of mercury in a glass vial against two wires to detonate the mine. Called the VS Mk 2-E, it, as well as the mechanically initiated VS Mk 2, is manufactured by *Valsella*

SpA in Brescia, Italy (a NATO alliance nation). Quite obviously, the merchants of death hold allegiance to no one.

Bouncing mines include the Czech PP-Mi-Sr, Soviet OZM-4, Yugoslav PROM-1 and PRC Type 69. Two types of Soviet magnetic limpet mines have been encountered—the SPM and Type 158 (a copy of the British Clam Mk 3).

A large variety of antitank mines have been employed against 101 Battalion's Casspirs: Soviet TM-57, TM-46, TMN-46 (anti-lift type), TM-62M and TMK-2 (shaped charge with tilt rod fuse); Yugoslav plastic TMA-2, 3, 4 and 5; Czech plastic PT-Mi-Ba-II/III and metallic PT-Mi-K; and Hungarian UKA-63. Older wooden box-type mines, such as the Czech PT-Mi-D and Soviet TMD-B and TM-44 are no longer used as they are easily detectable.

Anti-vehicular mines can be hazardous to anyone's health—including a SWAPO terrorist traveling on foot. Surprised by a team of Casspirs, a terr recently popped out of the bush and started racing down a bush track at full speed. Blazing away with everything they had, the Casspirs gave chase with little success. Lead poured out of every Browning and Hispano-Suiza in the team, to no avail. The terr just kept bounding down the trail. Finally, a 101 Battalion troopie shoved his M79 blooker (issued prior to the adoption of the Y2 40mm MGL) out the open top of the vehicle and fired a round. With more than a little luck, the M406 HE round struck the terrorist squarely in the back. He was carrying an antitank mine in his rucksack. They found only his shoes and feet.

Who is the real enemy in this tormented land? Clearly, the catalog of confiscated munitions points directly to the Soviet Union and its communist clones. And who are the terrorists? Most certainly not the South Africans, as dangerous political buffoons such as Jesse Jackson and Michael Dukakis have alleged. The devils in this matter are most assuredly those who sow the mines and hope to reap a bitter harvest of anguish and despair.

Regardless of world opinion and posturing politicians whose pompous pronouncements create little more than nebulous smoke screens behind which the communist terrorists continue to operate, the men of 101 Battalion and the SWATF will fight on. In the center of the battalion's compound stands a black marble monument with the names of all those struck down in battle. With no end to the conflict in sight (in the opinion of those who must bleed more than Linotype ink) the list will continue to grow. *Aut vincere aut mori* (conquer or die) remains the soldier's lot forever.

Originally appeared in *Soldier Of Fortune*
February 1989.

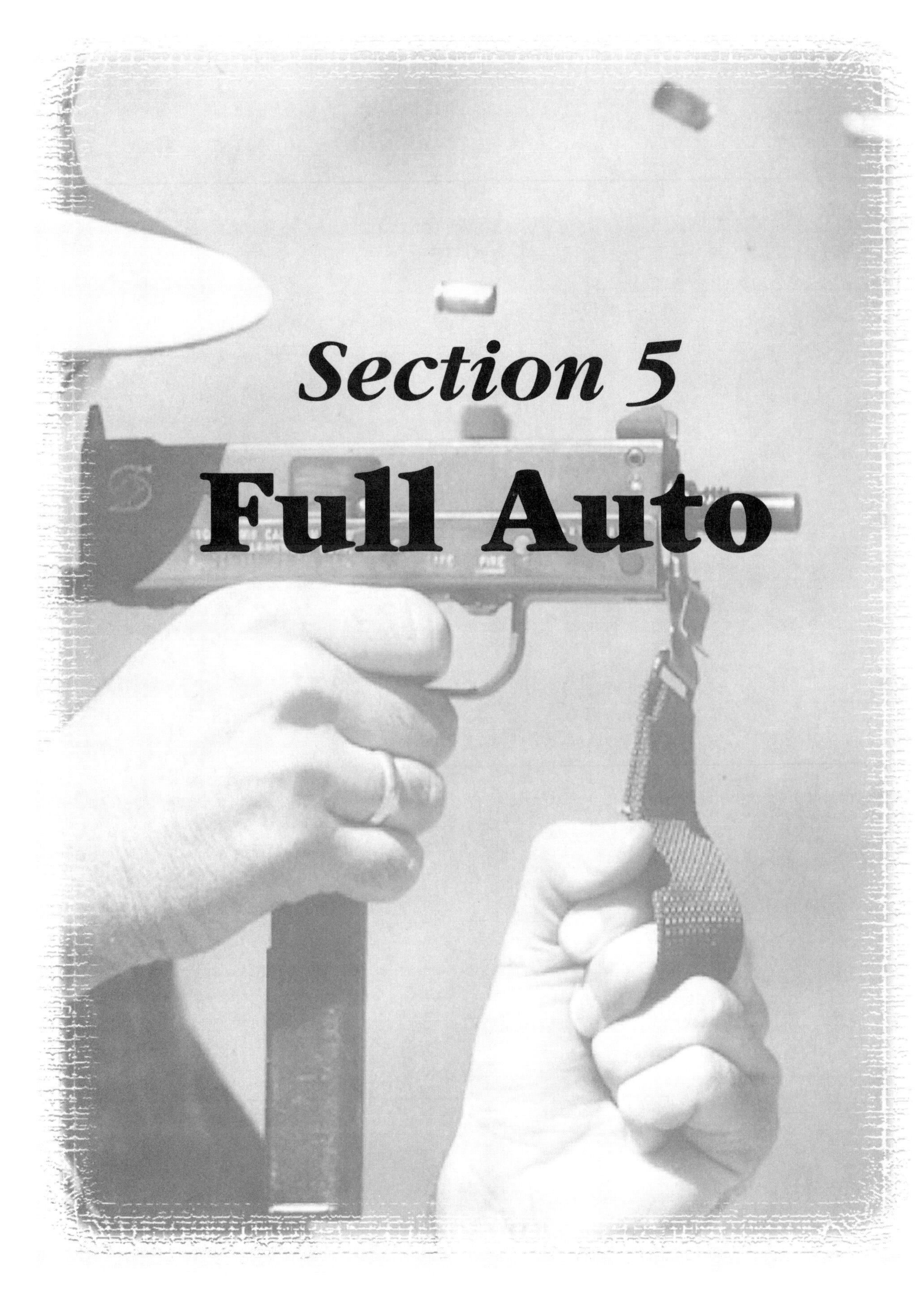

Section 5
Full Auto

UZI REBORN

Vector Arms' New Model HR4332

Mr. Machine Gun's 9-year-old grandson, Cody J. Kokalis, fires a burst to demonstrate the UZI's low recoil impulse.

A quarter of a century ago the UZI was, without doubt, the most widely distributed submachine gun in the Western world. In the interim, Heckler & Koch's MP5 series has largely become the submachine gun of choice for most military and law enforcement SpecOps groups throughout the Americas and Western Europe. Even the U.S. Secret Service, so long an advocate of the UZI recently turned them in for more modern hardware. Additionally, in recent years more and more police tactical teams are exchanging their pistol caliber submachine guns for short-barreled caliber 5.56x45mm carbines, such as the M4.

In the face of this, Vector Arms, Inc. has just introduced a new production series UZI submachine gun. Why? As an individual, the only MP5 available for you to buy is a registered-sear or registered-receiver gun made from a converted HK94 semiautomatic-only carbine. Factory MP5s are *all* dealers' samples-only and available exclusively to qualified law enforcement agencies and military organizations. Furthermore, every registered-sear MP5, without exception, that I ever encountered while teaching the SMG Operators Course at Gunsite Training Center failed before it reached 3,000 rounds. Yet these weapons are selling *(selling,* not asking) for $8,000 to $9,000. Vector Arms' new Model HR4332, is manufactured from Group Industries receivers and IMI (Israel Military Industries, Tel Aviv, Israel), FN (Fabrique Nationale d'Armes de Guerre, Herstal, Liege, Belgium) and LIW (Lyttelton Engineering Works Pty., Ltd., Pretoria, Republic of South Africa) components. It sells for only $2,995 and is a non-restricted-transfer Title II firearm. I don't think it requires a rocket scientist to determine that an older, but completely reliable, burp gun for less than three grand is a better buy than a cobbled-up, unreliable copy of a more modern design selling for three times as much.

Submachine guns—commonly defined as shoulder-mounted, selective-fire weapons chambered for pistol cartridges—were fielded in larger quantities during World War II than before or since. During this time frame most submachine guns fired from the open-bolt position by means of unlocked blowback. They were thus easier to manufacture than more complex locked-breech machine guns. In addition, both design and manufacturing techniques had advanced to a level that permitted fabrication using cost-effective welded and pinned sheet metal stampings together with synthetic furniture. In short, SMGs provided a great deal of firepower, cheaply.

For these reasons, the submachine gun's popularity continued during the decade following the war. In Czechoslovakia, the first small arm placed into series production after the war was Jaroslav Holecek's CZ447 submachine gun, which was designed at *Ceska zbrojovka* in Strakonice and adopted as the Model 48 in 1948. In the spring of 1950, the designation was changed to the Model 23 for the fixed-stock version and Model 25 for the folding-stock variant. Both were chambered for the 9mm Parabellum cartridge. When ComBloc standardization was imposed upon Czechoslovakia, the Models 24 and 26, respectively, were introduced in 1951, chambered for the Tokarev 7.62x25mm round.

MAJOR UZIEL'S FAMOUS SUB GUN

Shortly after this sequence of events in 1950 to 1951, Major Uziel Gal of the Israel Defense Force (IDF), assisted by other Israeli small arms technicians, designed what became the justifiably famous UZI submachine gun, chambered for the 9mm Parabellum round. Its single most salient characteristic, presumably taken from Holecek's SMG series, is a so-called wrap-around bolt design.

The 10.2-inch barrel extends well into the receiver, with a recess machined into the bolt body to accept the rear portion of the barrel. This requires an ejection port to be machined into the bolt body also to match that of the sheet metal receiver body. The primary advantage of this feature is compactness. At the moment of ignition, the telescoping bolt surrounds 3.75 inches of the rear portion of the barrel. Thus, by placing greater mass over the chamber—where the expanding propellant gases are generated—upward muzzle flip is moderated considerably. This system also alters the weapon's center of mass and accounts to a great extent for the UZI's excellent handling characteristics. Finally, this is an added safety factor in the event, albeit unlikely, of a blown case. The recessed breech face on the bolt has a fixed firing pin milled into its surface. A hole in the rear face of the rectangular bolt body retains the recoil spring assembly. The single-coil recoil spring and a red fiber buffer at the rear are captive on the steel guide rod. The fixed ejector, riveted to the receiver, rides in a slot milled into the bolt body under, and in back of, the breech face.

Another feature apparently taken from the Czech Model 23/24/25/26 series is the place-

TITLE II FIREARMS

The firearm reviewed in this article is a non-restricted-transfer Title II firearm. Title II firearms, which include machine guns, sound suppressors, and short-barreled shotguns and rifles are strictly controlled by the National Firearms Act (NFA) of 1934. There are almost a quarter of a million registered machine guns on the NFA log books. It is important to point out that during the 65-year history of the NFA no more than one or two registered machine guns have ever been used in the commission of a felony. This is a remarkable record matched by no other classification of firearms. "Non-restricted-transfer" means that if you, as an individual, live in a state that permits possession of machine guns, you can personally acquire the new Vector Arms UZI, providing you meet the federal requirements for an ATF Form 4 transfer (law enforcement certification, FBI fingerprint check, and $200 transfer tax).

ment of the magazine well within the pistol grip assembly. This is, in my opinion, somewhat controversial. To its credit, this (in addition to the telescoping bolt) leaves the point of balance directly above the grip, provides a firm support for the magazine and supposedly aids in rapid magazine changes, using the so-called principle of "hand finds hand." These advantages, real or imagined, are counterbalanced by sacrificing any attempt at a proper grip-to-frame angle, since to accommodate the magazine's required alignment for feeding, the grip must be vertical to the bore's axis. The most advanced human engineering I have ever encountered on a submachine gun, with regard to the grip-to-frame angle, is that of the Sterling.

UZI magazines, derived from those of the Italian Beretta Models 1938A and 38/42 submachine guns are of the two-position-feed type. The staggered-column, detachable, box-type Beretta magazines are generally conceded to be among the best ever designed. Less bolt energy is usually required to strip rounds from this type of magazine. It should also be noted that the top round in an UZI magazine is held at a slight angle with the nose up so that the primer does not align itself with the fixed firing pin until the cartridge enters the chamber. Single-position-feed magazines, such as those encountered with the German MP38/40, British Sten, U.S. M3 ("grease gun") and despicable MAC10 are more difficult to load without a tool and are sometimes not quite as reliable.

All UZI magazines are equipped with indicator holes on the sides so the rounds remaining can be seen at a glance. The standard UZI magazine holds 25 rounds. A 32-round magazine is also available and it is as reliable as the original. However, the 40-round magazine, fielded only shortly, was most decidedly not adequately reli-

UZI SMG. Left-side view with folding buttstock extended. (Inset) The all-steel folding stock is hinged at two points and collapses in an accordion-like manner under the receiver.

able for combat. This problem is generic to almost all large-capacity, spring-loaded, box magazines. If you have enough follower spring pressure to hold the last 10 or 20 rounds tightly up against the magazine body's feed lips, you will usually have too much pressure when a large-capacity magazine is loaded to its maximum extent. As a consequence, until the magazine has been partially emptied there is usually too much "stripping pressure." Thus, when the bolt drives forward in counter-recoil, it meets with excessive resistance as it attempts to strip the next round from the magazine. The bolt's forward velocity can be reduced to such an extent as a consequence that there might not be sufficient forward momentum to ignite the primer.

A special clip is available to join two UZI magazines together in a "L" configuration. This is useful because, when both magazines are loaded, the one not in use lies under the barrel and helps reduce muzzle climb. Then, when the first magazine has been removed and the new one rotated into the magazine well, the empty magazines lies back under the buttstock. This magazine clamp is available from Vector Arms for $8.

TRIPLE SAFETIES

The UZI is equipped with three safety systems, each independent of the others. The fire selector is located on the left side of the trigger assembly, just above the grip panel. When moved rearward to the "S" position, the trigger is blocked from moving by a curved extension on the front of the sliding selector bar. Pushed fully forward into the "A" position will permit full-auto fire, at a cyclic rate of approximately 550 to 600 rounds per minute. Sliding the selector to "R" (standing for Repetition, the British expression for single shot fire) will provide semiautomatic fire only. By design, the selector is usually quite difficult to manipulate, although over time it usually becomes easier to slide it forward or rearward. Furthermore, it's easily adjustable by tweaking the shallow V-shaped leaf spring that presses up against the slide's three detented positions. However, when in the operational area, most IDF personnel would

Rare vertical foregrip, made by IMI for Israeli Special Operations groups, substantially improves burst-fire hit probability.

An effective BFA was designed for the UZI by the Dutch munitions firm of NWM, although 9mm blank ammunition is now difficult to obtain.

The fire selector is located on the left side of the trigger assembly, just about the grip panel and has three positions for full-auto (A), semiauto (R), and safe (S).

slide the selector to full-auto and depend upon the grip safety.

The trigger is L-shaped with the short leg as the trigger itself and the long leg (or trigger bar) in a horizontal plane. It is spring-loaded and pivots on an axis pin at the junction of the two legs. Pulling the trigger depresses the trigger bar. At the end of the trigger is a pivoting disconnector, which holds down the sear when the trigger is pulled. The unusually wide sear has two prongs, between which lies the disconnector. In full-auto fire the disconnector moves downward and rotates the sear with it. The gun will fire until the trigger is released and the disconnector rises with the trigger bar. The sear spring pushes the sear into the path of the forward moving bolt body to hold it to the rear.

In semiautomatic-fire the disconnector is not only pulled down by the trigger bar but rotates about the curved extension on the front of the selector bar. This rotation brings the disconnector through the sear's two prongs and the sear then rises to impinge against the bolt and stop its forward travel. To fire another shot, the trigger must be released to permit the disconnector to rotate back up through the sear's two prongs and re-engage the sear. This is actually a very simple, but quite effective, method of operation.

Grip safeties were not uncommon on submachine guns. They were included on both the mediocre Danish Madsen M50 and the excellent French MAT49. But, a firm firing grip is absolutely essential to ensure reliable operation in high stress environments. However, this is just a matter of programming through intensive training. A lug projecting from the top of the spring-loaded grip safety's front bar on the right side of the trigger mechanism lies under the right sear prong and prevents it from moving down. When the grip safety is squeezed, its bar moves forward together with the lug permitting the sear to drop.

There are 13 small serrations on the underside of the receiver's top cover, next to the spring-loaded bolt-retracting slide. These serrations serve to lock the bolt if the retracting knob is accidentally released and an incomplete cocking motion results. Specifically, after the retracting handle and slide have come back 1.875-inch, they cannot go forward again until they have been retracted fully to the rear, a distance of 3.25 inches. This bolt blocking safety mechanism prevents accidental discharges if the weapon is jarred or dropped, as sometimes occurs when hastily off loading an armored personnel carrier in a combat zone. This is an excellent feature on an open-bolt burp gun.

The flip-type rear sight, equipped with large

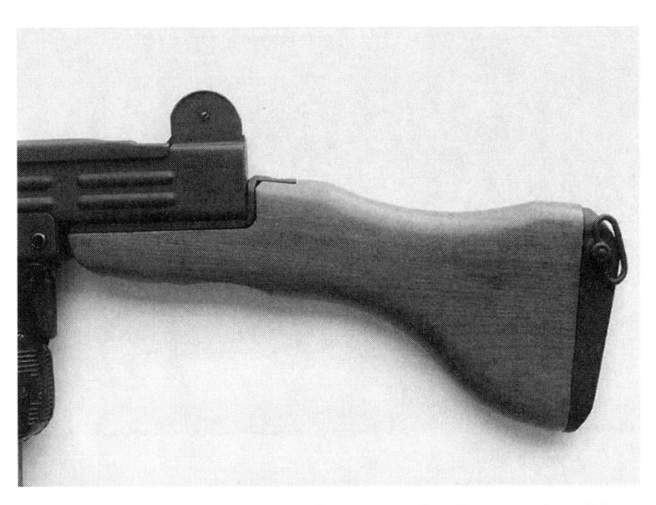

Early UZIs were issued with fixed wooden buttstocks of the quick-detachable-type. Fixed buttstocks provide a more stable firing platform for any shoulder-mounted weapon.

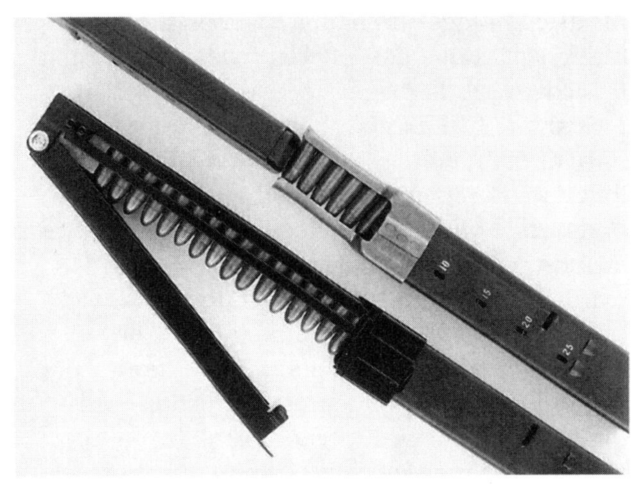

Two different magazine loaders are available for the UZI. (Top) Standard IDF-issue six-round loader is best used with another magazine to shove the rounds downward. (Bottom) A more sophisticated, after-market tray-style tool that loads 16 rounds at a time is considerably more efficient.

protective ears, has peep apertures marked for 100 and 200 meters. This is a bit optimistic, since submachine guns are CQB (Close-Quarter Battle) weapons most effectively deployed at distances well under 100 meters.

The front sight is a tapered post, also with more than adequate protective ears, mounted eccentrically on a threaded base. Both windage and elevation zero are adjusted at the front sight. As the post is offset, rotating the post for windage zero will also raise or lower it slightly. Elevation adjustments will also effect the horizontal zero to a slight extent. A half-turn of the front sight post will move the point of impact about 4 centimeters at 50 meters and as much as 64 centimeters horizontally. Submachine guns are best zeroed at 50 meters and employed at that distance and less. Remember, when you make zero adjustments at the front sight you must move the front sight post down to elevate the point of impact and to the left to move it to the right. The front sight should be adjusted with the IMI tool made specifically for that purpose and available from Vector Arms for only $7. As an adjunct to this, keep in mind that after the contact distance increases to farther than 15 meters, you should be seriously thinking about sliding the selector back to semi-auto, especially if you're shooting on the move.

While the extensive use of welded, sheet-metal stampings and high-impact plastic furniture resulted in a cost-effective weapon system (exactly what the Israelis needed in the early 1950s), the UZI exhibits a great deal of quality in both design and manufacture. As an example, longitudinal channels stamped into the receiver body's side walls serve effectively to trap debris, ensuring reliable operation under the most extreme conditions of high dust and sand, so often encountered in the Middle East. They also add structural support to the receiver body.

Vector Arms commences production with the Group Industries receiver blank (produced, marked and registered by Group Industries, Louisville, Kentucky, prior to the 16 May 1986 ban and purchased by Vector at auction in August 1995 during the dissolution of Group Industries), which must be trimmed and

STRIPPING THE UZI

First, remove the magazine. Squeeze the grip safety and retract the bolt and visually inspect the chamber to make certain it's empty. Allow the bolt to move forward into battery under control. Set the selector bar to "S." Depress the top cover's latch, just in front of the rear sight, and lifting it up at the rear, remove the top cover assembly from the receiver. Slide the bolt group rearward until it can be lifted out of the receiver. Pull the recoil spring, guide rod and buffer out the rear of the bolt body. Using a punch of the proper size drift out the extractor's retaining split pin.

Remove the extractor. This should be done every time the weapon is cleaned. Dirty extractors can account for serious stoppages. Rotate and withdraw the barrel nut from the receiver's threaded trunnion. Withdraw the barrel. Drift out the trigger group's captive retaining pin from the left side and swing the trigger group down and off the receiver. The magazines can be, and should be, disassembled in the conventional manner. No further disassembly should be attempted except by qualified armorers. After cleaning and subsequent lubrication, reassemble in the reverse order.

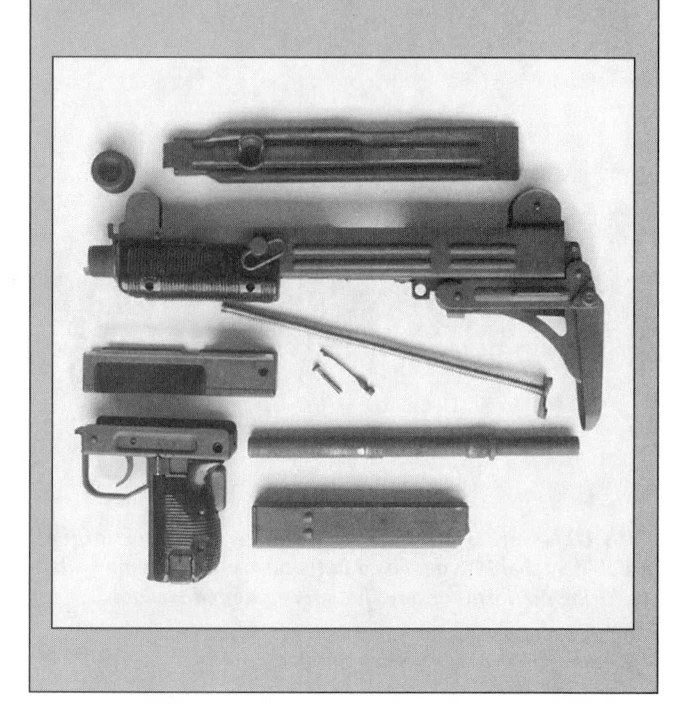

deburred. Three machined lugs are then TIG-welded to the bottom of the receiver. They are 1) the bayonet lug; 2) the handguard lug, into which the handguard retaining screws are threaded; and 3) the pistol grip/trigger group lug, which also holds the folding stock in place after it has been folded and retains the rigid wood buttstock when it is installed. After this the barrel sleeve (or trunnion), another machined part, is TIG-welded to the receiver in four places. This important component supports and indexes the barrel. The front end of the barrel sleeve, 3.75-inch in overall length, is threaded to accept the barrel locking nut. The machined backplate is then welded to the rear of the receiver.

Three stamped, sheet-metal components are subsequently welded to the receiver. The barrel nut latch housing, which holds the spring-loaded ratchet that prevents the barrel nut from vibrating loose, is TIG-welded under the front sight. Both the U-shaped front and rear sight saddles are spot-welded to the receiver. Two parts are riveted to the receiver: the fixed ejector and the front sling swivel.

THE HEAT TREATMENT

All of the receiver's fixed-parts manufacturing and receiver assembly is accomplished by Vector Arms, Inc. Welds and machining marks are polished to achieve an appearance far beyond the quality of any UZI ever manufactured for military use. After complete assembly, the receiver is vacuum heat-treated to minimize reciprocating bolt wear. A modern carburizing process is used. Carbon gas is induced into the steel in an oxygen-free environment to a depth of 16 thousandths (0.016-inch). After this seven-hour process, the receiver is quenched in oil to prevent warpage. At this time the receiver is too brittle at about 60C Rockwell Hardness. Subsequent tempering brings the receiver down to 44C Rockwell Hardness. This hardens the receiver's skin, while the core remains soft.

It should be remembered that neither IMI, FN or LIW ever heat-treated their UZI receivers. The UZI was essentially manufactured by them as a "throw-away" submachine gun. The bolt will, in fact, eventually wear two deep troughs in the bottom of these receivers as it reciprocates over a period of time. Vector Arms offers a repair service for damaged IMI receivers that includes subsequent heat-treating.

After installation of the 11 welded or riveted components and subsequent heat-treating, the receiver is then ready to accept the brand new, never-issued milspec parts kit. There are a total of 74 parts in the kit. Only about 5% of the parts came from either IMI or FN. The vast bulk of the parts were obtained from LIW in South Africa, whose manufacture of the UZI was never officially acknowledged. I have examined LIW-made UZIs in South Africa and they are substantially better than those produced by either IMI or FN. By any standards LIW (or Vektor, as it is now known) is a major player in the defense field.

In addition to their more or less clandestine production of the UZI, LIW produces the CP1 compact 9mm pistol, the Z88 and SP1/2 service pistols, the R4 rifle series, several different 20mm automatic cannons, the GS towed and G6 self-propelled 155mm artillery, a 35mm computerized anti-aircraft cannon system, the Rooikat Armored Fighting Vehicle, mortars of all types and arguably the finest GPMG ever manufactured—the caliber 7.62x51mm SS-77, and the superb Mini SS 5.56 SAW.

Production of the UZI at IMI, FN and LIW ceased long ago. With the exception of Vector Arms, the UZI is currently manufactured only by RH-ALAN in Zagreb, Croatia, as the ERO 9.

Unloaded weight of the UZI is 7.7 pounds (3.5kg). The 10.2-inch (260 mm) barrel has four grooves with a right-hand twist of one turn in 10 inches (254 mm). The chamber is slightly tapered as the weapon fires by means of *advanced primer ignition,* in which ignition of the percussion primer occurs while the bolt is still traveling forward and just before the round is completely chambered (almost *de rigueur* with unlocked blowback, open-bolt-firing SMGs). With the buttstock folded, the overall length is only 17.3 inches (440 mm). When extended, the overall length increases to 25.2 inches (640 mm).

The Vector Arms UZI carries an excellent, dark manganese phosphate ("Parkerized") finish. UZIs will also be encountered with a black baked enamel finish over phosphate.

Early UZIs were issued with fixed wooden buttstocks of the quick-detachable-type and with a distinctive configuration. The IDF actually used three different buttstock lengths. The Dutch contract called for a long, wooden buttstock. Original IMI wooden buttstocks, in the standard length only, with new wood and refinished hardware, are available from Vector Arms for $85. There's no doubt in my mind that a rigid buttstock provides a more stable firing platform for any shoulder-mounted weapon.

The subsequent all-steel folding stock is hinged at two points and collapses in an accordion-like manner under the receiver. To deploy it, simply slap the buttplate with the palm of the hand to release it from the bottom of the receiver and unfold it until it locks at both hinge points. To fold the stock, squeeze the back stock bars inward at the rear hinge point and fold the rear section upward. Then push in the locking stud on the left side of the front stock bars, at their hinge point at the rear of the receiver, and fold this section downward. Snap it back in place under the receiver. The Vector Arms UZI comes equipped with this folding stock, which is one of the most secure of its type.

Both the pistol grip panels and forearm are usually made of black, synthetic material. There is some air space between the forearm and the receiver to provide for a small amount of heat dissipation. A unique vertical foregrip is also available. Made by IMI for Israeli Special Operations groups and made of a dark gray, high-impact plastic, it fits all UZIs. In very limited supply, this rare vertical forearm is available from Vector Arms for $60. It helps to control the weapon significantly.

UZI ACCESSORIES

Vector Arms can also supply a wide range of other accessories for the UZI. The Israeli six-round magazine loader is available for only $8.

UZI SPECIFICATIONS

CALIBER . .9mm Parabellum (.22 LR and .45 ACP conversion kits available).

OPERATION . . . . .Unlocked blowback with advanced primer ignition. Fires from the open-bolt position. Full-auto and semiauto capability.

CYCLIC RATE . . . . .550 to 600 rpm.

FEED MECHANISM . . . . .25-, 32-, and 40-round staggered-column, two-position-feed, detachable, box-type magazines.

WEIGHT, empty . . . . .7.7 pounds (3.5 kg).

OVERALL LENGTH
with buttstock folded . . . . . .17.3 inches (440 mm);
with buttstock extended . . . .25.2 inches (640 mm).

BARREL . . . . .Four grooves with a right-hand twist of one turn in 10 inches (254 mm).

BARREL LENGTH . . . . .10.2 inches (260 mm).

SIGHTS . . . . .Tapered post, with protective ears, mounted eccentrically on a threaded base. Both windage and elevation zero are adjusted at the front sight. Flip-type rear sight, also with protective ears, with peep apertures marked for 100 and 200 meters.

FINISH . . . . .Phosphate ("Parkerizing").

FURNITURE . . . . .Grip panels and forearm: black, high-impact synthetic.

OPTIONAL ACCESSORIES . . . . .Vertical foregrip, spare parts and magazines, loading tools, bayonets, caliber-conversion kits, bayonets, screw-on sound suppressors, wooden stock, front sight adjustment tool and soft case.

PRICE . . . .$2,995.

MANUFACTURER . . . . .Vector Arms, Inc., 270 West 500 North, North Salt Lake, Utah 84054; Web site: www.vectorarms.com.

T&E SUMMARY: Battle-proven, open-bolt, blowback submachine gun that meets all conceivable standards of reliability. Heat-treated receiver exceeds durability of an UZI previously manufactured. The best UZI ever made, and available as a non-restricted-transfer Title II firearm.

Slip the loader over a magazine, drop in six rounds and then use the top end of another magazine, positioned sideways over the rounds, to shove the rounds downward and into the empty magazine. A more sophisticated, after-market tray-style tool that loads 16 rounds at a time can be obtained for $35. Spare 25-round magazines, made by LIW, are available for $15 each. Made by USA, 32-round magazines cost $29 each.

Vector Arms offers several caliber conversion kits for the UZI. The .45 ACP full-auto kit, made by IMI, sells for $210. IMI's semi-auto-only .22 LR conversion kit cost $195. A full-auto .22 LR kit is also available for $235. All of these kits include a bolt, barrel and one magazine. Extra 16-round .45 ACP magazines are $50 each. The 16-round .22 LR magazine costs $38. Both magazines were made by IMI.

Screw-on sound suppressors, with cutting-edge technology from Bruegger & Thomet of Switzerland—supplier to H&K in Germany and the German Army—are also available from Vector Arms. These suppressors provide greater sound pressure level reduction than any others designed for the UZI. The interior components have self-aligning interlocks that enable the operator to dismantle and reassemble the unit without alignment tools. Caliber conversion is possible by merely replacing the baffles and front cap. These suppressors screw directly onto the receiver's barrel sleeve, replacing the barrel nut. No barrel threading is required. The cost is $450 for the 9mm Parabellum unit and $475 for the .45 ACP suppressor. NFA rules apply.

The oddest accessory available for the UZI is a bayonet. There are two, almost identical except for manufacture. The IMI bayonet in new condition will cost you $95. One manufactured by LIW in South Africa sells for only $65. Both include the scabbard. Years ago, I asked two officers from the Golani Brigade of what use was a bayonet on the UZI. They indicated it was useful in herding prisoners. You go figure.

A heavy-duty, black Cordura case with a discreet rectangular configuration can be purchased for $69. Designed and manufactured by Blackhawk Industries, it features 3/8-inch closed-cell removable foam padding, wrap-around handle straps, a detachable, padded shoulder strap, three interior magazine pouches, a padded exterior pocket and 360-degree padded protection.

An excellent BFA (Blank Firing Attachment) was designed for the UZI by the Dutch munitions firm of *NWM*. While 9mm blank ammunition is difficult to obtain, Vector Arms will sell you the BFA for $25. It slips over the barrel and merely replaces the barrel nut. The IDF also fielded a grenade launcher for the UZI. It is no longer available.

I have fired many thousands of rounds through UZI submachine guns. It remains one of the most reliable designs ever fielded. The version being manufactured by Vector Arms is, without doubt, the finest UZI ever produced. It is currently the best buy in a nonrestricted-transfer sub gun. I can recommend it without reservations of any kind. The Vector Arms UZI is sold with a one-year warranty on materials and workmanship.

Originally appeared in *Soldier Of Fortune*
December 1999.

M60

The Great GPMG SNAFU

The M60 GPMG mounted on the M122 tripod.

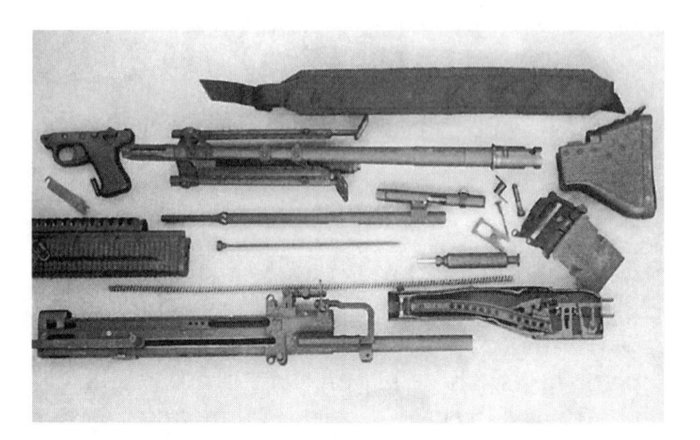

The M60 GPMG field-stripped.

The concept of the so-called "general purpose machine gun" is a failure. And as a prime example of the GPMG, the well-known M60 has proven to be *a very large failure*. It's a tale that deserves telling for all the Americans who humped and fired M60s in Vietnam and for those of all nationalities who are saddled with the gun today.

The GPMG concept emerged from German blitzkrieg tactics of World War II. Falling somewhere between medium and light machine guns, it is supposed to be light enough for use by squads in the assault and yet heavy enough to provide sustained support fire. It is not satisfactory in either task.

As fire and maneuver tactics have developed—with increasing emphasis on fast-moving units such as immediate-reaction battalions, fire forces or strike forces—the GPMG has proven too cumbersome for employment as a squad automatic weapon. Although equipped with quick-change barrels, it is still insubstantial for protracted support fire. The GPMG's ascendancy has been matched by an unfortunate decline in the science of machine-gun employment.

In its infancy on the battlefield, the machine gun was used incorrectly as a form of artillery, Since that time the machine gun has often been viewed as a mere bullet hose. The modern machine-gunner is usually capable of little more than pointing (not aiming) his weapon in the direction of the enemy and pulling the trigger. That gives rise to the argument that traditional tactical employment of machine guns no longer applies to the modern battlefield. That's just not so and most professionals could see it if only we had the proper weapons and trained people to employ them.

Do we have any call these days for protracted rates of fire along fixed lines? Emphatically yes, we do.

I was desperate for that potential recently in a small A camp in northern El Salvador. But my Ma Deuce crew didn't know enfilade (firing down the long axis of the target) from defilade (indirect fire on targets beyond the crew's field of view) and had never sandbagged a tripod in their lives. To make matters worse, I was too occupied with manning our sole M60, for which I had no tripod, to teach them.

Accurate defilade fire is not possible with only a bipod mount. It's also impossible to traverse and search with any degree of precision without a tripod's T&E mechanism. Finally, would you like to move during the final assault under the cover of overhead plunging fire delivered by a GPMG operated off the bipod? I wouldn't. Nor am I particularly fond of moving anywhere on the battlefield with an M60 as the only available machine gun.

The M60 was supposed to be the very model of a modern GPMG. We marched off to Vietnam eagerly clutching its rubber-covered forearm to our bosoms. It was the best. We dearly loved it, that we know, for the Ordnance Corps told us so. And then the shit hit the fan. The M60 fell into the ultimate crucible and the hype began to pale.

This gun was flawed from its very soul to its outer skin. It fires from the open-bolt position. That's fine. It is gas-operated with a unique constant-energy gas cut-off system. That's *not* fine.

The barrel's gas vent is eight inches from the muzzle. After the projectile passes this point, gas moves downward and strikes a recess milled around the outer circumference of the chromed piston. The gas passes through seven holes drilled into this recess and into the piston's interior. When sufficient energy has been imparted to the piston it begins to move rearward. The holes move away from the gas supply and cut it off. There is no adjustable gas regulator. In theory it is not required. Debris and fouling will hold up the piston's movement only until enough gas arrives to supply the required energy.

That ignores the demonstrable fact that increased fouling will usually decrease the cyclic rate of a gas-operated machine gun. The M60 provides no means by which the energy level can be increased to bring up the rate of fire. The system is supposed to be self-regulating and has often been referred to as a "constant volume" system. In reality, if the piston cannot be accelerated by the first impulse of gas, it may never receive enough energy to move. It happens all the time.

In the guerrilla-infested area of Ciudad Barrios in El Salvador, armorer Sam Allen and I once worked on six M60s being employed in combat. Fouling and corrosion inside the gas cylinders was so severe that four guns were completely inoperative with frozen pistons. Several of the pistons had to be removed with a punch and hammer. A little investigation revealed the gunners had been instructed by the American MTT (Mobile Training Team) from Panama *never to disassemble the gas system*. But the M60 gas system situation was bad even before the guns found their way to Central America.

Early M60s were equipped with gas-cylinder plugs that had small four-sided heads which vibrated loose during firing. A new plug with a larger hex-head was issued. *It* vibrated loose when firing. A slot head was filed across the end of the plug with safety wire run across it and around the top of the gas cylinder. The safety wire regularly slipped out of the slot and the plug vibrated off the cylinder. As a final solution, a hole was drilled through the plug's head to retain the safety wire. Then it was discovered that the gas-cylinder extension (threaded to the front of the gas cylinder) would *also vibrate loose* so the experts recommended retaining it with safety wire on the lock-washer. Unfortunately, these lock-washers break frequently. The gas nut, threaded to the rear of the cylinder, would also vibrate loose and back off against the gas tube. Finally, it became SOP to safety wire *its* lock-washer also. How many gunners (U.S. or Salvadoran) carry a $75 pair of safety-wire pliers and a spool of stainless-steel aircraft safety wire in their kits?

To make matters worse—and that's hard to do—the M60 piston can easily be reassembled backward. Believe me, if any part can be reinserted incorrectly on a military small arm, it will be—repeatedly. When that happens only the first round will fire as no gas will enter the system. Can you imagine lying in a ditch with your face in the mud to avoid incoming rounds while trying to remove the safety wire with your hands and teeth so you can reverse the piston? That's extremely unhealthy. It virtually means a qualified armorer must be present for this gun's gas system to be maintained or someone will die from a stupid mistake that's entirely too easy to make.

That's stupid. But the gun must have maintenance on the gas system and these components should be removed periodically with the wrenches provided on the combination tool. The tool's reamer is used to clean the barrel's gas vent after the gas-cylinder plug has been removed. *Always* replace the piston with its closed face to the rear. Remember that. Most M60 machine gunners don't.

The M60 gas cylinder is sweated and pinned to the barrel. That means when the barrel is changed an entirely new gas system comes into operation. No telling what might be wrong with it. The bipod is also attached to the barrel. Humping spare barrels means dealing with the unnecessary weight of another bipod. That's not the only problem with the M60 barrel and fixed-bipod design.

Each bipod leg includes a perforated sheet-metal skirt which stiffens the leg assembly and supposedly provides a heat-shield and handhold during barrel changes. It doesn't work, so a WWII-vintage asbestos mitten has been issued to protect the crews' hands during hot-barrel changes. Naturally, the mittens are mostly left in the supply room or the nearest ditch. A towel, cleaning rag or fatigue shirt does the job better. The rather flimsy carrying handle, which would make an adequate handhold for barrel changes, has been foolishly attached to the receiver—not the barrel—and cannot be used to facilitate changes.

There's more. The bipod legs are designed to be folded up against the barrel by pulling outward and compressing a lockspring. These springs break frequently and it's a common sight to see M60s being carried with one or both bipod legs dangling in the breeze. The bipod-leg plunger permits height adjustment through five positions from a minimum of 10 inches to a maximum of 14 inches. That's thoughtful but the bipod yoke is attached to the barrel directly behind the flash suppressor and that's *too far forward*.

While a bipod provides control over the weapon and produces smaller burst groups, positioning it up near the muzzle sharply restricts lateral maneuverability. And the yoke rotates about the barrel's axis only which means *no lateral pivot* without physically shifting the gun. Moving targets are difficult to address quickly under these design restrictions. Targets on the flanks cannot be engaged without lifting the weapon off the ground and re-positioning it. That's an awkward movement from the prone position with a 23-lb. machine gun.

A more sensible position for the bipod yoke would have been eight inches to the rear in back of the gas-port plug. With the bipod yoke

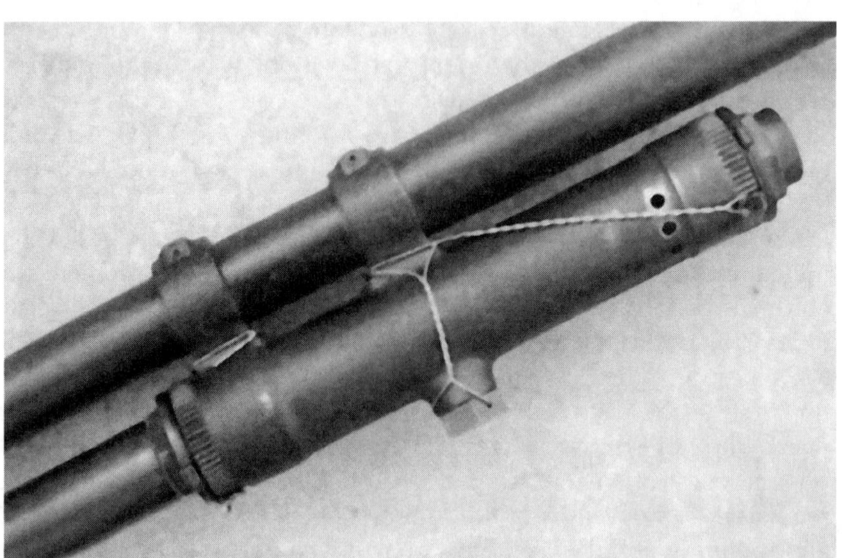

M60 gas system with stainless-steel safety wire on gas-port plug, gas-cylinder nut and extension washers.

located in its present position and with the bipod extended, the gun can be rested muzzle down in the ground. If it can be, it will be.

The M60 flash suppressor, threaded and roll-pinned to the muzzle, has five longitudinal ports. It performs its intended function adequately which is refreshing. There are some other interesting—if not perfect—points about the design of the M60.

The barrel is 22 inches long without the muzzle device. The rifling has four grooves with a right-hand twist of one turn in 11.5 inches. The bore and chamber are chrome-plated. In addition, there is a Stellite liner in the six inches of bore in front of the chamber. Stellite is an alloy made of cobalt, chromium, tungsten and molybdenum with traces of iron which retains its mechanical strength at high temperatures. It sounds good but there are drawbacks. The interference fit of the liner is critical as gas seeping between it and barrel interfaces can eventually blow out the liner. The Belgians and British discovered this the hard way and were unable to produce satisfactory Stellite liners for the MAG 58 GPMG. Stellite liners first appeared on U.S. Browning machine-gun barrels (both .30 and .50 caliber) after the Korean War. The liner should be inspected periodically. It must be free of chips and cracks and the liner expansion gap must be less than 1/8-inch, as measured with the M8 barrel-erosion gauge.

The M60 barrel is changed by removing the belt of ammo, retracting the bolt group and pushing the barrel catch (located on the trunnion block under the carrying handle) to the right while rotating the lock lever upward. Then withdraw the barrel. Trained crews need no more than eight seconds for a barrel change. Because of the bipod's location, many have stated that the chamber end of the barrel and the receiver/gas tube will fall in the mud during barrel changes in the field. I have never seen this happen. Men rarely take the time to change barrels when under fire.

A square-post front sight is sweated and pinned to the barrel. It's workable but far from optimum. While the sight is very sturdy, it is extremely high-profile and snags frequently when moving through dense brush. It cannot be adjusted for windage or elevation zero. In theory, the rear sight must be recalibrated every time a barrel is changed. In practice, no one bothers.

The folding rear sight of the M60 is mounted to the trunnion block in back of the carrying handle on a dovetail base above two ball bearings over springs. The sight must be rotated upward for use. For some strange reason, there is no battle sight in the horizontal position. Elevation can be adjusted from 300 to 1,100 meters but it's not easy. The elevation markings are small and difficult to discern even in the daylight. The open square notch is mounted to a spring-loaded slide release used for making major changes in elevation. A rotating knob is used for making fine adjustments. Four clicks on this knob equal a one-mil change in elevation. There is a windage knob on the left side of the sight but the index scale is just as bad or worse. The index marks are almost invisible to the naked eye. One click on the windage knob equals a one-mil change in deflection and the sight can be adjusted five mils right or left of zero. Night-vision equipment can also be mounted over the top cover.

Given all the problems with this sight and its inadequate scales, veteran gunners have come up with an alternative method of zeroing the M60. Pace off 25 meters and place a one-inch-square paster on a target. Flip up the rear sight and center the windage by eye-balling. Firing single shots only, adjust the elevation knob up or down until the shots impact on the paster. Pay no attention whatever to the graduated range scale. Then, without moving the adjustment knob, loosen the screw on the range scale. Slide the scale up or down until the rear sight is set for 500 meters and lock down the set screw. The M60 is now zeroed for all ranges.

Moving up on the gun, a forearm assembly includes a ventilated sheet-metal heat guard which covers the rear eight inches of the barrel. The underside surrounds the gas tube and is covered with rubber. Although the M60 can be fired in the hip-assault position using the extended left bipod leg for support, the forearm affords a more comfortable, reliable surface for this purpose. There is not too much trouble with this aspect of M60 design—until you have to remove the forearm assembly for cleaning. Gunners are taught to insert the nose of a cartridge into the latch hole at the bottom rear of the guard and depress the latch. Raising the rear end slightly, they should be able to withdraw the forearm assembly. It doesn't always work that way. This simple feature can be the only component on the M60 that can prove difficult to disassemble. The latch springs also break on occasion. The front sling swivel is attached to the left front of the forearm by two screws with lock washers. They will loosen and should be inspected often to keep gunners carrying the weapon on a sling from crushing their toes when the swivel detaches from the gun. On early forearm assemblies, the swivel bracket was mounted vertically, but this was modified by canting the bracket toward the rear on later specimens. The rear sling swivel is mounted on top of the buttstock. Surprise! That's exactly where it belongs. The black nylon "silent type" padded sling is the very best I have ever used on a GPMG or light machine gun. It's very comfortable even on long patrols—when you can find one to use.

The M60 receiver body is a bent sheet-metal pressing. It's riveted to a forged trunnion block. The gas tube is threaded and pinned to the trunnion block. The feed ramp is also attached to the trunnion block. In the fall of 1964, the feed-ramp angle was decreased to prevent the nose of an incoming cartridge from striking the primer of any live round in the chamber. A number of helicopter door guns had blown up because of this shortcoming. Two side-by-side machined guide rails are riveted to both the trunnion block and receiver body. Early receivers failed at about 50,000 rounds. The life span has been increased by the use of Bond-tite rivets on later guns.

Guide rails support and track the bolt group. The right guide rail is cut for the ejection port. There are three holes aft of the ejection port. They were designed as a case deflector, but the concept caused "spin-backs" into the receiver and was soon discarded. The non-reciprocating cocking lever's runway is cut into the receiver body's right side. The cocking lever is held to the receiver by a sheet-metal guide which, in turn, is retained by a screw and lock

M60 GPMG Specifications

CALIBER . . . . .7.62x51mm NATO

OPERATION . . . . .Gas. Constant-energy cut-off system. No adjustable regulator. Bolt/operating-rod mechanism based on those of the FG42—Lewis Gun.

CYCLIC RATE . . . . .550 rpm.

FEED MECHANISM . . . . .Belt-fed using M13 disintegrating links. Single-pawl system; actuator cam roller and feed arm based on those of the MG42.

WEIGHT, empty,
with bipod and buttstock . . . . .23 pounds.

OVERALL LENGTH . . . . .43.5 inches.

BARREL . . . . .Air-cooled, quick change type. Four grooves with a right-hand twist of one turn in 11.5 inches. Chrome-lined chamber and bore. Stellite liner in the 6 inches of bore forward of the chamber.

BARREL LENGTH . . . . .22 inches (without flash suppressor).

SIGHTS . . . . .High-profile, unprotected front; not adjustable for windage or elevation zero. Folding rear with open U-notch; adjustable for barrel zero and elevation from 300 to 1,100 meters (4 clicks = 1 mil); windage adjustments 5 mils right or left of zero (1 click = 1 mil).

ACCESSORIES . . . . .M122 tripod, sling, 100-rd. cloth bandoleers, spare-barrel carrying case, night-vision equipment, blank firing adapter, helicopter door-gun conversion components, and cleaning tools.

MANUFACTURER . . . . .Primary U.S. contractor: Saco Defense Systems Division Maremont Corporation, 291 North Street, Saco, Maine 04072, Australia; Ordnance Factory, Maribyrnong, P.O. Box 1, Ascot Vale, Victoria 3032. Republic of China: Kaohsung Arsenal, Taiwan.

STATUS . . . . .In service with the armed forces of Australia, El Salvador, Republic of Korea, Republic of China and U.S.A.

washer. It's yet another problem to plague gun crews and armorers.

These screws loosen with irritating regularity and they should be re-inserted with Super Strength Loc-Tite—when and if that's available. The cocking lever's handle is not of the folding type and it will bend or break if struck with a hard object, such as the heel of a boot. The front tripod studs are fixed to the trunnion block. The rear tripod bracket is welded to the receiver body. A machined guide rail for the buffer is welded to the bottom of the receiver body. Several of the M60s I have worked with in El Salvador showed evidence of welding repairs when rebuilt in 1975 and 1977. This is acceptable only if the receivers remain absolutely firm and straight.

In principle, the M60 operates using the bolt system of the WWI Lewis Gun. The operating rod carries a fixed cam yoke which rides in a cam slot cut into the bolt body. The bolt rotates on this cam yoke which is fitted with an anti-friction roller bearing. This is another short-sighted design fault. The bearing must be inspected for chips and excessive side play on its shaft. Burrs and gouges on the front of the cam yoke (they develop after just a few hundred rounds) must be removed with a Swiss file. The piston post is permanently attached to the front of the operating rod where it ends in the shape of a scraper with helical cuts which are supposed to remove fouling from the gas tube.

The action is short-stroke as the piston travel is limited to 2.375 inches. The operating rod is separated from the bolt group by pushing the rod away from the bolt and pivoting it upward. When reinstalling, position the rear of the operating-rod yoke against the rear firing-pin spool. Push on the op rod to compress the firing-pin spring and place the operating-rod yoke between the two firing-pin spools. The bolt body can also be assembled backward onto the operating rod although I have never personally witnessed that.

As the bolt enters the barrel socket, a cartridge has been deflected downward and into the chamber by the front cartridge guide, trunnion-block feed ramp and the barrel-socket feed ramp. The barrel-socket lead cams rotate the bolt clockwise as the two locking lugs enter into the socket-locking cam. And here's where the fun really begins.

The rotating action is quite violent and most of the forces are directed against the end of the lugs, especially the top lug which has the smallest surface area. After a few hundred rounds the ends of the locking lugs will begin to chip away, increasing wear on the barrel socket camways, which in turn places even greater stress on the locking lugs. That generally means *M60 bolts have to be discarded after 15,000 to 20,000 rounds*. The bearing surfaces on the barrel socket and bolt must be kept lubricated—and not with Break-Free CLP. Latest reports indicate that guns maintained with this product are exhibiting excessive wear. If not constantly shaken, the Teflon beads, upon which Break-Free's lubricating characteristics depend, will remain in the bottom of the container and never reach the weapon.

Chipped areas on the bolt lugs should be carefully rounded with a Swiss file. Some armorers have stated that this merely removes more metal, increasing the gaps between the lugs and barrel sockets further and accelerating wear. Not so. Allowing the sharp edges to remain focuses the points of stress and increases the crystallization process that causes erosion on the lug ends. In addition, if left unattended, these sloughed-up sharp edges on the upper lug will eventually start to cut into the bottom of the feed tray.

After the cartridge has been chambered, its base contacts the bolt face, compressing the ejector spring while the extractor snaps over the rim. As the bolt rotates, the firing pin—retained by its rear spool on the op rod's cam yoke—moves forward, assisted by the force of its compressed drive spring. I remain skeptical concerning the value of the firing-pin spring as the gun can be fired without it. The firing pin, as well as its spring and cup, are M60 components that can be reassembled backward. Remember Murphy's Law?

The correct sequence is firing pin to the front (with its two spools to the rear), followed by the cup with the spring inside and to the rear.

A small bolt-plug pin must first be removed before the bolt plug can be unscrewed to disassemble the firing pin components. If you're armed with an M60 or tasked with repairing them, always keep a small bag of these pins tied around your neck. They are constantly lost. Without the pin, the gun will fire only 40-50 rounds before the bolt plug rotates off and everything ceases to function. The cam-roller assembly can also be reassembled backward, but this is not often encountered as the bolt-plug pin cannot then be replaced. Be sure the cam roller is rotated topside when reinserting these components into the receiver.

After ignition, gas action drives the piston rearward about 13/16 of an inch before unlocking commences. Pressures have by this time dropped to a safe level. The delay also contributes to the gun's low cyclic rate. The bolt rotates counterclockwise and at full unlock, the op rod's cam yoke has once more fully compressed the firing-pin spring. As rearward travel continues, the empty case is withdrawn from the chamber and spun sideward to pivot about the extractor claw, ejecting out the port on the receiver's right side.

As the bolt and operating rod move rearward they compress the recoil spring. This drive spring is of multistrand construction. Machine-gun recoil springs are subjected to sudden loading which initiates shock waves that travel to the fixed end of the coil. These direct and reflected waves combine to produce a transient load that is extremely destructive to single-strand springs. Using multistrand wire can raise spring life by 10 times by avoidance of "wind-up" produced by the twisting movement from the axial load placed on a single-strand spring. A multistrand spring in the M60 seems smart but the spring drags along the bottom of the op rod's interior and flat spots develop which compromise the advantage gained through the use of multistrand wire. When these flat spots are noted, the spring should be discarded.

The head of the recoil spring's guide rod is press-fit and induction-brazed onto the rod. These heads will often loosen if improperly brazed—many of them are and should be inspected carefully.

The buffer's plunger rests in the guide rod's head. The original buffer employed a series of composition-fiber pads with a return spring. It can be identified by its stainless-steel exterior which is clearly marked, "DO NOT OIL INTERIOR." These buffers were withdrawn in 1965 after they caused more problems with this flawed GPMG.

As the fiber pads wore out, the gun's cyclic rate became more and more erratic. This caused serious difficulties with the solenoid-fired M60C helicopter guns. Although they had limit switches, their ammunition drive motors were constant speed. When the gun's cyclic rate dropped, ammunition backed up and caused a stoppage. The GAU-2A1B minigun eventually replaced the M60C because hard mounting outside the helicopter and sideways positioning, with consequent downward ejection, caused continual and insolvable difficulties.

The M60's new buffer is one of the few shining lights in the evolution of this machine gun. It's a hydraulic design and has a phosphate-finished exterior. Both buffers have a low coefficient of restitution and thus absorb considerable energy to both reduce the felt recoil transmitted to the operator's shoulder and lower the cyclic rate. The buffer, in conjunction with the operating mechanism's rather long "dwell" time, drives the M60's rate of fire down to a very proper 550 rpm.

The buffer is housed within the buttstock. Both are held to the rear of the receiver by the buffer yoke. FM 23-67 states that the buttstock should be detached from the buffer during disassembly by inserting the nose of a cartridge into the latch hole at the end of the butt and depressing the latch. I prefer to keep the stock and buffer together in the field and remove them as a unit from the receiver by withdrawing the buffer yoke. Neither of these components will normally require maintenance. The recoil spring, guide rod, operating rod and bolt group can be withdrawn from the rear of the receiver body after the buttstock and buffer have been removed.

It may be hard to believe I'm saying something positive at this point, but the M60 buttstock is well-designed and superior to that of the FN MAG 58 machine gun. Of sheet-metal construction, its outer surface is covered with the same rubber used on the forearm, top cover and pistol grip (Mil-Std-417). There is a hinged shoulder rest. Overall, an excellent application of human engineering.

In the very earliest stages of its development the M60 used the MG 42 feed system. Employing both inner and outer feed pawls, which were driven by a feed arm actuated by a roller on top of the bolt body, the belt was moved by the maximum energy level available to the system (just after the round had been fired and once more when it reached the end of its return motion). The pawls moved in opposite directions in two phases as the bolt moved back and forth. As the belt moved only half a pitch in each phase, less force was required for its acceleration across the feed tray.

The cam roller and feed arm were retained on the M60, but the inner and outer pawl system was discarded in favor of a single pawl. Forward motion of the bolt and actuator cam roller causes the feed arm to pivot to the right, forcing the front of the feed-arm lever to the left. The lever draws the feed-pawl plate assembly to the left as well, where the pawl arms drop over and engage the next cartridge for transport and remain there until the preceding round is stripped and fired.

As the bolt travels rearward, the actuator cam roller pivots the feed arm to the left and the feed-pawl assembly transports the next cartridge to the right onto the feed-tray groove. The round is forced down into the groove by the two spring-loaded cartridge guides in the top cover. The empty link is driven out the feed tray's port on the right by the new round. A spring-loaded retaining pawl on the feed tray holds the belt by grabbing onto the second round while the first round is in the tray groove.

While this mechanism has been much criticized by those who bewail the dropping of the German two-pawl system, it does work. There is sufficient reserve power to lift a 100-rd. belt vertically, albeit somewhat sluggishly when the gun is badly fouled. Few realistic scenarios require the gunner to stand and fire with an unsupported, free-hanging 100-rd. belt.

Once again, that doesn't mean the designers hit the mark. There are legitimate criticisms to be leveled at the M60 feed system. The top cover's outer shell is a thin aluminum stamping. The rear portion is covered with rubber in the area where the gunner places his support hand. The front portion is black anodized and the finish wears away much too quickly, leaving a silver mirror to shine in the gunner's—or the enemy's—eyes. I have toted many aerosol cans of flat-black stove paint to El Salvador to deal with that problem but your average grunt gunner is not going to have paint available.

Although it would have been a simple task to spring load the actuator cam roller in the same manner as on the FN MAG 58, it wasn't done. Thus, the top cover should be closed only after the bolt has been retracted to prevent damage to the feed arm and/or actuator roller. That's mentioned in training but troops in the field constantly slam the top cover down with the bolt forward. It would have been much more effective to engineer the flaw out of the gun.

The top cover's latch-lever assembly is rather fragile. While it looks professional to smash down the cover with your fist, it should be closed by pushing down with one hand and pivoting the latch lever back with the other.

There are four small holes on the left side of the receiver just above the trigger group—remnants of the old bandoleer holder which was riveted to the receiver which accommodated two types of large, but sturdy, 100-rd. pouches. One was fabricated from rubberized oilcloth, the other of canvas with a heavy metal zipper on one side. Unfortunately, the pouches were too large to permit the gun to left traverse on the M122 tripod. Additionally, the four alignment tabs on the holder would all too often catch belts fired without the pouch and momentarily shut down the weapon. The tabs also broke off with alarming frequency.

The modified bandoleer holder is attached to the feed tray at the place previously occupied by two anti-friction rollers. It will not accept the older pouches. They have been replaced by a cloth bandoleer into which fits a cardboard box containing 100 linked rounds. Two of these bandoleers are packed in the standard U.S. .30-cal. size ammo can. A cloth carrying strap is sewn to each bandoleer and a web strap is sewn to the top of the bandoleer for attachment to the gun—difficult to do anytime and next to impossible under stress. The flimsy cotton bandoleer itself will rot and rip to shreds in just a few days in the tropical bush. Yet, M60 belts should never be carried "Pancho Villa" style across the chest as was the custom in Vietnam and now in El Salvador. The M13 links rust quickly and cartridges often slip out of the link's extractor-groove tab resulting in stoppages. Very macho, but very stupid.

To load the M60, place the selector on "F" and then retract the bolt. Slide the cocking handle forward and set the selector to "S". Raise the top cover. Place the ammo belt (links up) on the feed tray with the first round in the feed-tray groove. Close the cover and move the selector back to "F" to fire.

I have logged thousands of rounds through M60s of every configuration (except the new M60E3)—including some "chopped" versions. When you can keep the gun in repair, ergonomic characteristics are excellent. Coupled with a cyclic rate of only 550 rpm and low felt recoil. the basic design can be fired comfortably in all positions: prone, kneeling, hip assault and standing. Whenever possible, the M60 should be fired behind cover and concealment from the prone position off the bipod or tripod. The sling should be employed with the hip-assault position, which should be reserved for the final close with the enemy when the gunner is a member of the assaulting group. The standing position should be used for emergencies only. Gunners must exercise fire discipline and bursts should normally be limited to 3 or 4 rounds at 4- to 5-second intervals, with target reacquisition between each burst group. The barrel should be changed every 15 minutes. To prevent the enemy from returning fire or to establish initial fire superiority, the interval between bursts can be shortened to 2-3 seconds. The barrel should then be changed every five minutes. The cyclic rate of fire should be employed only to win the fire fight in the final moments of the assault. It can be maintained for no more than 60 seconds before changing the barrel. But, if the shit is really flying, forget about changing the barrel and worry about damage due to overheating back in the barracks.

It may seem incredible given what I've said so far, but the M60's accuracy potential is high. It cannot be faulted in this area. Although the sights are deficient, trained operators can expect superior hit probability.

There are a couple of unique feed systems for the M60 besides the standard belt. Interestingly, they were not designed by Americans. Using the old-style feed tray and bandoleer holder, the Australians developed two ammunition containers. A nylon-coated, aluminum belt box holds 40 rounds and can be left mounted on the side of the gun. The weapon can be fed by loose belts with the assault pack still in place. Holes in the rear of the box enable the gunner to see how many rounds remain. The box's lid also serves as a belt-stop pawl to retain the exposed rounds on the magazine's platform. The exposed rounds can be clipped to those on an external belt already in the feed mechanism.

A waterproof PVC-coated nylon bandoleer was also designed. It holds 50 rounds and several can be joined together. These bandoleers are packed three to a can and can be re-used by feeding the belt from the bandoleer's mouth or torn open along a seam to instantly expose the entire belt.

To disassemble the feed system, raise the cover assembly and, using the nose of a bullet, unhook the hinge-pin latch from the right side of the cover's hinge pin and pull the latch out from the left. Pull the hinge pin out from the right side and pull the cover off the receiver. Pull out the torsion helical spring and remove the feed tray. For thorough cleaning, depress the feed-arm (called the cam) retainer plunger with the tip of the recoil spring guide rod and lift the feed-arm assembly away from the cover. To remove the feed-lever assembly, push the clip away from its pivot post and lift the lever, clip and single-coil spring out of the cover assembly. No further disassembly is required for normal cleaning. Reassemble in the reverse sequence but do not snap the single-coil spring into the cover cutout until you have placed all three lever-assembly components on the pivot post.

That brings us to the M60 trigger mechanism and one of the most devilish problems with the gun's design. The trigger mechanism was loosely patterned after that of the MG 42. But alterations from the original concept have resulted in some serious difficulties. The MG 42 trigger mechanism incorporated a unique spring-loaded sear trip attached to the top of the trigger. When the trigger is pulled, the sear trip descends, allowing the front of the sear to rise while the rear end is lowered to release the bolt group. When the trigger is released, the sear trip rises up and the rear end of the sear is lowered even farther. The sear trip now projects into the bolt body's path of travel. The bolt then shoves the sear trip rearward. That frees the front end of the sear and permits the sear spring to drive the rear end of the sear upward to engage the bolt's bent. Full-face engagement is achieved in one sudden, sharp movement as the bolt starts forward after its overrun. This system reduces wear and chipping of the mating surfaces to a minimum. Unfortunately, the sear-trip mechanism was not used on the M60 and both the sear and the sear notch cut into the operating rod can wear quickly. Armorers are constantly ordering and installing replacements for these parts.

The pistol-grip trigger housing is covered with rubber which provides a comfortable nonslip surface at a vital area. The selector has only two positions. There is no provision for semiautomatic fire, but with a cyclic rate of only 550 rpm, gunners can be easily trained to fire single shots. The selector is a rotated cross-pin type. When the selector lever is moved up to "F" (fire), a cutout in the cross-pin rotates into position under the sear extension. When the selector is moved down to "S" (safe), an uncut portion of the pin is rotated into position to block the sear extension's downward movement. It's reliable. Early change levers were difficult to manipulate and their surface area was increased to afford a larger purchase for the thumb of the firing hand. Unfortunately, they stopped short of other needed improvements in the trigger mechanism.

The trigger-housing group is held to the receiver by a rear holding notch which rests in recess at the bottom of the receiver and a front trigger-housing pin (interchangeable with the sear pin). These two pins are retained by a leaf spring on the right side. Originally these leaf springs were inserted from the bottom, but they fell off too often, eventually leaving the gunner without a trigger group. As it is most difficult to fire the M60 without its trigger mechanism, these leaf springs were redesigned and they are now inserted from the top. It wasn't enough.

U.S. Marines flush the enemy from a hillside in Vietnam. ***(Photo: DOD)***

These leaf springs are troublesome to remove when you want to disassemble the gun and they are also prone to breakage. To disassemble the trigger-housing group, first remove it from the receiver by pressing inward on the front of the leaf spring and rotate the front end down to clear it from the trigger-housing pin. Pull forward to disengage it from the sear pin's notch. Remove the trigger-housing pin by pushing it to the left. Slide the trigger housing forward and rotate down away from the receiver.

Detailed disassembly of the trigger-housing group is as follows. Depress the sear and remove the sear pin by pushing it to the left. Remove the sear, sear plunger and spring. Remove the trigger pin by pulling it from the right. Remove the trigger through the top of the trigger housing.

When replacing the trigger, make certain its spring is hooked under the channel surface in the housing. Reinsert the trigger pin from the right. Drop in the sear-plunger spring and then the sear plunger. If you reverse this sequence, the sear-plunger spring will be exposed. The gun will fire, but very shortly the spring will bend and the sear may no longer engage the op rod's bent resulting in a runaway.

I've seen this a number of times including once when I had a runaway gun during a demonstration of the kneeling position for the Atlacatl Battalion gunners. It caused me to fall backward flat on my ass. To shut down the gun, grab the belt and twist it. That's not easy from the kneeling position but it works.

You're heading for the next potential problem with reassembly of the M60. You should replace the sear with the shoulder up and to the rear. Unfortunately, the sear can be reassembled backward which will not permit the bolt to remain retracted. Murphy's Law applies. Remember that the sear pin and trigger-housing pin must be reinserted from the left side so the leaf spring can be properly reinstalled.

Despite problems such as those I've described, the M60 has found its way off the ground and into the air. A commonly encountered variant is the M60D. This version is seen in El Salvador mounted in the doors of the UH-1H helicopters. The buttstock has been replaced by spade grips with triggers. This is connected to a sear-activator assembly (which replaces the normal trigger housing) by a steel rod. The selector remains a cross-pin type, but it is pushed to the right for "safe" and to the left for "fire," rather than rotated. The sear, plunger and spring remain the same. A rubber boot protects the rear end of the receiver body. The bipod and carrying handle are sometimes removed. An aerial ring sight usually replaces the folding-leaf rear sight. The Salvadoran Air Force has recently installed some twin mounts of indigenous design on the UH-1H to increase its firepower. In my opinion, the M60's cyclic rate is too low for this role.

All this doesn't mean the M60 machine gun should be scrapped. If America could afford to eat such military mistakes, we'd have gotten rid of this gun long ago. There is some solution on the horizon for our fighting men—and our allies—who may carry an M60 into future fights. Many of the M60's more glaring deficiencies have been redressed by its latest version—the M60E3.

Designed by George F. Curtis, Maremont's Manager of Advanced Development

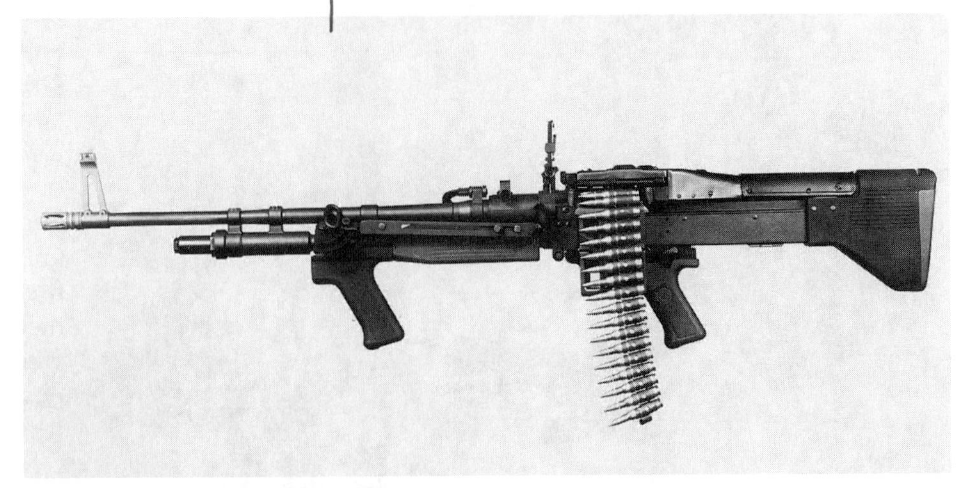

The Maremont M60E3 GPMG is far better than the older model.

Engineering, the Lightweight M60 weighs only 18 lbs. A light bipod has been mounted to the receiver. The forearm has been redesigned and now ends in a pistol grip. Forearm, trigger housing and buttstock are now fabricated from high-impact, glass-reinforced nylon. There is a heat shield between the forearm and barrel. The barrel has been lightened (a 15 5/8-inch assault barrel is also available) and incorporates an M16-style flash suppressor.

The carrying handle has been moved to the barrel. The trigger assembly has a winter trigger guard for use in arctic environments. The sear has a double notch to prevent runaway fire. The selector is ambidextrous. The rear sight is available with an optional peep aperture and the front sight is adjustable for both windage and elevation zero. The hinged shoulder rest has been removed from the butt plate. The feed system can be charged with a starter tab-type belt after the top cover has been closed. The cam activator roller is now spring-loaded—the top can be closed with the bolt either forward or retracted.

Most important of all, the gas system has been revamped which means the piston can no longer be placed in backward and interlocking cylinder nuts eliminate the requirement for safety wire. All these improvements are multiplied in value by a 60-percent parts commonality with the older M60.

The M60E3 has been adopted by the USMC. They have ordered 1,600 units to date. One thousand of their old M60s will be shipped to Maremont for retrofitting to the new M60E3 specifications. The Marines plan to retrofit another 4,000 older M60s themselves with conversion kits obtained from Maremont.

Fifty M60E3s have been purchased by the U.S. Navy SEALs and another 300 by the Navy itself. The Navy also plans to convert M60s into the M60E3 configuration by use of the Maremont kits. Various other U.S. government agencies have purchased the M60E3. Officials at Ft. Benning have fired the M60E3 and Maremont says they were impressed—especially when it was employed in conjunction with the 7.62x51mm NATO SLAP round.

The M60E3 has recently been demonstrated in Central America. El Salvador could easily retrofit to the M60E3 through their facilities at the Maestranza (Central Ordnance Depot).

It's a nice scramble but it's a day late and a dollar short. The U.S. Army and Marine Corps are totally committed to the M249 SAW at the squad level. At 18 lbs., the M60E3 is certainly targeted for the squad automatic role. It will never make it.

Now that we have adopted the fine M249 SAW and the infantry once more has a true lightweight squad weapon, it's time to reconsider the machine gun's sustained-fire-support role. There are some attractive options.

Maremont has developed a new lightweight .50-cal. M2 Browning machine gun that offers a quick-change barrel and fixed headspace. The venerable M1917A1 Browning water-cooled medium machine gun, rechambered for the 7.62x51mm NATO cartridge, is another proven alternative. Both have greater merit than our most likely course of action: shuffling the M60 upstairs to the company or battalion heavy weapons' section.

The M60's performance in Vietnam was a disaster of almost major proportions. Those who ranted and raved about the M16's failings may have served their country better by focusing some of their rage on this ill-conceived machine gun. At least now it's time for a change and it looks like that change is coming.

Originally appeared in *Soldier Of Fortune*
August 1985.

MAC ATTACK

Poor Boy's SMG has Indigestion

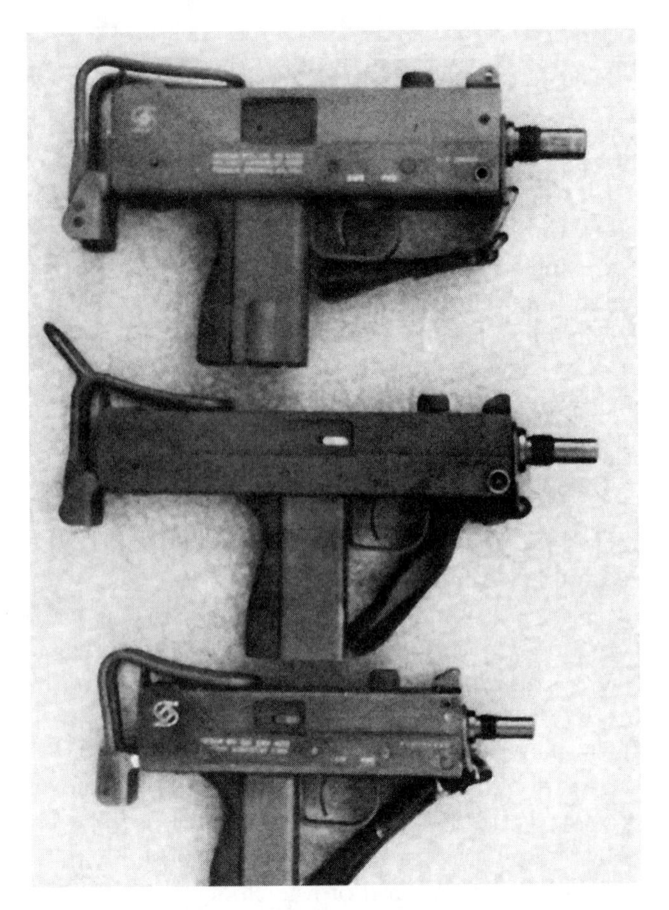

MAC lineup: Original Ingram's M10 in .45 ACP (top), S.W.D., Inc. M11 in 9mm Parabellum (center) and a rare example of the M11 in .380 ACP (bottom). It is stamped 9MM AUTO rather than .380 ACP.

Cowboys nailed villains and fence posts with the Colt .45 Peacemaker. FBI agents will forever be connected to Colonel Thompson's submachine gun. And, although Hollywood heroes have destroyed evil from *McQ* through the *Raid on Entebbe* with the MAC 10, in the real world, such as the streets of Miami, it rests today in $500 alligator briefcases next to bags of cocaine and dirty money. The poor boy's submachine gun refuses to die.

Gordon B. Ingram has been designing submachine guns since shortly after World War II. At that time the M3/A1 Grease Gun was the standard issue U.S. SMG. Allowing for a possible M4, Ingram started his design nomenclature at M5, and came up with a thoroughly conventional weapon with a wooden buttstock, tubular receiver and barrel casing and adapted to the Reising 12-round magazine. It was never put in production.

The Model 6—in external configuration more than a little like the Thompson—was manufactured by the Police Ordnance Company of Los Angeles and was purchased by the Cuban Navy (Batista regime), the Peruvian Army and the U.S. Constabulary in Puerto Rico. A prototype Model 7 was developed, identical to the M6, except that it fired from the closed-bolt position. Another modified version of the M6, called the M8, was produced in very small numbers in Thailand in 1954. Ingram again modified the M6 in 1959, adding a retractable M3-type stock and other improvements and calling it the Model 9. Ingram Model 6 submachine guns are still occasionally seen in U.S. collector circles. They sell for $600 to $800.

Ingram began work on the M10 in 1964. The first prototype was chambered for the 9mm Parabellum cartridge and used Sten magazines. In 1966 the U.S. Army obtained prototype number two and tests were conducted at Frankford Arsenal. While on his way to Vietnam to demonstrate counterinsurgency weapons manufactured by his firm, the Sionics Company, Mitchell L. Werbell III contacted Ingram to obtain M10 9mm number three and a .45 ACP version with sound suppressor. Shortly thereafter, Sionics obtained the manufacturing rights to both the M10 and the new M11, chambered for the .380 ACP cartridge. Ingram became chief engineer at Sionics in 1969. Early guns are marked with the plant's location at that time, Powder Springs, Ga. A very few MAC M11s escaped from this factory marked "9MM AUTO" (for 9mm Kurz = .380 ACP) and into the hands of collectors. Most MAC M11 lower receiver bodies were altered by a shallow mill cut over the "9MM" and restamped ".380."

In 1970 a group of New York investors obtained controlling interest and moved the manufacturing facility to Marietta, Ga. The company was renamed "Military Armament Corporation" (MAC) and both Werbell and Ingram were shunted aside. The MAC M10 was hyped as the replacement for the U.S. Government's Colt M1911A1 .45 ACP pistol. This was not to be. Thank God.

Without Werbell to tout the weapon by firing suppressed versions into phone books in the offices of Manhattan investment-banking firms, MAC declared bankruptcy under Chapter 11 receivership in 1975. In 1976 the inventory was auctioned off to pay corporate debts. It included 2,500 M10/9mm SMGs, 6,400 M10/.45 ACP SMGs, 175 M11/.380 ACP SMGs, 1,000 M10/9mm suppressors, 875 M10/.45 ACP suppressors and 50 M11/.380 ACP suppressors. In the late 1970s so-called "pre-auction" MAC M10 submachine guns in both 9mm Parabellum and .45 ACP could be purchased from Class 3 dealers for about $85, having been sold at the auction for less than $20 each.

By 1978 the MAC was back in production by RPB Industries, Inc. of Atlanta, Ga. While M10 and M11 submachine guns were manufactured, RPB focused on the domestic market and great emphasis was placed on semiautomatic-only pistol and carbine versions. Conversion to full-auto proved simple and became popular with drug pushers and pimps from Florida to California. The BATF shut down the RPB operation several years ago.

There was another auction. Now Class 3 dealers, their "expertise" purchased each year with a $200 license fee, hawk "pre-auction" RPBs at local gun shows for $400 to $600. Hilarious.

Leatherwood Enterprises, now operating under the name Military Armament Corp. currently manufactures a semiauto-only .45 ACP/9mm Parabellum pistol version called the MAC M10A1 They filed for bankruptcy a year ago. Modified versions of the MAC M10/M11 are now manufactured by S.W.D., Inc. One of them, the M11/9mm submachine gun, is the subject of this column's test and evaluation.

Firing the Mac M11, caliber .380 ACP. Five- to six-round bursts are the norm due to the high cyclic rate—over 1,200 rpm.

More than 20 countries purchased the original MAC M10/11 Ingram series submachine gun, mostly in very small quantities (50 to 1,000), for use by special operations groups and assassination teams. Among them were: Argentina, Bolivia, Brazil, Guatemala, Indonesia, Israel, Jordan, Korea, Mexico, Portugal, Saudi Arabia, Spain, Thailand, Venezuela and Yugoslavia. What exactly were they buying?

Blowback operated, both the M10 and M11 fire from the open-bolt position. Their single most salient feature is compactness, achieved by means of a rectangular telescoping bolt copied from the early Czech designs. With the stock removed, the M10's overall length is 10.5 inches and that of the M11 but 8.75 inches. The barrel lengths are 5.75 and 5.06 inches, respectively, and all barrels are externally threaded to accept the Sionics-type sound suppressor. The M10 weighs 6.25 pounds, empty, and the M11 only 3.5 pounds.

The wrap-around bolt's firing pin and face are located far back to permit the greater part of the bolt body to envelop the barrel. A spring-loaded extractor is held in place with a roll-pin—impossible to remove without a punch which usually destroys the pin in the attempt.

The recoil spring and guide rod travel through a hole drilled the length of the bolt body. And guess what? They're retained by yet another roll-pin. The ejector, a steel rod, moves through a parallel channel and projects through the bolt face during the recoil momentum. Both the ejector and guide rod are attached to a steel backplate covered by a synthetic rubber buffer.

Protruding through the top of the upper receiver is the retracting knob. Like the Thompson M1921/28 series, a U-notch has been cut in the cocking knob to clear the line of sight. When the bolt is closed, the retracting knob can be rotated 90 degrees in either direction to lock the bolt. Don't ask me why. I've been told this feature prevents an opponent who has grabbed the weapon from shooting you. But, such a scenario is little more than a novice's fevered nightmare.

Both upper and lower receivers are fabricated from stamped sheet metal and are substantial. They are phosphate finished. The barrels are permanently attached to the upper receiver body by another roll-pin.

A hand strap is attached to the barrel just in front of the upper receiver by a pivoting sling swivel. It's supposed to counter muzzle climb and prevent the support hand from sliding forward in front of the muzzle. In use, it's noisy and almost impossible to insert two fingers into the strap's loop under the time constraints of a stress environment. In any event, the strap is usually disregarded, as these weapons are commonly fired with the Sionics sound suppressor in place, which serves as a forearm when equipped with a neoprene or Nomex cover.

The front sight's protective ears, a single U-shaped pressing, is welded to the upper receiver. On the MAC M10, the front sight itself is nothing more than a bent strip from the bottom of the protective ears' stamping. A bit more elaborate on the MAC M11, the round front sight post is adjustable for elevation zero by 180-degree movements and held in place by an allen-head screw.

The upper and lower receivers are connected by a pin at the muzzle end. Nothing prevents this pin from failing out on the MAC M10. A spring-loaded receiver pin catch was added to the MAC M11 to prevent this misfortune. The lower receiver body contains the serial number (hence the unrestricted sale of MAC upper receivers), trigger mechanism and magazine well. A hole in the extended portion of the backplate (welded to the end of the lower receiver) serves as a rear sight aperture.

The magazine well is located in the pistol grip and the housing is welded to the lower receiver. While this provides a solid support for the magazines and permits them to be changed rapidly, even in the dark, it requires a grip-to-frame angle of 90 degrees, far more awkward than that found on submachine guns like the Sterling or Beretta Model 12S. A plastic grip extender is screwed to the rear of the magazine well. The magazine catch release is located in an undesirable position at the bottom of the pistol grip.

After analysis by high-speed photography it was determined that the feed angle on the MAC

M10/9mm was incorrect. The angle was altered on the MAC M10/.45 ACP, but production proceeded on the 9mm version without modification and this model is subject to unreliable feeding. *Caveat emptor*. The feed ramp on these submachine guns is nothing more than a bent piece of sheet metal spot welded to the lower receiver.

The MAC M10/.45 ACP uses the M3 Grease Gun magazine, which is altered to mate with the three points of contact in the magazine well. This single-position-feed, 30-round magazine is, without doubt, one of the worst submachine magazines ever designed and produced. But it was cheap and available. The MAC M10/9mm accepts a slightly modified Walther MPL 32-round magazine. This wedge-shaped, two-position feed magazine is excellent. MAC M11/.380 ACP magazines are of two capacities: 16 and 32 rounds, both single-position feed. A standard M3 magazine loader (stamped as were the magazines with the Cobray logo) was available for the MAC M10/.45 ACP. A scaled-down version was made for the MAC M11/.380 ACP.

The trigger mechanism is housed within the lower receiver. The trigger itself is L-shaped. When the trigger is pulled with the selector set to full-auto, the long portion of the L, an extension bar, drops and a spring-loaded catch mounted to the front of the bar holds down the sear until the trigger is released. When the selector is set to semiautomatic, a shaft, connected to the selector, rotates and cams back a tripping lever under the sear catch. The head of the tripping lever lifts up at the same time. As the bolt moves forward it strikes against the tripping head and drives it forward and down. The end under the sear catch forces the catch back off the sear and the sear rises to stop the bolt. The trigger must be released to move the sear catch back over the sear. The safety moves a bar under the sear to prevent its rotation downward.

The selector lever is located in an inept position on the left side of the lower receiver, forward of the trigger guard, and can be rotated continuously in either direction. On the MAC M10 the lever is detented in two positions: "SEMI" and "FULL." There are no detents on the MAC M11 and you must visually inspect the selector lever's position. The safety is located on the right side of the lower receiver directly above the trigger guard and the rearmost position is marked "SAFE." Slide the catch forward to "FIRE."

One could live with all of the MAC's minor idiosyncrasies were it not for its major flaw: The cyclic rate in every version is close to 1,200 rpm. Submachine guns should ideally fire between 500-600 rpm. A bullet hose serves only the ends defined by movie producers. Only the most highly trained operators can muster the trigger discipline required to produce consistent two-to-three-shot bursts with a MAC submachine gun. Hit probability decreases as the length of the burst increases.

The new M11/9mm submachine gun is the same breed of cat with some different spots. At 11.25 inches without the stock, it's actually 3/4-

MAC M10 Specifications

CALIBER .... 9mm Parabellum.

OPERATION .... Unlocked blowback. Fires from the open-bolt position. Selective fire: semiautomatic or full-auto modes.

CYCLIC RATE .... 1,200 rpm.

FEED MECHANISM .... 32-rd. staggered box; single-position feed. Zytel construction.

WEIGHT, empty .... 3.75 pounds.

LENGTH
stock retracted .... 13 inches
stock extended .... 23 inches.

BARREL LENGTH .... 5.25 inches.

SIGHTS .... Protected, fixed front post; fixed aperture rear set for 100 meters.

ACCESSORIES .... Complete with cleaning rod, wrist strap, magazine loader and one magazine.

PRICE .... $218 to Class 3 dealers.

MANUFACTURER S.W.D., Inc., 1872 Marietta Blvd., Atlanta, GA 30318.

inch longer than the MAC M10. But, at 3.75 pounds, empty, it's only 1/4-pound heavier than a MAC M11. The maximum width is 1.36 inches, identical to that of the MAC M11.

There are other differences. The firing pin is no longer milled into the bolt face, except on the earliest specimens. A sheet-metal stamping, it now rocks back and forth on a steel pin and can be easily removed. While the cocking knob can no longer be rotated 90 degrees to lock the action in the closed position, doing so with the bolt retracted, just past fully cocked, will lock the action in the open position. There is now a receiver pin lock within a hollow receiver pin—in the manner of the M60's top cover. The hand strap's loop has been enlarged and a less noisy Dot fastener is employed.

Most of the other changes are found down in the lower receiver group. The sear catch spring has been altered to another configuration. Positive stops dimpled into the lower receiver body now prevent the selector lever from rotating more than 180 degrees. The "FULL" marking has been changed to "SMG." A U-shaped loop has been welded to the bottom of the butt, which can now be rotated out of the closed position without pressing inward. The magazine release catch is no longer knurled and is more difficult to manipulate than ever before. The magazine well is now welded to the trigger guard, not the lower receiver body. Dropping the weapon will surely bend the pistol grip assembly away from the lower receiver.

The most dramatic change is the adoption of a Zytel nylon 32-round magazine body and follower. Zytel is glass-reinforced plastic, impervious to water and petroleum-based products. It should be ideal for this application. However, the design has gone backward. The Walther's two-position feed system has been dropped in favor of a single-position feed configuration that resulted in many stoppages. That's not a surprising result considering that it does the same thing in the Sten and MP40 submachine guns.

Disassembly procedures remain as before. Remove the magazine and clear the weapon. Remove the receiver pin and separate the upper and lower receivers. Pull the bolt to the rear and pull out the cocking knob. The bolt assembly can then be withdrawn from the rear of the upper receiver. Depress the stock latch button and withdraw the stock assembly from the lower receiver. No further disassembly is usually required. Reassemble in the reverse order.

We fired 1,500 rounds through our test specimen and the Zytel magazine began to fail after 800 rounds. The magazine body started to bulge, the feed lips spread and rounds began popping out of the magazine at inappropriate moments. The magazine body is simply too thin. The test ammunition included reloads, PMC 115-grain FMJ, Federal 115-gr. JHP and Czech military ball surplus ammunition. There were no other stoppages except those attributable to the magazine.

But after 1,200 rounds the bolt's sear notch and its corresponding contact surface on the sear are peened severely, as is the feed ramp where it is overridden by the bolt. None of

S.W.D., Inc. M11/9mm submachine gun field-stripped.

these components have been properly heat-treated. A runaway gun was the final result. Now you just insert a fresh magazine, pull the trigger and wait for all 32 rounds to cycle through the weapon. In addition, the buttstock now collapses in your face with each burst sequence and the receiver pin lock usually drops to the ground.

The cyclic rate and all other handling characteristics remain the same. If you like to throw lead all over the scenery and listen to sounds akin to Hitler's Zipper (the MG42 GPMG), then be my guest. Class 3 dealers can purchase an M11/9mm SMG for $218 complete with one magazine, loader, hand strap and cleaning rod. As for me, these ghastly little guns hold no allure. In my opinion, they have already done far too much to blacken the image of full-auto weapons collecting to ever receive any praise from me.

An upper receiver unit chambered for the caliber .380 ACP cartridge is also available from S.W.D., Inc. It uses the same 32-rd. Zytel magazine and the cyclic rate is reported to be considerably lower. It was not tested.

Originally appeared in *Soldier Of Fortune*
January 1986.

All-Out Shoot Out

Knob Creek's Full-Auto Reunion

Model 1909 Argentine Maxim, caliber 7.65mm, with solid brass water jacket and brass fixtures.

Full-size reproduction of Gatling "Camel Gun" chambered for the 9mm Parabellum cartridge with 18-round strip feed.

At least 50 machine guns are on the line, blazing away at close to their respective cyclic rates. They attack all the senses. Eyes sting and water from swirling clouds of gun smoke and exploding dynamite. Nostrils clog and burn from acrid cordite. You can taste the burnt gunpowder on your tongue as it grinds between your teeth. And, of course, the noise—it's deafening even with ear protection. Exciting, but my psyche remains strangely unaffected. The basic feverish ingredient of a firefight is missing—incoming rounds.

All these sensual stimuli are being generated by a most innocuous group. This is "The Machine Gun Shoot," held twice a year at the Knob Creek Range in West Point, Kentucky, about 20 miles south of Louisville. Machine-gun enthusiasts and collectors from across the country gather at this former World War II U.S. Navy test range to bust as many caps as they can in two days. Wheel-lock and Walker Dragoon collectors never more than stare with glazed eyes at their prizes, protected from all harm in glass display cases. But, no matter how rare and unique, automatic weapons collectors invariably want to shoot their rattle guns, straining them until they redline. If spare parts are not available, they will machine them.

Machine gunners flippantly refer to their gigs as "shoots." The Knob Creek party is, without doubt, the world's largest civilian assemblage of legally owned machine guns. By any standards, the inventory at last October's shoot was impressive. The participants hauled more than 130 full-auto weapons onto the firing line.

There was an assortment of more than two dozen assault/battle rifles. Almost half of these were either M16s in one form or another (including an unusual belt-fed variant) or FAL derivatives. The others included Ruger AC556s, StG(MP)-44s, Kalashnikovs, M-14 E2s, an HK33, HK51 (short-barreled 7.62X51mm NATO), FN CAL (unsuccessful predecessor to the FNC, complete with three-shot burst control) and a rare first model FG-42 (Fallschirmjäger Gewehr, 1942). Unfortunately, the FG-42 broke its firing pin before I could lay back on the trigger.

There were almost three dozen submachine guns. Although Thompsons (8), Stens (4), UZIs (3) and MP-40s (3) dominated, there were also four of that despicable Doper's Delight—the MAC 10. Others included one each of the following: Walther MPK; Reising Model 50; Sterling; Lanchester; MP-28, II Schmeisser; Soviet PPSh-41; S&W Model 76; American 180 (.22 LR); Madsen Model 50; Franklin Arsenal Model 9 (9mm SMG in M16 clothes) and Yugoslavian M61(j) Skorpian.

But machine-gun shoots belong to the belt guns and their magazine-fed corollaries. There were no less than 72. Richard Jordan Gatling invented and produced the first successful rapid-fire guns. Two full-scale reproductions of his famous series were rolled into action: the Model 1875 .45-70 Gatling, complete with Accles 100-round drum and field carriage, and a Model 1876 Camel Gun with tripod, chambered for the 9mm Parabellum cartridge. Both were manufactured by Thunder Valley Machine Co. They are exact replicas in every detail, cost a great deal, but operate flawlessly. The Model 1875 with field carriage will extract $12,750 out of your savings account.

Moving closer to World War I, there was an intriguing array of rare pieces which included the French Model 1914 Hotchkiss (8mm Lebel); Model 1904 Colt Maxim (.30-03); Model 1909 Argentine Maxim (a Model 1898 in caliber 7.65mm Argentine) with solid brass water jacket, feed block, fusee cover, spade grips and other fixtures; BSA Lewis Gun; and German Maxims in both the sled-mounted '08 and bipod-mounted '08/15 configurations.

Extremely rare Portuguese Vickers, caliber .303 British.

Rare first Model German World War II FG-42 (Fallschirmjäger Gewehr, 1942), caliber 7.92mm, with original mount and ZF-4 scope. This LMG/assault rifle was used by German airborne troops.

The inter-war period was represented by a rare Portuguese Vickers (caliber .303 British); Colt Monitor (compact, but unsuccessful, version of the BAR); two Colt Commercial Model 38B, .30-caliber, water-cooled Browning machine guns with spade grips; and three Czech ZB Vz26 light machine guns (from which the famed Bren Gun was derived). Many ZB Vz26s served in the German army during WWII and one of the specimens carried Waffenamt stempels.

Man-portable anti-tank rifles were popular until the middle of WWII, when tank armor improved to the extent that it could no longer be penetrated by such weapons. Two 20mm semiautomatic anti-tank rifles from this era, an S18/1000 Solothurn and Finnish Lahti Model 1939, were also present. Once not subject to Federal registration, these guns sold in the 1950s and early 1960s for $189.50 and $99.95 respectively. Their value has increased twentyfold.

Most abundant at the Knob Creek Shoot were the machine guns of World War II. The .30-caliber Browning series was represented by four M1917/A1 water-cooled MGs, three AN-M2 aircraft guns (fitted with spade grips and attached to M2 tripods) and seven M1919A4/A6 or M37 LMGs. There were both a BAR and FN Model D version.

King of any machine-gun mountain is the .50-caliber Browning. Ten were on the line the entire weekend, although four were attached to a completely restored Quad .50 towed mount. Placed in deuce-and-a-half trucks in Vietnam, these Quad .50s rained certain death on Charlie when employed in airfield perimeter defense. I fired a belt of 50 through each gun, and watching 200 of these 700-grain pills dump their energy 400 yards downrange would make any 97-pound weakling feel omnipotent.

One of Ma Deuce's keepers fired 2,200 rounds after sunset through an already shot-out barrel, stopping only to load belts. At 800 rounds the barrel commenced to glow. After the entire 2,200-round sequence, the barrel was cherry-red from muzzle to barrel extension and hot enough to boil C-rats.

A Russian 12.7mm Degtyarev heavy machine gun proved bush league to Ma Deuce's line-up, as did the interesting JAP Type 96 (caliber 6.5mm Arisaka) and Type 99 (caliber 7.7mm rimless) light machine guns. Hitler's zippers were also present in great strength: nine MG-42s and six MG-34s. When you run nickel ammo through tired guns, the result is inevitable. Most of these German wonders were down at one time or another during the weekend and one MG-42 blew off its bipod, excavating a four-foot trench in front of the muzzle. These guys won't blink an eyelash when they have to cough up $3,500 to $5,000 for an MG-42 or MG-34, but ask them to pay more than five cents a round for ammo and they cry crocodile tears. You can expect up to a 40 percent dud rate on 50-year-old ball. Firing with a cyclic rate of 1,800 rpm per barrel was a German WWII MG-81Z twin-barreled aircraft gun. Produced by Mauser in caliber 7.92mm, it has one common receiver and trigger. Operating in a manner similar to the MG-34, by means of gas-assisted short recoil, some of these guns were re-chambered after the war to 7.62x51mm NATO for use in helicopters.

There were but nine postwar machine guns: three M60s, two L4A3 Bren Guns (caliber 7.62X51mm NATO), two FN Minimis (M249 SAW), a G.E. 7.62mm minigun sitting on a Browning M2 tripod and an HK21. Firing a belt through the HK21 reinforced my impressions from El Salvador that this painful pile driver belongs in no army's inventory.

Firepower like this goes begging for pyrotechnic targets and Knob Creek Range provides some dramatic devices. Dynamite (as much as 40 sticks) is attached to cars, trucks, refrigerators, water heaters, bathtubs, pyramids of tires and barrels with an alcohol, sawdust and oil mixture. The results are often spectacular (especially at night) and sometimes amusing. With 50 or more belt guns raining destruction on the targets no one can be sure he hit anything, so of course everyone assumes he hit everything.

There's a great deal to interest both those with a historical and technical penchant at the Knob Creek shoot. You can learn far more about a particular weapon's salient features from watching it operate than you can in the musty confines of a museum. But don't come looking to see examples of fire discipline. These buckos are blasters. Bursts are limited only by the length of a belt or the capacity of a magazine. What you'll see at Knob Creek is nothing more or less than a weekend of innocent, albeit awesome, fun by the most law-abiding group of gun owners in the world. In the 53-year history of machine-gun control under the National Firearms Act of 1934, no more than one or two felonies have been committed with registered automatic weapons.

Originally appeared in *Soldier Of Fortune*
April 1987.

STONER'S SUPER 63

Legendary Weapon of Navy SEALs

When the cradle adapter is used to fire the Stoner 63A from the M2 Browning machine gun tripod, the group dispersion is very low. However, there is no evidence the weapon was ever employed in this manner, as weapons of .30 caliber or larger provide greater range and penetration and are to be preferred when weight is not an overriding consideration.

Produced only in small quantities, this 250-round drum-type belt carrier—developed at China Lake, California—was too clumsy and heavy.

The Viet Cong (VC) called them devils with green faces; their middle names were almost always Stoner; they evolved from U.S. Navy Underwater Demolition Teams (UDTs). They're the U.S. Navy's Sea, Air, Land commandos (SEALs).

In Vietnam they capitalized on the element of surprise, superb intelligence assets and fire superiority to raid VC strongholds, stage ambushes, capture prisoners and supplies and generally create havoc in Charlie's rear areas. Most often, they gained superiority of fire with their Stoner 63(A) machine guns.

Although few in number and unfortunately never viewed as anything more than a local tactical asset, three SEALs won the Medal of Honor during the Vietnam War. Naval special warfare units suffered only 48 KIA while inflicting several thousand confirmed kills on the VC. The only group to ever employ the Stoner 63(A) in significant quantities, the SEALs created a mystique about this fascinating weapon system far out of proportion to the small number manufactured.

An ex-Marine, Eugene M. Stoner is arguably our nation's most prolific and imaginative modern military small arms designer. While associated with Armalite as chief engineer, Stoner was responsible for the design of the AR15/M16, the caliber 7.62x51mm NATO assault rifle, the AR-10 battle rifle (small quantities of which were sold to Nicaragua, Sudan, Burma and Portugal), the AR-5 caliber .22 Hornet bolt-action rifle adopted by the U.S. Air Force as the MA-1 survival rifle and the blow-back-operated AR-7 "Explorer" takedown rifle. Stoner's prototype projects included a bolt-action sniping rifle (AR-1), a semiautomatic 7.62x51mm rifle (AR-3) and a 12 gauge shotgun with an aluminum bore and receiver (AR-9).

After leaving Armalite, Stoner developed what was at that time a revolutionary concept for a weapons system to be built around certain basic components that could be transformed into a rifle, carbine or various machine gun configurations by fitting different barrels, feed mechanisms, trigger systems and other parts to the basic assembly.

Through his acquaintance with Howard Carson, who was in charge of the West Coast plant of Cadillac Gage in Costa Mesa, California (where Armalite was located), he convinced the company's president, Russell Baker, that the project was feasible. Cadillac Gage established a small arms development center in Costa Mesa, to which Stoner brought his two principal aides at Armalite, Robert Fremont and James L. Sullivan (who later went on to design the Ultimax 100 SAW for Chartered Industries of Singapore).

A STONER IS BORN

The first working prototype was completed in 1962. Called the Stoner M69W (for no other reason that when turned upside down it reads the same—obviously to symbolize the receiver's inversion to assemble different configurations), it was chambered for the 7.62x51mm NATO cartridge. A second prototype was fabricated and named the Stoner 62. Only these two specimens were produced before it was decided to focus instead on the 5.56x45mm NATO cartridge, as it appeared the smaller round would soon predominate the small arms arena.

At that time the name was again changed to Stoner 63; No more than 200 guns were produced at the Costa Mesa facility when manufacture was transferred to the Cadillac Gage (a subsidiary of the Ex-Cello-O Corporation) plant in Warren, Michigan. By 1967, NWM (*Nederlandsche Wapen-En Munitiefabriek)* de Kruitboorn N.V. of 's-Hertogenbosch, Netherlands, was licensed to produce the Stoner 63 system with rights to sell the weapon in all countries of the world outside of the United States, Canada, and Mexico. NWM never fabricated more than a few prototypes. By the end of the decade the Stoner 63's all too brief production life had all but ended.

The complete Stoner 63 system was demonstrated to the U.S. Marines at Quantico in August and September of 1963. They were impressed and favorably inclined toward the rifle and light machine gun configurations. What happened next was an example of inter-service military politics and deceit at its very worst.

The Army Material Command, upon whom the Marine Corps depended for logistical support, offered to perform the trials on the Stoner system at their test facilities. As they were with Stoner's AR-10, U.S. Army Ordnance personnel were predisposed against the Stoner 63 from day one. In any event, toward the end of the Stoner 63 trials, they were already bound and determined to adopt the bullpup-configuration Rodman Laboratory, Rock Island Arsenal, Squad Automatic Weapon (SAW) chambered for the 6mm XM732 cartridge developed at Frankford Arsenal.

By an incredibly strange coincidence the Stoner 63 trials were scheduled to be held at Rodman Laboratory. Equally suspicious was the fact that one of the Stoner 63 test project managers was also involved in the 6mm Rodman Laboratory bullpup program up to his ears. The Marines were informed that any small arms system, including of course the Stoner 63, should be able to function reliably with ammunition exhibiting an extremely wide range of port pressures. This was, of course, untrue, unrealistic and blatantly unfair and thus the Stoner 63's Mean Rounds Between Stoppages (MRBS) and Mean Rounds Between Failures (MRBF)—critical criteria in the test and evaluation of any military small arms system—were "demonstrated" to be unacceptable.

The tracer ammunition used to sabotage the Stoner 63 trials was of such low port pressure that it would not even function reliably in the M16. Stoner was eventually invited to submit the Stoner 63 chambered for the 6mm XM732 cartridge. He wisely advised Cadillac Gage to decline.

SIX GUNS IN ONE

The Stoner 63 system was composed of six weapons, all constructed from the basic receiver group: 1) fixed machine gun, 2) light machine gun (LMG), belt-fed, 3) automatic rifle (AR), magazine-fed, 4) medium machine gun (MMG), 5) assault rifle and 6) carbine.

The belt-fed fixed machine gun was the lightest (10.4 lbs.) and smallest (30.4 inches in overall length) weapon of its type ever produced. Intended to be mounted in either vehicles or helicopters, either singly or in tandem, it could be fired by hand or, with the trigger group removed, remotely actuated through a trigger-linkage by a solenoid, cable or pneumatic system. The weight of both the gun and its ammunition are not overriding considerations when they are mounted in a vehicle or aircraft. Weapons of .30 caliber or larger provide greater range and penetration. To my knowledge, this configuration of the Stoner 63 system was never employed in combat.

The MMG was little more than the left-hand, belt-fed light machine gun mounted on a tripod with the buttstock removed. A cradle adapter was fabricated to attach the gun to the U.S. M2/122 tripod (originally designed for the M1918A4 .30 caliber Browning machine gun). The body of the adapter was machined from aluminum, then black anodized. A standard M2 pintle and T&E mechanism were attached. A steel pin on the front of the adapter fits into a hole drilled into the underside of the U-shaped barrel bracket attached to the gun's gas cylinder. Steel, spring-loaded hooks and latches, screwed

Firing Stoner 63A with left-hand feed mechanism and 150-round drum-type belt carrier. There is very little perceived recoil and hit probability is quite high if the operator exercises proper fire discipline.

and pinned to the rear end of the adapter, were fixed to steel pins welded to the receiver sides. These adapters are quite rare and there is no evidence the Stoner was ever employed as an MMG—undoubtedly for the same reasons that precluded its use as a fixed gun. With the tripod and cradle adapter, the total weight in this configuration is 29 lbs.

The magazine-fed automatic rifle, often referred to as the "Bren" configuration, weighs only 10.2 lbs. Because it is top-fed, both the front and rear sights are offset to the left. As a consequence, in addition to the magazine-well with rear sight that replaces the belt feed mechanism, a special barrel is required. There was little interest in this configuration either—a pity as magazine-fed LMGs, battle-proven in the case of the Bren, provide a substantial number of advantages over belt guns in certain tactical roles.

The belt-fed LMG is the Stoner 63 configuration most often associated with the U.S. Navy SEALs. Weighing only 11.9 lbs., empty (compared to 15.5 lbs. for the M249 SAW), with an overall length of 40.25 inches, it could be fired from the prone position with a bipod, offhand or from the hip. The latter position was employed all too frequently by the SEALs who would send long bursts downrange with poor hit probability.

As all of the above configurations are fed ammunition from the top, it is required that the basic receiver group be oriented with the gas cylinder to the bottom so that the reciprocating parts do not interfere with the feed mechanisms. They all fire from the open-bolt position and in full-auto mode only (the selector lever is inoperative in these configurations).

Both the assault rifle and carbine fire from the closed-bolt position and are bottom-fed. Thus the receiver group must be inverted with the gas cylinder above the barrel. They are both selective-fire weapons. The rifle, used to some extent by the SEALs in Vietnam, weighs 7.9 lbs., empty, with an overall length identical to the LMG. To convert the assault rifle to a carbine it is only required to replace the 20-inch rifle barrel with the 15.7-inch carbine barrel and change the rigid buttstock for a folding type. This drops the weight to 7.7 lbs. and reduces the overall length to 26.75 inches with the stock folded.

Early factory literature shows the Stoner 63 with all wood furniture. Black and white spacers between the rubber buttpad and stock provide a somewhat sporting appearance. However, the wooden pistol grips and buttstocks were quickly replaced by black polycarbonate components. The handguard remained wood, but was painted black.

RUSH TO VIETNAM

By the time the Stoner 63 was rushed into the Vietnam War by the SEALs, it was barely beyond the prototype stage. Battlefield experience and tests by the Marines, Army and Air Force indicated a number of modifications were required. Shortly after serial number 2,000 was produced, substantial alterations were introduced and the name was changed to Stoner 63A.

Improvements included a one-piece buttstock pad, a 17-4 PH stainless steel gas cylinder, polycarbonate forearm with sling swivel, increased barrel life (from 8,000 to 30,000 rounds) through additional hardening processes, a sling attaching ring at the front end of the carrying handle rod to permit carry at the "ready" position, an investment-cast feed tray, an improved feed cover, a spring-loaded actuator roller to prevent damage to the feed mechanism if the feed cover is closed with the bolt group forward, a longer retracting handle, improved bipod and smaller ammunition boxes. The rifle and carbine's magazine-well was extended and flared to facilitate magazine insertion and the cocking handle was placed on top of the receiver to permit charging with either hand. A wire folding stock with polyvinyl chloride coating and only 14 parts replaced the carbine's previous polycarbonate folding buttstock (all told nine different types of folding stocks were produced at one time or another in varying quantities).

There were also a number of major alterations. The "SAFE" position was removed from the selector lever and a safety mechanism of the

STONER 63A METHOD OF OPERATION

Stoner 63A machine guns are gas operated, rotary-locking, air-cooled belt- or magazine-fed and fire from the open-bolt position to improve cooling and reduce the possibility of cook-offs. The mechanism is driven by a conventional long-stroke piston (which means that the piston and bolt group travel rearward a distance somewhat greater than the length of the loaded cartridge). Using the bolt's movement to generate energy to lift and feed belted ammunition requires sustained pressure on the piston and increased momentum of the reciprocating parts which would not be provided by the short impulsive blow produced in a direct gas impingement system like that of the M16.

When the trigger is pulled the piston/bolt group flies forward and a round is stripped from the belt or magazine. The bolt head has seven symmetrical locking lugs which distribute the firing load evenly around the bolt head and barrel socket. Attached to the piston extension, the bolt carrier has a curved cam path in which the bolt's cam pin (retained by the firing pin) moves. Final forward movement of the piston rotates the bolt through 22 1/2 degrees clockwise and the locking lugs engage behind abutments in the barrel socket

When the carrier cap with the actuator roller (for the belt feed mechanism) at the rear end of the bolt carrier is rotated 180 degrees—required to assemble the bottom-fed rifle and carbine configurations—it also rotates the spring retracted firing pin by the same amount and orients a small notch on the firing pin to the top. There is a thick steel flat washer in front of a series of Belleville washers in the bolt carrier's hollow interior. This washer, called the firing pin collar, has a tab which engages a channel in the "bottom" (when the receiver is inverted for the rifle/carbine configuration) of the bolt carrier's interior to preserve its orientation. The hole in the center of this unusual washer is cut flat at the top (called a "D" hole). The notch on the spring-loaded firing pin permits it to move rearward past the "D" hole in the firing pin collar an additional sixty five-thousandths of an inch so that it may be driven forward by the hammer. When oriented downward—for firing from the open-bolt position—the firing pin is held forward by that amount and when the bolt rotates into battery and moves into the carrier, the firing pin protrudes from the bolt face to strike the primer. All of this is remarkably ingenious.

Unknown to most, an antibounce device, consisting of a tungsten carbide rod, approximately 4 inches in length, rides within the piston extension's hollow interior and moves forward and aft during the recoil and counter-recoil cycles. This reduces bolt bounce during locking and eliminates the possibility of primer ignition out of battery during closed-bolt firing of the rifle/carbine.

Upon firing, gas following the bullet up the bore is tapped through the barrel's gas port (about 46 inches from the chamber) and into the gas cylinder to drive the piston rearward. This movement is translated to the bolt carrier. There is about 0.2-inch of free travel, to permit gas pressure to drop to safe level, before the carrier's cam path rotates the bolt 22 1/2 degrees counter-clockwise to unlock. There is no pitch on the locking lugs and so no primary extraction occurs during unlocking. The claw extractor, which has a pivoting fulcrum at its rear end, is retained in the bolt head by a spring-loaded pin. Seated deeply, almost in line with the bolt axis, it alone provides extremely efficient extraction of the empty case as the harder the bolt pulls the deeper the extractor claw bites into the case's extractor groove. There is an angled channel on the opposite side of the bolt head through which the ejector passes as the bolt group moves rearward. The spring-loaded ejector is fixed to the front feed-mechanism trigger housing bracket by a roll pin. After extraction and ejection, the carrier continues rearward to compress the recoil spring on its guide rod (located in the receiver body to permit removal and changing of the buttstock for the different configurations).

Another unique feature is the buffer system within the bolt carrier. In front of the carrier cap are a steel shim and a set of 27 saucer-shaped Belleville washers which absorb piston energy by deforming into a flat plate when the bolt carrier strikes the receiver's end cap. When they return to their original shape they release a surge of strain energy which throws the reciprocating parts forward in counter-recoil with a speed only slightly less than the original recoil velocity. These Belleville washers, oriented in opposing sets of three, prevent the gun from literally beating itself to death. Depending upon the cyclic rates employed and the port pressure of the ammunition used, they should last between 40,000 and 50,000 rounds before replacement is required.

As the piston and bolt group move forward, another round is stripped from the feed mechanism and the cycle repeats until the trigger is released or the ammunition supply is exhausted.

Belt movement is produced by the actuator roller riding in the channeled feed arm. This spring-loaded feed arm rests under the hinged top cover and is pivoted at its rear end. As the bolt reciprocates, the front end of the feed arm moves across the feed tray and operates a lever attached to a single set of spring-loaded feed pawls. The feed pawls move a cartridge and link over the feed tray's spring-loaded stop pawl from where they are next positioned onto the slotted feedway and held firmly in place by a spring-loaded steel plate in the top cover. After the cartridge has been pushed forward out of its link, the empty link is pushed toward the link ejection port which has a spring-loaded dust cover.

M1 Garand/M-14 type installed in a slot in the front of the trigger guard. As the new trigger guard was welded to the trigger housing and could not be removed, a winter trigger attachment was eventually designed. The selector lever's "SEMI" and "AUTO" positions were placed in 180 degree juxtaposition to preclude confusion under stress.

A carbon-relief groove was cut inside the gas cylinder to permit firing over longer periods of time without maintenance. A three-position gas regulator was added to the machine gun barrel's gas block. A spring-loaded ejection-port dust cover was added.

To understand the importance of these changes and the brilliance of the overall design, a detailed examination of the Stoner 63A's salient features is required. The basic component group consists of the receiver, gas cylinder, bolt and piston group, recoil spring group and trigger housing.

The receiver body is a long rectangular steel pressing with support structures, the gas cylinder and other appurtenances (i.e., trigger housing and feed mechanism retaining brackets, tripod studs, forearm bracket and barrel latching mechanism) welded in place. The rear portion holds the piston extension and bolt group. The front portion holds the piston and barrel and is ventilated to reduce weight and improve heat dispersion. The ejection port is on the right side when the receiver is inverted for the rifle and carbine and on the left side in the machine gun configurations. The "Endurion"® finish appears to be some type of black baked enamel over phosphate.

The open-bolt-firing trigger mechanism is quite simple. There are four so-called trigger pins. The front pin retains either a flapper-type magazine catch/release (used with the rifle/carbine and left-hand-feed 150-round drum-type belt carrier), a full dust cover (used with the top-fed AR or vertically mounted ammo box on either belt-fed system) or a half size dust cover (used with the right-hand-feed bottom box carrier). The next two pins hold the timer and hammer, both of which components are deleted in the open-bolt guns. The fourth pin is the trigger's axis shaft. The spring-loaded sear pivots on the selector lever's axis shaft. The selector lever's position is of no consequence when firing from the open bolt. The sliding safety blocks all trigger movement when pushed to the rear. The rear of the trigger housing serves as the receiver's end cap and holds the buttstock. The black polycarbonate pistol grip has a small checkered area on each side and is flared at the bottom to prevent the operator's hand from sliding downward. There is a storage compartment in the grip, sealed by a hinged cover with a spring-loaded hatch.

The hooked, steel retracting handle is normally mounted along the right side of the receiver. Non-reciprocating, it has 24 lightening holes and when pulled rearward engages a projection on the piston to draw the piston/bolt group back to the cocked position. After retraction of the operating group it should be pushed fully forward so a flat-spring latch riveted to the front end can engage a slotted plate welded to the front of the receiver. It is somewhat difficult to operate when the right-hand feed mechanism with bottom box carrier is installed, so a special slotted forearm with a bottom cocking handle was devised. SEALs in Vietnam often added an extended handle to this apparatus.

FIVE-BARREL SYSTEM

Five different barrels are associated with the Stoner 63A system. In addition to the rifle, carbine and AR barrel (with offset front sight), there are two types of machine gun barrels—a standard heavy barrel and a short "commando" tube. The standard machine gun and AR heavy barrels are 20 inches in length (bolt face to muzzle). Add another 1.67 inches for the flash hider.

The fluted, so-called "commando" barrel with a length of 15.7 inches was sometimes employed by the SEALs. However, the gas port is almost at the muzzle and since port pressure drops to zero as soon as the bullet leaves the barrel very little energy reserve is retained. Even though the port diameter was increased, this merely resulted in faster initial acceleration of the piston and operation was never totally reliable. (Note: This problem was also encountered by FN when they developed the first short-barreled "para" FAL. They were eventually forced to lengthen the barrel.)

All Stoner 63(A) barrels are of the quick-change type. The front of the barrel rests on a U-shaped barrel bracket attached to the gas cylinder. A spring-loaded (with two nested coil springs) barrel latch on the receiver drives a steel pin into a hole in the barrel socket to hold the barrel firmly in place. When the bolt is retracted, the barrel can be removed by pressing down on the latch and pulling the barrel forward.

The front sight assemblies of these barrels, as well as the bayonet lug, are mounted to the gas block and feature a round post with protective ears, adjustable for windage and elevation zero. The bird-cage flash hider has six oval ports. There are carrying handles on the AR and standard machine gun barrels that can be snapped into any one of three positions. The black-painted wooden handles are attached to a steel rod by a roll pin. The entire assembly can be removed from the barrel, if desired.

There are three split-ring valves on the gas plugs of these barrels. While they appear similar to those on the M16's bolt, they are not interchangeable. Furthermore, they are static and serve only to seal the gas cylinder, while those on the M16 are dynamic (i.e., they reciprocate with the bolt) and are thus much more subject to wear and damage.

The rifle, carbine and AR barrels have no gas regulator as they are magazine-fed and do not require the energy reserve levels of belt-fed mechanisms. The standard machine gun barrel has a three-position regulator. It can be adjusted by inserting the nose of a bullet into the hole over the regulator's lock detent and pushing down on the detent. Then rotate to the desired setting. The slowest rate of fire is obtained when the narrowest indicator notch is set over the detent. The cyclic rate is supposed to vary from 700 to 1,000 rpm depending upon the gas regulator setting, port pressure of the ammunition, ambient temperature, lubrication, degree of fouling and a few other variables. When set to the lowest setting, *SOF*'s test specimen fired at about 715 rpm. The middle gas regulator position increased the rate of fire to approximately 830 rpm. When the maximum amount of gas was thrown into the system, the cyclic rate increased to only 865 rpm. Generally, the regulator should be left at the middle setting as this provides reliable performance at a reasonable rate of fire that will not induce excessive wear.

Most Stoner 63(A) barrels have six grooves with a right-hand twist of one turn in 12 inches to stabilize the 55-grain M193 ball projectile commonly used during this time frame. However, after NWM obtained the license to produce the Stoner 63A weapons, experiments were conducted with heavier bullets and some NWM-manufactured barrels will be encountered marked "200MM," which indicates a twist of one turn in 8 inches.

Ball, armor-piercing and tracer ammunition was manufactured for these tests by *Industriewerke Karlsruhe (IWK)*, Germany—the post-1945 title of *Deutsche Waffen und Munitionswerke* (DWM). The ball ammunition is headstamped "NWM 67 5,56." The brass, Berdan-primed case has three flash holes. There is a lime green primer annulus and the propellant charge consists of 24.8 grains, nominal, of a spherical ball powder. The bullet weighs 77 grains, nominal, and has a flat base, lead core, slender ogive and cupro-nickel, clad steel jacket. There is a sharp crimping groove around the projectile and it is stake-crimped to the case mouth. Nothing is known about this bullet's wound ballistics potential, but it is an amazing precursor to the Belgian SS109 and U.S. M885—predating these rounds by almost two decades.

During this same time frame, Cadillac Gage distributed ammunition with heavy bullets to the law enforcement agencies (principally in the Michigan area) who had purchased Stoner 63As, in one configuration or another. This ammunition appears to have come from two sources. Some is headstamped "IVI 69" *(Industries Valcartier* Incorporated, Quebec City, Canada—1969) and features a 68-grain boattail bullet with lead core, copper alloy jacket and crimping cannelure. The Boxer-primed, brass case contains 26.5 grains, nominal, of a spherical ball powder. Federal also supplied small lots (headstamped "FC 66") with a similar bullet. Some of the Federal lots have "Colt" stamped on the box, apparently indicating use for the Colt machine gun. Very little else is known about this ammunition.

APERTURES & AMMO BOXES

Stoner 63(A) LMGs were initially equipped with a left-hand feed mechanism, followed later by a right-hand feed mechanism, which is almost the former's mirror image except for an additional stop pawl on the side of the feed tray. (Note: Early Stoner 63 bolt carriers will not operate the right-hand feed mechanism).

Rear sights are mounted on the top covers of the belt-fed LMGs. These folding-leaf, peep aperture (0.06 inches in diameter) sights are elevation-scale-graduated in 100-meter increments from 200 to 1,100 meters. Both windage and elevation can be adjusted in 1/4-mil increments. When the sight is folded down, a 200-meter battle-sight aperture with a diameter of 0.09 inches is exposed. Rifle and carbine rear sights—with large ventilated protective ears—are simple flip-type apertures with settings for 0-300 meters and 300-500 meters. They can be adjusted for windage or elevation in one-minute of angle increments.

Early ammunition boxes for the Stoner 63 held 150 linked rounds. Made of ribbed plastic, they have a tab that permits them to be attached to the side of the left-hand feed tray. The very first specimens were olive drab in color and indicate manufacture in Costa Mesa, California. They were followed by black boxes of the same capacity manufactured by Cadillac Gage in Warren, Michigan. Stoner 63A ammunition boxes were also manufactured in Warren of black plastic, but hold only 100 linked rounds as the gun was unbalanced when the larger box was attached. They can either be attached to the left-hand feed tray or held in a bottom box carrier when the right-hand feed mechanism is employed.

Several drum-type belt carriers were designed for the left-hand feed mechanism. Most common was a 150-round anodized aluminum container. The left-hand feed mechanism with this 150-round drum-type belt carrier

Field Stripping a Stoner

For all its complexity, disassembly procedures for field stripping the Stoner 63A are relatively simple. Engage the safety. Lift the top cover and remove the belt. With the bolt retracted, inspect the chamber and make sure the weapon is clear. Close the top cover. Disengage the safety and move the reciprocating parts forward, under control.

Push the captive take-down pin to the right as far as it will go. Pivot the trigger housing group downward. Withdraw the recoil spring and guide rod out the rear of the receiver. Lift the muzzle and drop out the piston/bolt carrier group. Unscrew and remove the trigger housing pivot pin assembly and separate this group from the receiver. Unscrew and remove the feed mechanism pivot pin assembly. Push forward on the top cover latch. Raise the top cover and remove it and the feed tray from the receiver. Unscrew and remove the forearm pivot pin assembly and remove the forearm. Depress the barrel latch and remove the barrel. Pull back on the retracting handle until its lug and disassembly notch are aligned with the notch on the receiver. Pull out and down on the retracting handle to separate it from the receiver.

Push against the base of the firing pin until it is flush with the back carrier cap assembly. While holding the firing pin in this position, rotate the carrier cap assembly (with the piston oriented down and to the front) a quarter turn counter-clockwise while also pressing down on the actuator roller until it is in line with the bottom of the bolt carrier. Use a work bench as a fulcrum as it is difficult to rotate the carrier cap assembly. Remove the firing pin. Push the bolt into the carrier and remove its cam pin. Withdraw the bolt from the carrier.

Under no circumstances should you attempt to separate the carrier cap assembly from the bolt carrier body. The Belleville washers inside the carrier were installed by means of a hydraulic press and calibration gauges. If you drive out the roll pin and two notched pins retaining the carrier cap assembly, you can kiss your ass goodbye, as you'll never be able to reassemble these parts.

Push out the stock retaining pin to separate the buttstock from the trigger housing group. To disassemble the trigger housing group, you must first push in the lock plate tang on the left side of the housing and slide the lock plate forward. Push out the trigger housing cover pin and remove the trigger housing cover. Push out the trigger pins and remove the trigger, sear and sear spring. Remove the lock plate. Remove the safety by sliding it to the rear and dropping it down through the trigger guard. When re-assembling these parts, the trigger pins must be aligned perfectly so their notches will engage the lock plate.

The gas regulator can be removed by rotating it until the flat side exposes the roll pin. Then push the regulator out from the left side.

After cleaning, lubricate and reassemble in reverse order. Do not lubricate the gas cylinder, piston head or gas regulator.

Note: Do not "play" with the take-down pin while there is a belt in the feed mechanism and the bolt is retracted. If it is pushed out by mistake, with the 150-round belt carrier attached, the receiver and trigger housing will separate just enough so the bolt will override the sear, but not enough for the recoil spring and guide rod to fly out the rear. The result will be an uncontrolled runaway burst. A SEAL was killed in Vietnam while he sat in a boat with the muzzle pointed at his chest and accidentally pushed out the take-down pin. Any number of weapons will fire if this procedure is followed and this is hardly a defect in the Stoner 63A. Nevertheless, three modifications were made to the subsequent XM207E1 prototype-only Stoner to prevent accidental discharge of this type.

attached predominated in use over the right-hand feed with bottom box carrier (the Stoner 63A was called the Mk 23 Mod 0 by the Navy when in this latter configuration) in Vietnam by a ratio of almost 10 to one. A 250-round drum-type belt carrier was developed at China Lake, California. Only a small quantity were manufactured as they were too heavy and bulky. RPD belt carriers were also sometimes jury-rigged to the Stoner by enterprising SEALs.

Stoner links are marked "S-63 BRW" and are scaled-down versions of the U.S. M13 push-through link for the M60 GPMG. M27 links, designed for the M249 SAW (FN Minimi), are similar, but will not perform with complete reliability in the Stoner 63(A) as their angle of pitch is slightly different.

Early Stoner 63(A) 30-round magazines featured steel bodies with black oxide steel followers and weigh 8 ounces. Later magazines of this type had chrome-plated steel followers. An aluminum-bodied magazine was developed that cut the weight to 4 ounces. There was also a 20-round magazine, but it is rarely encountered.

As they were mostly hip-shooting blasters, U.S. Navy SEALs rarely employed the Stoner bipod. Non-locking Stoner 63 bipods cannot be correctly attached to the Stoner 63A's gas tube as it is of larger wall diameter than the earlier model. Stoner 63A bipods can be locked in either the open or closed positions. Fabricated from stamped sheet-metal with numerous distinctive lightening holes, the Stoner 63A bipod can be adjusted in command height from 9 3/4 to 14 inches, albeit not easily. It does not pivot and the gun must be lifted to engage flanking targets.

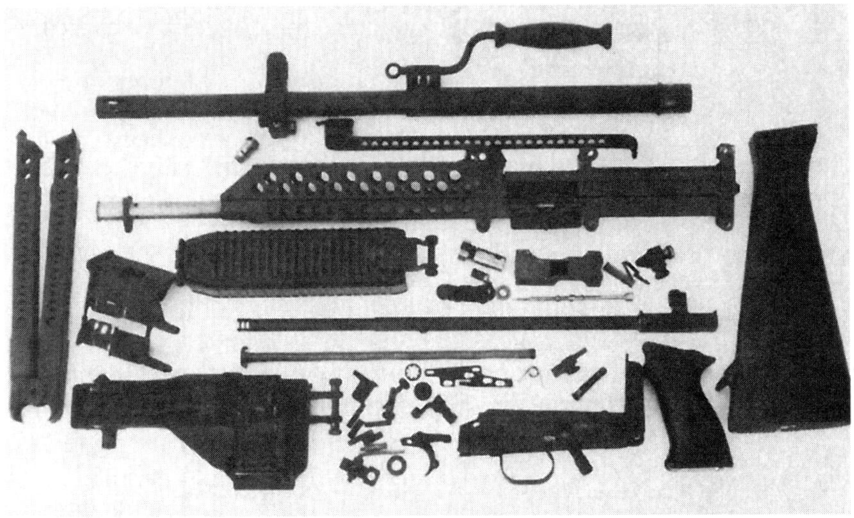

THE WELL-DRESSED STONER

Stoner 63(A) accessories are as varied as the system itself. Either series produced or in prototype form only, in addition to those accouterments already mentioned, there was a blank firing attachment (BFA), winter trigger, asbestos-lined spare barrel bag, complete cleaning kit, 40mm grenade launcher, various types of slings and a bewildering array of magazine and belt box web pouches. SEALs frequently carried Stoner belts either in M14 magazine pouches or across the chest, "Pancho Villa" style, between two layers of T-shirts.

Three different bayonets can be attached to the Stoner 63(A). Most mundane is the standard U.S. M7 for the M16 series. In 1970, Eickhorn of Solingen, West Germany, together with NWM, developed a bayonet for the Stoner 63A system with wire-cutting capability. The clipped-point Bowie blade, complete with sawteeth fetish, uses the same wire-cutter concept employed on the ComBloc AKM bayonet. It can be identified by the figure of a squirrel over the NWM logo stamped on the blade, a ribbed, black plastic grip and a black plastic scabbard with a gray leather leg thong. Most esoteric of all is a Stoner 63A bayonet manufactured in very small quantity by SIG. It is of typically Swiss design with a black plastic scabbard and olive drab web front. Its highly polished blade will attract every sniper within a 1,000-meter radius and is endemic of a nation devoid of all battlefield experience since 1815.

Stoner 63A LMG, completely disassembled. Operators should never disassemble the bolt carrier components as shown here as the Belleville washers are installed with a hydraulic press.

STONER REBORN

One final version of the Stoner system remains to be mentioned. In 1966 Great American Arms Co. Inc. (now known as Navy Arms, Inc.) in Ridgefield, New Jersey, advertised a semiautomatic-only version to be known as the Stoner 66. The retail price was $199.50. While this was just $10 more than the semiauto-only Colt AR-15, the project was dropped after Cadillac Gage purportedly assembled no more than a handful.

SOF's test and evaluation of our Stoner 63A test specimen was entirely satisfactory. Several thousand rounds were fired using South Korean PMC M193-type ball ammunition (headstamped "80.05 5.56 PS"). There was one dead primer and one empty case spin-back into the receiver when using the 150-round drum-type belt carrier. There were no other stoppages. Group dispersion of tripod-mounted bursts were exceptionally small. Accuracy and hit probability were both high when the bipod was employed. When fired off-hand, accuracy was acceptable, if the bursts were held to two or three shots.

No more than approximately 3,600 Stoner 63(A)s were ever produced. Compare this to 50 million Kalashnikovs, 9 million M16s or even the only 7,500 German WWII FG42s manufactured and you start to have some perspective of the amazing mystique developed around a weapon system that never actually reached major series production. Stoner 63(A)s remained in SEAL inventory until about 1983. Thousands were torched by the U.S. government. With the exception of maybe two or three specimens, the few in collectors' hands in this country are restricted-transfer Class 3 dealers' samples, most manufactured by Cadillac Gage, exported to NWM in Holland and then imported back to the United States.

BETTER THAN M249?

Prior to the 19 May 1986 deadline, C. Reed Knight, Jr.—well-known for his sound suppressors and designer of the recently introduced and quite revolutionary Colt M1 America/Model 2000 pistol—formed, welded, finished and registered with the BATF, 100 Stoner 63A receivers from sheet-metal flats and fixtures obtained from Cadillac Gage. Manufactured to exact Stoner 63A specifications, all of these receivers are "non-restricted transfer" firearms, i.e., they can be legally registered to qualified individuals. Together with the new parts obtained from Cadillac Gage and NWM, they are available in any of the six Stoner 63A configurations. One of these units was used in *SOF*'s test and evaluation of the Stoner 63A system. Every single component was in brand new, "in-the-box" condition when we received it. They are not inexpensive and a complete weapon will set you back up to five figures. For further information contact Knight's Armament Company.

Clever is certainly an appropriate adjective for the Stoner 63A—in some areas perhaps too clever. While it successfully prevents an 11.9 pound belt-fed machine gun from hammering itself to death, the unique buffer system in the bolt carrier is not tamper-proof. And, while factory literature asserted that 81.3% of the interchangeable components were common to two or more configurations, the number of bits and pieces required to change from a rifle or carbine to a machine gun or to go from right-hand feed to left-hand feed would cause any supply sergeant to swoon. Furthermore, its complexity requires meticulous maintenance, probably more than the average grunt is capable of mastering.

Yet, in the hands of an experienced operator, it's a superb lightweight and reliable instrument. While comparisons with the M249 SAW are unfair, as they are separated in developmental time by almost a quarter-century, the Stoner 63A is more than 3 1/2 pounds lighter, every bit as reliable and quite a bit more accurate. Had the rifle version ever entered mass production, with supplemental manufacture of the machine gun, as anticipated, it would have inevitably been simplified and refined. If forced to choose, I would personally select the Stoner 63A LMG with left-hand feed and the 150-round drum-type belt carrier over the M249 SAW if for no other reason than its lighter weight—although I feel the M249 will eventually serve with distinction and provides greater simplicity.

During the 1980s Eugene Stoner marched forward to design the Ares caliber 5.56x45mm SAW. Using features and components from a number of different weapons, the so-called Stoner 86's modular design emphasizes simplicity and reliability. Weighing only 10.85 pounds, empty, it is adaptable to both belt and magazine feeds. Unfortunately, the parent company recently divested itself of Ares and the project has been all but terminated. Once again, one of America's most brilliant small arms designers finds himself in the wrong place at the wrong time, a victim of the duplicity of others.

Originally appeared in *Soldier Of Fortune*
July 1991.

RUSSIA'S BUFFALO BURP GUN

New Bizon SMG Conceived by Kalashnikov, Designed by Dragunov

Featuring 60% parts interchangeability with the AK100 series for cost-effective manufacture, the new Russian Bizon SMG provides a lightweight, compact envelope with excellent handling characteristics and a low recoil impulse, controllable rate of fire, more than acceptable accuracy and superb hit probability.

Bizon SMG has the usual noisy AK sheet-metal selector. The uppermost position is "safe." The middle position (marked with a Cyrillic "AB") provides full-auto fire, and the lowest position places the mechanism in the semiautomatic firing mode.

Although no major military organization has adopted a new submachine gun in decades, they continue to hold significant interest for military special operations units (especially when provided with sound suppressors) and law enforcement agencies that contain special response teams within their organizational structures.

By definition, a submachine gun is a firearm chambered for pistol-caliber ammunition that is capable of full-auto fire and most often is equipped with a shoulder stock.

As they usually operate by means of unlocked blowback and thus avoid the complexities of locked-breech, gas-operated systems, SMGs are generally suited to low-cost, fast production in great numbers. No country has more experience in the design and mass production of submachine guns than Russia, as they manufactured and fielded close to 10 million caliber 7.62x25mm PPSh 41 SMGs during the "Great Patriotic War" (World War II). The military application of the pistol-caliber submachine gun was brought to its tactical zenith by the Soviet army during this time frame. Although widely used by communist forces during the Korean War as well, the PPSh 41, and submachine guns in general, reached military obsolescence by the late 1950s.

During my recent trip to Russia to participate in the celebration of Mikhail T. Kalashnikov's 75th birthday, I was presented with the opportunity to test and evaluate a new submachine gun designed by a project team headed by his son, Victor Kalashnikov.

Bizon Specifications

CALIBER .... 9x18mm Makarov, 9x18mm Makarov High Impulse and 9x19mm Parabellum (requires magazine change).

OPERATION .... Unlocked blowback. Semiautomatic and full-auto fire modes. Fires from the closed-bolt position.

CYCLIC RATE .... 650-700 rpm.

FEED .... Fiberglass-reinforced polyamide 64-round helical-type magazine.

WEIGHT
empty .... 5.45 pounds (2.47 kg);
without magazine .... 4.6 pounds (2.1 kg).

LENGTH, overall
With stock unfolded .... 26 inches (660mm);
with buttstock folded ..... 16.7 inches (425mm).

BARREL .... Four-groove with right-hand twist of one turn in 9.4 inches (240mm).

SIGHTS .... Flip-type rear sight with protective ears and two open square-notches with 50- and 100-meter elevation settings. Round, post-type front sight with protective hood; adjustable for both windage and elevation zero.

FINISH .... Black phosphate.

FURNITURE .... Black fiberglass-reinforced polyamide pistol grip.

MANUFACTURER .... Izhmash, 3 Derjabin Str., 426006 Izhevk, Russia.
STATUS .... Proceeding to series production.

T&E SUMMARY: A clever design featuring 60% parts interchangeability with the AK100 series for cost-effective manufacture. A lightweight, compact envelope with excellent handling characteristics and a low recoil impulse, controllable rate of fire, more than acceptable accuracy and superb hit probability.

Alexi Dragunov, youngest son of Evgeni F. Dragunov, who designed the highly regarded SVD sniper rifle, was also a member of the design team that developed the "Bizon" (Bison) submachine gun.

The Bizon was developed at the state-owned Izhmash factory. Located 1,130 kilometers east of Moscow in the western part of the Urals, Izhevsk, formerly a closed city, has a population of 730,000. Its small-arms factory was established in 1807, making it the third such facility established in Russia. (Tula Arsenal was founded in 1712 and that of Sestroretsk—not far from St. Petersburg—in 1722, although the latter no longer produces small arms and is now a machine tool factory.)

Firing from the closed-bolt position (to enhance the weapon's accuracy potential) by means of unlocked, pure blowback, the Bizon has several unusual features.

Most immediately noticeable is what appears to be a grenade launcher mounted under an extremely short AK. Neither supposition would be correct. Although the Bizon features 60% parts interchangeability with the AKS-74, it is chambered for the standard 9x18mm Makarov cartridge, as well as the new high impulse round in this caliber. Furthermore, what appears to be a grenade launcher is in reality the Bizon's unique 64-round, helical-type magazine. Although reminiscent of the U.S. Calico magazine, all cartridges are oriented nose forward in the Bizon magazine and, I was told, rounds cannot be loaded incorrectly. Why 64 rounds? Apparently for no other reason than the fact that this number is a multiple of 16 and 9x18mm Makarov cartridges are packaged 16 to the box.

The prototype magazine used in *Soldier Of Fortune's* test and evaluation of the Bizon SMG actually held 67 rounds. Its body was also fabricated from aluminum tubing. Production series Bizon magazines will be made from fiberglass-reinforced, thermosetting (this indicates that heat is used to cure the resin) polyamide (epoxy-based resin) of the type used on the furniture of the new AK100 series rifles.

Injection-molded polyamides are super industrial-strength synthetics well known for their resistance to high temperatures, corrosion, wear, chemicals and radiation. Lighter than steel, they have a higher tensile strength than aluminum.

Loading this magazine is quite simple. Insert a round and turn the loading bar at the front end of the magazine one click counterclockwise (with the front end facing you). Release the bar after loading has been completed. Hooks on top of the front end of the magazine grab on pins under the front sight. The rear end of the magazine engages a paddle-type, spring-loaded, magazine catch/release of the Kalashnikov type located in front of the trigger guard.

SUB-GUN SYSTEMS

The Bizon provides potential users with an extremely lightweight and compact envelope.

High Impulse Makarov

The 9x18mm Makarov cartridge is thought to be derived from a 1936 German project, the 9mm Ultra.

Neither should be confused with the post-World War II German 9mm Ultra. There are subtle dimensional differences among all three and they are not interchangeable.

A pipsqueak by U.S. standards, the 9mm Makarov lies midway between the .380 ACP and 9mm Parabellum in both size and wound ballistics potential. At 17,000 psi, chamber pressures are low enough to permit unlocked blowback operation (Note: A few unlocked blowback-operated pistols, such as the Spanish Astra 400/600 series, have been chambered for more powerful cartridges—in this instance the 9mm Bayard Long and 9mm Parabellum, respectively).

The rimless, straight-sided 9mm Makarov case is 18.05mm in overall length. Russian production of this cartridge features a copper-washed steel case loaded with 3.5 grains of a short, extruded tubular kernel powder. The squat, round-nose Full Metal (steel) Jacket (FMJ) bullet has a lead core and weighs 95 grains with a diameter of 9.27mm and a length of only 11.4mm. Muzzle velocities range from 951 to 1,033 fps.

In response to criticism concerning its effectiveness and a request by Russian law enforcement agencies for a projectile that would defeat body armor, a new 9mm Makarov High Impulse round was developed. Until now very little was known about this cartridge. While in Russia, I not only fired it extensively in both pistols and submachine guns, but was able to examine it in detail.

The standard copper-washed steel 9x18mm Makarov case is used. The one I examined carried a headstamp of 539 93. It is Berdan-primed with two flash holes. When assembled, this ammunition is provided with red primer annulus and case mouth sealants.

The FMJ bullet is most unusual in its configuration, which is that of a truncated cone with a rounded tip. It weighs 85.5 grains and has a steel core. The propellant is a flattened-ball type similar in size and shape to Winchester 231. The charge weight is 7.8 grains, nominal. With a bullet lighter by 10%, this charge generates muzzle velocities of 1,410 fps.

While designed to penetrate body armor, the trade off will undoubtedly be excessive penetration in unprotected human targets as a consequence of its higher velocity and shoulder-stabilized truncated-cone bullet which will not yaw in soft tissue. (Yawing 3 to 4 degrees in either direction—a characteristic of most FMJ round-nose handgun bullets— decreases their penetration in soft tissue.)

9x18mm Makarov ammunition: on the left, standard cartridge with 95-grain round-nose FMJ bullet; right, new high impulse round with 85.5-grain truncated-cone bullet with rounded tip.

Weight, without the magazine is only 4.6 pounds (2.1 kg). With an empty magazine installed, the weight of the Bizon goes up to 5.45 pounds (2.47 kg). With the butt-stock folded, the length of the weapon is only 16.7 inches (425mm). With the stock unfolded, the overall length is increased to 26 inches (660mm).

The pinned and riveted sheet-metal receiver body is essentially that of the AKS-74 with the front end chopped off and modified, as there is no gas system. The usual noisy AK sheet-metal selector lever is located above the trigger on the right side of the receiver. The uppermost position is "safe" and in this position the lever physically impinges against the bolt's integral retracting handle to block the bolt's rearward travel while a projection on its shaft blocks the trigger. The middle position (marked with a Cyrillic "AB") provides full-auto fire and the lowest position places the mechanism in the semiautomatic firing mode.

This is the most logical sequence, because sweeping the selector lever downward under stress will invariably place it in the bottom, or semiauto, position. Thus the operator must consciously move the bar back upward to obtain full-auto fire. That's just what we want, since much of the time (but not always) shoulder-mounted submachine guns and assault rifles should be fired single-shot to enhance hit probability.

The Bizon's trigger mechanism is exactly that of the AKM/AKS-74, which is, in turn, based upon that of the U.S. .30 M1 Garand rifle. Its major components are a hammer with two hooks, a trigger with an extension that is the main sear and a spring-loaded auxiliary sear. When the hammer is cocked, it is held back by the main sear. When the trigger is pulled, the main sear moves forward off its hook on the left side of the hammer and the hammer rotates up and forward to strike the firing pin, driven by its multiple-strand spring. When the bolt moves back and rolls the hammer down, it's caught by the auxiliary sear.

In semiautomatic fire, since the trigger is already rearward, another round cannot be fired. When the trigger is released, the main sear moves back to catch the hammer as it is released by the auxiliary sear. As control of the hammer is restored to the main sear, provided the bolt group is fully forward, pulling the trigger will fire another shot.

There is also an auto safety sear on the right side of the trigger mechanism. Its tip protrudes from a slot in the right receiver rail. Until it is disconnected by the bolt in its final forward movement, it continues to engage the hammer.

When the selector lever is set to full-auto, the projection on the selector lever shaft that blocks trigger movement when the lever is in the uppermost, or "safe" position, prevents the auxiliary sear from rotating upward to catch the hammer. Thus as long as the trigger is pressed, ammunition is available and the bolt goes fully forward to trip the auto safety sear, the weapon will continue to fire.

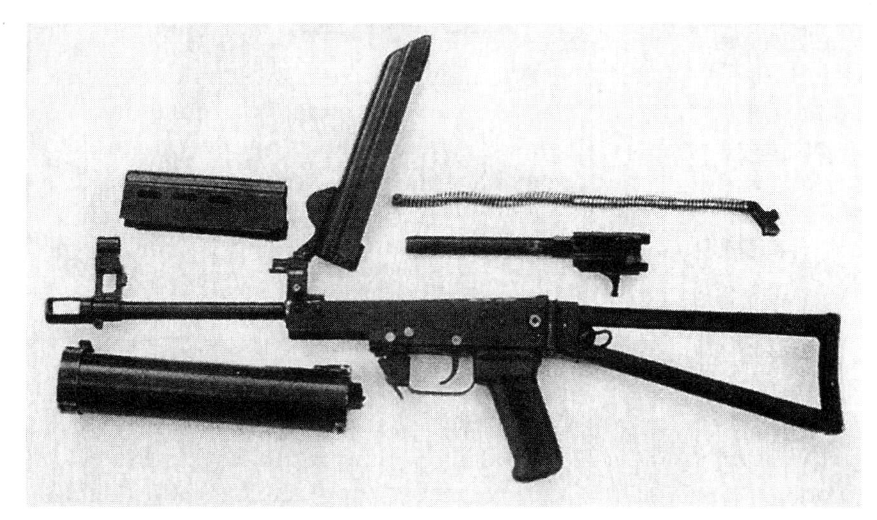

Russian Bizon SMG, fieldstripped

RUSSIAN ORIGINALITY

There is also a five-component "anti-bounce" device attached to the trigger mechanism that serves as a mechanical drag on the hammer to delay firing just enough until the bolt has settled completely into battery. Originally thought by some to be a "rate reducer," this feature was introduced with the AKM because its thin (1mm) sheet-metal receiver body had different bounce characteristics than the heavy, milled, drop-forged AK-47 receiver it replaced.

The Bizon's bolt has a spring-loaded, floating firing pin. Externally, the bolt carrier resembles that of the AK, with the usual integral retracting handle, but without a piston and the piston extension plugged with a steel insert. The recoil spring and guide rod are those of the AK.

The excellent AKS-74 folding buttstock has been used on the Bizon SMG. Fabricated from U-shaped, stamped sheet-metal struts and assembled by welding, it was a decided improvement over the down-folder stock used on the AK-47/AKM series and copied directly from the German World War II MP-40 SMG. It folds to the left, and is held open by a spring-loaded, catch/release button at the rear of the receiver on the left side. The stock is held closed by a spring-loaded hook on the left-forward end of the receiver. This is one of the simplest, most robust and rigid buttstocks I have ever used.

The pistol grip is exactly that of the AK100 series and is made of black fiberglass-reinforced polyamide. The stamped sheet-metal receiver cover has also been taken from the AK. It has been altered only at the front end. The rear sight module is permanently attached to the top cover. A hinge at the front of the rear sight base holds the top cover permanently to the receiver body. When the top cover is opened and rotated upward, a pin holding the top handguard in place is withdrawn so that the handguard can be removed. The handguard is a sheet-metal stamping with three elongate ventilation holes on each side. The Bizon's magazine serves as a bottom handguard.

The rear sight, a flip-type with protective ears, has two open square-notches with 50- and 100-meter elevation settings. The front sight has been taken from the SVD sniper rifle. A round post-type, its bottom end is threaded and it can be rotated up or down to adjust elevation zero. The post is protected by a hood with a hole in the top to insert the elevation adjustment tool. The hood and post unit are fitted to the sight base by means of a dovetail and can be drifted right or left to provide windage zero.

The Bizon's four-groove barrel has a right-hand twist of one turn in 240mm (9.4 inches). The muzzle device has a large rectangular port on each side of upper dead center. It has little effect on the flash signature but does moderate the weapon's muzzle jump to some extent. Its principal purpose is to protect the muzzle and magazine from damage.

One of the Bizon's most interesting design features is its short recoil stroke. When standard 9x18mm Makarov ammunition is fired in this weapon, the bolt does not impinge against the rear of the receiver body. This ammunition will produce a cyclic rate of 700 rpm. Makarov High Impulse ammunition will drive the bolt rearward until it just barely strikes against the rear of the receiver. A cyclic rate of only 650 to 680 rpm is the result. With no more than a change of magazines, the Bizon SMG can be converted to fire 9x19mm Parabellum ammunition.

This short recoil stroke also decreases muzzle jump and substantially increases controllability, accuracy potential and hit probability. Although it should never be fired this way, except for a dog-and-pony show, I watched Alexi Dragunov dump an entire 64-round burst into an 18-inch group from a distance of 50 meters. Very impressive.

Disassembly is incredibly simple. Remove the magazine and then clear the chamber. Push in the serrated end of the recoil spring guide, which protrudes from a cutout in the end of the receiver cover. Lift up the receiver cover. Remove the top handguard. Lift out the recoil spring and guide rod with the bolt. Separate the bolt from the recoil spring and guide rod. No further fieldstripping is necessary or recommended. Reassemble in the reverse order.

My test firing of the Bizon SMG indicated that it has excellent handling characteristics with a low recoil impulse, controllable rate of fire, more than acceptable accuracy and superb hit probability. There were no stoppages of any kind during *SOF's* test and evaluation.

Commonality of many parts with the AK provides the Bizon with even greater cost effectiveness than normally encountered with the generally economical submachine-gun genre. I can envision a bright future for the Bizon amid the crowded field of burp guns and so-called "personal defense weapons," especially when a 9x19mm Parabellum magazine is snapped in place.

Originally appeared in *Soldier Of Fortune*
May 1995.

RUSSIAN KLIN/KEDR SUBGUNS

Russian KLIN submachine gun, shown here with the 30-round magazine inserted and the 20-round magazine to the left, can be fired with the stock folded, but only in an emergency and only at extremely close ranges.

Sound-suppressed version of the Russian KEDR submachine gun. For law enforcement tactical teams involved in the "dynamic entry" of a building, the ability to hear voice commands is a primary reason for employment of the sound suppressor.

Nobody should know more about submachine guns (SMGs)—shoulder-mounted, full-auto, pistol-caliber weapons—than the Russians. This genre reached its absolute zenith during World War II. The Germans produced an estimated 1 million MP40s and the British manufactured a total of 4,184,237 Sten SMGs (at a cost of only 3 to 5 pounds sterling per gun). But the Russians turned out more than 10 million caliber 7.62x25mm PPSh41s, rivaling the caliber 7.62x54R Mosin Nagant M1891 bolt-action service rifle in the numbers manufactured and issued.

As they usually operate by means of unlocked blowback and thus avoid the complexities of locked-breech, gas-operated systems, SMGs are generally suited to low-cost, fast production in great numbers. The Soviet army was the first to take what had previously been a special-purpose small arm with limited applications and turn it into a tactical weapon. Russian soldiers armed with PPSh41s rode into battle like a swarm of ants clinging to their T34 tanks.

After the Korean War, which was fought by both sides with World War II weaponry, the military applications for the submachine gun fell off sharply. No major military organization has adopted a submachine gun in substantial quantities in at least three decades. However, they still have viable applications for law enforcement agencies and, when fitted with sound suppressors, for military special operations units.

During my recent trip to Russia to participate in the celebration of Mikhail T. Kalashnikov's 75th birthday, I was presented with the opportunity to test and evaluate a new submachine gun designed by the late Evgeni Feodorovich Dragunov who developed the well-known caliber 7.62x54R SVD (*Samozariyadnaya snaiperskaya vintovka Dragunova*) semiautomatic sniper rifle, adopted by the Soviet Army in 1963. Dragunov commenced the design of his submachine gun in 1969 and died before it was fully executed. The project team who developed the design into a production-series weapon was headed by his eldest son and my very good friend, Mikhail E. Dragunov.

The *"KLIN"* (which was named by the Ministry of Internal Affairs and in the Russian language means "wedge") submachine gun is chambered for both the standard 9x18mm Makarov cartridge and the new high-impulse round in this caliber.

The slightly different *"KEDR"* (which means "cedar" in Russian and is also an acronym for Konstruktsia Evgeni DRagunova) is chambered only for the standard 9x18mm Makarov round but is also available in a sound-suppressed version.

KLIN submachine gun with stock extended. Recently adopted by the Russian national police, it is one of the world's lightest and most compact burp guns.

Operating by means of unlocked blowback and firing from the closed-bolt position, the KLIN/KEDR family of submachine guns have a 1mm-thick stamped sheet-metal receiver that features pinned, riveted and welded construction. This is a very compact and lightweight design. Production series specimens carry a black phosphate finish. Weight of the KEDR, empty is only 1.4 kg (about 3.1 pounds). The KLIN is just slightly more. This is somewhat lighter than even the Czech Scorpion. With the stock folded, the overall length of KEDR is just 303mm (about 12 inches). With the stock extended, the length increases to only 540mm (21.26 inches). The KLIN is 2mm longer. Cyclic rates of fire are 800 to 850 rpm with standard 9x18mm Makarov ammunition and 1,030 to 1,200 rpm when high-impulse Makarov ammunition is fired in the KLIN SMG.

The barrel for both weapons is 120mm (4.7 inches) in length. The bore has four grooves with a right-hand twist of one turn in 240mm (9.45 inches). The KLIN barrel has been moved forward in the receiver 2mm so that the feed ramp and receiver body could be altered to provide reliable feeding of the high-impulse cartridge's truncated cone bullet. A conventional feed ramp which is integral to the receiver is all that is required on the KEDR to feed the standard 9x18mm Makarov cartridge with its round-nose bullet. A projection added to the center of the KLIN's feed ramp lifts the round upward to meet two lips (one on each side wall of the receiver) added to the rear of the barrel. This modification pivots the cartridge upward for alignment with the chamber.

The KLIN's barrel has another unusual feature that has been designed to safely accommodate the new 9x18mm Makarov high-impulse round. Its chamber has three helical grooves. After ignition, expanding gases swell the case and drive it into the helical grooves to increase the coefficient of friction and retard extraction until pressures drop to a safe level. The Chinese Type 64 pistol I fired seven years ago on the mainland had a chamber with four helical grooves. Chambered for the odd 7.62x17mm cartridge which has a muzzle velocity about 150 fps greater than that of the .32 ACP round, the Type 64's slide, which has slightly less mass than that of the Walther PPK, required some retardation to operate safely.

With checkered sides and three finger grooves, the injection-molded, one-piece grip is attached to the receiver by a threaded bolt. A sheet-metal cap on the bottom of the grip covers the bolt's access hole. The prototype grip on the weapon I fired was made in the model shop and looked like wood, but was machined plastic.

The front sight base is integral with the front receiver cap which also contains the barrel support, feed ramp, spring-loaded magazine catch/release, magazine support structure within the receiver and also the sound suppressor's mounting brackets. The round post-type front sight has protective ears and is adjustable for both elevation and windage zero in the manner of the Kalashnikov rifle series.

The rear sight is quite innovative. When Evgeni Dragunov fired the Stechkin machine

Sound-suppressed version of the KEDR submachine gun, fieldstripped

KLIN/KEDR SMG Specifications

CALIBER .... .9x18mm Makarov and 9x18mm Makarov High-Impulse (KLIN only).

OPERATION .... .Unlocked blowback. Semiautomatic and full-auto fire modes. Fires from the closed-bolt position.

CYCLIC RATE
KEDR .... .800 to 850 rpm;
KLIN .... .1,030 to 1,200 rpm.

FEED .... .20- or 30-round staggered-column detachable box magazines of the two-position-feed type.

WEIGHT, empty ....
KEDR .... .3.1 pounds (1.4 kg);
KLIN .... .1.41 kg.

LENGTH, overall .... KEDR with stock unfolded, 21.26 inches (540mm); with buttstock folded about 12 inches (303mm). KLIN is 2mm longer.

BARREL .... .Four-groove with right-hand twist of one turn in 9.45 inches (240mm).

BARREL LENGTH .... .4.7 inches (120mm).

SIGHTS .... .When the shoulder stock is extended, a peep aperture with a 1.6mm-diameter is exposed; when the stock is folded over the top of the receiver, it rotates the spring-loaded sight assembly forward to expose an open U-notch sight. Round, post-type front sight with protective ears; adjustable for both windage and elevation zero in the manner of the Kalashnikov rifle series.

FINISH .... .Black phosphate.

FURNITURE .... .Injection-molded plastic pistol grip with checkered sides and three finger grooves.

STATUS .Adopted by Russian national police and commencing series production.

T&E SUMMARY: Among the smallest and most compact of all currently produced submachine guns. Reliable and cost-effective. Provides a sound-suppressed option and several ammunition alternatives. An abundance of innovative features: helical-grooved chamber, unique rear sight system and a modular trigger assembly.

pistol he noted that the rear sight, while adequate for close-range shooting, did not match the weapon's potential at longer ranges when the buttstock/holster was employed as a shoulder mount. As a consequence, he designed the KLIN/KEDR so that when the shoulder stock is extended, a peep aperture with a 1.6mm-diameter is exposed. When the stock is folded over the top of the receiver, it rotates the spring-loaded sight assembly forward to expose an open U-notch sight for firing at close range without use of the shoulder stock.

A notch at the rear of the front sight base engages a spring-loaded catch/release at the butt end of stock to retain it when it is folded on top of the receiver. Push forward on the elongated, checkered catch/release button to release the stock so it can be rotated rearward to the extended position. A spring-loaded catch/release on the front end of the stock locks it in the extended position on the rear receiver cap. The buttplate, a machined forging, is welded to the stamped sheet-metal stock.

The stamped sheet-metal trigger guard has been welded to the receiver at the front and rear. Equipped with a pivoting locking bar, the stamped sheet-metal receiver top-cover has a forward cutout on the right side for the ejection port.

A machined drop-forging, the bolt body has an integral cocking piece on the left side. Its floating firing pin has no spring. There is a spring-loaded claw extractor on the breechface. The fixed ejector is pinned in place on the left receiver wall. The bolt reciprocates on rails inside the receiver. Two sets of cutouts in the receiver rails at the rear permit removal of the bolt group during disassembly.

There is a single-coil recoil spring and conventional guide rod. The rear end of the guide rod is attached to a plate with a pin that enters a hole in the rear of the bolt body when the bolt reaches the end of its recoil stroke. This prevents the bolt group from jumping off the rails because of the disassembly cutouts at the rear.

Another of the KLIN/KEDR's innovative features is the modular trigger assembly, which is reminiscent and an expansion of the drop-in hammer and sear subassembly of the Russian TT-33 (Tokarev) caliber 7.62x25mm pistol. The KLIN/KEDR trigger pack consists of the trigger housing and all the required components: trigger, trigger spring, hammer, hammer axis pin, hammer spring and strut, auto safety sear and its axis pin and the disconnector.

This modular trigger mechanism is held in the receiver at the rear by a pin at each end of a bushing that holds the rear end of the hammer spring in place. The front end of the trigger mechanism is held in place by the selector lever shaft, which also has a trigger block on it. A protrusion on the selector shaft activates the spring-loaded bolt hold-open after the last shot has been fired. In the *safe* position, the bolt hold-open pin also locks the bolt in the forward position. Cyrillic selector markings on the receiver are, from top to bottom: "AB" for full-auto, "OD" for semiauto and the character pi and "P" (the letter R in English) for *safe*. This is not the same sequence used with the AK series, which is, top to bottom: safe, full-auto and semiauto. In the KLIN/KEDR design the bolt does not contact the auto safety sear during semiauto fire.

Magazines of either 20- or 30-round capacity are available for the KLIN/KEDR. These staggered-column, detachable box magazines are of the two-position-feed type. They have stamped sheet-metal, punch-welded bodies with sheet-metal followers, floorplates and retaining plates. They are disassembled in the conventional manner. To withdraw a magazine from the magazine-well, press the magazine catch/release button inward from the left side.

Disassembly procedures are easily mastered for the KLIN/KEDR series. First, remove the magazine and then inspect the chamber. If it is folded, extend the shoulder stock fully. Turn the disassembly lever on top of the sheet-metal receiver cover in either direction to disengage the rear of the lever from its slot in the rear receiver cap. Lift off the receiver cover. Push the guide rod and recoil spring assembly forward and rotate the rear end of the guide rod upward. Separate the recoil spring and guide rod from the hole in the rear of the bolt. Push the bolt group to the rear and lift it out of the receiver. Turn the selector lever until its nose is at 12 o'clock and pull it out of the receiver. Lift up the trigger pack from the front, twist it slightly to disengage the rear retaining pins and lift the trigger pack out of the receiver. Although the trigger pack can also be easily disassembled, this will not usually be required at the operator's level. Reassemble in the reverse order.

The sound-suppressed version of the KEDR SMG can also be easily disassembled for maintenance. This is a two-part sound suppressor system. The barrel has 10 ports, about 1 inch to the rear of the muzzle, arranged in two sets each of two and three ports. The port diameter is 3.8mm. Over the ported barrel is a collar which interfaces with brackets on the front receiver cap assembly. This is followed by tightly wrapped stainless steel mesh held in place by a coil of thick wire. At the muzzle end is another collar to which a conventional baffle-type muzzle suppressor is attached by means of interrupted threads. A smooth steel housing 74mm (2.9 inches) in length, which serves as an expansion chamber, covers the collars and mesh-wrapped, ported barrel.

The muzzle suppressor measures 136mm (5.4 inches) in overall length. External diameter of both the muzzle suppressor and the expansion chamber over the ported barrel is 32mm (1.24 inches). Both housings are machined from steel bar stock. The muzzle suppressor housing has knurling in the center.

The muzzle suppressor can be disassembled for maintenance by removal of its front cap. Simply depress its springloaded retaining pin and turn off the front cap.

This unit drops the sound signature by only a modest 20 decibels and should be more appropriately termed a "sound moderator." Complete silence is not necessarily golden. To the layman, the total drop in sound pressure level as measured by the logarithmic ratio known as the decibel (dB) is the bottom line in assessing any sound suppressor, or "silencer" as they are known to fiction writers. To professional users, such as military special operations units and law enforcement special reaction teams, other parameters often outweigh the requirement for the absolute obtainable reduction in sound pressure level.

For those who must dance through the elaborate and complex choreography involved in the "dynamic entry" of a building infested with terrorists or drug dealers, the ability to hear voice commands is a primary reason for employment of a sound suppressor. Size, weight, durability, smoke and flash signatures, ease of maintenance and sometimes special capabilities—in addition to sound reduction—make up the spectrum of attributes desired in sound suppressors by the armed professionals dressed in Ninja black.

AWC Systems Technology's TAC NINE muzzle-type sound suppressor features a 30 dB drop in sound pressure level, but adds 9 inches to the length of the weapon. This is 3.6 inches more than the KEDR's sound suppressor and it weighs considerably more. The KEDR suppressor drops the sound level below the pain threshold, about 141 dB, and for most of those engaged in CQB this is sufficient.

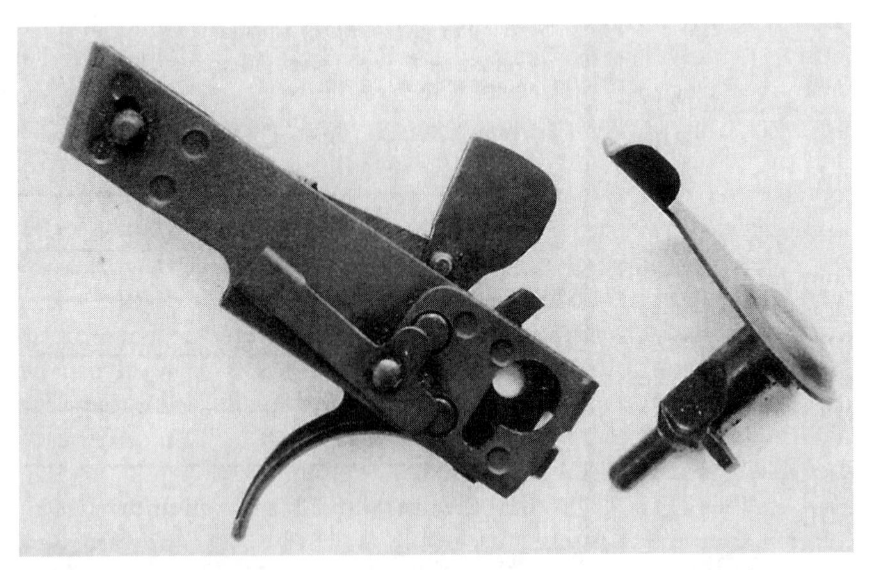

KLIN/KEDR modular trigger mechanism and selector lever. The assembly is held in the receiver at the rear by a pin at each end of a bushing that holds the rear end of the hammer spring in place.

In 1994 the KLIN SMG was pitted against a number of other submachine guns, including a Tula Arsenal design called the "Kypariss" or PP-90, during the Russian police trials. The KLIN emerged as the clear winner and in that year was officially adopted by the Russian national police.

It was cold, very cold, the day I test fired the KLIN and both versions of the KEDR submachine guns. Russians and their weapon systems thrive on this type of weather. Throughout the day many thousands of rounds were fired through several specimens of the three variants. There were no stoppages of any kind.

Strict fire discipline is a mandatory requirement with this category of weaponry. Even though submachine-gun magazines usually have almost double the capacity of any semiautomatic pistol, full-auto bursts should be restricted to only two rounds whenever the design permits, as any subsequent shots in an extended burst will impact high and often over the target. Furthermore, in my opinion, burst fire should be limited to ranges under 15 meters in most instances. At 15 meters and beyond, operators will increase hit probability by firing in the semiautomatic mode. Head shots, at any distance, should be attempted only with the selector lever set to fire semiautomatic.

All of my firing with the KLIN/KEDR submachine guns was conducted with the stock extended and from a strong shoulder mount. At ranges beyond 15 meters, the semi-automatic accuracy potential was substantially greater than any handgun. This is to be expected, especially so since this design fires from the closed-bolt position. These weapons should be fired with the stock folded only in an emergency and only at extremely close ranges. I found the length-of-pull (the distance between the front edge of the center of the trigger to the center of the buttplate) to be somewhat too short for me. I would like to see it lengthened by about 2 inches. The Russian police have made the same criticism.

At ranges beyond 15 meters, I found the sound-suppressed version of the KEDR provided the greatest hit probability. Firing standard 9x18mm Makarov ammunition, it required no great skill to fire consistent two-shot bursts from either of the KEDR variants as the 800 to 850 rpm cyclic rate generated by this ammunition closely approximates that of the MP5. When firing the KLIN with Makarov high-impulse ammunition, experienced operators can expect to fire three-shot bursts at best.

The KLIN/KEDR series of submachine guns are among the smallest and most compact of this genre, rivaled only in this area by the Czech Scorpion. Reliable and cost-effective, they provide law enforcement agencies and military special operations units with a sound-suppressed option and several ammunition alternatives. Employing the helical-grooved chamber, it would be no great trick to chamber these submachine guns for the 9x19mm Parabellum cartridge. In fact, using this principle I fired blowback-operated Makarov pistols chambered for the 9mm Para round. But that's a story for another issue of *Fighting Firearms*.

Originally appeared in *Fighting Firearms*
Summer 1995.

Section 6

Sniper Rifles

ComBloc Sniper Rifles

Crosshairs of the Warsaw Pact

Dragunov-copy M76 upgrades the Kalashnikov system to minimum sniping quality.

Yugoslavian M76 rifle field-stripped.

While American use of telescopic sights on military rifles dates from the Civil War, optically equipped service rifles did not reach prominence in Europe until World War I when highly trained German sharpshooters helped decimate the Russian horde along a thousand-mile line from the Baltic to the Black Sea. The Russian army at that time possessed no sniping capability. But by 1941 Germans dropped like flies before Red Army snipers who helped prevent the blitzkrieg from overrunning Mother Russia.

During WW II, Russian snipers were equipped with the M1891/30 Mosin Nagant bolt-action rifle fitted with either the four power "PE" scope or 3.5-power "PU" scope. Although still employed by some Third World client states of the Soviet Union, these have generally been replaced by the Dragunov system. To determine accuracy potential, hit probability and overall quality of ComBloc sniping systems, an M1891/30 Mosin Nagant with "PU" scope and a Yugoslavian M-76 Dragunov derivative were selected by *SOF* for test and evaluation.

The M1891 rifle was developed by Colonel S.I. Mosin of the Russian Artillery who designed the action, and the Belgian Nagant brothers who were responsible for the magazine system. The complex and unusual bolt consists of three components, excluding the firing pin: the body with the bolt handle, the recessed bolt head with two locking lugs and extractor, and a coupling unit which also serves as a bolt guide.

Locking lugs cam laterally into recesses on each side of the receiver ring instead of vertically as in the case of the Mauser. The bolt handle on sniper versions is turned down. Cocking occurs when the cocking piece is held back by the sear as the bolt is opened. The safety is simple but clumsy. The cocking piece must be pulled rearward against the firing pin's spring tension and turned counterclockwise about an eighth turn.

Model 1891/30 rifles differ from the original M1891 only by virtue of a shorter barrel (28.7 inches versus 31.6 inches), a rounded rather than hexagonal receiver and a rear sight marked in meters instead of archines (one archine is approximately 28 inches). With a weight of 8.7 pounds without scope, sling or bayonet—standard for this period—and an overall length of 48.1 inches, the M1891/30 rifle is awkward to carry and was supplemented by carbine versions throughout its production history. To my knowledge, Mosin Nagant carbines were never fitted with scopes.

Because of the large-rimmed 7.62x54R cartridge, the M1891/30 magazine has a peculiar feed interrupter which holds down the second round, permitting the top cartridge to feed and move forward without pressure from below. This magazine catch helps to eliminate jams and is not released until final rotation of the bolt. For purposes of unloading and maintenance, the magazine floorplate can be released by a spring-loaded catch. Magazine capacity is only five rounds. The rifle is loaded from five-round clips, but they cannot be used with the scope in place.

Almost full-length, the buttstock ends with a metal cap just 3 1/2 inches from the muzzle. Of poor grade, the wood furniture on these rifles is usually stained and varnished or painted. Two metal barrel bands retain the top handguard. A cleaning rod fits into a groove on the buttstock's underside. A web sling is attached to leather straps that run through two slots in the buttstock in a manner peculiar to this rifle series. There is a finger groove on either side of the stock just under the receiver. A recoil crossbolt at this location all too frequently loosens and permits the barreled action to move about in the stock. This does nothing to maximize accuracy potential.

Sight radius of the M1891/30 is 24 1/2 inches. A round, post-type front sight is threaded into the protective hood which can be drifted in its dovetail on the barrel for windage zero. The post can be rotated by inserting a tool through the top of the hood for elevation zero. A sliding tangent rear sight carries an open U-notch and can be adjusted for elevation only from 100 to 2,000 meters—a typical fantasy of this era.

The so-called "PU" 3.5-power scope is the most prevalent and worst of the two. Only 6 3/4 inches long, it sits too far forward on the receiver and causes neck strain when used for prolonged periods of time. The reticle pattern has been copied from that used by the German military since World War I. It consists of a single, thick, pointed post at the bottom of the field of view with thick horizontal sidebars. This format excels in subdued light and offers faster target acquisition than standard crosshairs. But since there is

Mosin Nagant M1891/30 Dragunov Sniper Rifle

CALIBER . . . . .7.62x54R Russian rimmed.

OPERATION . . . . .Manually operated turn-bolt.

FEED MECHANISM . . . . .5-round box magazine with feed interruptor.

WEIGHT, empty . . . . .8.7 pounds without scope, sling or bayonet.

LENGTH, overall . . . . .48.1 inches.

BARREL LENGTH . . . . .28.7 inches.

SIGHTS . . . . .Round, post-type front with protective hood; adjustable for windage and elevation zero. Sliding tangent type rear with open U-notch; adjustable for elevation only in 50-meter increments from 100 to 2,000 meters. Sight radius 24.5 inches

OPTICS . . . . .4X PE or more prevalent 3.5X PU scope mounted to left side of receiver. PU scope reticle pattern pointed post on bottom with thick, horizontal sidebars.

MANUFACTURER . . . . .Numerous Soviet arsenals and PRC. Many rebuilt in late 1950s at Polish arsenals.

STATUS . . . . .No longer manufactured. Obsolete in ComBloc armies. Replaced by Dragunov series. Still encountered in Third World ComBloc client states.

The Mosin Nagant M1891/30 sniper rifle, replaced by the Dragunov series, is obsolete in all but a few Third World ComBloc client states.

3.5X "PU" scope on Mosin Nagant M1891/30 sniper rifle.

Russian crest and serial number on 3.5X "PU" scope.

no range-finding capability the sniper must estimate the range and rotate the external elevation knob on top of the scope which is adjustable in 100-meter increments from zero to 1,300 meters. I don't know why the supposed effective range of this rifle is reduced from 2,000 to 1,300 meters by the addition of a telescopic sight.

Another external knob on the left side permits windage adjustments. The scope mount is attached to the left side of the receiver. Although it can be adjusted for initial zero by means of two opposing set screws, it is not designed to be detached by the operator. Inconvenient, but acceptable. The later four-power "PE" scope, also mounted on the left side of the receiver, was longer and permitted the sniper to maintain a normal shooting position when aligning his eye with the ocular.

Our test specimen was one of a large number rebuilt in Polish arsenals during the late 1950s. The stock had been repaired in three places. The receiver, bolt, magazine floorplate and checkered buttplate had been stamped with new serial numbers. The bore was in fair condition only, so the barrel had obviously not been replaced. The metal parts, except for the bolt group which remains in the white, were reblued. Trigger pull weight was a consistent but spongy 7 pounds.

Without doubt, the Russian 7.62x54R cartridge remains the last of the old-fashioned, large-capacity, rimmed-case, long-range rounds to survive in the service of any major military power. Both the Soviet Dragunov sniping rifle and the PK series of current Soviet general purpose machine guns are chambered for this round. With the 150-grain light ball projectile, it's in the same class as the .30-06.

Three lots of 7.62x54R ammunition were used to test the M1891/30 Mosin Nagant sniper rifle. All were Berdan primed and corrosive. Two of the lots were Russian light ball. Both contained 49 to 50.5 grains of extruded tubular kernel type (IMR) powder and copper-clad steel cases. The first lot was headstamped "60 45" (factory code 60, 1945). This ammunition used the M1908 projectile (Type "L") which is a full metal jacketed (FMJ) spire-point bullet with a hollow base and no color code. It left the 28.7-inch barrel at an average velocity of 2,968 fps. Extraction was somewhat sticky. The other lot, headstamped "188" (factory code 188, no date) and captured on Grenada, is the current Russian light ball. It features an FMJ boattail bullet with a mild steel core (Type "LPS") that is patterned after their armor-piercing round and carries a silver color code on the tip. It produced an average velocity of 2,833 fps, but a standard deviation of only 12 fps, which is excellent for military ammunition.

A substantial quantity of Egyptian ammunition was also fired through this rifle. The headstamp has Arabic characters which transliterate into "Misr" (the Arabic name for Egypt) and Factory No. 10. These are also Berdan primed, corrosive, and the cases are brass. A heavy 180-grain boattail FMJ bullet is propelled by 47 grains of European-type cut-sheet flakes. These Israeli battlefield pick-ups are really Egyptian junk. Forty percent have dead primers and when they do go off, hangfires are numerous. Those cartridges that ignited without incident have an average velocity of only 2,578 fps (expected because of the heavier bullet) and an unacceptable standard deviation of 90 fps.

With regard to accuracy, current Russian light ball turned in the best performances with 3 MOA, mediocre at best. Older Russian ammunition and the Egyptian heavy ball never shot better than 3.6 MOA. Remember, however, that our test specimen's barreled receiver floated about in the stock with every shot fired. Consistent head shots would not be possible with this rifle at even 200 meters. The antiquated M1891/30 Mosin Nagant rifle simply cannot cut the mustard as a modern sniper rifle.

By the late 1960s, these rifles had been replaced in the Soviet armed forces by the Dragunov system. The Dragunov rifle is based on the Kalashnikov system and has an awesome reputation. Our test specimen was the Yugoslavian M76 derivative of the Dragunov manufactured by Zavodi Crvena Zastava, Kragujevac. It will soon be available in the U.S. in caliber 7.62x51mm NATO. A PRC version of the Russian Dragunov, in caliber 7.62x54R, will soon be imported by Keng's Firearms Specialty.

Gas operated, the Dragunov has a Kalashnikov-type bolt and carrier but a separate short-stroke piston. Long-stroke operation results in movement of a heavy mass which jars the weapon and decreases the accuracy potential. The Dragunov's light piston, driven rearward by gases moving through the gas port, transfers its energy to the bolt carrier. There is no piston extension and the carrier moves back alone to rotate the bolt running in its cam path and unlock the action. The carrier and bolt continue to move rearward and compress the recoil spring which then drives the bolt group forward to strip a round from the magazine and lock the two-lug rotary bolt into its recesses in the receiver walls before the next shot is fired. The Yugoslavs have, for some reason, opted to retain the Kalashnikov's original longstroke method of operation with the piston and extension attached to the bolt carrier. The M76 differs in other significant details from the Soviet SVD (Samozaryadnaya Vintovka Dragunova) and Rumanian FPK (an almost exact duplicate of the Russian SVD), although disassembly/assembly procedures are identical to any other Kalashnikov and will not be described here.

Before WWII the Yugoslavs adopted the German 7.92x57mm cartridge for use in their military rifles and machine guns and until recently the standard sniping system was a rebuilt, accurized Mauser 98k fitted with a telescopic sight. This caliber has been retained for both machine guns and the M76 sniper rifle. With its rimless case it constitutes a better choice than the 7.62x54R while retaining equivalent ballistic potential.

At 11.6 pounds, empty, but with scope and mount, and an overall length of 45.4 inches, the M76 is not exactly a lightweight scout rifle. But snipers don't usually move about as much as other straight legs and this is an acceptable tradeoff for an enhanced accuracy potential and reduction in felt recoil.

A forged and milled receiver, like that of the AK-47, is partially responsible for this heft. This receiver and all other metal parts have been salt blued with a black finish, except the bolt and piston group and regulator which have been left in the white. The gas block and receiver on my specimen are slightly plum colored, indicating improper bluing techniques.

A simple, three-position regulator has been fitted to the gas block, the normal position being "1." It produces the least felt recoil, expelling most of the gases into the atmosphere before they act upon the piston head. Soviet SVDs have two-position regulators. The regulator can be adjusted only after the gas tube has been removed. Except for a protruding tab to lock the regulator, the gas tube is that of the Kalashnikov series, but with four gas vents on the bottom.

Since there is no provision for full-auto fire, the Kalashnikov trigger mechanism has been simplified. It consists of a trigger, sear, disconnector, hammer, hammer and trigger pins and multiple strand spring. The safety sear and its coil spring, which permit the hammer to fall only after it has been tripped by the bolt carrier when the bolt locks, have also been retained. While the trigger pull weight on my specimen is only 3.25 pounds, the trigger must be pulled along a seemingly infinite course of travel before let-off. The standard Kalashnikov safety lever has only two positions. With the lever set in the upper position to safe and a magazine in place, the bolt carrier can be retracted only enough to inspect the chamber for a loaded round. Stiff and exceptionally difficult to manipulate, the selector lever makes the usual loud AK noises.

A moderately effective muzzle device has been attached to the M76's four-groove, 22-inch barrel. Flash suppression is no better or worse than any number of other attachments of this type. Muzzle climb is moderated slightly by location of the five elongate vents with three on top and two at the bottom. Nothing to rave about here. Four interrupted threads at the rear of the muzzle device will presumably accommodate rifle grenades.

The rearmost portion of the muzzle device serves as the front sight mount and the entire assembly is pinned to the barrel at this point. The round, post-type front sight is guarded by heavy protective ears. Threaded to a steel pin which can be drifted right or left, the front sight can be adjusted for both elevation and windage zero. The rear sight is a sliding tangent type with the usual open U-notch. It can be adjusted for elevation only from zero to 1,000 meters in 100-meter increments. There is no battle position.

CHRONOGRAPH RESULTS

Instrumentation: Oehler Model 33 Chronotach with Skyscreen detectors positioned 10 feet from muzzle. Ambient temperature: 74 degrees F. All readings in feet per second at elevation of 1,080 feet above sea level. All projectiles military with full metal jackets (FMJ). 7.62x54R ammunition fired through Mosin Nagant M1891/30 rifle with 28.7-inch barrel. 7.92x57mm ammunition fired through Yugoslavian M76 rifle with 22-inch barrel.

7.62x54R	**Low velocity**	**High velocity**	**Extreme spread**	**Average**	**Standard Deviation**
Russian 150-gr. Type "L"	2,939	3,003	64	2,968	25
Russian 150-gr. Type "LPS"	2,818	2,853	35	2,833	12
Egyptian 180-gr. boattail	2,378	2,713	335	2,578	90
7.92x57mm					
Canadian WWII 154-gr. spitzer	2,691	2,816	125	2,759	43
German 1938 156-gr. AP tracer	2,642	2,679	37	2,665	12
Portuguese 198-gr. boattail	2,393	2,443	50	2,421	16

There is a bayonet lug on the bottom of the front sight mount, a strange anachronism. Bayonets serve no useful purpose on such military small arms as sniper rifles and submachine guns. A Soviet-style AKM wire-cutter bayonet is now standard issue in the Yugoslavian armed forces. Both the scabbard and bayonet are of the second model type: plastic- rather than rubber-insulated steel scabbard and steel instead of plastic pommel as found on the first model AKM bayonets. The Yugoslavian version is distinguished by its black scabbard and grip panels.

Of all-steel construction, the staggered box-type magazine holds 10 rounds. The magazine's follower is raised to hold the bolt group back after the last shot has been fired in the manner of the PRC Type 63 20-round magazine. But there is no hold-open stop on the receiver and when the empty magazine is removed, the bolt group will immediately fly forward. This is foolish, and serves only to increase the difficulty encountered in removing the magazine.

The magazine release catch, directly in front of the trigger, is a spring-loaded flapper and magazines must be inserted from the front and rolled back to engage the catch.

Soviet and Rumanian Dragunovs have skeletonized, laminated-wood buttstocks with the front portion as a pistol grip. The Yugoslavs have opted for a more conventional configuration. A blond, tight-grained wood, lightly stained and oiled, has been used for the buttstock, pistol grip and handguards. A 1/2-inch-thick, solid black rubber recoil pad has been attached to the rear swivel under the buttstock and the front swivel fitted to the gas block.

All very interesting, but the M76's most intriguing component is its telescopic sight. The ON M76 scope is patterned after the PSO-1 optical sight found on the Soviet SVD, with one important difference. The PSO-1 scope employs a small battery-operated internal light bulb to illuminate the reticle during operation under subdued light conditions. The Yugoslavs have eliminated both bulb and battery and substituted tritium illumination. At low light levels the reticle pattern glows green. A nuclear-powered betalight is far superior in both longevity and maintenance to the cruder system used on the PSO-1.

Both PSO-1 and ON M76 scopes also have an infrared detector, often referred to as a "Metascope." A small knob on the ON M76's right side can be rotated forward to flip an internal green filter up into the scope's optical path. Infrared sources, such as small arms night sights and searchlights mounted on MBTs, can then be detected as an orange glob in the scope. Future ON M76 scopes will omit this feature since military technology has moved away from IR emitters.

Optically, the ON M76 scope is a conventional, fixed four-power telescope of military grade with an aluminum housing about 91/2 inches in overall length. The actual field of view is 5 degrees 10 feet. A threaded, removable sunshade is attached at the objective end. A long, soft rubber eyecup of the proper length to obtain the required eye relief can be slipped over the ocular end. The eyecup has four pinholes to minimize fogging and prevent suction. In addition to the infrared detector knob there are external controls for elevation (top of scope) and windage (left side). Elevation can be adjusted from zero to 1,200 meters in 100- and 50-meter increments. Calibration matches the trajectory of 7.92x57mm Yugoslavian M49 ball (standard issue), M70 (tracer) and M75 ball (match grade) ammunition.

The range-finding reticle pattern is, in modified form, that of the Russian PSO-1 scope which, in turn, copies the principle used in the PGO-7 and PGO-7V optical sights on the RPG-7 rocket launcher. It's quite simple, requiring a minimum of training, but very effective. To the bottom left of the vertical crosshair is a baseline marked "1.75," the average height of a European in meters. Above the baseline are seven short ascending steps, moving upward from the left. Every other step is marked (8, 6, 4 and 2 corresponding to 800, 600, 400 and 200 meters). Simply place the feet of your target on the baseline and match the top of your target's head with the appropriate step. This is the distance to the target. All that remains (if the rifle and scope have been zeroed) is to dial the equivalent number on the elevation knob.

The mount and rings are black anodized aluminum. Two circular holes in the mount permit use of the iron sights while the scope is in place. The scope can be quickly detached from the single side rail and riveted to the receiver's left side by a locking lever on the mount.

YUGOSLAVIAN M76 DRAGUNOV SNIPER RIFLE

CALIBER 7.92x57mm

OPERATION . . . . .Gas with three-position regulator and conventional Kalashnikov piston and extension attached to bolt carrier. Locking by means of rotary two-lug bolt. Fire from the closed-bolt position. Auto safety sear, but semiautomatic fire only.

FEED MECHANISM . . . . .10-round staggered box type, detachable magazine.

WEIGHT, empty . . . . .11.6 pounds with scope and mount.

LENGTH, overall . . . . .45.4 inches.

BARREL . . . . .Four-groove with right-hand twist of one turn in 9.6 inches.

BARREL LENGTH . . . . .22 inches.

SIGHTS . . . . .Round, post-type front with protective ears; adjustable for windage and elevation zero. Sliding tangent type with open U-notch; adjustable for elevation only in 100 increments from 0 to 1,000 meters.

OPTICS for ON M76 4X scope . . . .side rail mounted with aluminum rings and mount (quick detachable); Tritium illuminated range-finding (200 to 800 meters) reticle pattern; metascope for infrared detection.

ACCESSORIES . . . . .Wirecutter bayonet, rifle grenades, sling, passive optical night sight [PN5X80(j)], carrying case and optical maintenance kit for ON M76 scope.

MANUFACTURER . . . . .Zavodi Crvena Zastava, Kragujevac, Yugoslavia.

STATUS . . . . .In service with Yugoslavian armed forces and several unspecified Third World countries.

Repeated removal and installation during our test and evaluation did not alter the unit's zero. When detached, the scope and mount are carried in a leather and canvas pouch that also contains a camel's hair brush, optical cleaning cloth and a combination tool with screwdrivers for adjusting the scope.

A passive night sight, called the PN5X80(j), can also be installed on the M76. It has a five-power magnification and a 10-degree field of view. Using rechargeable batteries it is reported to be effective against human targets up to 700 meters and vehicles up to 2,000 meters.

Although its case is far too long by today's standards, the German 7.92x57mm cartridge still performs with excellence in machine guns and sniper rifles. Adopted by the German army in 1888, it is one of the most widely distributed military cartridges in history.

Ammunition from three different sources was tested in the M76. At one time surplus WWII Canadian ammunition in this caliber was commonly available, and with good reason—particularly for reloaders—since their brass cases (headstamped "7.92MM44") are boxer primed, but corrosive. The bullet used was a copy of the German S ball, a 154-grain FMJ spitzer. With 47 grains of IMR-type powder, this ammunition left the M76's barrel with an average velocity of 2,759 fps and a standard deviation of 43 fps.

We also fired some ancient German ammunition, headstamped "PS*7 38" (manufactured by Polte, Magdeburg, in 1938). Berdan primed and corrosive, its brass cases are filled with 44 grains of European-type cut-sheet flakes. The black-tipped, 156-grain boattail FMJ bullet and red primer annulus indicate this to be armor piercing tracer. By now it only traces 50 percent of the time and for just a short distance, but like fine wine its accuracy performance only seems to have improved with age as the average velocity was 2,665 fps with an astounding standard deviation of only 12 fps.

Portuguese ammunition of 1971 vintage ("FNM 71-20") is Berdan primed, but noncorrosive. Its 198-grain FMJ boattail bullet duplicates the German so-called s.S ball. Propelled by 44 grains of European-type cut-sheet flakes, this ammunition produced an average velocity of 2,421 fps with a standard deviation of only 12 fps.

The German ammunition produced the best accuracy with 2.6 MOA. The 198-grain Portuguese projectile did not fare well in the M76 Dragunov and shot no better than 5 MOA. I had no access to the special ammunition supposedly issued to ComBloc snipers. Assuming it to be of match grade, I think we could easily get the M76 down to 1.5 MOA. That would permit head shots out to 300 meters and chest shots up to 450 meters in the hands of a moderately trained rifleman. Most sniper kills I have observed were made at less than 500 meters.

Testing for this report required two essential pieces of equipment: a high-quality spotting scope and a sturdy bench. Steiner's 24x80 military field telescope, weighing only 44 ounces, offered exceptional clarity and brightness. Waterproof and dust resistant, it's covered with rugged, frost resistant NATO green rubber armor and has passed a rugged shock vibration test. All the glass surfaces are coated against glare and extreme light conditions. This 24-power telescope has a field of view of 105 feet at 1,000 yards which equals two degrees. The fiber reinforced polycarbonate housing exhibits negligible temperature expansion from minus 40 C to plus 70 C. An adjustable ocular ring permits eyeglass wearers to see the full field of view. An objective lens will accept an optional sunshade. A compact Porro lens system helps hold internal light loss to a minimum. Suggested retail price is $599. Optional accessories are available. For further information contact Pioneer Marketing and Research Inc.

We used a sturdy, portable shooting bench manufactured by Armor Metal Products. Built like the proverbial brick latrine, the bench top is a massive 1 1/2-inch-thick plywood unit with 10 plies oriented in all directions, which gives outstanding resistance to deflection throughout the surface while keeping the weight manageable. Top area dimensions are 29x43 1/2 inches with an 11 1/2-inch inset (right- or left-handed). The assembled bench stands 33 inches high on heavy-wall galvanized pipe legs. The leg mounting brackets are heavy 1/4x5x5-inch steel plates with welded, precisely angled leg sockets which are through-bolted to the bench top. Once the brackets have been installed, field setup is simply a matter of screwing in the legs.

Three- and four-leg models are available. I used the combo model, weighing 65 pounds. That's heavy, but compared to the flimsy portable units, it's an acceptable tradeoff. The three-leg model costs $119.95, the four-leg is $129.95 and the combo model is $139.95, and that's not too heavy at all for quality that locks your crosshairs on the target.

Originally appeared in *Soldier Of Fortune*
August 1986.

Jeff Cooper fires the Steyr Scout Rifle, which was built to his specifications, from the classic standing position. Note optional 5-round sidesaddle-type spare cartridge carrier on the right side of the buttstock.

STEYR SCOUT RIFLE

A Gun for All Seasons

While longer than the caliber 5.56x45 Steyr AUG bullpup, the Steyr Scout Rifle is chambered for the considerably more effective 7.62x51mm NATO (.308 Winchester) cartridge.

Jeff Cooper owns the 'Scout Rifle" concept, lock, stock and barrel. In 1983, he defined the "general purpose rifle" as a conveniently portable, individually operated firearm, capable of striking a single decisive blow, on a live target up to 200 kilos in weight, at any distance at which the operator can shoot with the precision necessary to place a shot in a vital area of the target." To Cooper a "scout" was one man moving from cover to cover, operating alone, and highly trained in all the fieldcraft arts.

The envelope was prescribed by Cooper with a maximum length of one meter (39.37 inches) and a total weight, empty, no greater than three kilograms (approximately 6.6 pounds). The rifle resulting from these parameters was to be convenient, powerful (whatever that means), accurate, rugged, versatile and aesthetically pleasing.

SCOUT HISTORY

An early commercial precursor to the modern scout rifle was the Mannlicher Schoenauer Model 1903 Carbine chambered for the 6.5x54mm cartridge. There have also been numerous short, lightweight military bolt-action rifles chambered for full-size cartridges. The British No. 5 MkI "Jungle Carbine" is an example. There were an almost infinite number of carbines based upon the '98 Mauser action and even earlier Mauser designs. Examples include the Spanish and Argentine M91 Carbine, Belgian M89 Lightened Carbine, Spanish M95 Carbine, Swedish M94/14 Carbine, Argentine Model 1909 Cavalry Carbine, FN

Dutch Police Carbine, Iranian Models 98/29 and 49 Short Rifles, German Model 33/40 and so on. Most of them exhibited an unacceptable flash signature and increased recoil.

Furthermore, the turn-bolt, except in the hands of a dedicated and highly trained sniper, is an anachronism on today's battlefield. Today's military scout would be far better equipped to meet with, and destroy, the enemy in "shoot and scoot" scenarios armed with an M4, Steyr AUG or Kalashnikov than a short-barreled bolt-action. The scout rifle is a superb instrument for the game field, but not combat. I have personally hunted in Africa using a Scout Rifle with great success on impala at ranges out to 300 yards.

Until now all scout rifles have been custom-made from a variety of turn-bolt short actions, such as the Remington Model 600, Czech Brno ZKK and the superb Sako L579 medium-length action. Most have been chambered for the .308 Winchester cartridge. Some "pseudo" scouts have been chambered for the .30-06 round using standard-length actions like the pre-64 Winchester Model 70. Other cartridges presumably compatible with the scout rifle concept include the 7mm-08, .350 Remington Magnum, .35 Whelen and 6.5mm Remington Magnum.

Over seven years in development and designed using specifications provided by none other than Jeff Cooper himself, Steyr-Mannlicher AG has just introduced a production series scout rifle with several dramatic innovations. Distributed by GSI, Inc., the new Steyr Scout Rifle falls clearly within the envelope configuration originally mandated by Cooper.

With two buttstock spacers, the overall length of the Steyr Scout Rifle is 39.57 inches. The weight, empty, with two five-round magazines and the scope with its mounts is 6.93 pounds. Initially chambered for the .308 Winchester (7.62x51mm NATO) cartridge, other

Steyr Scout Rifle Specifications

CALIBER . . . . .7.62x51mm NATO (.308 Winchester)

OPERATION . . . . .Bolt-action. Four front locking lugs. Unique safety ring in the barrel's lock bushing shields extractor from excessive pressures. 70-degree bolt lift with an additional 20 degrees for locking of the firing pin. Roller-type tang safety with three positions: Fire, Loading and Safe.

FEED MECHANISM . . . . .Detachable, double-detent, staggered-column, 5- or 10-round box-type magazines, injection-molded from a tough, black synthetic called "Grivory."

WEIGHT .Empty, with two five-round magazines and the scope with its mounts: 6.93 pounds.

BARREL . . . . .Four-groove with a right-hand twist of one turn in 12 inches. Hammer-forged and fluted to cut weight and add rigidity.

BARREL LENGTH . . . . .19 inches.

OPTICAL SIGHTS . . . . .Leupold 2.5x28mm Scout Scope with heavy duplex reticle pattern.

EMERGENCY SIGHTS . . . . .Spring-loaded flip-up, ghost-ring-type polymer rear sight adjustable for elevation zero only. Spring-loaded, front-sight post, with a vertical white bar, installed in the polymer receiver front cap, can be adjusted for windage zero.

FURNITURE . . . . .Gray synthetic Zytel (an ABS glass reinforced polymer) stock, with a non-skid texture and an integral lightweight folding bipod. Rail on the underside of the forearm to attach accessories. Equipped with two spacers which provide a length-of-pull of 13.58 inches.

SUGGESTED RETAIL PRICE . . . .$2,595, complete with Leupold Scout Scope, two five-round magazines, leather three-point scout sling and a carrying case.

MANUFACTURER . . . . . .Steyr-Mannlicher AG, Postfach 1000, Steyr, Austria.

IMPORTER . . . . .GSI, Inc., P.O. Box 129, Trussville, AL 35173-0129.

SCOPE MANUFACTURER . . . . .Leupold & Stevens, Inc., P.O. Box 688, Beaverton, OR 97075.

T&E SUMMARY: First production series example of Jeff Cooper's scout rifle concept. Superb medium-game rifle. Excellent accuracy. Long-eye-relief scope provides both fast target acquisition and visual command of the tactical frontal area. Lightweight and compact. Highly recommended.

rounds such as the 7mm-08 Remington and .243 Winchester are possible sometime in the future. Barrel length is only 19 inches with four grooves of rifling and a right-hand twist of one turn in 12 inches. Although thin-walled, the hammer-forged barrel has been fluted to cut weight and add rigidity.

The barrel's method of attachment to the receiver is unique. First, there is an integral cone machined into the chamber-end of the barrel. In front of this is an aluminum ring (longitudinally slotted for tightening) with an opposing interior cone on each end. Another counteracting ring with a cone goes in front of the aluminum ring. This is followed by a locking nut. Using a special wrench, the locking nut must be torqued to Steyr specifications by a trained armorer only. A rectangular steel locking wedge on an aluminum block attached to the underside of the receiver further secures the barrel. The barrel is completely free-floating and the barreled-action is held to, and bedded in the stock by means of two aluminum pillars in the stock.

A steel lock bushing (or barrel extension) at the chamber end of the barrel permits the use of an aluminum receiver, an important factor in keeping the complete package under 7 pounds. Made from a 6061 T6 aluminum alloy extrusion, the receiver has been black hard-anodized. With the bolt locked in battery into the lock bushing at the end of the barrel, the receiver, in essence, serves only as a framework for the barrel and bolt in a manner reminiscent of the World War II German MG42 General Purpose Machine Gun.

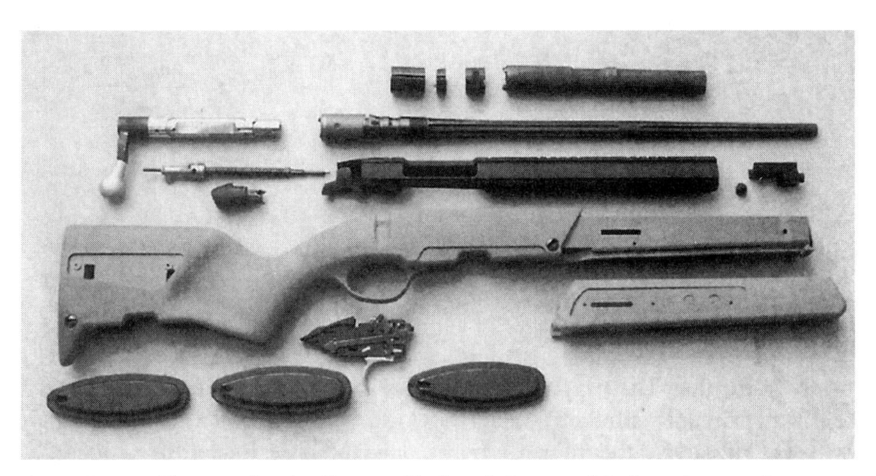

Steyr Scout rifle completely disassembled and shown with three buttstock spacers and the armourer's wrench used to remove the barrel from the receiver.

SCOPES, SIGHTS, AND THE SBS

To accommodate mounting of the long-eye-relief Leupold Scout Scope and the emergency "iron" sights, the receiver extrusion is unusually long: 15.67 inches (398 mm). Installed in a recess at the rear end of the receiver is a spring-loaded flip-up, ghost-ring-type polymer rear sight which is adjustable for elevation zero only. The spring-loaded, front-sight post, with a vertical white bar, has been installed in a polymer receiver front cap and can be adjusted for windage zero. Just slide a serrated polymer bar forward slightly and the front sight post will pop up. The integral scope rail on top of the receiver has twelve mounting points and is configured to MIL-STD-1913. In other words, it is a so-called "Picatinny Rail."

The new Steyr Scout Rifle uses the SBS (Safe Bolt System) action introduced in 1996. With four front locking lugs (the Steyr SSG featured rear locking lugs), an important feature of the SBS is a safety ring that is an integral part of the lock bushing. This safety ring rotates with the bolt and shields the extractor, so that escaping gases can only go down the barrel, instead of pushing outward against the extractor. There are also two standard gas ports on the bolt body to accommodate the safe escape of gas from a pierced primer. A cut-out on the bolt body at the ejection port (when the bolt is closed) and six grooves around the body were designed to reduce weight.

However, a groove around the ejection port area of the body is present for anti-debris and anti-icing purposes. The bolt body has been nickel-plated because this finish is tough and corrosion-resistant and it also reduces the coefficient of friction. The two rear locking lugs are smaller than the front lugs. To enhance the bolt lift motion, there are dual opposed cocking cams within the bolt body. The classic Mannlicher-Schoenauer butterknife bolt handle's lift is 70 degrees. An additional 20 degrees is utilized for the downward locking motion described below. An indicator pin protrudes from the rear of the bolt shroud when the firing pin is cocked.

The SBS bolt assembly is much easier to disassemble than that of the Steyr SSG. Just depress the disassembly button on the left side of the

Leupold Scout Scopes

During World War II Germany fielded some K98k rifles with the long-eye-relief ZF41 1.5X scope, which mounted on a side rail machined into the rear sight base of specially prepared rifles. It was not popular with the troops and many were simply discarded on the battlefield.

Nonetheless, an important and unusual attribute of the Scout Rifle is a forward-mounted scope with an eye relief of about 10 inches. Cooper has argued the case for a low-power, long-eye-relief scope with great conviction. He was convinced that it permitted the shooter to see both the entire area in front of him, as well as the cross hairs printed on the target as long as the scope's magnification remained under three power to prevent a great disparity between the view perceived by each of the two eyes. And, although the technique is rarely used in the field, a long-eye relief scope is particularly suited for rapid snap shooting.

He also argued that it permitted low mounting. This is important since a peep-aperture "ghost ring" rear emergency sight is another key ingredient in the scout rifle package. Keeping the scope's line of sight as close as possible to that of the iron sights is important because it permits the buttstock's comb to provide a proper cheek-weld with either sight.

Until recently, the only low-magnification, long-eye-relief scope available for this purpose was the 2.75X fixed power Burris Scout Scope The concept was correct but the execution was seriously flawed. In my circle—where the scout rifle prevails—the Burris scope was noted principally for its poor quality and tendency to fail at the most inappropriate time. Leupold & Stevens, Inc. has introduced a 2.5X28mm (actual magnification: 2.3X) Scout Scope that quite frankly blows Burris out of the water.

Leupold's fixed-power scopes have been the purest form of their incomparable craftsmanship since they began manufacturing scopes just after World War II. Leupold's Scout Scope provides 9.3 inches of eye relief. The Steyr Scout Rifle scope is equipped with a heavy duplex reticle pattern as per Jeff Cooper's specifications.

Using a 1-inch tube, the overall length of the Leupold Scout scope is 10.1 inches. The weight is only 7.5 ounces. Both the ocular and objective diameters are 1.36 inches. At 100 yards, the elevation and windage adjustment range is 110 inches. One click of adjustment equals 1 minute of angle. Both the windage and elevation knob caps are marked "Steyr Scout Rifle." The field of view at 100 yards is 22 feet.

Excellent image quality is a characteristic of all Leupold scopes. This is possible only because Leupold uses material and techniques that minimize chromatic aberration, distortion and curvature of the field to an extent not achieved by any other scope manufacturer in the United States. More top benchrest shooters, professional hunting guides, custom rifle makers and elite military and law enforcement special operation units use Leupold scopes than any other make.

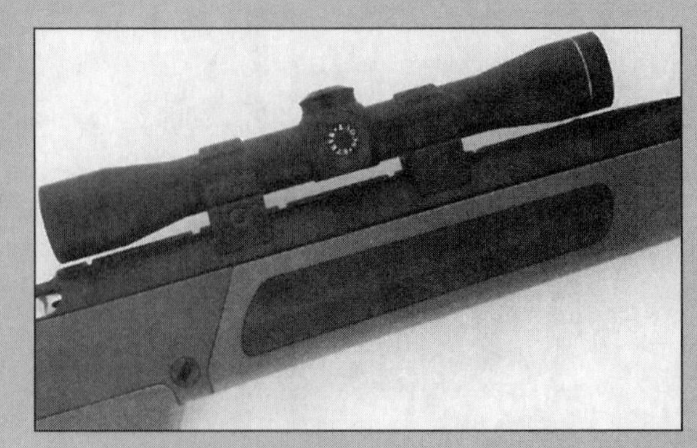

Leupold's 2.5X28mm Scout Scope provides 9.3 inches of eye relief and unparalleled optical quality.

shroud and then rotate the shroud approximately 1/4-inch clockwise. The shroud, firing pin assembly and cocking cam ring can then be withdrawn from the rear of the bolt body. Reassemble in the reverse order.

The ambidextrous roller-type tang safety on the trigger mechanism has three positions: Fire, Loading and Safe. When the roller is rotated all the way forward a red dot becomes visible and the rifle is ready to fire. The adjustable (by trained armorers only) trigger has been set at the factory to provide a trigger pull weight of 3.5 to 4.0 pounds. All SBS rifles use the same triggers. As a consequence, the nickel-plated trigger on the Steyr Scout Rifle is smooth on the front edge and serrated at the rear so it can be more easily pressed forward when it is installed in a single-set mechanism available only in Europe. The trigger mechanism itself is, in principle, similar to that of the Steyr SSG. However, the method by which the bolt is removed from the receiver required changes to the mechanism.

When the roller-type tang safety is rotated rearward to the middle position, a white dot is exposed and the trigger is blocked. However, the bolt can be manipulated and thus the rifle loaded or unloaded with the trigger and sear blocked. In conjunction with this, it should also be noted that if the magazine is lowered approximately a 1/4-inch to a drop-lock position, the bolt can be cycled and a single round inserted and chambered by hand.

When the safety is rotated all the way rearward, a white dot becomes visible and spring-loaded gray safety catch pops upward. In this position, the trigger remains blocked and the bolt cannot be rotated. Furthermore, the bolt handle can now be rotated downward another 20 degrees to remain locked in that position and block the firing pin. The gray safety catch must be depressed downward to rotate the tang safety forward.

To remove the bolt from the receiver, unlock it while it is in the loading position with the trigger blocked. Then rotate the safety all the way rearward and the bolt can be withdrawn. During re-installation, the trigger and firing pin remain blocked until the roller-type tang safety is rotated forward.

FUNCTION DICTATES FORM

The gray synthetic Zytel (an ABS glass reinforced polymer) stock, with a non-skid texture, has an integral lightweight folding bipod and a rail on the underside of the forearm to attach accessories. The rifle comes equipped with two spacers which provide a length-of-pull of 13.58 inches. Removal, or installation of additional spacers, offers a length-of-pull ranging from 12.68 to 16 inches. Extra 0.45-inch spacers are available as an optional accessory. The heel of the buttstock is rounded to prevent snagging. The stock has been molded with filler material

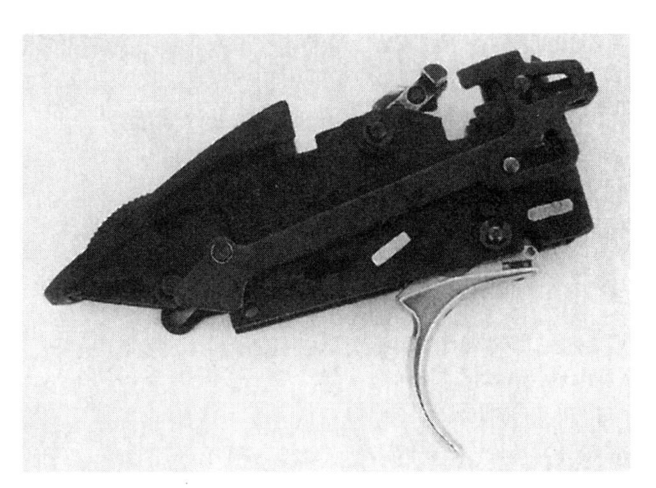

Steyr Scout Rifle trigger assembly. Note the ambidextrous roller-type tang safety.

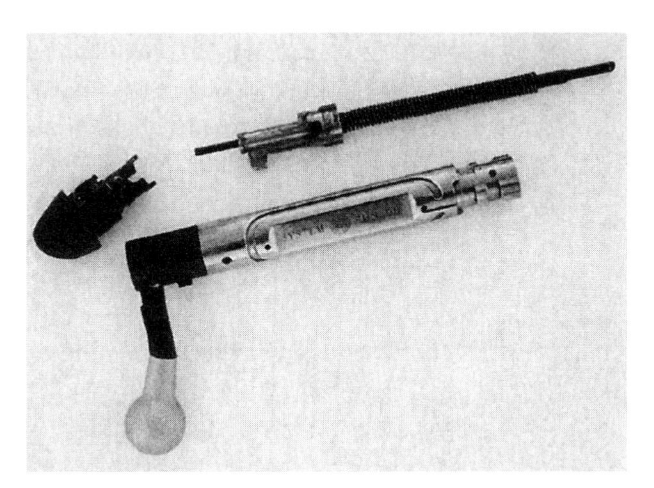

Steyr Scout Rifle bolt group is much easier to disassemble than that of the Steyr SSG sniper rifle.

in back of the trigger guard to prevent slamming against the finger during recoil.

The underside of the stock's butt end is noticeably undercut and in this area is a compartment for storage of the spare magazine. In front of this is a storage area for a cleaning kit. Both 10-round magazines and an adaptor kit with a shroud to protect them are available options. The double-detent, detachable, staggered-column, box-type magazines, whether of 5- or 10-round capacity, are injection-molded from a tough, black synthetic called "Grivory." The magazine catch-release system is similar to that found on the Steyr SSG. There are black removable panels on each side of the butt end of the stock and on the bipod legs. They can be replaced with inserts of other material for cosmetic reasons, or the one on the right side of the buttstock can be substituted for a sidesaddle-type spare cartridge carrier.

To deploy the integral bipod, depress the polymer lever to the rear of the accessory rail on the underside of the stock's forearm area and then pivot the legs to their locked and completely extended position. The command height (the distance from the ground to the center line of the barrel's axis) is approximately 11 inches. With the bipod deployed, the rifle can be rotated 15 degrees in either direction. Bipods on hunting rifles are useful to a) check zero in the field, b) shoot game from an ambush position, and c) a convenient rest for the rifle during break periods.

A so-called "CW" or three-point sling is another important ingredient in Jeff Cooper's Scout Rifle concept and three-point, flush-mounted sling sockets on the stock are provided for this purpose. The CW sling is named after Cooper's Guatemalan friend, Carlos Widmann. Widmann showed Cooper an old military rifle with a sling loop attached to the front of the trigger guard. This permits a sling to be secured forward of the base of support without resorting to the more complex U.S. military-type target sling. The rifle was most probably a Lee Metford or SMLE as the British have employed this sling position since 1895. Eventually the concept was improved by Cooper and his associates by adding an additional sling segment between the looped firing position and. the rear sling swivel. In this latter configuration, a leather sling manufactured by Turner Saddlery is supplied with the Steyr Scout Rifle.

Jeff Cooper has quite justifiably criticized "the benchrest mentality" with its obsession for accuracy, stating emphatically that it has set practical field shooting backwards. He has gone so far to state that group size can be a fallacious measurement. And yet, the Steyr engineers have provided us with a Scout Rifle that clearly has the potential for sub-MOA accuracy with match-grade ammunition. My personal results with the Steyr Scout Rifle in a benchrest environment produced 3/4-MOA groups.

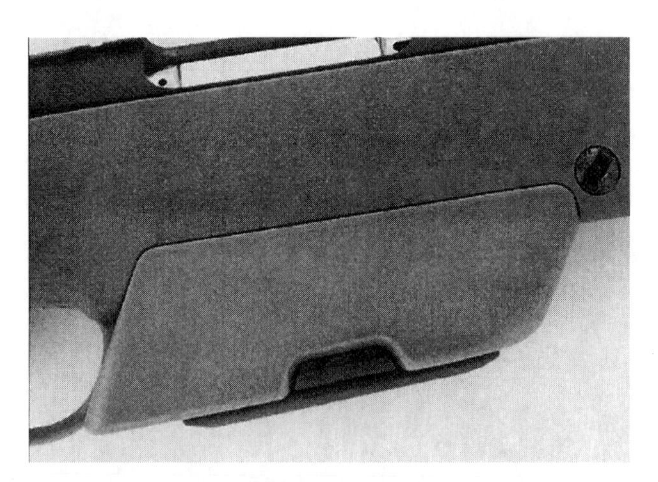

Both 10-round magazines and an adaptor kit with a shroud to protect them are available options.

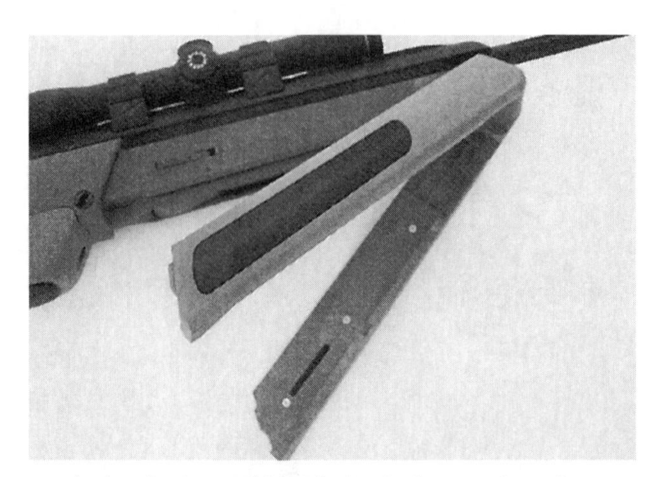

To deploy the integral bipod, simply depress the polymer locking lever and then pivot the legs to their locked and completely extended position.

Many believe that barrel lengths under 20 inches will significantly reduce the muzzle velocity. Although the propellant used is an important factor in this equation, short barrels do not necessarily mean lower velocities. Chronograph results obtained during a recent Steyr Scout Rifle writers' seminar held at the NRA Whittington Center in Raton, New Mexico were as follows: Federal 168-grain BTHP 308M, 2,600 fps; Hornady 165-grain Lite Mag, 2,785 fps; Hornady 150-grain Lite Mag, 2,834 fps; and American Eagle 150-grain FMJ, 2,855 fps. While fired at an elevation of approximately 6,500 feet asl, nevertheless, these muzzle velocities are very close, and in some instances slightly higher than results expected of longer barrels, albeit in denser air at lower elevations.

Others are even more convinced that thin-walled barrels will invariably overheat during long firing sequences with a consequent change of impact downrange, or at least noticeable "vertical stringing" of shot groups. Our tests at Whittington Center indicate that this is not the case.

Jeff Cooper has stated that an essential element of the scout rifle is *handiness*. And, Steyr's new Scout Rifle is certainly that and much more. Cooper believes that the scout rifle is not just a short range rifle, but can also be deployed in three different hunting scenarios: very quick target acquisition, i.e. snap shooting; reasonably quick target acquisition in the normal game hunting environment; and slow, or firing from ambush as at water holes or tree stands. Made to his standards, the Steyr Scout Rifle meets all of Jeff Cooper's specifications. I personally have only one criticism of the pre-production series samples I have fired. The action is a bit too stiff for my tastes.

The Steyr Scout Rifle carries a suggested retail price of $2,595, complete with Leupold Scout Scope, two five-round magazines, leather three-point scout sling and a carrying case. Production series specimens should be available by March, 1998.

Originally appeared in *Soldier Of Fortune*
February 1998.

GALIL'S NEW SNIPER RIFLE

Israelis Take After Sovs and Design Mass-Produced Sniper System

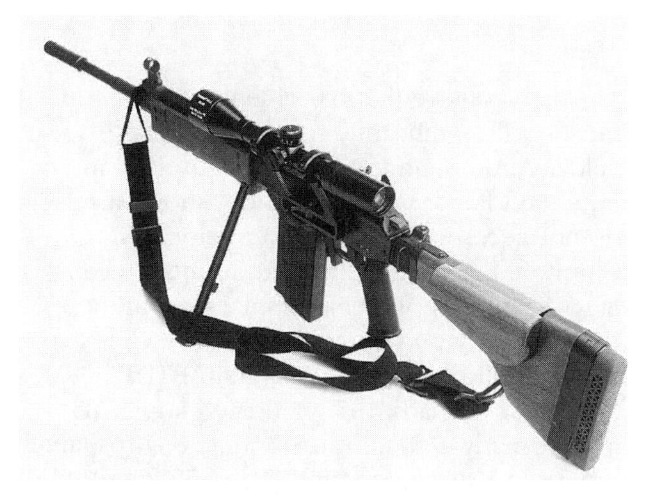

Nimrod 6X40mm scope attached to Galil Sniper Rifle.

Galil Sniper Rifle field stripped and shown with Ciener sound suppressor, scope and all accessories.

Throughout history snipers and their equipment have held the military's interest only during periods of armed conflict. Between wars they languish in a limbo buried under higher priorities. But Israel's state of siege never ends, so it stands to reason that sniping would occupy no small role in Israel Defense Forces (IDF) tactics. Yet until recently their snipers have made do with little more than exhausted M14s—junkyard remnants of American military hopes left rotting in the jungles of Southeast Asia.

Working in close support with the army, an Israel Military Industries (IMI) weapons division team commenced development of an indigenous sniping system in 1980. Taking note of Russia's success at mass producing the Dragunov, large-scale series production without serious compromise of the accuracy potential was a key objective. The resulting rifle, while not totally satisfactory in my opinion, is, nonetheless, a sincere effort to comply with the user's requirements.

Starting with the Galil rifle, itself an extensive modification of the Kalashnikov system (early prototypes were in fact assembled with Finnish Valmet M62 receivers), IMI's response to the infantry's need to strike at multiple targets both quickly and accurately checks in at 18.3 pounds, complete with scope, bipod and loaded magazine. While Israeli soldiers are well-known for their ability to hump with awesome loads, this is still far too heavy. Overall length with the stock unfolded and the muzzle brake installed is approximately 43 inches.

Built around the Galil's heavy forged receiver, chambering is for the 7.62x51mm NATO cartridge. Since snipers will employ match-grade ammunition, adding in essence a different cartridge to the pipeline, the belted .300 Winchester Magnum would have been a superior choice, as it provides less variation in point of impact at long ranges.

Firing from the closed-bolt position, the Galil is gas-operated without an adjustable regulator. When the trigger is pulled, the hammer drives the firing pin forward to strike the primer. The bolt has been fitted with a strong firing pin spring to prevent premature ignition of more sensitive commercial primers—an especially important feature as the auto safety sear has been eliminated from this semiautomatic rifle.

After ignition of the primer, a portion of the propellent gases passes through the barrel vent into the gas block pinned to the barrel. Gas enters the cylinder (to which a small spring has been attached to secure its retention during reassembly) and drives the piston rearward. The piston is hard-chrome plated for ease of maintenance. A notched ring in back of the piston head provides a reduced bearing surface and permits excess gas blow-by, which is vented into the atmosphere out six ports in the gas cylinder. The bolt carrier is permanently attached to the piston. After a short amount of free travel, during which time the gas pressure drops to a safe level, the carrier's cam slot engages the bolt's cam pin and the bolt is rotated and unlocked as the carrier moves rearward.

Primary extraction occurs as the bolt is rotated. Empty-case ejection is typically violent. Cases are severely dented by the ejector (milled into the left receiver rail) and thrown to the right and front as much as 40 feet (an undesirable characteristic with regard to position disclosure). At this time, the recoil spring is compressed and its return energy drives the carrier forward to strip another round from the magazine and chamber it.

The Galil's hammer spring is made of multistrand cable. Both the trigger and sear springs are fabricated from conventional single-strand wire. The two-stage trigger on *SOF*'s test specimen breaks cleanly at 3.25 pounds. But the right-side selector lever is the same stamped sheet-metal bar common to all Kalashnikovs and every bit as noisy. Something should be done about this.

The top position, marked "S," is safe and blocks upward rotation of the trigger bar. In this position, the bolt can be retracted only far enough to inspect for a chambered round. There is also a thumb-operated selector switch on the left side. By means of a two-piece hinged bar inside the receiver, the rearmost position of this selector is safe, and pushing forward with the thumb will place the weapon in the firing mode, marked "F."

Taken from the Hungarian AKM/AMD-65 series, the Galil's gray-plastic pistol grip exhibits excellent human engineering. Of more than adequate length, with a sharp bottom flare to prevent the hand from slipping, the grip has been attached to the receiver at precisely the correct grip-to-frame angle.

Protected by the front of the trigger guard, the spring-loaded, flapper-type magazine latch must be inserted from the front and rolled back to engage the latch. Two tough, all-steel, ribbed 25-round magazines are issued with each rifle. Both the magazines and the receiver are finished with black baked enamel over phosphate. All other steel components (except the piston) are phosphate finished.

Carrying handle and bayonet stud on the ARM have been deleted. The retracting handle remains attached to the bolt carrier and bent upright to permit cocking with either hand—a useful feature.

No small portion of this rifle's horrendous weight is consumed by the 20-inch, heavy, stepped barrel. Its four grooves twist to the right with a turn of 1:12 inches. A faster 1:10 inch twist would have offered greater bullet stabilization at ranges approaching 1,000 meters.

A large 4-inch muzzle brake has been threaded to the barrel. It has three rows of exhaust ports of five holes each positioned to the rear of four transverse compensator cuts arranged two abreast. It can be rotated offset to the right or left to accommodate right- or left-handed shooters (unfortunately, to no avail). It's quite effective but retained by an allen-head set screw which will surely loosen and disappear in the field.

Two-piece wooden handguards, without the ARM's longitudinal grooves, are attached to the barrel and receiver by screws through a hole at the rear of the gas block and into the bottom, front portion of the receiver. The bipod, attached to the gas block on the ARM, has been moved 9.5 inches to the rear and mounted to the end of the forearm assembly to avoid interference with the barrel's vibration pattern. While this location enhances the operator's ability to quickly engage targets on the flanks without lifting the rifle off the ground, it has an unfavorable effect on accuracy. Stored under the handguards, this sturdy, supposedly adjustable bipod, unlike that of the ARM, cannot be used to cut wire or open beer bottles. That's of small consequence, but one of the legs on our test specimen refused to retract. Command height can be varied from approximately eight up to 10 inches, with two intermediate positions.

The buttstock can be folded to the right for transport, reducing the rifle's length of 33.6 inches. Although this feature would usually be undesirable when accuracy potential must be maximized, the Galil's rugged stock latch is every bit as rigid as a fixed stock. The clumsy-looking wooden buttstock has an adjustable spring-loaded cheekpiece. Locked by a slotted screw on the right side which slowly backs off during firing sequences, the operator soon finds his eye well above the scope's ocular. The rubber recoil pad can also be adjusted for height.

The rear end of the Galil's recoil-spring guide rod, which serves as a retainer for the sheet-metal receiver cover, is extended to ease disassembly and lock the cover more securely to the receiver body. This is especially important as the rear sight has been mounted on the receiver cover. While this does not provide the rigidity offered by the receiver-mounted rear sight of the Kalashnikov series, the trade-off is a longer sight radius of 19 inches.

The rear sight is a flip-up peep style with 300- and 500-meter apertures. The hooded front-sight post is adjustable for windage and elevation zero. Elevation adjustments are by means of the UZI front-sight tool. Windage adjustments are made by loosening and tightening the two opposing screws which move the entire front-sight assembly in its dovetail on the gas block. The front sight hood forms an additional aiming circle just within the rear-sight aperture to further assist sight alignment and speed target acquisition.

The Galil's tritium (betalight) night sights set for 100 meters have also been retained. To use, at dusk or night, the front betalight is flipped up to expose a luminous vertical bar which is aligned between the two rear luminous dots. When the rear tritium sight is flipped up for use, the rear peep sights must be placed in an offset position, midway between the two apertures. But all of these are, at best, for emergency use only, since the heart of any sniper system is its optical unit.

Knowing full well that mounting a scope on a Galil (or Kalashnikov or FN FAL) sheet-metal receiver will result in unacceptable vertical dispersion, IMI has wisely welded a dovetail base to the receiver's left wall. Interface with the optical sight is by means of a sturdy, all-steel, quick-release mount. While its heavy construction and latchwork seem to ensure maintenance of zero through repeated removal, it offsets the scope to the left, which prevents left-handed shooting. One advantage of this setup is that the iron sights remain unobstructed.

The milspec Nimrod scope mounted on the Galil sniper rifle is manufactured in Japan by a subsidiary of KOOR Industries, an Israeli firm. It has a fixed magnification of six power, an ideal compromise and more reliable than any variable-power scope. Objective and ocular diameters are 40mm and 32mm, respectively. The field of view is 19.6 meters at 300 meters (3 degrees 45 feet). The eye relief is about 3 inches. Heavy bars are superimposed over the crosshairs on the right, left and bottom. There are two auxiliary crosses, one for aiming at 900 and 1,000 meters, the other for high-trajectory ammunition.

Range estimation with the reticle pattern duplicates that of the Dragunov/RPG-7 optical sights. At the bottom of the field of view is a baseline below five short steps. The step closest to the baseline is marked "10" for 1,000 meters, while the farthest is marked "2" for 200 meters. The three steps in between correspond to 800, 600 and 400 meters in ascending order. Just align

Galil Sniper Rifle Specifications

CALIBER . . . . 7.62x51mm NATO

OPERATION . . . . .Gas with no regulator and conventional Kalashnikov piston and extension attached to bolt carrier. Locking by means of rotary two-lug bolt. Fire from the closed-bolt position. Semiautomatic fire only. No auto safety sear.

FEED MECHANISM . . . . 25-round staggered box-type detachable magazine.

WEIGHT . . . . 18.3 pounds, complete with scope, bipod, sling and loaded magazine.

LENGTH
stock unfolded . . . . 43 inches.
stock folded . . . . 33.6 inches.

BARREL . . . . .Four-groove with a right-hand twist of one turn in 12 inches.

SIGHTS . . . . .Round, post-type front with protective hood; adjustable for windage and elevation zero. Flip-type rear with protective ears; 300- and 500-meter peep apertures. Also has flip-up tritium night sights: vertical bar front and double-dot rear.

OPTICS . . . . .Nimrod 6X40mm scope, side-rail mounted with steel rings and mount (quick detachable). Ocular diameter of 32mm with eye relief of approximately 3 inches. Double cross-hair pattern with reticle rangefinder of Dragunov/RPG type. Range drum calibrated from 200 to 800 meters in 50-meter increments. Windage drum with 5 mils of adjustment right or left in 1/2-mil increments.

ACCESSORIES . . . . .Complete with foam-lined, fitted drop case; four-piece cleaning rod; cleaning kit; sling; two magazines and Nimrod scope with rubber eye cup, protective caps, amber and neutral density ocular filters.

MANUFACTURER . . . . .Israel Military Industries, P.O. Box 1044, Ramat Ha Sharon, Israel.

U.S. DISTRIBUTOR . . . . .Action Arms, Ltd., P.O. Box 9573, Philadelphia, PA 19124.

PRICE . . . . .Suggested retail: $3,995, complete. Limited production and availability.

STATUS . .Recently adopted by Israel Defense Forces.

the target's groin with the baseline and match the top of his head with the appropriate step. Dial the correct distance into the range drum on top of the scope (calibrated in 50-meter clicks from 200 to 800 meters for 7.62X51mm M118 match ammunition) and fire away. The windage drum, located on the scope's left side, provides five mils of adjustment to the right or left in 1/2-mil increments. This method is simple, quick, reliable, adequately accurate and requires a minimum of instruction.

A constantly centered reticle pattern has been achieved by inversion of the lenses. Each scope is equipped with protective caps, a rubber eyecup and two ocular filters: amber for overcast light and neutral density for extreme brightness. Night vision equipment can also be incorporated. The scope tube is black anodized aluminum. This is an excellent military optical system which meets the user's requirements at all levels.

Each rifle is also equipped with a fitted, foam-filled drop case with nylon carrying handles and sling, four-piece cleaning rod with brass tip and the standard IDF cleaning kit consisting of a tan, plastic container with plastic oil bottle, cotton-rope pull-through and nylon bristle brush. A wide, black nylon sling of sufficient length for carry at waist height, in the IDF fashion, is attached to the buttstock sling stud with a sturdy steel spring-hook that rotates 360 degrees and to a hole in the front end of the gas block by black nylon cord.

Our test specimen was equipped with a sound suppressor manufactured by Jonathan Arthur Ciener. User maintainable, this suppressor is 17 inches in overall length, with a 1.5-inch diameter phosphate-finished outer tube and a weight of approximately 2.5 pounds. Thus, complete with the suppressor and a loaded magazine, the Galil sniper rifle checks in at 20.8 pounds.

Maximum performance with this rifle is supposedly achieved with Lake City's M118 match ammunition (so-called " Special Ball") and its 173-grain Full Metal Jacket (FMJ) projectile. This is an unfortunate choice, as Lake City Arsenal has failed to produce consistent M118 match-grade ammo in more than a decade. Instead we selected Federal's match ammunition (M308) with its fine 168-gr. Jacketed Hollow Point (JHP) bullet for our test. It departed from the Galil's heavy barrel with an average velocity of 2,622 fps. Ciener's suppressor dropped this by only 68 fps with no change in the point of impact or loss in accuracy.

Our best group at 100 yards was 1.6 inches. However, most groups hovered around 2 inches under minimal wind conditions. Although no semiautomatic rifle can ever reach the accuracy potential of a turn-bolt, this is mediocre performance at best, as M14 rifles, at half the weight, can be tuned to achieve 1 MOA groups (although this will not last for more than 800 to 1,000 rounds). Since the Galil's forged receiver is quite rigid, the optics of high quality, the scope-to-rifle interface apparently secure and the buttstock adequately stable, the problem almost certainly lies with the barrel.

It seems to be of no better quality than IMI's .30-caliber Browning machine gun barrels. Substitute a Douglas Number 1 Contour Premium barrel and this rifle would turn 1 MOA all day long. Douglas uses 4140 chrome moly steel barrel blanks which are carefully heat-treated before they are bored, reamed and button-rifled. Equally important, in caliber 7.62x51mm NATO the bores are cut to .309-inch, as subsequent stress relieving not only collapses all radial stress but will spring-back the bore to the correct groove dimension. No doubt about it, the very best match-grade barrels hail from either the U.S., Austria or West Germany.

Over the course of our test, more than 300 rounds were fired and we experienced two stove-pipes and several failures to feed when the bolt was retracted by hand. There were no other stoppages of any kind. As expected, felt recoil was almost imperceptible, but firing from any position other than the prone is not practical with this beast.

Although it's as quiet as any modern .30-caliber rifle sound suppressor (since the .30-caliber projectile leaves the muzzle at a velocity above the speed of sound, the downrange "crack" is not eliminated), Ciener's unit, when fitted to a Galil, is badly in need of a gas-relief valve similar to that fitted to the original Sionics suppressor fielded during the Vietnam War. When you contain propellent gases in a suppressor tube, they will eventually migrate either forward or rearward. Most will be exhausted into the atmosphere from the muzzle end of the suppressor. Back pressure will, however, drive some rearward. In a Kalashnikov-type rifle they will invariably exit out the square cut at the end of the sheet-metal receiver cover, directly into the shooter's face. This would prove more than just irritating to snipers in a combat environment.

Military snipers require both high first-round hit probability and the ability to fire rapid succeeding shots. Bolt-action rifles cannot meet this latter criterion. The Galil sniper rifle is based upon a reliable, battle-proven system. It shows great promise, but we need to shave off at least five pounds, screw in a more accurate barrel and effect some other minor modifications. This can be accomplished without sacrificing the potential for mass production.

It has just recently been accepted for service by the IDF. In very limited quantity, the Galil sniper system is being imported by Action Arms, Ltd. with a suggested retail price of $3,995, complete.

Originally appeared in *Soldier Of Fortune*
April 1987.

THE DEFINITIVE SNIPING RIFLE

Accuracy International's Incredible AW/AWP

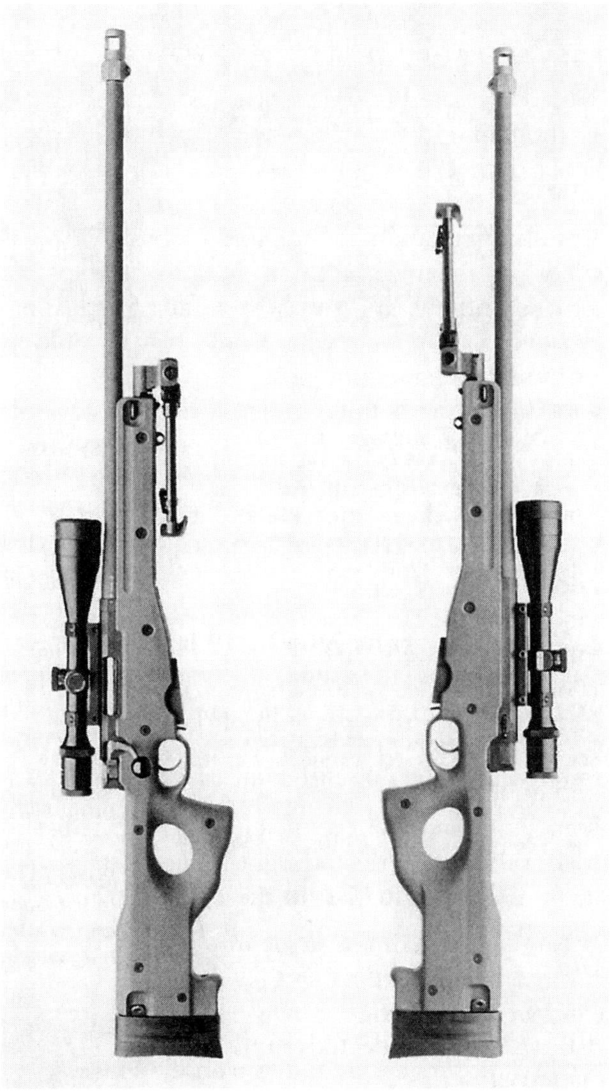

At left, the right-side view of AW sniper rifle. Bolt lift upon opening is only 60 degrees. Total bolt throw is 4.2 inches, which allows the operator to maintain his cheek-weld on the stock while manipulating the bolt and thus observe the target during reloading. At right, the left-side view of the AW sniper rifle with Hensoldt "Sight 90" fixed power 10X42mm scope. *(Photo: Accuracy International, Inc.)*

Soldiers, usually expert shots operating from concealment, who pick off individual enemy targets, have, since the late 18th century, been called snipers in the British army. The word "snipe" is derived from the Middle English "snype," probably of Scandinavian origin.

During the Civil War, both Union and Confederate troops employed large-caliber percussion rifles, weighing up to 40 pounds and often equipped with scopes of high magnification and limited field of view, to record kills at distances considered incredible even by today's standards. They were called "sharpshooters."

Over the last decade the term "sniping" has taken on a sinister tone with definite terrorist, or at least criminal implications. As a consequence, it is now fashionable to refer to law enforcement personnel equipped with scoped rifles as "countersnipers" or by the even more politically correct term, "selected marksmen." This tautological silliness aside, rifles with optical sights permit trained police marksmen to obtain the precise target discrimination so often required in hostage situations. Countersnipers are commonly an integral component in the composition of law enforcement special reaction teams.

Despite their successful deployment by both sides during World War II, with the advent of the Cold War and the tactical concept of "fire and movement" and the fast-moving armored warfare that was anticipated in central Europe, interest in military sniping declined sharply. However, by the mid-1970s the experience of Vietnam and other low-intensity conflicts encouraged renewed interest in sniping.

A police selected marksman is not the equivalent of a military sniper. A sniper on the battlefield has three equally significant roles. They are: 1) to kill selected enemy, such as commanders, snipers, weapons crews, helicopter crews and special operations personnel at ranges from 300 to 600 meters; 2) to provide harassment fire up to about 1,000 meters intended to damage equipment and inhibit enemy troop movement; and, 3) to observe and report information about enemy troop and vehicle movements and to sometimes act as forward observers to direct mortar and artillery fire and tactical air support.

The modern military sniper will often be equipped with night vision equipment, a laser rangefinder, IR laser aiming module, communications gear, a thermal imaging device and a navigation system in addition to his rifle and optical sight. These latter are most important and should be of the highest possible quality and accuracy potential.

COOPER'S MISSION

In 1982 Malcolm Cooper was requested by the British special force community to design a new sniper rifle. Cooper's credentials in the area of competitive rifle shooting, in which he participated from 1962 to 1991, are impeccable. A member of the British national team from 1970 to 1991, Cooper has 156 international medals, which include two Olympic gold medals, eight world championships and 12 world records. Trained as an engineer, he started Accuracy International in 1978 and involved himself in various activities within the firearms industry, including development of a match rifle for international competition.

Recognizing early on that the British special forces community did not have specialist support in this area and that neither his target rifle nor any other rifle available at that time could meet the Mission Essential Need Statement (MENS), Cooper sat down with a design team who formulated two lists. One list comprised all of the failings of current designs, while the other outlined all of the desired features—which included user friendliness, ease of maintenance, first-shot hit capability, total and long-lived reliability, interchangeability of sights without loss of zero, durability, and safety and interchangeability of all components without use of special tools or gunsmithing techniques.

After a competitive evaluation, Accuracy International was awarded a contract to supply 1,238 rifles to the British armed forces in 1984. The special forces and police version was known as the PM (Precision Magazine). As adopted by the Ministry of Defense (MoD), with emergency iron sights, a flash hider and different optical system, the rifle's nomenclature was L96. Export sales of these rifles to infantry, special operations groups and police—including sound-suppressed versions optimized for covert and counterterrorist operations which use subsonic ammunition at ranges out to 300 meters—in 19 other countries brought the total sales to 2,000 units. Production of the PM and L96 ceased in 1990 when the second generation AW (Arctic Warfare) rifle was introduced.

In 1983 the Swedish MoD invited nine small arms manufacturers to participate in a test and evaluation process intended to provide the Swedish armed forces with a sniper weapon system. Since the Swedish army did not have snipers at that time, the selection and evaluation of both equipment and operational concepts, drawing largely upon the experience of the British army and USMC, took seven years.

In 1990, a contract was awarded to Accuracy International for 1,105 AW rifles, which are designated the PSG 90 by Sweden. Subsequent orders from the Belgian, Irish, New Zealand, Canadian and Omani armies and several police agencies and special operations groups from half a dozen countries have brought the total produced to more than 3,000.

NATO APPROVED

Soldier Of Fortune was provided with both an AW rifle and its law enforcement variant, the

AWP (Arctic Warfare Police—first introduced in 1993), for an extensive test and evaluation.

All of the Accuracy International series rifles are fully NATO certified. Chambered for the 7.62x51mm NATO cartridge, the AW rifle weighs about 14.3 pounds (6.5kg), empty, but with bipod, scope and magazine. The overall length is 46.4 inches (1,179mm) with 1.2 inches (30mm) of spacers. Also provided are 10mm and 40mm spacers to adjust the length of pull to suit the user. The six-groove barrel measures 26.5 inches (673mm) with muzzle brake and is available with user-specified twist rates (1:10" or 1:12"). Its muzzle brake/flash hider has an integral mount for an emergency front sight. It is commonly referred to as the "green rifle" because the barrel and receiver have been finished with a green epoxy paint and the structural-filled nylon stock furniture is also green in color.

The stock is a chassis system based on benchrest rail gun technology. In other words, this rifle has an aluminum-alloy frame, or chassis, to which the other components are attached. The two-piece, thumb-hole-type stock panels are made from reinforced nylon produced on a huge injection-molding machine and are present only to make the rifle ergonomically acceptable. They are held together by eight allen-head screws and washers.

The massive action body is bolted and epoxy-bonded to the frame. This completely removes any interface between the action body and the stock panels, while greatly increasing the moment of inertia (or stiffness) of the barrel support system. This system also permits the stainless steel barrel to be totally free-floating.

The three-lug bolt locks into a ring which is sandwiched between a shoulder in the receiver and the end of the barrel. This ring has been designed to take up all of the wear, so that when the headspace becomes excessive, only the ring needs to be replaced to bring the headspace back to normal. This should not be necessary until more than 15,000 rounds have been fired (perhaps two barrel changes).

All of the lock-up system—the barrel spigot (threaded portion of the barrel), case-head protrusion, locking ring, and action and bolt bodies—on Accuracy International rifles are held to very tight tolerances, which permit complete interchangeability between bolts and barrels without the need for "breeching" to get the correct headspace.

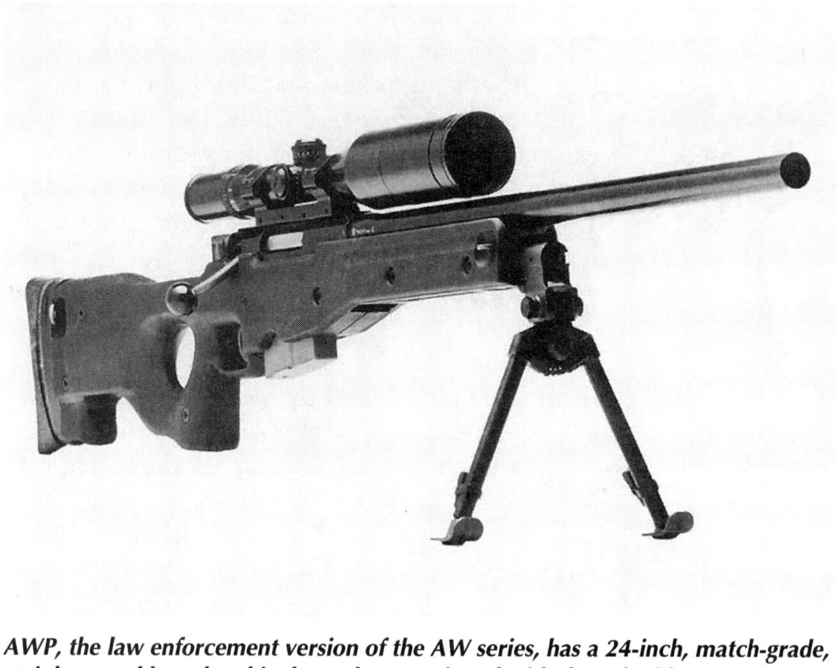

The bolt's three forward lugs are supplemented by an emergency fourth lug at the bolt handle. Bolt lift upon opening is 60 degrees (as compared to 90 degrees on a '98 Mauser action). Total bolt throw is 4.2 inches (107mm), which allows the operator to maintain his cheek-weld on the stock while manipulating the bolt, and thus observe the target during reloading.

A large extractor, which is reminiscent of that found on the Austrian Steyr SSG rifle and gets its strength from its length rather than an auxiliary coil spring, and the spring-loaded "plunger" or "bump" type ejector were designed for covert

AWP, the law enforcement version of the AW series, has a 24-inch, match-grade, stainless steel barrel and is shown here equipped with the Schmidt & Bender variable power PM3-12Vx50M scope. *(Photo: Accuracy International, Inc.)*

bolt manipulation. No matter how slowly the bolt is operated, the empty case will still clear the action.

The receiver is milled from solid barstock, working off the bolt-way hole. An integral, proprietary sight rail has been milled into the top of the receiver. Gas protection is offered by minimum action cuts and tight tolerances. Gas ports are away from the shooter's face.

A Winchester-style safety is mounted on the right side of the bolt shroud and is designed as an antifreeze, anti-dirt mechanism. In the rearward position, the safety catch withdraws the firing pin and locks the bolt in its closed position, thus ensuring safety during airborne operations. When pushed forward into the middle position, the bolt can be manipulated to withdraw a round from the chamber, but the trigger remains blocked. Push the serrated knob all the way forward to fire the rifle.

COCKING PIECE

The cocking piece is also close to that of the Winchester Model 70. The firing pin extends past the shroud. In this manner it serves as both a cocking indicator and a measuring device for firing pin lift. This is the distance the firing pin is pulled to the rear by the cocking piece. The ease with which the safety can be manipulated is a function of the firing pin lift. The bolt's gas protection is by shroud deflection. The firing pin cannot protrude until the bolt is fully closed. Grooves on the bolt body serve a dual purpose of lightening the bolt and providing an area for the accumulation of debris.

Removal of the cocking piece and spring-loaded firing pin from the bolt body is quite simple. With the bolt cocked and the safety lever set to "half safe," just depress the spring-loaded bolt location pin on the inner edge of the shroud and rotate the shroud clockwise until this subassembly can be separated from the bolt body.

The antifreeze and anti-dirt trigger mechanism is of the two-stage type familiar to U.S. military high-power shooters. The trigger pull weight can be adjusted to between 3.5 and 4.4 pounds. *SOF*'s test specimens were set to the lowest pull weight.

Sniper rifles should be equipped with a bipod, and the Accuracy International series are all provided with the very best—a modified, quick-detachable Parker Hale LM6 model. This is nothing more nor less than a scaled-down version of the excellent Bren LMG MkI bipod. Its adjustable legs permit the command height (the distance from the ground to the center line of the barrel's axis) to be varied from 8.5 to 12 inches. The head can be swiveled and canted approximately 15 degrees in either direction without altering the leg position. The bipod attaches to a spigot in the front of the stock panels that has a spring-loaded, heavy-duty catch/release.

There are five sling attachment points on the rifle: two at the butt end (right and left sides) and three up front (left, right and bottom). The olive-drab (AW) or black (AWP) nylon web sling is 1.5 inches wide and has spring-loaded steel hooks on each end for attachment to the stock's sling points. This sling has enough nylon buckles and brass loops to serve as either a carrying or shooting support sling. In addition, there is a competition-type, adjustable, aluminum-alloy hand-stop with a sixth sling point on the underside of the stock's forearm area that moves in a slot on an aluminum assembly containing the three front sling points and the bipod's spigot.

Detachable box magazines are another essential of a successful sniper weapon system. Both the AW and AWP rifles are equipped with 10-round, staggered-column detachable box-type magazines with steel, phosphated bodies and floorplates, and plastic followers. A ballistite (blank cartridge) magazine is also available. It can be identified by the floorplate, which has been painted red, and by the red plastic follower which will accept only blank rounds.

Other accessories include a set of seven different-sized allen-head wrenches—used to disassemble almost every component on the rifle—a rubber muzzle cap and lens cleaning equipment. Two different types of cleaning rods are available: a Parker Hale, single-piece, plastic-coated rod with a chamber guide and both a jag

AW/AWP Sniper Rifle Specifications

CALIBER . . . . . . 7.62x51mm NATO (.243 Winchester optional for AWP).

OPERATION . . . . . . Bolt-action. Three-lug bolt with an emergency fourth lug at the bolt handle. 60-degree bolt lift. Three-position safety on right side of bolt shroud.

FEED MECHANISM . . . . . . Detachable, 10-round, staggered-column, box-type magazine. Ballistite magazines also available for training exercises.

WEIGHT, empty, with bipod, scope and magazine
AW . . . . . . 14.3 pounds (6.5 kg);
AWP . . . . . . 15.5 pounds (7 kg).

LENGTH, overall, with 30mm butt spacer
AW . . . . . . 46.4 inches (1,179mm);
AWP . . . . . . 43.88 inches (1,114mm).

BARREL . . . . . . Six-groove. AWP: match grade stainless steel, manufactured by Maddco Arms, Australia, or Border Barrels, Scotland. AW barrel available with either 1:10" or 1:12" twist.

BARREL LENGTH
AW . . . . . . 26.5 inches (673mm) with muzzle brake;
AWP . . . . . . 24 inches (610mm) with inverted muzzle crown.

OPTICAL SIGHTS . . . . . . Hensoldt "Sight 90" 10x42mm; Schmidt & Bender PM10X42M, PM6X42M or variable power PM3-12Vx50M; Leupold Vari-X III 3.5X10 Tactical or Mark 4 series: M1-10X, M1-16X and M3-10X.

EMERGENCY SIGHTS . . . . . . AW only. Round front post; adjustable for elevation zero only; assembly bolts directly to the muzzle brake housing. Two rear sights: rotating disc aperture-type with apertures for elevations from 200 to 600 meters in 100-meter increments (Swedish army issue) or flip-up aperture-type with a single battle aperture for ranges from zero to 400 meters only (Belgian army issue).

STOCK . . . . . . Chassis system based on bench-rest rail gun technology: aluminum-alloy chassis to which the action body is bolted and epoxy bonded. Two-piece, thumb-hole-type stock panels made of reinforced nylon (AW: green; AWP: black) and held together by allen-head screws and washers.

FINISH . . . . . . Barrel and receiver: green (AW) or black (AWP) epoxy paint.

PRICE . . . . . . AW or AWP rifle complete with bipod, butt spacers, one magazine and scope mount with rings, but without scope: $2,999. Emergency sights included with AW rifle. Sound-suppressed AWS rifle: $3,295. Metal carrying case, $268; soft carrying case: $109; sling, $17; extra magazines, $23; cleaning kit, $25; tool kit, $7.50.

MANUFACTURER . . . . . . Accuracy International Ltd., P.O. Box 81, Portsmouth, Hampshire, England, PO3 5SJ.

T&E SUMMARY: Ultimate turnbolt sniper weapon system. Police model optimized for Federal 308M with guaranteed sub-1/2 MOA accuracy. Superiority over Remington 700 types includes detachable box magazines and a robust, milspec action with many desirable features. Fitted with excellent Parker Hale bipod.

tip and bronze bore brush; or a military-type, six-piece, sectioned steel rod with an eyelet tip and bore brush—all in a camouflage-pattern cloth bag. While a chamber guide is important for cleaning precision rifles, I do not like jag tips and I feel that abrasive material can embed itself in plastic-coated rods.

An action/scope cover is also provided with the AW rifle. As it is Swedish army-issue, it is reversible with a Swedish camouflage-pattern cloth on one side and white plastic on the other. A foam rubber housing inside the cover protects the scope from damage. The allen-wrench set, jag tip and bore brush are housed inside an olive drab plastic box that contains two extra-long screws for the buttplate extensions and a small bottle of Break Free CLP. I would discard the latter, because unless it is shaken continuously, its lubricating Teflon beads will settle to the bottom of the dispenser bottle. All of this, including a box of Parker Hale bore patches and the rifle with its scope, can be stored in an olive-drab, heavy-gauge, aluminum carrying case with steel furniture. A black foam strip around the inside edge of the case's lid gives rain protection, but it is not completely waterproof.

HEART AND SOUL

If the heart of any SWS is the rifle itself and its inherent accuracy potential, then its soul must certainly be the optical sight with which it is equipped. Both the AW and AWP rifles are available with a wide array of excellent glass.

The AW rifle supplied to us for test and evaluation was fitted with a Hensoldt "Sight 90" fixed power 10x42mm scope, housed in a 1-inch tube. It has a 3/4-minute Mil Dot reticle pattern that is tritium-illuminated for night shooting. The focusing ocular has +/– 1 diopter adjustment forward of the focusing ring. The bullet drop compensator (BDC) on the elevation knob is calibrated for NATO military ball ammunition and is rendered in meters. There are +/– 13 clicks of windage adjustment available. Both knobs provide .2 Mil Rad clicks which means that one click provides 2 centimeters of adjustment at 100 meters. This is a high-quality optical sight.

Three equally high-quality Schmidt & Bender scopes are also available: the fixed power PM10x42M and PM6x42M, and the variable power PM3-12Vx50M. This latter scope has a fixed-position reticle pattern with the result that, while there is no change in the point of impact throughout the range of power settings, the reticle gets bigger as you power up.

Accuracy International mounts, fabricated from aluminum-alloy, are available with either 1-inch or 30mm rings. The mount slides onto the receiver's integral grooves and is held in place by three allen-head screws.

As the AW rifle is essentially a military sniper system, it is provided with emergency iron sights. The front sight assembly bolts directly to the muzzle brake housing. The round front post, adjustable for elevation zero only, rests in a machined aluminum alloy block with protective ears.

There are two types of emergency rear sights. The one provided with the Swedish contract PSG 90 is a rather bulky, rotating disc aperture-type with apertures for elevations from 200 to 600 meters in 100-meter increments. It is adjustable for windage zero only. These sights can be pre-zeroed and either stored in the pouch provided or left on the rifle. However, if the scope's position has to be adjusted so the operator can acquire a proper cheek-weld on the stock within the eye-relief parameters provided by the scope design, this latter option is precluded, at least with regard to the rear sight.

The emergency rear sight provided for the Belgian army contract is a flip-up aperture-type with a single aperture intended for use as a defensive battle sight at ranges from zero to 400 meters only.

All of this is more than a little interesting, but it bears small relevance to U.S. law enforcement personnel and American shooters in general. The AW's long range capability (800 meters or more) is a consequence not only of its long and lightweight barrel but of the fact that the internal barrel configuration has been designed to optimize the performance of European 7.62x51mm NATO ammunition, which has a Full Metal Jacket (FMJ) with enhanced ballistic shape and higher muzzle velocity than Federal 308M. With match-grade sniper ammunition, this rifle is guaranteed to shoot 3/4 MOA groups.

Trials have commenced in the United Kingdom to develop "sniper-grade" ammunition that conforms to NATO specifications but has improved, external ballistics enabling supersonic flight out to 1,100 meters. This would advance the effective supersonic range of the 7.62x51mm NATO cartridge by about 300 meters. There is also a project now in Sweden to develop a saboted round for the AW rifle that will increase hit probability at ranges of 1,000 meters or more.

However, the AWP rifle has a chamber throat and leade optimized for the Federal 308M cartridge that features the superb Sierra 168-grain BTHP Matchking bullet. Developed for 300-meter shooting in international matches, this remarkable Boat-Tail Hollow-Point bullet has been winning competitions ever since it was introduced in 1959. It was used by the gold medal winner in the 1968 Olympics and set a new Wimbledon record at 200-15 X's in 1983 at Camp Perry. It remains far and away the most popular bullet among law-enforcement selected marksmen in the United States.

The AWP's six-groove, 24-inch (610mm), heavy, match-grade stainless steel barrel is connected to the receiver via a large-diameter, long-threaded portion at the chamber end. This permits low torquing values to be used on the barrel. This, in turn, eliminates any possibility of stress being transferred to the barrel, which could compromise the first shot hit probability. No muzzle device is fitted to the AWP barrel. To protect it from damage, the barrel crown has been countersunk. A deep, inverted muzzle crown of this type is an important feature on a precision rifle. If the muzzle is not completely squared, propellant gases will escape prema-

AWP rifle with its heavy match barrel, Leupold mark 4 M3-10X scope and Federal 308M ammunition will consistently shoot sub-1/2 MOA groups. *(Photo: Accuracy International, Inc.)*

turely and tip the projectile's base as it departs from the muzzle.

AWP barrels are currently manufactured by two well-known barrel makers: Neville Madden of Maddco Arms, Australia, who apprenticed under Pat McMillan in the early 1970s; and Jeff Kolbe of Border Barrels, Scotland. Kolbe barrels were fitted to the majority of rifles on the winning British Palma Match team in 1992 at Raton, New Mexico. Maddco barrels are button-rifled and Border barrels are fabricated by the cut rifling process. These barrels are guaranteed to deliver sub 1/2 MOA accuracy. They increase the weight of the system to about 15.5 pounds (7kg), empty, but complete with scope, bipod and magazine. Overall length of this rifle is 43.88 inches (1,114mm).

There are only a few other differences between the AW and AWP rifles. AWP stock panels and the barreled action are colored black. A multi-adjustable buttplate for shooting from the awkward positions often required of urban police selected marksmen is also available for the AWP, as is an adjustable hand-stop and optional bipod fitting for more height or for parapet shooting. An alternative chambering, .243 Winchester, can also be provided.

In Europe, the AWP rifle is usually fitted with the Schmidt & Bender variable power PM3-12Vx50M scope. However, in the United States the Leupold Mark 4 series reigns supreme with police agencies and within the special operations community. Justifiably so, as this is the flagship in Leupold's product line and represents the highest possible milspec quality in a production series optical sight.

All Mark 4 scopes feature 30mm diameter main tubes which are machined from a solid piece of 6061-T aircraft aluminum. A 30mm tube with a one-eighth-inch wall thickness still has almost 30% more cross-sectional area inside the tube than most 1-inch tubes with their much thinner walls. Once this additional area is available, the erector tube inside the scope body (which carries all lenses except the ocular and objective lenses) and its lenses can be increased in size to transmit more light and thus yield greater resolution and a brighter image. Furthermore, this heavy 30mm housing is more shock-resistant than any 1-inch tube.

All lenses are coated on all internal and external surfaces with Multicoat 4, Leupold's exclusive multicoating process. This, in addition to a computer-designed optical system, results in edge-to-edge sharpness, precise resolution, minimal distortion and optimum low light visibility. All Mark 4 scopes are tested for water resistance in a reduced-pressure, hot-water immersion tank before leaving the factory and are completely waterproof.

I chose to attach the Leupold Mark 4 M3-10X scope—developed for the U.S. Navy

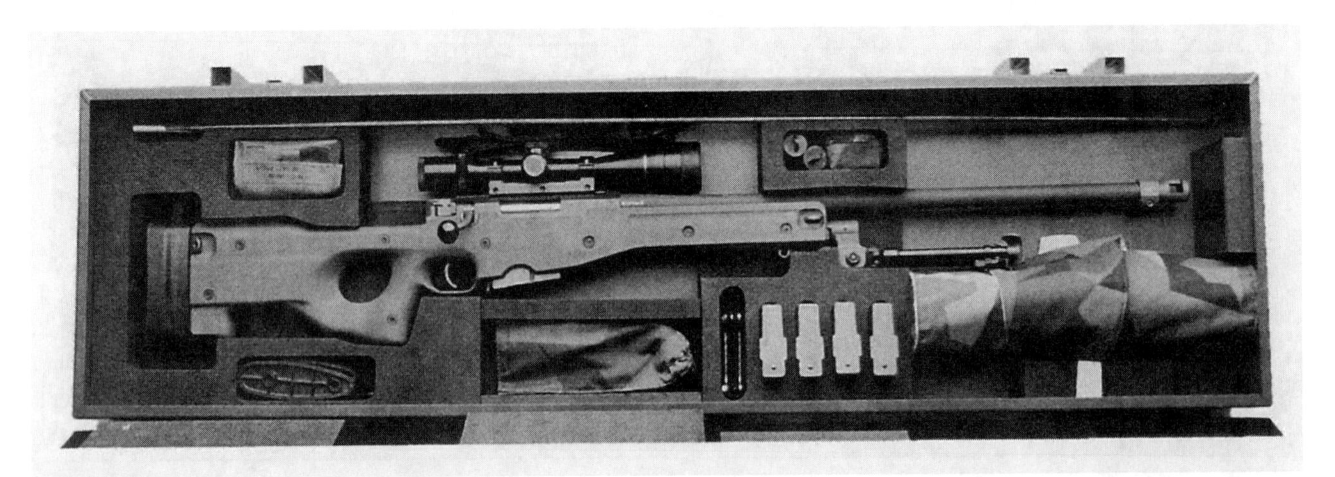

Metal carrying case contains complete AW sniper weapon system which, in addition to the scoped rifle with bipod, includes: cleaning equipment, muzzle caps, emergency sights, scope/action cover, wrenches, spare magazines and ballistite magazines. ***(Photo: Accuracy International, Inc.)***

SEALs and used on the U.S. Army's M24 SWS—on the AWP rifle for test and evaluation. This fixed 10-power scope has a duplex reticle pattern (Target Dot and 3/4-minute Mil Dot reticle patterns are also available in the Mark 4 series). At 100 yards the thin section opening (distance from one heavy post to the other, either horizontally or vertically) is 14.6 inches. The Mark 4 M3-10X scope has an elevation adjustment system that is used as a BDC. The dial is calibrated for bullet drop increments from 100 to 1,000 meters in 100-meter increments, all in less than one complete revolution of the adjustment. Elevation adjustment resolution is in one-minute clicks on the Mark 4 M3-10X scope and in 1/4-minute clicks on the Mark 4 M1 scopes.

SIGHTING IN

After sighting in, and depending upon the range at which the scope is zeroed, the bullet drop dial is then set to that range. We zeroed this rifle/scope combination at 100 meters with Federal 308M, and thus the dial was set at the 100-meter mark. BDC dials are provided with the Mark 4 M3-10X scope for the following bullets: Federal 308M with 168-grain Sierra BTHP, 7.62x51mm NATO M118 Match with 173-grain FMJ, 220-grain .300 Winchester Magnum, 180-grain .30-06 and 55-grain 5.56x45mm NATO. The scope is also provided with a sunshade and Butler Creek flip-up lens covers.

Windage adjustments have 1/2-minute clicks. This adjustment dial can be reset to zero once the scope is sighted in on the rifle. There is a third adjustment knob on the left side of the Mark 4 series scopes. This is for eliminating parallax, and these scopes can be set parallax-free at any distance from 15 meters to infinity when the shooter is in the firing position. It has limiting stops with the two extreme positions symbolized by the infinity mark and the largest dot. Parallax occurs when the primary image of an objective lens does not coincide with the reticle.

The ocular can be focused by backing the eyepiece away from the lockring and then screwing it in or out until the reticle pattern is sharp and crisp.

The M3 comes with screw-on caps for its three adjustment knobs, which have a lower profile than the large, oversized so-called "ergonomic" knobs of the Mark 4 M1 (both fixed 10X or 16X magnifications are available) scopes. Optimum eye relief of the Mark 4 M3-10X is 3.4 inches. At 100 yards the field of view is about 11 feet.

For those who operate in a more confined urban environment, the Leupold Vari-X III 3.5X10 Tactical variable power scope—originally developed for the FBI's HRT—is probably a better choice. At 100 yards, its field of view when powered down to 3.5X is almost 30 feet. It's also almost half the price of any of the Mark 4 scopes. All of these excellent law enforcement scopes carry an unconditional warranty against defects in materials or workmanship, without time limits and even if you are not the original owner.

This combination—the AWP rifle with its heavy match barrel, Leupold Mark 4 M3-10X scope and Federal 308M ammunition—will print 0.4 MOA groups for me and any other experienced operator any day of the week. This is, without question, the most accurate SWS I have ever fired. No semiautomatic rifle can match this accuracy potential.

Subjected to grueling tests in several countries—endurance firings of more than 10,000 rounds during the Swedish trials using the same barrel and with no maintenance were without malfunctions, misfeed, parts breakage or loss of zero—the AW/AWP rifles are without peer among turnbolt sniping systems. It is certainly superior to the M24 SWS in several important areas: accuracy, detachable box magazines (which the M24 does not have) and bipod (the M24 is equipped with a Harris bipod).

Suggested retail price of either the AW or AWP rifle complete with bipod, butt spacers, chamber guide for a cleaning rod, sling, allen-head wrench set and storage box, muzzle cover, two magazines and scope mount with rings—but without a scope—is $2,999. Emergency iron sights are included with the AW rifle. The sound-suppressed AWS version with integral barrel and suppressor costs $3,295. Accessory prices are as follows: metal carrying case, $268; soft carrying case (DPM or black), $109; sling, $17; cleaning kit, $25; tool kit, $7.50; extra magazines, $23 each.

Recent introduction of the AWP rifle in the United States at a reasonable price for such incredible precision should ensure its dominance in U.S. law enforcement and special operations circles well into the foreseeable future.

Originally appeared in *Soldier Of Fortune*
April 1994.

After subjecting the barreled action to deep cryogenic stress relief, the Steyr SSG-PIIK consistently shot 0.4 MOA groups off the bench at both 100 and 200 meters.

SHORT BARREL, LONG RANGE

Steyr's Compact Sniper

Original Steyr SSG 69 (top), most often seen with a Kahles scope, was equipped with a 26-inch medium-weight barrel, iron sights and a green ABS Cycolac stock. The spigot for attaching a Parker-Hale bipod has been permanently fixed to the front of this specimen's stock. Steyr SSG-PIIK (bottom) has a heavy 20-inch barrel, no iron sights, a black ABS Cycolac stock and is marketed with the Hensoldt 10X scope.

One of the very best of the production series sniper rifles ever, since its inception, the bolt-action Steyr Mannlicher *SSG 69 (Scharf-SchuetzenGewehr*—Sharpshooter's Rifle, Model of 1969—the year of its adoption by the Austrian army) was first introduced to American gun owners by Jeff Cooper in *Soldier Of Fortune* more than 18 years ago. His impressions were distinctly favorable. I have worked with this rifle on several occasions while training the Special Reaction Team *(Equipo de Reaccion Especial* or ERE) of the *Policia Nacional* in El Salvador. My reaction was also overwhelmingly positive, although the rifle has several idiosyncrasies peculiar to the design.

The Steyr SSG 69 (now referred to as the "PI") has continued to evolve over the years. The first significant modification appeared with the SSG-PII (*Polizei*—Police Model II) which was first imported to the United States in 1984. This version, designed for law-enforcement applications, substituted a large knob-type bolt handle for the "butterknife" bolt handle so characteristic of modern Steyr Mannlicher rifles. In addition, the SSG 69's iron sights were deleted, the Cycolac stock was changed from green to black and a heavier barrel was installed. Two years ago the SSG-PIIK *(Kurz*—short) was introduced. It differs from the standard SSG-PII only by its 20-inch barrel. *SOF* was recently sent two PIIK rifles, together with the new 10x42mm Hensoldt scopes they are now equipped with for the police market, by the exclusive U.S. distributor, GSI Inc. These rifles are based upon the Steyr Mannlicher SL (Super

Light) sporting rifle which was first exposed to the U.S. market in 1967.

The SSG-PIIK's 20-inch (508 mm) heavy, free-floating barrel with match crown has four lands and grooves with a right-hand twist of one turn in 12 inches (305 mm). Overall length of this rifle is 38.5 inches (978 mm). The rifle alone with an empty magazine weighs 8.7 pounds (3.68 kg). With the Hensoldt scope attached the weight increases to 10 pounds (4.54 kg). Add approximately another 1.5 pounds if you deploy with the aluminum-legged Parker Hale bipod. This is about what you would expect for a heavy-barreled sniper weapon system. As a comparison, the AW sniper rifle with a 20-inch barrel, scope, bipod and empty magazine tips the scales at about 15.75 pounds. If forced to fire snap shots off the shoulder from the standing position at close ranges (25 to 50 yards), I would much prefer the Steyr SSG PIIK, as it is a significant 4.25 pounds lighter than the AW system.

All SSG 69 series barrels are manufactured by cold hammer forging, a process originally developed by Steyr. The billet is fixed on a mandrel with the rifling raised in relief. A series of hammers force the rifling onto the bore and simultaneously form the barrel's exterior contours, work-hardening both the bore and exterior surface. Steyr does not remove the oval-shaped hammer marks on the exterior surface and this lends a distinctive appearance to their rifle barrels.

While Steyr's sporting rifles mate the barrels and receivers by the standard threading method, the SSG 69 series is mated by a shrink-fit process that involves fitting a cold barrel to a heated receiver. After cooling, a 10-ton press would be required to separate the two components. All of this is intended to enhance the barrel's harmonics and improve the inherent accuracy potential. The barreled actions are phosphate-finished and rest in either a green or black ABS Cycolac synthetic stock which can be adjusted for length-of-pull by adding or removing buttpad spacers. There is a fully rotating sling swivel on the front of the stock and a fixed sling attachment point on

Steyr SSG-PIIK Specifications

CALIBER .....7.62x51mm NATO (.308 Winchester).

OPERATION ....Bolt action. Two-piece bolt with a body and bolt-handle sleeve. Six rotating, rear locking lugs, symmetrically arranged in pairs on the bolt handle sleeve. 60-degree bolt lift. Two-position safety on right side just to the rear of the bolt handle.

FEED MECHANISM ....Detachable, 5-round rotary magazine of Schoenauer design.

WEIGHT, empty, with bipod, scope and magazine: ....approximately 11.5 pounds.

LENGTH, overall ....38.5 inches (978 mm).

BARREL ....Four-groove with a right-hand twist of one turn in 12 inches (305 mm) and match crown.

BARREL LENGTH ....20 inches (508 mm).

OPTICAL SIGHTS ....Hensoldt 10X42mm ZF500 or ZF800 milspec scopes, both with 3/4-minute Mil Dot reticle patterns.

IRON SIGHTS ....None.

FINISH ....Phosphate.

FURNITURE ....Green or black ABS Cycolac synthetic stock which can be adjusted for Length-of-pull by adding or removing buttpad spacers.

SUGGESTED RETAIL PRICE ....$2,195 with one magazine; Hensoldt ZF500 or ZF800 scope: $1,195; extra magazines: $42 each; green or black web sling, $36.

MANUFACTURER ....Steyr-Mannlicher AG, Postfach 1000, Steyr, Austria.

U.S. DISTRIBUTOR ....GSI Inc., 108 Morrow Avenue, Trussville, AL 35173.

T&E SUMMARY: Accuracy potential, after inexpensive deep cryogenic stress relief, edges very close to far more expensive systems. An excellent choice for law enforcement agencies who desire maximum performance capability without the risk of bankrupting their budgets. Lighter than many other heavy-barrel sniper weapon systems. Compact envelope—highly maneuverable in crowded urban environments.

the left side of the butt. A green or web sling is available for $36.

An approximately 8-inch-long slot, or rail, in the stock's forearm area permits installation of a sliding handstop/sling swivel unit to which can be attached the superb, but expensive Parker-Hale bipod. This is nothing more nor less than a scaled-down version of the excellent Bren LMG MkI bipod. Its adjustable legs permit the command height (the distance from the ground to the center line of the barrel's axis) to be varied from 8.5 to 12 inches. The head can be swiveled and canted approximately 15 degrees in either direction without altering the leg position. The bipod, which has a spring-loaded, heavy-duty catch/release, attaches to a spigot on the front of the handstop. A new version, available from Brownell's features lightweight aluminum-alloy legs.

The receiver is machined from a single piece of steel with a massive recoil lug at the rear. Bolt lift is a very short 60 degrees. The bolt itself is of two-piece construction with a body and bolt-handle sleeve. On this short sleeve are the six rotating locking lugs, symmetrically arranged in pairs. Rear locking of this type permits shorter bolt travel than the front-locking '98 Mauser system; but, in theory, is less desirable since the entire bolt, not just the head, is compressed when locked in battery and the receiver weakened somewhat by the cut-out on the right side in front of the locking shoulders. Steyr has

Maximizing a Rifle's Accuracy Potential

Both of our Steyr SSG-PIIK test specimens came with computer-generated factory test targets indicating a 3-shot accuracy potential of about 0.7 MOA at 100 meters. That's not bad for production series rifles, but these groups were undoubtedly fired with the rifle clamped in a rigid fixture and in an indoor tunnel. In an attempt to improve on this we subjected both barreled actions to a process called deep cryogenic stress relief, which until recently was known only to precision high-power shooters. The accuracy potential of a rifle barrel is dependent, along with other factors, upon the concept of residual stress. These stresses are present in any piece of cast or forged steel and more are introduced when a barrel is machined, bored, formed and heat treated As these stresses are uneven, when a barrel is heated or cooled it will warp off the bore's axis. This phenomenon, called "warping an arc," can significantly increase the group dispersion down range.

Relief of these stresses can take place if the barrel is brought to an equal temperature—both surface and core—and then cycled through a wide range of temperatures. If the rate of temperature change is maintained at a slow enough pace, thermal compression and expansion occur evenly from the core to the surface and internal stresses are released, resulting in a homogeneously stabilized barrel. The process by which this is accomplished takes more than several days as the barrel is taken by precise computer control to –310 F, held for up to 60 hours then raised to +310 F and slowly brought back to room temperature. Due to recent technical developments, the deep cryogenic tempering process is now relatively inexpensive ($49.50 per barrel or barreled action). Furthermore, modern computer control has brought the process to a consistency not possible in the past.

Barrels subjected to this process develop more uniform, refined microstructure with far greater density as a consequence of carbide fillers precipitated during the process. These carbides fill interstices in the steel, leaving a much denser more coherent structure with far greater wear resistance. Thus, in addition to increasing accuracy, deep cryogenic stress relief can provide barrels with an exceptional increase in durability with barrel life extended by as much as 300%.

We had this work done by Cryo-Maxx. The results? Consistent 0.4 MOA groups fired by both myself and Alex Pappas the other *SOF* staff member participating in these tests, from the bench off a sand-bag rest at both 100 and 200 meters with Federal 308M. To me this really demonstrates that deep cryogenic stress relief will significantly improve the accuracy of an already accurate rifle.

THE SNIPER'S OPTIC

Hensoldt Optishe Werke AG of Wetzlar, Germany is part of the Zeiss optical group and a highly regarded manufacturer of milspec rifle scopes. They have recently introduced two new scopes, the ZF500 for ranges out to 500 meters and the ZF800 for distances up to 800 meters, both with 3/4-minute Mil Dot reticle patterns.

Mil Dots were developed by the USMC in the late 1970s to assist Marine Corps snipers in estimating distances. It is now the standard reticle pattern with all branches of the U.S. Armed Forces. The term "Mil Dot" comes from "mil"—a unit of angular measurement used in artillery and machine gunnery and equal to 1/6400 of a complete revolution—and the fact that the dots are spaced in 1-mil increments on the crosshairs. It should be made clear that the dots themselves are not measured in mil increments, but rather in increments of MOA (minutes of angle, or arc). The dots are actually oval-shaped (with the long axis oriented in the vertical position on the vertical crosshair and in the horizontal position on the horizontal crosshair). In addition, the dots are actually a 1/4 mil in length (slightly longer than 3/4 MOA). In any event, the distance between the dots is 3/4 mil and the center-to-center distance between them is exactly 1 mil as is the distance from the top (or bottom) of one dot to the top (or bottom) of the dot above or below (or to the right or left). There are also four thick posts at the edges of the field of view: The posts are almost 1 mil thick (0.98 mil = 3.36 MOA). The distance from the center of the crosshairs out to the beginning of any of the four posts is exactly 5 mils.

The distance to a target in yards can be calculated using the Mil Dot formula:

$$\frac{\text{Height or width of target (in yards) X 1,000}}{\text{Height or width of target (in mils)}}$$

Both the Hensoldt ZF500 (with which our test specimens were equipped) and ZF80 have an elevation adjustment that is used as a BDC (Bullet Drop Compensator). The elevation dial on the ZF500 is calibrated for bullet drop increments of the Sierra 168-grain BTHP bullet as loaded in Federal 308M ammunition from 100 to 500 meters in 100-meter increments with 50-meter marks in between in less than one complete revolution of the adjustment knob. Elevation adjustment resolution is in 1/4-minute clicks all the way out to 500 meters (with 1/2-minute clicks on the ZF800 out to 800 meters). Windage adjustments have 1/2-minute clicks. The windage dial can be reset to zero once the scope is sighted in on the rifle.

Hensoldt 10X ZF500 scope features a 3/4-minute Mil Dot reticle pattern.

Both of these fine scopes provide fixed 10 power magnification with a 42mm entrance pupil and an exit pupil of 4.2 mm. Eye relief is 2.7 inches (69 mm).

The one-piece aluminum tube has a center diameter of 1 inch. The objective diameter is 48 mm and the ocular diameter is 38 mm. The tube has a tough, non-glare black matte finish. Overall length is 12.6 inches (320 mm), with a weight of 17.6 ounces (500 grams). The field of view is 4.2 meters at 100 meters. The suggested retail price of either model is $1,325.

Optional flip-up lens covers are also available.

compensated for this by both strengthening and lengthening the receiver so the barrel is seated with its chamber within the receiver walls. The bolt face is recessed for the cartridge head and is fitted with a spring-loaded "bump"-type ejector. The bolt has two gas escape ports.

The extractor is of the Italian Carcano type and can be easily removed. The one-piece firing pin has a shorted threaded section toward the rear end onto which is attached the double-headed cocking cam. The firing pin's coil mainspring is located between this cocking cam and a collar at the front end. A small set screw on the cocking cam prevents the firing pin from rotating. Dual opposed cocking cam recesses at the rear of the bolt handle's sleeve engage the cam surfaces on the cocking cam when it rests in a slot at the rear of the bolt body. Everything is held together by the bolt sleeve which also puts the firing pin under tension when it is cocked by turning the bolt handle upward. When cocked, the rear of the firing pin extends through a hole in the bolt sleeve to serve as a cocking indicator. Complete disassembly of the bolt for replacement or cleaning of the striker assembly, although periodically required, is somewhat complex and best left to those with an SSG 69 armorer's manual. The firing pin is quite long and spares are advisable in my opinion, although their protrusion must be initially adjusted by an armorer.

Both single and double-set triggers are available. While most are well-advised to opt for the single trigger, which can be adjusted for length and pull-weight, I prefer the double-set system. The trigger mechanism is built into a cast aluminum-alloy housing that fits in a recess at the bottom of the receiver and is retained by two pins. The double-set trigger system can be adjusted to a pull-weight of only a few ounces. To employ this system, first pull back the rear trigger until the mechanism is cocked, then merely touching the front trigger releases the cocked rear trigger, which then releases the sear and firing pin. Optionally, the rifle can be fired by only pulling the front trigger, which will then provide a pull-weight of between 3 to 4 pounds. A small set screw between the two triggers can be used to adjust the pull-weight of the set trigger only. Turning the set screw out will increase the pull-weight.

The trigger housing also holds the bolt-stop mechanism. To remove the bolt from the receiver, pull the front trigger as far as possible and withdraw the bolt.

The grooved and serrated manual safety button is located on the right side of the receiver, just to the rear of the bolt handle where it is quite easy to reach. Moving it to the rear exposing a white dot locks the sear and prevents the bolt handle from being lifted. Pushing forward exposes a red dot and permits firing the rifle. No middle position that locks the sear or trigger but allows bolt-manipulation for unloading is required as the Steyr SSG 69 magazines are easily removed.

The Steyr SSG 69 detachable, 5-round rotary magazine is based upon the design of Otto Schoenauer. Except for its springs, the principal components are fabricated from injection-molded plastic. Inside is a spring-tensioned rotary spool. The rear face of the magazine is made of clear plastic so the cartridge heads can be counted. Spring-loaded latches at each side of the magazine at the bottom engage notches in the magazine-well and permit the magazine to be inserted and withdrawn with ease. At one time, a 10-round magazine, which projected below the stock, was also available to military and law enforcement agencies. It was dropped from production in 1990. Regardless of their capacity, magazines are the weakest link in the Steyr SSG 69 system, in my opinion. While strong enough for almost all law-enforcement applications, they are not milspec by any stretch of the imagination, and many of the ERE's magazines that I examined in the equipment-hostile environment of El Salvador were damaged to one degree or another. Extra 5-round magazines carry a suggested retail price of $42.

The trigger guard and magazine-well of these rifles is a complex one-piece, injection-molded black plastic component. Screws at each end are threaded into the receiver and retain the barreled action in the stock.

For our test and evaluation of the Steyr SSG-PIIK rifles we employed both Federal and Lapua ammunition. The Federal 308M cartridge features the superbly accurate Sierra 168-grain BTHP Matchking bullet. After deep cryogenic stress relieving of both barreled actions, we shot consistent 0.4 MOA to 0.5 MOA groups at both 100 and 200 meters with this ammunition. That's excellent accuracy for a production series rifle. The Sierra 168-grain BTHP Matchking bullet remains far and away the most popular bullet among law-enforcement selected marksmen in the United States. But, is it really the best choice from a wound ballistics perspective?

The hollow point cavity in this competition target bullet does not guarantee the type of consistent, early expansion exhibited by hollow points and soft points designed for use on living tissue targets. Research recently conducted by Lucien C. Haag and reported in the *Wound Ballistics Review* (Vol. 2, No 2), the journal of the International Wound Ballistics Association, revealed that these bullets frequently fail to expand in tissue simulant even after as much as 6 inches of penetration and with close-range impact velocities. Dr. Martin L. Fackler, in the same issue of the journal, reported that these bullets will commonly break up and fragment after 7 inches of penetration. His experiments indicated that when the cavity was increased in diameter to 0.055-inch by a drill, reliable expansion was obtained after penetrating less than 1 inch of 10% gelatin or muscle. Recent controlled testing has demonstrated that this alteration degrades accuracy by no more than 0.1 MOA—an insignificant amount. Winchester is reportedly in the process of developing their own bullet for law enforcement use that will provide both match-grade accuracy and reliable performance in tissue.

Three different types of Lapua match-grade .308 Winchester ammunition were tested: 167-grain Jacketed Hollow Point, 170-grain Full Metal Jacket (FMJ) and 185-grain FMJ. The 167-grain bullet, which closely approximates the Sierra 168-grain BTHP Matchking will shoot from 0.5 to 0.6 MOA on a reliable basis. The Lapua 170-grain bullet will provide accuracy very close to that, increasing the group size by no more than 0.1 MOA. The 185-grain bullet would shoot no better than 0.8 to 0.9 MOA. That's to be expected as the Steyr SSG PIIK barrel twist at one turn in 12 inches is too slow to completely stabilize this heavier bullet. We need a faster 1:10-inch twist for .30 caliber bullets weighing more than 170 grains.

With a suggested retail price of $2,195, the Steyr SSG-PIIK lies between low-end systems such as the Ruger or Savage and pricey custom rifles or the AW from Accuracy International. Its accuracy potential, after inexpensive deep cryogenic stress relief has been performed, edges very close to that of the costly AW and custom shop systems, which are usually based upon the Remington 700 action. It remains an excellent choice for law-enforcement agencies who desire maximum performance capability without the risk of bankrupting their budgets. Its compact envelope provides optimal maneuverability in crowded urban environments. Finally, and of no small importance to the user, the Steyr SSG-PIIK is significantly lighter than other sniper weapon systems with similar accuracy potential.

Originally appeared in *Soldier Of Fortune*, August 1996.

20-MIKE-MIKE MAYHEM

Aerotek's Shoulder-Fired Sniping Cannon

At 800 meters the Aerotek 20x82mm Anti-Materiel Rifle will shoot to about 1.1 MOA. Its excellent accuracy potential is a consequence of a number of important factors, including bolt-action operation, the superb GA1 automatic cannon barrel and high quality ammunition from PMP. Note the long eye relief provided by the Lynx 8x42mm scope.

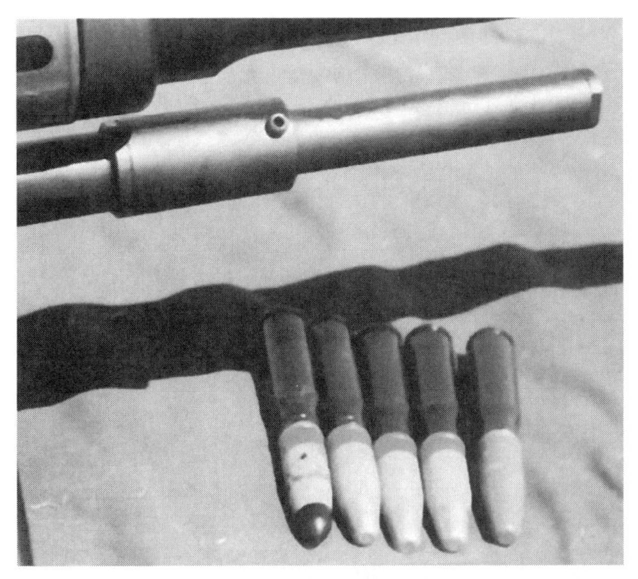

An important element in the Aerotek 20x82mm Anti-Materiel Rifle's recoil management system is the oil-filled hydraulic damper which absorbs most of the recoil momentum generated at the initiation of the recoil stroke.

The first time I began to press the trigger on the South African Aerotek 20x82mm Anti-Materiel Rifle, I had a momentary fantasy of my arm being torn from its socket. However, only the 55 gallon drum 400 meters downrange separated from its platform on the ground when it was struck dead center, as in reality this incredible new sniper weapon system has less perceived recoil than the Barrett M82A1 Light .50.

Designed by my good friend, Tony Neophytou, who also designed the revolutionary Neostead combat shotgun, this innovative new weapon is produced by Aerotek, CSIR. While there are now many .50-caliber (either .50 Browning, aka 12.7x99mm, or Soviet 12.7x108mm) rifles to choose from, the 20x82mm automatic cannon round provides a wider variety of projectile types. Although it was developed as an anti-materiel system for deployment against such targets as communication masts, power lines, radar installations, parked aircraft, missile deployment installations, refineries, security guard towers, satellite dishes, ground-to-air weapons, machine-gun bunkers, vehicles and both sea and land mines, it can be effectively employed as an anti-personnel weapon also.

The Aerotek 20x82mm Anti-Materiel Rifle weighs 57.33 pounds (26 kg). This breaks up into two manpacks of 24.3 pounds (11 kg) and 33.1 pounds (15 kg), respectively. Some .50-caliber sniper weapon systems weigh almost as much. The overall assembled length is 70.7 inches (1,795 mm). The barrel length is 39.4 inches (1,000mm).

Neophytou commenced development in August 1995 and from concept to firing the project took only four and a half months. His previous work on helicopter turrets in the area of recoil reduction facilitated the rapid progress. To further simplify development, the pre-existing GA1 automatic cannon's barrel, bolt head and barrel extension were utilized.

For reasons of accuracy potential, reliability, cost effectiveness and weight reduction, a bolt-action system was employed. There are six locking lugs, in two sets of three each on the left and right sides of the bolt body with primary extraction of the empty cases.

The three-round, single-column, detachable box-type magazine has two slots at the rear in the interior of the magazine body to prevent rounds from moving forward until they're chambered and thus protect the projectile's sensitive nose fuze. The magazine is located on the left side of the receiver because the barrel, action and magazine recoil together and a bottom-feed magazine would have complicated the design. None of the human interface components: buttstock, pistol grip, scope and bipod, travel with the recoiling parts. The barreled action and magazine move rearward in the frame about 100 mm.

HOW DO YOU SPELL (RECOIL) RELIEF?

The heart of this amazing rifle is the unique recoil management system which permits a 20x82mm automatic cannon round to be fired from an operator's shoulder. It consists of three elements. First of which is the large muzzle

20x82mm Ammunition

This cannon cartridge was developed by Mauser in the early 1930s for their MG 151/20 aircraft gun, which was originally chambered for a 15 mm case but was later scaled up to 20 mm at the request of the Luftwaffe. It is similar to the 20x80RB Oerlikon round, but has a pronounced bottleneck and a full-diameter extractor rim. After World War II, the 20x82mm cartridge was used principally by France and the Republic of South Africa. In South Africa it is manufactured by PMP (Pretoria Metal Pressings).

Automatic cannon ammunition differs from that of heavy machine guns mostly by virtue of their respective projectile types. Most cannon projectiles are shells (i.e. hollow projectiles which contain explosive material that is initiated by an impact fuze at the target or at some earlier point during the trajectory by a time-delay mechanism). Most heavy machine gun projectiles are of the solid ball type with lead or steel cores which sometimes have hollowed-out areas at the base or nose that contain tracer or incendiary elements, respectively. This distinction between heavy machine-gun and cannon ammunition is somewhat blurred by an early Soviet 12.7x108mm antiaircraft round with a projectile that contained both high explosive and incendiary composition, 14.5x114mm HEI ammunition and by the more recent Norwegian .50 caliber BMG, multi-purpose Raufoss ammunition whose bullet is filled with a hard core and a loading of RDX explosive.

South African 20x82mm ammunition is available in five types: nose-fuzed High Explosive Incendiary (HEI) and High Explosive Incendiary Tracer (HEIT) filled with 107 grains of Hexal P30, base-fuzed Semi Armor Piercing High Explosive Incendiary (SAPHEI) filled with 92 grains of Hexal P18, and both dummy-fuzed Target Practice (TP) and Target Practice Tracer (TPT).

Muzzle velocity of this ammunition is about 2,360 fps (720 m/s). The arming distance of the fuze is 8 meters, minimum. Tracing time of the HEIT and TPT rounds is two seconds, minimum. Soft iron driving bands are used to insure long barrel life and to prevent fouling. The weight of the projectile (or shell), filled and fuzed, is about 1,690 grains.

brake. It has an angle of deflection of about 130 degrees on each side to avoid disturbing the operator.

The mechanical spring buffer system, with two co-axial helical-coil springs, operates during both the recoil and counter-recoil strokes of the barreled action and magazine. It is supplemented by an oil-filled hydraulic damper. At the beginning of the recoil stroke, when the recoil velocity is high, the hydraulic damper is utilized to absorb most of the recoil momentum. At the end of the recoil stroke, when the recoil velocity is lowest, the hydraulic damper falls away and the spring buffer system, which is now compressed to its maximum extent, is used to absorb the remaining recoil energy. There is no restriction on deployment modes. The buffer/damper system is designed to be effective whether the gun is fired from the prone position or a hard mount.

The barreled action's maximum recoil distance is less than 4 inches and is dependent upon the atmospheric temperature, because the hydraulic system uses oil. As a consequence, the standard damper fitted to the rifle will cover operation in environments ranging from 15 degrees F to 140 degrees F. However, an optional damper is available to cover operational temperatures from minus 15 degrees F to 95 degrees F. This overlap in temperature ranges permits moving from a training or assembly area to the operational area without changing dampers. Equipped with milspec O-rings, the unit is designed to be maintainable by trained armorers.

The frame (or more correctly, chassis) of this rifle is a stamped sheet-metal pressing made from 4130 steel. There is a monopod at the rear end of the rifle which is part of the carrying

This is quite a payload in a small package and it is significantly superior to .50-caliber ball for anti-personnel applications, because it disperses numerous fragments into an area pattern often referred to as the "effective casualty radius." This is defined as the distance from the point of detonation at which a certain percentage of exposed personnel will be casualties. Unfortunately, there is no consensus as to what constitutes the proper percentage of casualties to be used in the definition. Some authorities and governments use 20% and others up to 50%. Furthermore, the orientation of the projectile's long axis at the moment of detonation will affect its wounding capability. Finally, in most instances this information, even if it has been determined, is classified.

Nevertheless, 20x82mm HEI ammunition can wreak awesome destruction upon unprotected troops. In addition, the SAPHEI round will defeat 15mm of NATO-standard armor plate at zero degrees angle of incidence from a range of 100 meters (and about 10mm at 1,000 meters) and can thus be employed with considerable effectiveness against lightly armored vehicles, boats, sandbagged bunkers and buildings.

In general, nose-fuzed rounds are more suitable for "thin-skinned" targets, such as fuel storage tanks and aircraft. Base-fuzed rounds are less sensitive and often have a longer delay after impact before full detonation. This makes them more suitable for use against armored personnel carriers and other targets with more substantial armor protection.

20x82mm ammunition provides quite a payload in a small package and is significantly superior to .50-caliber ball for anti-personnel applications. From left to right: Target Practice, High-Explosive Incendiary and base-fuzed Semi Armor Piercing High-Explosive Incendiary.
(Photo courtesy Aerotek.)

handle at the bottom of the frame. It can be adjusted for height or disregarded by folding the carrying handle against the frame. The butt-stock, also a stamped sheet-metal pressing, is a separate component and contains the bolt stop. The quick-detachable bipod can be adjusted for command height and rotated 15 degrees on either side of center.

The modular trigger mechanism can be removed from the bottom of the frame. Trigger pull weight is adjustable without removal of the trigger group. A crossbolt-type safety is part of the trigger housing. The pistol grip to the rear of the trigger group is that of the South African R4 infantry rifle (a licensed and highly enhanced version of the Israeli Galil).

There is an accuracy interface collar on the frame that permits the barrel to be securely fitted to it. The collar can be tightened or loosened with the head of a cartridge case. The GA1 barrels supplied by LIW and PMP ammunition help to provide exceptionally high accuracy. Fired from a stable bench this rifle will generate 10-inch groups at 875 yards (800 meters). That's about 1.1 MOA.

Only one revolution of the scope mount's elevation adjustment wheel covers ranges from 150 to 1,500 meters. A quick-release type with levers on each side, the mount can effortlessly be removed from the rifle. The mount is attached to a collar fitted to the non-recoiling frame. Secured to the mount with 30mm rings is a Lynx 8X42mm scope, which is made in Japan and features a standard NATO crosshair reticle pattern and, most important, long eye relief.

The Aerotek Anti-Materiel Rifle can be easily converted to the still prevalent 14.5x114mm round. All that is required is a new barrel and muzzle brake, bolt head, magazine and scope mount. The recoil momentum of this round is very similar to that of the 20x82mm cartridge. The 14.5x114mm scope mount is calibrated for ranges from 300 to 2,300 meters. The 20x82mm cartridge will fit in the 14.5x114mm magazine, but the reverse does not hold true.

The most obvious historical precedent to the Aerotek rifle is the German World War I Mauser "T" rifle (Tankgewehr 1918) which was developed in 1917 and fielded a year later. It was in essence a scaled-up, single-shot version of the Mauser Model 1898 bolt-action rifle chambered for the 13x92mmSR round developed specifically for it and a machine gun that was never adopted. The semi-rimmed case held 200 grains of nitrocellulose. Its 965-grain boat-tailed spitzer bullet had a steel core with a lead filler and gilding metal jacket. It could penetrate 22mm of armored plate at 100 meters. The recoil was ferocious and it was not popular with the German troops.

But that was then, and this is now, and after almost 80 years designers have learned how to cope with the problem of the excessive recoil generated by man-portable rifles chambered for 20mm automatic cannon rounds. While I cannot say the Aerotek design is a "pleasant" rifle to shoot, it doesn't seem to exhibit any more perceived recoil than my .505 Gibbs bolt-action hunting rifle. However, in addition to its sophisticated recoil management system, the weight

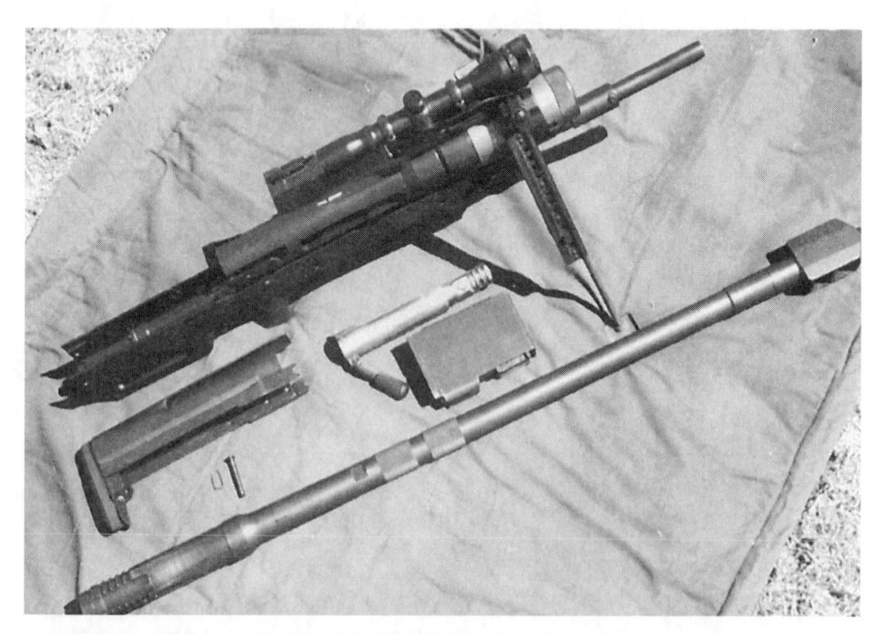

Aerotek 20x82mm Anti-Materiel Rifle, fieldstripped.

Aerotek Anti-Materiel Rifle Specifications

CALIBER . . . . .20x82mm automatic cannon or 14.5x114mm heavy machine gun.

OPERATION . . . . .Bolt-action with six locking lugs, in two sets of three each on the left and right sides of the bolt body and featuring primary extraction of the empty cases; recoil-management system with barreled action moving approximately 100 mm rearward in the frame during the recoil strike.

FEED MECHANISM . . . . .Three-round, single-column, detachable box-type magazine, on the left side of the receiver body, with two slots at the rear in the interior of the magazine body to prevent rounds from moving forward until they're chambered and thus protect the projectile' s sensitive nose fuze.

WEIGHT, empty . . . . .57.33 pounds (26 kg). This breaks up into two manpacks of 24.3 pounds (11 kg) and 33.1 pounds (15 kg), respectively.

LENGTH, overall . . . . .70.7 inches (1,795 mm).

BARREL . . . . .20x82mm—eight-groove with a right-hand twist of one turn in 22 inches (550 mm).

BARREL LENGTH .1,000 mm.

SIGHTS . . . . .Lynx 7X42mm optical sight with long eye relief.

MANUFACTURER . . . . .Aerotek, CSIR, P.O. Box 395, Pretoria 0001, South Africa.

T&E SUMMARY: Designed for deployment against thin-skinned targets and light armored vehicles, with anti-personnel applications as well; recoil impulse effectively moderated by sophisticated recoil-management system; accuracy potential approaches 1.1 MOA at 800 meters; reliable bolt-action operation.

required to moderate a cannon cartridge's recoil impulse is an important consideration to those who must tote these beasts on the front lines. Fifty-seven pounds, even when broken into two loads, is not an inconsiderate amount to be added to the already overburdened modern soldier's load.

Aerotek's only competition at this time is the Croatian RT-20 sniper rifle chambered for the 20x110mm Hispano Suiza HS 404 round. It is a single-shot, bolt-action type with a bullpup configuration. It uses a peculiar "recoilless" system to compensate for the recoil impulse generated by this cartridge. About half-way down the barrel is a ported gas block which directs propellant gases down a tube to be vented into the atmosphere above and behind the bolt group. As a consequence, a rear hazardous area common to all recoilless weapons requires the operator to assume an unorthodox firing position and keep clear of obstacles at the rear. The standard version weighs the same as the Aerotek. There is supposed to be a lightweight model that weighs "only" 40 pounds. While I never personally encountered this weapon during several assignments to Bosnia, it appears that the Aerotek design is superior in both concept and execution.

I predict the trend to develop and field anti-materiel rifles chambered for automatic cannon cartridges will continue. The only limiting factor on those introduced to date are the optical sights. When engaging man-sized targets at ranges approaching and exceeding 1,000 meters, a scope with a minimum magnification of 16 power, such as the Leupold Mark 4 M1 16X40mm, is necessary in my opinion.

Originally appeared in *Soldier Of Fortune* April 1997.

As Far As You Can See

The 1,000-Yard Chandler Rifle

CHANDLER SNIPER RIFLE SPECIFICATIONS

CALIBER . . . . .7.62x51mm NATO (.308 Winchester).

OPERATION . . . . .Bolt-action. Two-lug bolt. Two-position safety on the right side of the bolt sleeve.

FEED MECHANISM . . . . .Five-round, staggered-column integral magazine with steel hinged floorplate.

WEIGHT, empty . . . . . .Approximately 13-15 pounds, depending upon optical equipment and stock.

LENGTH, overall . . . . .46.5 inches.

BARREL .Six-groove, heavy, match-grade, stainless steel, with a right-hand twist of one turn in 12 inches. Manufactured by Hart.

BARREL LENGTH . . . . .26 inches.

OPTICAL SIGHTS . . . . .Leupold, Mark 4 M1 10X or Tactical series with Mil-Dot reticle pattern.

FINISH . . . . barreland receiver: Black oxide.

FURNITURE . . . . .McMillan fiberglass stock; either USMC-type, A2 or A3.

MANUFACTURER . . . . .Iron Brigade Armory Ltd., 100 Radcliffe Circle, Jacksonville, NC 28546.

T&E SUMMARY: Built to exceed USMC specifications. Most accurate long-range system ever tested. Expensive and somewhat heavy.

Most often, writers for the popular gun press determine the accuracy potential of a centerfire rifle by sitting at a bench and firing groups into paper targets at a range of 100 yards. Some very mediocre rifle and scope combinations can look pretty good at that distance. In the case of a so-called sniper rifle, you need to shoot at ranges of 500 yards, and beyond, from actual field shooting positions in order to separate the few real contenders from the all-too-common pretenders.

The Chandler Sniper Rifle, made by Iron Brigade Armory Ltd., is a genuine 1,000-yard rifle built to exceed USMC specifications. I know because I own one and recently returned from running it through Clint Smith's ringer during Thunder Ranch's demanding Precision Rifle 2 course. I am convinced that Lieutenant Colonel Norman A. Chandler, USMC, (Ret.), has designed and assembles the very best military/law-enforcement long-range sniper rifle in the world, bar none. However, be advised that there is no free lunch and rifles of this type are neither compact, nor lightweight, nor cheap. Overall length of my Chandler Sniper Rifle is 46.5 inches. These rifles weigh from 13 to 15 pounds, depending upon the scope and stock selected.

Chandler uses a Hart barrel, Remington M700 receiver, McMillan stock and several different Leupold scopes. The barrels are all 26 inches in length. During the late 1970s, research conducted by the inner loop in the USMC sniper community clearly demonstrated that

26 inches provided a more complete propellant burn and a significantly reduced flash signature. Stating that airborne operations were a design parameter, the Marine Corps adopted a barrel length of only 24 inches. This has proven to be a mistake in all regards.

Made from milspec 416R stainless steel, these Hart barrels are button rifled to a dimension of 0.3082-inch and then lapped with a plug to remove high spots. They feature six grooves with a right-hand twist of one turn in 12 inches. After installation on the receiver, the barrels and actions are provided with a black-oxide finish. As you would expect with match-grade equipment the barrel has a countersunk muzzle crown, as the slightest damage to the crown can significantly degrade accuracy.

Iron Brigade Armory uses only Remington Model 700 actions, which are highly modified. All Chandler Super Grade Sniper Rifles have the clip slot for the M14 charger guide milled into the rear of the receiver's ejection port. The Chandler scope mount is designed to interface with this clip slot to provide greater rigidity. The Chandler bases are made oversize and then hand fit to the clip slot. Remington uses 6x48 screws to attach a scope base to the M700 receiver. Chandler enlarges these to 8x40 to further strengthen the scope-base-to-receiver interface. Leupold's quick-release scope rings have a spindle projecting below the ring that goes into the scope base. Chandler industrial-oven-brazes (at 850 degrees F.) the Leupold scope rings to their base. Incredibly,

THE REMINGTON M700 ACTION

The Remington M700 action was introduced in 1962 and is almost identical to the Model 721/722 actions it replaced. The receiver is machined from round bar stock and has the same dimensions its entire length. The right side of the bridge extends past the magazine well to prevent the bolt from binding. The bolt body is machined from a steel cylinder. The bolt head, with two solid locking logs, is brazed onto the bolt body. The low-profile bolt handle has also been brazed to the bolt body at the rear. This is one of the safest actions ever designed as the bolt extends about 0.150-inch ahead of the locking lugs and the barrel's breech face is recessed to receive the bolt with minimum clearance around its circumference. Furthermore, the bolt head's face is recessed for the case head.

The extractor, often criticized yet rarely failing, is a thin C-type flat spring which has a lip pressed into its inside curve. It interfaces with a shallow groove in the inside of the rim forming the bolt face recess. The ejector, which is a spring-loaded plunger in a hole along the perimeter of the bolt face recess, is retained by a cross-pin. The ejector prevents the extractor from rotating in its recess. The one-piece firing pin is spring-loaded and attached at its head to the bolt sleeve by a cross-pin. The bolt sleeve threads into the rear of the bolt body. Lock time is very fast, since the firing pin's total length of travel is no more than 0.300-inch. A flat steel stamping under the left locking lug raceway serves as the bolt stop. Pushing the bolt-stop release (a bent spring-steel metal strip sliding on the trigger and safety pivot pins) up pivots the bolt stop down so the bolt can be removed from the action.

The trigger and safety group is attached to the underside of the receiver by the boltstop and sear pins. The sear, a thin piece of hardened steel located in the top right side of the trigger housing and in the trigger housing opening in the receiver, pivots on the sear pin in front of the housing and is tensioned by a small coil spring. The trigger also pivots on an axis pin. When pulled, it releases the trigger connector. The trigger over-travel screw contacts the trigger connector. The weight of pull adjustment screw is just below this screw. A third screw adjusts trigger-sear engagement.

A bent steel stamping on the right side serves as the safety lever. When the action is cocked, the safety can be pulled all the way back and both the firing pin and bolt are locked. To fire the rifle push the safety all the way forward.

Reach Out and Touch Someone with a Magic Bullet

Moly-coated bullets now dominate precision high power rifle competition, but have not yet made a significant impact in the law enforcement arena. Including myself, only three students used moly-coated ammunition in Thunder Ranch's PR2 course. All of it was provided by Black Hills Ammunition. Black Hills loads .223 with 52-grain and 68-grain BTHP (boat-tail hollow-point) moly-coated bullets and .308 Win match-grade ammunition with either a 168-grain or 175-grain BTHP. I chose the 175-grain BTHP for its potential to reach out at long ranges.

This process involves impact-plating bullets with molybdenum disulphide and a protective coating of wax. Moly-coating produces a protective surface in the barrel with a number of important benefits.

Moly-coated bullets will reduce both chamber pressure and muzzle velocity because of the reduction in the coefficient of friction. Increasing the propellant charge weight will bring the muzzle velocity back up. When fired at identical velocities, moly-coated bullets will provide flatter trajectories at long range than untreated bullets. Accuracy is also significantly enhanced. Many shooters are reporting an improvement of 10-20% with match-grade projectiles. This is partially a consequence of the improved uniformity of muzzle velocity. The standard deviation is normally reduced by about 15%. Especially important is the incredible almost twofold increase in barrel life. Barrels that will normally maintain match-grade accuracy up to 3,000 to 3,500 rounds will last at least 6,000 rounds when moly-coated bullets are fired through them exclusively.

Out to 500 yards, my elevation come-ups with the Black Hills 175-grain moly-coated ammunition were 0.5 to 2 minutes higher than those firing 168-grain ammunition. But, at 500 yards, the 175-grain moly-coated bullet starts to get serious and my come-ups started to drop well below the others. At 1,000 yards, the USMC come-up for M118 173-grain ammunition is 39.5 minutes. Mine is 35.5 MOA, a full 5 minutes below the M118 173-grain projectile. That's an impressive difference. At 500 yards, firing from the prone position—not a bench—the Black Hills 175-grain moly-coated round printed a five-shot 0.5 MOA group with the Chandler Sniper Rifle and Leupold 4.5-14x40mm Tactical scope. A 2.5-inch group at 500 yards from a field shooting position is phenomenal and the only comment required on the rifle-scope-ammunition combination.

Metal fouling is also significantly reduced and less maintenance is required. While students firing standard ammunition had to punch their barrels after every ten rounds, I did not clean the bore until the end of each day, after more than 100 rounds had been fired. However, the cleaning procedure differs radically from the norm, and is essentially chemical rather than mechanical. First pass the cleaning rod through the chamber guide and out the muzzle. Then attach an undersize bronze bore brush (caliber .270 for a .308 barrel) and wrap a clean patch around it. Wet the patch with Rem-Oil or Shooters' Choice FP-10 lubricant and pull-push two times to remove the copper fouling. Replace the patch and cover it with J-B Cleaning Compound. Again, pull-push 10 times. Then two passes with an oiled patch, followed by two more passes with another oiled patch and two passes with a dry patch. All of the cleaning materials required can be obtained from Brownell's, Inc.

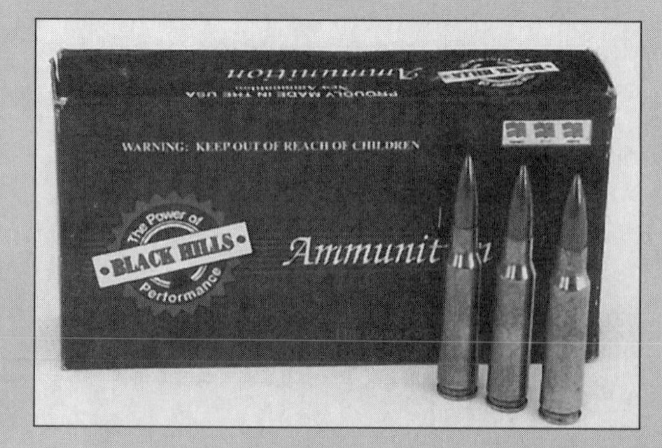

tests have indicated that this alone can add 0.5 MOA to the system's accuracy potential. The one-piece scope base is made from 4130 ordnance steel.

My Chandler scope mount has 20-minute angle-bored rings which were bored after the rings were brazed to the base. Chandler's new system is to put the 20 minutes of angle on the base. Either way, this gives you the ability to elevate the scope to 1,000 yards and beyond (actually out to 1,400 yards on my Chandler rifle/scope combination), but still have positive clicks at 100 yards. By this means, the scope's erector tube is operating at its optimum range of adjustment at the long ranges. In addition, the Remington receiver is milled by first indexing the front receiver ring and then milling down the rear bridge approximately 125 thousandths lower than the front receiver ring. This is somewhat lower than the Remington standard for this dimension.

Hart installs their barrels into the Remington M700 receivers for Chandler. They square the receiver before installation of the barrel. Hart also laps the bolt lugs and sets the headspace. At the very end of the production cycle, Chandler's armorers coat the bolt with Tetra gun grease, a polymer derivative, and hand work the action. The magazine follower is also altered to permit positive feeding of five rounds.

The USMC has been using standard Remington triggers for more than 30 years and so does Chandler. The trigger group is completely disassembled and tuned. The trigger is adjusted for a pull weight of 3.75 pounds, which Chandler feels is just about the lightest sear release you should have for a combat environment where the adrenaline is likely to be pumping. Badger Ordnance of Rapid City, South Dakota, makes the one-piece steel trigger guard with a wide trigger bow to protect the trigger and a hinged steel floorplate exactly to Chandler's specifications. Badger Ordnance also provides the new scope base and the heavy recoil lug.

Chandler Sniper Rifles are bedded twice. The first, or heavy, bedding is accomplished with titanium devcon. The rifle is then disassembled and skin-coat bedded with MarineTex, a high-strength, low-shrinkage two-part epoxy. The barrels are, of course, completely free floating.

I selected one of the superb McMillan A3 sniper stocks with a woodland camouflage pattern for my Chandler Sniper Rifle.

The outer shell of McMillan's A3 stock is made by a hand-laminated, pressure-cured process. It is laminated from about 130 pieces of 8-ounce fiberglass cloth, giving a finished shell wall thickness of six to 12 layers. The thickness is greater in high stress areas on the stock and less in other areas to reduce the overall weight. The interior is filled with different fiberglass compounds consisting of engineering-grade epoxy resins,

Built to exceed USMC specifications, the Chandler Sniper Rifle is the most accurate long-range sniper system SOF has ever tested.

LEUPOLD 4.5-14X40MM TACTICAL SCOPE

In the United States, Leupold & Stevens have an absolute lock on the military and law-enforcement scope market. Justifiably so, as their Mark 4 and Tactical scopes exhibit superb milspec performance, complete reliability, and unmatched ruggedness and optical excellence.

I chose to install a Leupold Vari-X III 4.5-14X40mm variable-power Tactical scope with adjustable objective on my Chandler Super Grade Sniper Rifle. During Thunder Ranch's Precision Rifle 2 course you will shoot at ranges varying from 35 yards in Thunderville, out to 1,000 yards. A variable-power scope is ideal for ranges of this extreme. I used the 4.5X power setting for urban scenarios and then powered up to 14X for long range shooting where precise target definition was essential. The objective lens is provided with settings for 50, 100, 200, 400 yards and infinity.

All exterior and interior lenses on this scope are treated with Multicoat 4 anti-reflective coating. This, in addition to a computer-designed optical system, results in edge-to-edge sharpness, precise resolution, minimal distortion and optimum low-light visibility. All of Leupold's military and law-enforcement scopes are 100% waterproof. In addition, all of the scopes in Leupold's Tactical series are economically priced and you can purchase one for almost half the price of a Mark 4 scope, which is the flagship in Leupold's product line.

It was my initial assumption that scopes with an elevation-adjustment system that is used as a BDC (Bullet Drop Compensator), such as the Mark 4 M3-6X and M3-10X, with replaceable elevation dials calibrated for an assortment of bullet types, would be superior to standard elevation adjustment knobs in the high stress environment of combat. My personal experience has demonstrated this to be an invalid hypothesis for several reasons. First of all, bullet designs are evolving rapidly and Leupold does not provide elevation dials for bullets like the 175-grain moly-coated BTHP (although an unmarked elevation dial is available for self-calibration). Furthermore, Leupold's Mark 4 M3 scopes have 1/2-minute adjustment increments. This is too coarse for really long range shooting. Remember, 1 MOA is the equivalent of 10 inches at 1,000 yards.

The 4.5-14X40mm Tactical scope has both windage and elevation adjustment knobs that feature audible and tactile feedback from one-minute numbered divisions with 1/4-minute click-stops clearly marked between each one-minute division. The ability to make 1/4-minute adjustments is an important attribute. Total elevation travel on this scope is 63 minutes and each complete revolution raises or lowers the point of impact by 15 minutes. The elevation knob also has a horizontal scale that is used to keep track of the number of revolutions that the dial has been turned. In addition, there is a built-in anti-backlash system that guarantees repeatable accuracy from click to click, and back again. Most snipers use an elevation adjustment system such as this by zeroing the rifle and scope at specific ranges and writing the elevation adjustment settings on a range card attached to the rifle's buttstock. The setting to which the elevation adjustment settings on a range card attached to the rifle's buttstock. The setting to which the elevation adjustment knob must be rotated for a specific distance is usually referred to as a "come up" by those who move in this elite loop. Mounted on the Chandler Sniper Rifle the Leupold 4.5-14X40mm Tactical scope, when zeroed at 100 yards, still has 7.5 minutes of depression available. The windage-adjustment knob provides approximately 30 minutes of lateral adjustment to both the left and right for a total of 63 minutes.

My Leupold 4.5-14X40mm Tactical scope is equipped with a Mil-Dot reticle pattern. Mil-dots were developed by the USMC in the late 1970s to assist Marine Corps snipers in estimating distances. It is now the standard reticle pattern with all branches of the U.S. Armed Forces. The term "mil-dot" comes from "mil"—a unit of angular measurement used in artillery and machine gunnery and equal to 1/6,400 of a complete revolution—and the fact that the dots are spaced in 1 mil increments on the crosshairs. It should be made clear that the dots themselves are *not* measured in mil increments, but rather in increments of MOA. Premier Reticles (who make these reticles for Leupold & Stevens) uses wire (or "mechanical") crosshairs onto which the dots are applied wet. Because of this, the dots cannot be made circular and are, thus, oval-shaped (with the long axis oriented in the vertical position on the vertical crosshair and in the horizontal position on the horizontal crosshair). In this particular instance, the dots are actually a 1/4-mil in length (slightly longer than 3/4 MOA). In any event, the distance between the dots is 3/4-mil and the center-to-center distance between them is exactly 1 mil as is the distance from the top (or bottom) of one dot to the top (or bottom) of the dot above or below (or to the right or left). There are also four thick posts at the edges of the field of view.

The formula for using the mil-dot system is:

$$\frac{\text{Height or width of target (in yards)} \times 1{,}000}{\text{Height or width of target (in mils)}} = \text{Distance (in yards)}$$

All of my Leupold scopes are equipped with Butler Creek (available at Michaels of Oregon Co.) lens caps. They are the best available. For tactical rifles avoid the see-through types. Butler Creek also has a glint screen system built into a front lens cap that I have installed on the Leupold 4.5-14X40mm Tactical scope. The KillFlash system used by Butler Creek is a honeycomb glint-suppressing screen with a highly specialized coating that eliminates the reflection from the surfaces of lenses. The loss of light transmission is only 15%.

On the battlefield, glint kills. Moshe Dayan, the famous Israeli general, lost his eye from the bullet of a sniper who caught the glare of his binoculars. During World War II, in the Battle of Stalingrad, Russia's top sniper, Vasili Zaitsev, won his famous three-day duel with Germany's ace, Major Zossen, by looking for—and targeting—the reflection from the German's scope.

chopped fiberglass strands, micro-balloons and other materials and slightly lighter than the A2 sniper stock. No polyesters or phenolics (both less expensive than epoxy resins) are used. There are no hollow spaces. All of McMillan's tactical stocks are filled solid. Inletting for the barreled action is done on CNC machinery to tolerances approximating a 0.001-inch. This yields a completed stock which is stable in environmental conditions ranging from -60 degrees F. to +240 degrees F. and is totally water-impervious.

McMillan's famous A2 stock was originally designed in 1988 at the request of a federal law-enforcement agency and a U.S. military organization. It represents a high-point in the application of human engineering to stock design. Both the width of the forearm and the angle of the pistol grip (about 8-9 degrees off vertical) were determined by using experienced shooters pretending to hold a rifle in the prone position. The A3 model was designed in 1995 as a somewhat lighter version of the A2. The A3 stock weighs 3.25 pounds—exactly 1 pound lighter than the A2. It features a shallowed forend which is 1/3-inch shorter than the A2 stock, and this also provides it with a trimmer appearance. Its integral adjustable fiberglass cheek piece accounts for a half pound of the weight shaving and provides lateral, as well as vertical, adjustment. Finishes available include black, gray and desert, woodland and urban camouflage patterns. The stock is equipped with QD-type sling swivels. This is the tactical rifle stock against which all others must be measured. The A3 stock is equipped with a half-inch Pachmayr recoil pad.

Marines don't use bipods, but I do. I had McMillan install an aluminum-alloy accessory rail to accept a sliding handstop and the superb new Parker-Hale bipod with lightweight aluminum-alloy legs. This is nothing more nor less than a scaled-down version of the excellent Bren LMG MkI bipod. Its adjustable legs permit the command height (the distance from the ground to the center line of the barrel's axis) to be varied from 8.5 to 12 inches. The head can be swiveled and canted approximately 15 degrees in either direction without altering the leg position. The bipod, which has a spring-loaded, heavy-duty catch/release, attaches to a spigot on the front of the handstop.

All Chandler sniper rifles are shipped in a Pelican hard case. Other collateral equipment includes a black or brown National Match leather sling made by Turner Saddlery. Their slings are the standard by which others must be judged and the vast majority of competition centerfire match shooters use slings manufactured by Turner Saddlery. In addition, you will also find in the Pelican case, as part of the Chandler sniper rifle package, a sniper's log book, a takedown tool and Craftsman torque wrench for the trigger guard screws (which should be tampered with a little as possible), and a plastic-coated cleaning rod with a 44-inch shaft made by the famous J. Dewey Mfg. Co., Inc.

Quality at this level is not inexpensive. Prices of the Chandler Sniper Rifle vary according to the scope selected. For further information contact Iron Brigade Armory Ltd. Together with his brother, Roy F. Chandler, Norm Chandler has authored a highly-regarded five-volume series on Marine Corps sniping, entitled *Death From Afar*, and a fascinating authorized biographical memoir of Carlos Hathcock, entitled *White Feather*. All of these books are available from Iron Brigade Armory, Ltd.

Originally appeared in *Soldier Of Fortune*
May 1998.

Section 7
Rifles

THE M16A2

The Final Verdict

M16A2: Small arms scandal or military masterpiece? America's battle rifle for the end of the 20th century must be the biggest firearms story of the '80s.

Two cases inches apart show high burst rate that gives tight full-auto groups. Ejection was smart, and failed only once in 6,000 rd.

One last high-pitched report bent the afternoon dust-devils as a bullet struck the berm at 250 yards.

"Well, that's it."

After three days in parching hundred-degree wind, the four testers didn't even cheer. Besides, we still had to fire for group, blow one last half-mag through the chronometer, and measure barrel erosion.

We weren't happy. But we were relieved. Over 6,000 rounds had gone down the spout, and the biggest civilian test of the free world's most important new small arm was done.

M16A2: the military technology story of the '80s.

Trying to beat SOF's *experts into print, virtually every gun magazine or military-oriented medium has whipped out a story of one sort or another on the son of the much-maligned M-16 service rifle. Others have told you how it feels, how it looks, how it shoots . . . But every other article has been based on the easily-available, semi-auto AR-15A2, or firing demos staged by Colt or the U.S. military.*

That's just not good enough for a weapon that's destined to accompany our fighting men into combat for at least the next generation or two. Such superficial treatment has created a storm of controversy over the quality of the M16A2 rifle that's devisive, counter-productive and unprofessional. With this in-depth test and evaluation report, we hope to correct all that once and for all.

SOF*'s Peter G. Kokalis—formerly of U.S. Army Technical Intelligence—bought the first M16A2 delivered to a civilian. At our direction, the world-renowned author and weapons scientist was told to conduct a duplication of the Department of Defense official 6,000-round* *endurance test. Ammo was gathered, magazines cleaned, a Black Canyon, Ariz., range reserved for three days.*

Every day, firing began when the range opened, and the assault rifle's roar stopped just long enough for Mil-spec cooling between cycles. The desert summer tested weapon and technical team to the limit. Over $2,000 worth of ammunition was stuffed into 20- and 30-round magazines. Although it was time-consuming and expensive, we felt the test had to be done.

To avoid bias and single-source problems with our M16A2 test, SOF *sent Bill Guthrie, director of editorial research, to observe the test and verify results. Guthrie checked every malfunction, read velocity from the chronograph and performed all barrel erosion measurements. No piece of the body of data that made up the report came from one tester. The 'A2 test is too important.*

This vital report is the longest military small arms article ever printed in Soldier Of Fortune. *This exhaustive technical account of the private-sector test of America's new battle rifle is presented uncut, with full scientific data, so* SOF *readers will know what DOD knows about the M16A2.*

—The Eds

•••••

Born in controversy, America's M16 service rifle seems destined to live—at least for some time to come—in that same angry clamor. Muzzle-energy fetishists and armchair experts, like angry bulldogs, just won't let go of their belief that the M16 will never be anything more than a plastic toy. Fortunately for the American fighting man and all the allies who will eventually end up with the M16A2 rifle, the facts don't support such conclusions. This study should finally demonstrate that for *SOF* readers. It's the first full-scale, Mil-spec endurance test of the M16A2 conducted by an independent civilian authority. By 1978 most of the M16A1 rifles in the U.S. inventory had been worn out from use as training weapons. Many had fired more than 50,000 rounds. The need for new weapons was apparent and urgent. Something was required to bridge the supposed 20-year gap when the millennium would commence with something as wonderful as a fully-perfected H&K G11 with its 4.7mm caseless ammunition. Enter the M16 PIP (Product Improved Program): A joint venture among Colt, the Marine Corps and Army which commenced in 1979.

Within nine months Colt submitted its first proposal on the improved rifle. The M16A2 was approved for service use by the USMC in September, 1982. It was type-classified (adopted) by the U.S. Army in November, 1982. The first 1,500 rifles were delivered to the USMC Marksmanship Training Unit (MTU), Quantico, Va., in January, 1984. Major deliveries to the U.S. Army will start by mid-1985. Canada has adopted the M16A2 to replace its aging 7.62mm FN FALs. The 81,500 Canadian M16A2 rifles will be manufactured by Diemaco, Inc., Kitchener, Ontario, with full-auto instead of three-round burst. Canadian 'A2s will have hammer-forged barrels and 30-round plastic magazines. A good source alleges the Canadian government will pay $1,300 per gun for the special run.

The M16A2 rifle retains the same method of operation as its predecessor. After firing a round, the projectile passes the gas port permitting gas to flow back through a stainless steel tube and a so-called bolt carrier key into the hollow interior of the bolt carrier. As the carrier moves rearward, a cam slot cut into the carrier turns the bolt's cam pin, which causes the bolt to rotate clockwise, freeing the eight locking lugs from their abutments in the barrel extension. The carrier's momentum draws the bolt rearward at a slightly reduced velocity.

There is no primary extraction and the extractor withdraws the cartridge from the chamber. The spring-loaded bump-type ejector emerges from the left of the bolt face and rotates the empty case, after it has cleared the chamber, around the extractor claw and out the ejection port of the upper receiver body. The bolt carrier assembly continues rearward, compressing the recoil spring and cocking the hammer.

The buffer and recoil spring return the carrier and a fresh round is stripped from the magazine. All forward bolt motion stops after the round is chambered. The carrier continues forward to contact the rear face of the barrel extension and its cam slot turns the cam pin which rotates the bolt and its lugs anti-clockwise into the locked position.

Direct gas action without a piston was taken from the Swedish Ljungman AG42 rifle. The M16's trigger mechanism is based on that of the M1 Garand.

The new M16A2 will be marketed in no less than six versions for the international military market. The rifle (Model 701) as adopted by the U.S. military has a 20-inch barrel. This barrel length is also available with a lightweight barrel (Model 703) or with the old M16 rear sight (Model 711). The carbine version (Model 723) has a 14.5-inch barrel and a sliding buttstock. The M16A2 Commando (Model 733) has an 11.5-inch barrel with sliding buttstock. Finally, a squad-level support weapon (HBAR-Model 741) is offered with a heavy barrel and M60 bipod.

A significant number of needed modifications have been incorporated into the M16A2 rifle. The M16A1's famous bird-cage flash suppressor remains, sans the sixth port on the bottom which was deleted to slightly reduce muzzle climb during burst-fire and diminish position disclosure when firing from the prone in desert climates. There has been no change in the flash characteristics. The lock washer used to retain the muzzle device has been replaced by a set of peel-washers so that the flash suppressor can be rotated either to the right or left, for right- or left-handed shooters, respectively.

The visible portion of the barrel is much thicker (0.73-inches in diameter). It tapers back to the old diameter just under the handguards past the gas port. While many will assume this was done to improve handling characteristics and/or accuracy potential, the real reason is a reflection of grunt mentality. Too many snuffies were using the M16 as a crowbar and bending the barrel just forward of the front sight. Three carefully-added ounces have doubled or tripled the barrel's stiffness.

The new handguards resemble those of the M16A1 carbine, the Commando model and early factory literature depictions of both the M16 rifle and the Colt HBAR. As the upper and lower halves are identical, spare parts inventories can be reduced accordingly. They are more comfortable than the previous triangular cross-section handguards. Improved handguards and heavier barrel have increased the rifle's sustained fire capability by raising the probable cook-off limit by 20 rounds to 160 rounds. The handguards are retained by a tapered slip ring, which was already in the system 14 years ago on the XM177E2. Joy in the barracks. No longer will you need four screwdrivers to pry off the handguards.

The pistol grip now has deep longitudinal grooves along its rear face and a finger swell one inch below the trigger guard. It's interchangeable with that of the M16A1 and, in fact, Colt seems to have exhausted their inventory of the older pistol grips since the new style was installed on all 4,000 M16A1 rifles recently sold to El Salvador.

The front sight-post is square and adjusted only for initial zero which is now done at the Colt factory. This change was requested by Marine

Post adjustment is unchanged, but with rear-sight elevation it is used only for original zero.

Army Wants M16A3

The U.S. Army is currently exploring modifications of the newly type-classified M16A2 rifle to accept optical sighting equipment. And it wants the often-criticized carrying handle removed.

Scopes mounted on the carrying handle do not permit a proper cheek-weld (spot-weld) on the M16 buttstock. And the Army says the luggage-type handle on top of the rifle makes it too easy for troops to avoid carrying the weapon combat-ready at all times. Project managers in the Department of the Army now see removal of the handle and installation of an optical sight with emergency iron sights as the answer to these problems.

Both collimator (devices that take advantage of human eyes' natural alignment by giving the conventional sighting eye an illuminated dot which projects to point of impact in the weak eye's field of view) and conventional telescopic sights are being tested. Scopes of the low-power 1.5x Steyr AUG type and more sophisticated concepts—such as the British Sight Unit, Small Arms, Trilux (SUSAT) are all in the running.

The 4x SUSAT scope would be an attractive choice. Presently fitted to the Enfield Individual Weapon and Light Support Weapon, SUSAT represents the results of 25 years of experience and refinement, Designed by RARDE (Royal Armament Research and Development Establishment), SUSAT has performed admirably during reliability tests in harsh environments during the UK Ordnance Board trials. The sight is fitted with a small acrylic aiming pointer—illuminated by a tritium-activated light source—to allow use under poor light conditions.

Latest word is that the Cobra Scope military optical sight (manufactured by Swarovski Optik, Hall, Tirol, Austria) is heavily favored for the Army's replacement for the 'A2. The Cobra Scope is a conventional, light-weight, compact, waterproof, shockproof rifle scope available with a variety of reticles. Unlike the SUSAT there is no tritium element, and it doesn't need batteries.

The Army intends to provide night-vision capability as well. But we'll have to wait to see what form that will take.

If the current tests demonstrate significantly increased hit potential (which I'm sure they will) the Army variant "M16A3" will soon become a reality. M16A1 rifles due for overhaul will be brought up to the required Army specs. Any M16A2 rifles delivered before adoption of the third-generation variant would be retrofitted. All of the other modifications incorporated in the M16A2 will be retained.

M16A3 is our unofficial designation for this development. We chose it because it represents a logical numerical relationship to old models in the type-classification system the U.S. military has previously used. We don't have any idea what the '16 variant the Army is pressing for might officially be titled after type-classification, but a Colt Firearms source recently denied any knowledge of an officially designated M16A3 variant.

Maj. Gen. Wm. G. Carson, Jr., Director of Material Branch, HQ, USMC, accepts M16A2 rifle from Guy C. Shafer, Executive Vice President of Colt Industries at 12 April 1984 dedication ceremony. *(Photo: Colt Industries)*

marksmanship experts who believe it offers an improved sight picture.

The new rear sight was also requested by the Marines, who first saw it on the old M16 light machine gun. It is a flip-type peep with two apertures calibrated for M855 ammunition. Flipped forward, the large aperture (0.197 inches in diameter), marked "0-2," is brought into view for ranges out to 200 meters, low light levels or moving targets. When firing within this range, the elevation knob should be set to "8/3" with the sight base at its lowest position. Flipped back, the peep sight brings the small aperture (0.70 inches in diameter) into view for ranges of 300 to 800 meters. A small indicator line matches up with the windage calibration lines on the back of the sight base. One audible click of the windage knob moves the sight 0.4 minute right or left. With the small aperture in use, the elevation knob should be set at the range required: 8/3 low for 300m, 4 for 400m, 5 for 500m, 7 for 700, and 8/3 high for 800m. Audible clicks between the main settings will raise or lower the elevation by approximately one minute of angle. (Elevation is graduated to fit the M855 projectile's trajectory path.) The new M16A2 rear sight is easily adjusted and will be appreciated by riflemen trained in its proper use. I hope it will help to re-establish the rifle marksmanship training so badly needed in the U.S. Army. As things stand right now, the Marines produce the only riflemen who are regularly trained to make full and effective use of the vastly improved sighting system.

The M16A2 upper receiver casting has also incorporated a case deflector to the rear of the ejection port which throws empty cases clear of left-handed shooters. In semi-automatic fire the ejection pattern has been altered five or six degrees forward by this deflection hump.

The spring-loaded retaining catch on the ejection port's dust cover has been strengthened and enlarged. The cover's latch-pin is no longer retained by the slip-washer that so often disappeared at the wrong time. The first 2,000 M16A2 rifles were delivered with the old-style dust cover.

The forward bolt assist has been changed from a casting to a screw machine component with a button-like shape and concentric rings. I have never once personally used or observed anyone using the forward bolt assist in a battlefield. When I inquired of Colt officials why this feature was retained, I was informed that no one requested its removal. And so mutations often pass into the orderly cycle of evolution.

A most discussed (and often disputed) feature of the M16A2 is the three-shot burst control consisting of a simple and durable ratchet with an over-running clutch. Each cycle of the hammer turns the spring 60 degrees until after 180 degrees the hammer falls from the auto sear to the trigger sear, holding it in place. The nine-component mechanism is of the interrupted type, which means it picks up the count wherever it left off. Thus the first in this cycle after a magazine change may be either one, two or three rounds.

This system is less complex and uses fewer components than the intricate ratchet-counting device fitted to Heckler & Koch weapons. The H&K burst control holds the sear off the hammer until the burst has been fired; any interruption (such as an empty magazine) starts a new count and releasing the trigger resets the counter. The M16A2 three-shot burst control requires a heavier disconnector and springs. Thus trigger pull weight is about one pound heavier. Because of the burst control's rotating cam, trigger-weight will increase with each pull until the cycle is completed. My M16A2 starts with a relatively drag-free pull of 7.5 pounds at stage one and ends with 9.0 pounds at stage three. While this may twitch the ultra-sensitive trigger-fingers of match shooters, it will not be noticed in the heat of combat.

Despite arguments to the contrary, from a military perspective three-shot burst controls are righteous devices. They allow the shooter to forget about fire discipline and concentrate on sight alignment and target acquisition. That's handy in the adrenalin rush of a fire fight where the brain is busy with other things.

The selector lever location and markings remain as before, on the left side of the lower receiver, except the "AUTO" position to the rear has been replaced by "BURST." The selector markings, "SAFE," "SEMI" and "BURST," are now stamped on the right of the receiver also and the selector shaft notched, for the benefit of left-handed shooters.

The trigger pull length has been changed to a dimension determined to be ideal by the Human Engineering Laboratory at Aberdeen Proving Ground. The lower receiver casting has been subtly altered and strengthened in the rear to accommodate the stronger buttstock—fabricated from foam-filled, high-impact plastic designed to resist fracture from rifle grenade launching (trendy once again with development of the bullet-trap types) and buttstroking Parris Island training dummies or body-armored enemy soldiers. The buttstock has been lengthened by 5/8-inch. The buttplate's edges are no longer rounded and the entire surface, not just the butt-trap, is deeply checkered.

The weight trade-off for these modifications is a modest increase from 7.0 lbs., empty (for the M16A1—the M16 without buttplate trap or forward bolt assist weighed only 6.7 lbs.), to 7.5 lbs.

All that makes the M16A2 a handier, more comfortable, stronger weapon, but those things aren't the final measure of a service rifle's effectiveness. What about performance: reliability, durability, accuracy and hit potential, handling characteristics and claims of accelerated bore erosion? To settle the issue once and forever, *SOF* decided to conduct its own 6,000 round U.S. Government Military Specification Test (MIL-R-63997) on a new, strictly-stock specimen of the Colt M16A2 rifle.

MIL-R-63997 is a 48-page document. Much of it is devoted to a detailed description of the military specification for each component of the rifle, defect evaluations, interchangeability tests, inspection lot tests and the individual rifle test.

A total of nine malfunctions (attributable to the rifle) is permitted in the 6,000-round endurance test. The firing procedure consists of 50 120-round cycles. Intervals between cycles are determined by the cooling of the barrel, which must be such that it is capable of being held by the bare hand. Each cycle is fired in the following sequence: 30 rounds burst-fire with one three-shot burst every five-to-eight seconds; 30 rounds burst-fire with one three-shot burst every two-to-five seconds; and 60 rounds semiautomatic at a rate of 10 to 30 rounds per minute. During the 18th and 36th cycles, the semiautomatic fire is conducted with the rifle held in the hands not touching the shoulder and without restraining the normal recoil of the rifle.

MIL-R-63997 specifies that the ammunition used is to be M193 ball conforming to MIL-C-9963 (ball propellant), as M855 ammunition is not yet available in sufficient quantities. In an effort to accelerate what had been rumored to be rapid barrel erosion, *SOF* substituted 1,100

M16A2 Specifications

CALIBER . . . . .5.56mm NATO—M855 ball and M856 tracer ammunition.

OPERATION . . . . .Gas—direct, no piston; rotary bolt with 8 locking lugs; semiautomatic and 3-shot burst fire.

CYCLIC RATE .700-900 rpm.

MAGAZINE . . . . . .20- and 30-rd. staggered box type.

WEIGHT, empty .7.5 pounds.

LENGTH, overall . . . . .39.6 inches.

BARREL . . . . .6-groove, right-hand twist with 1 turn in 7 inches.

BARREL LENGTH . . . . .20.0 inches.

SIGHTS . . . . .Front: protected square post adjustable for zero; Rear: flip-type peep with two apertures, 0.197-inch for 1-200m and 0.070-inch for 300-800m, knob adjustable for windage and elevation, calibrated for M855 ammunition.

ACCESSORIES . . . . .Bayonet, telescopic sights, bipod and blank adapter; will accept M203 40mm grenade launcher.

MANUFACTURER . . . . .Colt Industries, Firearms Division, P.O. Box 1868, Hartford, CT .06101.

STATUS . . . . .Adopted for service by the armed forces of United States and Canada.

rounds of ammunition using IMR-type propellants. Fired in sequence, our inventory was as follows: 4,300 rounds of PMC M193, 500 rounds of Samson (IMI) FMJ (M193), 100 rounds of Lake City '77 (M193), 100 rounds of AAC '83 (Yugoslav ammo brought back from El Salvador—IMR type powder) and 1,000 rounds of reloaded ammunition (55-gr Hornady FMJ bullet with 22.5-gr. IMR 8208M—a light charge designed to test the gas system's power reserve).

Our test site was the Black Canyon Shooting Range located in Arizona's Lower Sonoran Desert just north of Phoenix. Temperatures ranged from 91 degrees Fahrenheit to 112 degrees Fahrenheit. The test was conducted over a period of three days as the barrel cooled slowly (as did the four test participants). Two thousand rounds were fired each day.

Before beginning firing and after every 1,000 rounds we observed carefully controlled measurement procedures. The rifle was cleaned thoroughly. A U.S. Military M16 Barrel Erosion Gauge was inserted to determine the progressive erosion of lands at the throat of the barrel (increase in freebore). Thirty rounds were chronographed with an Oehler Model 33 Chronotach and Skyscreen III detectors to measure any loss in velocity associated with increasing barrel erosion. The rifle was fired from the bench for group to determine the amount of loss in accuracy potential that would be expected with deterioration of the barrel.

During the first 4,000 rounds a Steiner 4X24 NATO rifle scope was attached to the rifle with a Swan M16A1A2 Universal Top Scope Base. The Swan base accepts both Weaver and NATO STANAG military scope fastenings. It features a special bolt and self-locking thumbscrew that assures repeated positioning and zero hold and a wide-angle "TV-screen" iron sight see-through. The rugged Steiner scope uses the German military reticle pattern dating from WWI. It consists of a single, thick, pointed post at the bottom of the field of view with horizontal side bars and stadia lines. Although never popular in this country, this format excels in subdued light and offers faster target acquisition than Standard crosshairs

After 6,000 rounds the barrel erosion gauge dropped a total of only 3.6 percent closer to the reject line. No visible signs of barrel erosion could be detected with a fiber-optic probe even after the final 1,100 rounds had been fired using IMR type propellants. Although linear extrapolation is not possible—since erosion will eventually become asymptotic—I estimate useful barrel life to be in excess of 30,000 rounds. But, this is a function of numerous uncontrollable parameters, such as the amount of tracer fired.

The average velocity of the PMC M193 ammunition dropped from 3,340 fps to 3,305 fps over its 4,300-round run (these higher-than-specified velocities are a function of the high ambient temperatures prevailing during the test) This is not a statistically significant decline as it is very close to the standard deviation measured for this ammunition.

The M16A2 maintained two minutes of arc (MOA) at 100 meters throughout the test. MIL-R-63997 specifies an accuracy potential of no more than 1.4 MOA, but firing is done inside an enclosed tunnel with the rifle held in a fixture and the distance from the muzzle to the target measured with exactness. The *SOF* test was shoulder-fired under conditions of high, erratic winds and tremendous heat mirage, both of which magnified sighting errors and shooter fatigue.

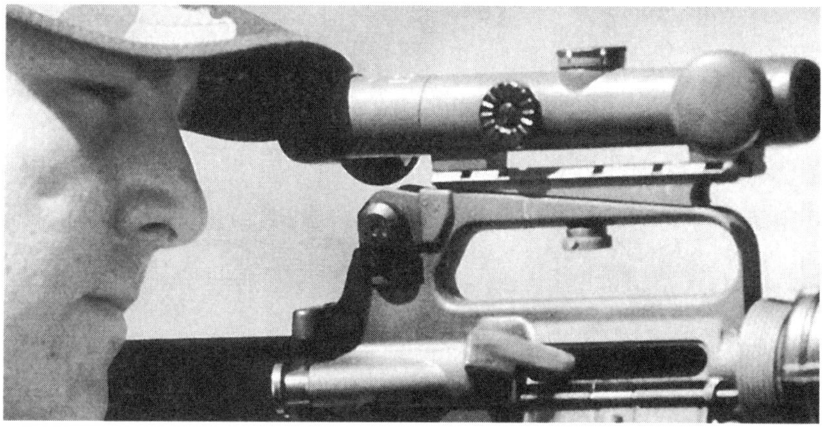

Steiner Scope lends accuracy, but head-high posture exaggerates recoil.

M16: Fact and Fiction

The M16 (originally called the AR-15) is a creature of Eugene Stoner, then-chief engineer for Armalite, a division of Fairchild Aircraft. The AR-15 was essentially a scaled-down version of Stoner's AR-10. A small quantity of AR-15 rifles were delivered to Ft. Benning for test and evaluation against the M14 on 31 March 1958. In a simulation of combat environments, the M16 proved to be almost three times as reliable as the M14. But Gen. Maxwell Taylor, the Army Chief of Staff, vetoed any further CONARC development of the AR-15 in favor of continued procurement of the M14. Fairchild, disenchanted with the AR-15 program, sold the entire AR-15 package to Colt in December 1959.

The AR-15 was tested in Vietnam by the Defense Department in the summer of 1962, under the code-name Project AGILE. The AGILE report was more than enthusiastic, as great claims were made for the .223 cartridge's killing power, and the improved handling, reliability, durability and ease of maintenance of the AR-15 over the M14. A favorable cost-effectiveness report followed from the DOD Comptroller's Office.

The anti-M14 group now had an alternative to rally around. A number of Pentagon agencies entered the fray and began comparing the AR-15 and M14. A comparative evaluation between the two rifles was held at Aberdeen Proving Ground late in 1962. The results were ambivalent.

A comparative lethality and wound-ballistics test at Edgewood Arsenal stated that the earlier Project AGILE report of the .223's killing power was a gross exaggeration. The official Army reply to Secretary of Defense McNamara's order for the comparative examination of the two rifle systems flatly concluded that ". . . only the M14 is acceptable for general use in the U.S. Army . . ."

But too much evidence pointed to an opposite conclusion. An Army Inspector-General's investigation decided that the Army had rigged some of the tests against the AR-15. As a consequence McNamara terminated procurement of the M14 rifle on 23 January 1963 and announced a "onetime buy" of 85,000 AR-15 rifles for the Army and 19,000 for the Air Force. An entirely new weapon system called SPIW (Special Purpose Infantry Weapon, firing small-caliber cartridges using steel-flechette multiple projectiles imbedded in plastic sabots) was believed to be at hand.

SPIW never materialized, and intensification of the war by 1965 caused Gen. Westmoreland to request the M16A1 rifle for all ground combat elements in Vietnam. Procurement was accomplished by August 1966. In December of that year, the U.S. Army type-classified the M16A1 rifle and it replaced all .30-caliber rifles in its inventory, except those eventually retained for use as sniper rifles.

By the spring of 1967, Colt's bed of M16 roses started to rot as reports of widespread malfunctions in combat began to appear in media hungry for every tainted tidbit about the war in Vietnam. The press gleefully printed melodramatic letters supposedly written by GIs whose comrades had fallen dead next to their jammed M16s.

And there were real problems.

Foremost was the change in midstream from a cartridge using an IMR propellant to a ball powder. Innuendoes of intrigue were leveled against the powder manufacturer, Olin Winchester. In truth, ball propellants generally burn cooler than extruded IMR-type powders, extending barrel life. This is no small consideration for modern, lightweight assault rifles with full-auto capability. The M16 upper and lower receiver bodies are fabricated from T6 aluminum, not steel, a far superior heat reservoir. The tradeoff—and there always is one—is that ball propellants generate more carbon residue which, of course, accelerates fouling of the gas system.

Ease of maintenance had been overstressed by both Colt and the Army. Proper cleaning equipment was not issued with the rifle and the troops were not impressed with its importance. In the humid jungles of Southeast Asia that was a lethal error of judgment. Subsequently, M16A1 chambers and bores were chrome-lined: another corrective refinement.

Early on, some M16s would occasionally fire out of battery as a result of excessive bolt carrier bounce. That bounce caused the same problem when the Soviets switched from the heavy, forged and milled receiver body of the AK-47 to the lighter sheet-metal receiver of the AKM. The Russian solution was a five-component device which acts as a drag on the hammer and has been erroneously called a "rate reducer" by Western experts. Research by Colt produced a new buffer which eliminated the potentially dangerous carrier bounce. An unexpected additional benefit was a reduction in the cyclic rate.

Original bolts could be installed with the extractor to the left, and that made the Black Stick futilely attempt to eject spent cases through the portless left side. But small arms systems mature with experience and one more fix was re-machining the cam pin and bolt to prevent misassembly. Designers and soldiers alike have been plagued by the M16 magazine. Sometimes troops overloaded them, but their flimsy, almost throw-away design and construction has troubled the system to this day.

With the single exception of a still-inadequate magazine, the M16's woes are now almost two decades in the past. Military authorities long ago conceded the M16 to be one of the world's best assault rifles, every bit the equal of the ComBloc Kalashnikov and its superior in accuracy potential.

But controversy haunts the M16. Charges of unreliability and inaccuracy, supported mostly by rumor or innuendo, dog the original microcaliber assault rifle. And for the M16's critics the *bete noire* is still the 5.56mm NATO cartridge.

On 28 October 1980, NATO approved the standardization of a second small caliber cartridge for use within the alliance (STANAG 4172), based on the Belgian SS109 5.56mm ammunition. Three different calibers were represented in the NATO trials: the British 4.85mm (never a serious contender), the German caseless 4.7mm (withdrawn because of cook-off problems) and three 5.56mm projectiles (SS109, U.S. XM777 and M193 for control).

Three considerations were paramount: increased long-range effectiveness, semi-armor-piercing potential and fear that high velocity projectiles of the M193 type might be restrained by international agreements. The first two are desirable attributes. The so-called "humanitarian" parameter—a decrease in the tumbling and breakup characteristic of the M193—is a grotesque charade. The Russians drop poison gas on the Afghans and biological bombs on the Cambodians and NATO worries about decreasing the inhumanity of mankind's most inhumane activity—the perpetual conflict of nations.

The SS109 projectile, with its more sharply tapered form (ogive), greater weight (62 gr.) and hardened-steel penetrator frontal core, offers armor penetration superior to the M80 7.62mm NATO projectile at greater ranges, boring right through the three NATO penetrations targets (3.5mm of mild steel plate at 640 meters, and the West German and U.S. steel helmets at 1,150 and 1,300 meters, respectively). As the use of body armor on the battlefield is expected to increase, this looms as an important quality. The improved ballistic stability which yields greater long-range capability through improved wind drift performance has not been offset by the expected marginal loss in lethality. Extensive tests (still classified) at Aberdeen Proving Ground indicate the M855 (SS109) projectile has greater lethality than the M193 at ranges past 100 meters.

M193 ammunition performs admirably in the new M16A2 with increased long-range potential and little effect on tumbling and breakup, since the change in barrel twist (from 1:12 inches to 1:7 inches) has less effect on ballistic stability (yaw) than does the bullet's geometry. In fact, the barrel twist was principally changed to stabilize the new, much longer M856 (L110) tracer projectile which started to keyhole in less than 100 meters through 1:12-inch barrels and to yield a ballistic match between the ball and tracer cartridges.

This is welcome news since Lake City Arsenal is apparently having real trouble producing the more complex M855 (SS109) projectile—so we probably won't be seeing it in quantity for some time to come (the 200-rd. plastic assault packs for the M249 SAW will initially be supplied by FN). Obviously, important work is still afoot in re-engineering the '16's cartridge, and the weapon's capacity to handle it.

Systems mature by correcting problems, and—as it is with men or nations—that takes time. Stoner's system and its cartridge have been progressively adapted to each other, to the battle environment, and to user need. The M16 entered the arena of war as one of the most reliable and effective tools of conflict, and decades of testing, user feedback, and controlled development have only changed it for the better.

M16A2 proved comfortable for both left-handed testers.

Portions of the historical segment of the text were obtained from the following sources:

Ezell. E.C. 1983. *Small Arms of the World*; Harrisburg. PA. Stackpole Books. 12th edition. 894 pp.

Kirby. W. K. 1972. *The M16 Rifle Controversy*: Oregon. Portland State Univ. unpub. Master's thesis. 63 pp.

Stevens. RB. 1983. *U.S. Rifle M14 from John Garand to the M21*: Toronto, Canada. Collector Grade Publications. 340 pp.

In calculations based on physical measurements, the least precise measurement determines the number of significant figures that may be used. Since range distances are always given, and measured, to one significant figure (i.e., 100m, not 100.0m) only, MOA figures measured to the tenth are stretching the point a bit. Extending this information to hundredth's (1.83, etc.) is an absurd assumption of nonexistent precision. Two MOA is excellent for a modern military assault rifle and will beat, by a considerable margin, any one of the more than 30 million Kalashnikovs manufactured.

So much for barrel erosion.

Another rumor about the M16 series of weapons is its reputation for mechanical unreliability. Of course, this has never been established by any scientific or even systematic test, yet the myth persists. There were three stoppages attributable to the test rifle: two failures to fire (light hits) and one failure to extract.

There were 24 failures to feed. But in every instance, the bolt overrode the cartridge base, dented the case severely with the locking lugs and failed to chamber the round. Called a "bolt-over-base" stoppage, it is always a result of faulty magazines.

We had only nine mags with which to conduct the entire 6,000-rd. test and the swirling dust played hell with them. Yet, the bolt-over-base malfunctions ceased when defective magazines were isolated and withdrawn from the test cycle.

As a practical matter, our test indicated shooters should never load M16 magazines to more than capacity and stick with Colt-manufactured mags rather than substitutes such as Adventure Line models.

Further, to avoid magazine stoppages, disassemble all mags; clean them thoroughly, and lightly lubricate the follower spring after you have stretched it by hand. If a magazine continues to cause problems, trash it or sell it at a gun show to someone you don't know. A new 30-round magazine with an improved magazine follower is now in production, according to Colt, but something more drastic than a change in magazine followers is required. The M16A2 needs and deserves an entirely new magazine system.

Incidentally, after these bolt-over-base failures the rifle fed, fired, extracted and ejected every bent cartridge when it was recycled from a reliable magazine.

More findings from our *SOF* test:

Sometime after round 4,000 three of the four rear retaining tabs on the aluminum heat shields under the handguards broke away. It did not affect the rifle's operation or handling characteristics. The rifle was almost too hot to hold at the end of every cycle in the test. Also, at the end of the test, the plastic nub at the end of the buffer was splayed and cracked and appeared close to failure.

SOF Senior Editor Bill Guthrie and I were both able to consistently place all three shots of a burst into a standard military silhouette target at 50 meters from the standing position without a sling. This is the maximum distance point targets should be engaged with three-shot bursts. Riflemen should be trained and ordered to fire their rifles semiautomatic only. Now that we once more have a true SAW at the section level—the M249—the rifleman's requirement to provide intensive supporting fire is diminished and three-shot bursts should be restricted to close-range—like "they're coming over the wall"—emergency use only.

Two of the test participants, Jason Kokalis and Patrick Martin, are left-handed. After more than 1,000 rounds apiece, they learned to appreciate the case deflector installed for their 20 percent of the world's population.

The new, all-checkered buttplate met with mixed reaction. Two testers felt it improved the handling characteristics, prevented stock slippage and increased stability, especially in the three-shot burst sequences. Two felt it cut too sharply into shoulders protected only by a T-shirt because of the broiling desert sun.

At first, this cutting effect was exaggerated by high perceived recoil. Surprised at this in the heavier weapon, we continued to suffer superficial bruising and abrasions, until we removed the Steiner Scope. Rugged, compact and well-adapted to battlefield precision shooting, the

Steiner requires a raised-head shooting posture which accentuates recoil.

Lower head position allows more secure stock-weld (or spot-weld), since the cheek must be lifted off the stock when aiming through any scope mounted on an M16's carrying handle. This may also be attributable to the lengthened stock which allows a more comfortable stock-weld for most shooters.

Acclaim for the new rear-adjustable sight system was universal. Vertical adjustment is what is most often needed for field marksmanship and the old 'A1's elevation by turning the front sight never was acceptable. Neither is holdover: Kentucky windage went out with the Kentucky rifle.

In general, reaction to the M16A2 was positive—even ecstatic. Everyone felt it was a quantum leap forward. USMC Commandant Gen. Paul X. Kelley has stated, "We think that in the M16A2 we have come as close to perfection as you can come in a rifle." I think he is correct.

A civilian semiauto version of the M16A2 is now available. Called the AR-15A2 Sporter, it differs from its military counterpart in the following ways. The old style rear sights have been retained. There is no forward bolt assist or case deflector. It is being issued with the older dust cover. The geometry of the rear portion of the lower receiver remains the same as before. There is, of course, no selective fire option. As with the earlier AR-15 semi-auto rifle, the lower and upper receivers are connected at the forward end by two screws instead of a push-pin—a condescension to the BATF, as it supposedly inhibits attaching a full-auto lower receiver to an AR-15 upper receiver assembly.

Originally appeared in *Soldier Of Fortune*
January 1985.

Heckler & Koch's Revolutionary G-11

H&K's revolutionary caseless gun looks like it came straight from Buck Rogers.

Caseless Gun Highlights ADPA's International Symposium on Small Arms

George F. Curtis, manager, Advanced Development Engineering for Maremont Corp. and designer of new M60 lightweight assault machine gun, caliber 7.62mm NATO. MG weighs only 18.5 lbs.; bipod mounted to receiver instead of barrel; amibidextrous safety; foregrip for improved controllability; adjustable front sight; double sear-notch for safety; winter trigger guard, and gun can be cocked after feed cover is closed.

Watching the Heckler & Koch G-11 caseless gun being fired is an experience like no other. Totally unorthodox in appearance, with an exterior seemingly composed entirely of plastic, it is a Buck Rogers Ray Gun of the first order. The scope tube (which features a 1X optical sight) and the scope mount itself seem to be integral with, and part of, the receiver molding. They also serve as the carrying handle. The completely in-line stock. the absence of protrusions, and the overall ultrastreamlined configuration at first almost offend the senses. "Surely," I caught myself saying, "this cannot be a real assault rifle."

The moment it is actually fired, the brain is jarred by another startling, yet quite obvious observation: There is no ejection port! No cases fly outward from either side of the weapon! It is, after all, what they said it was—a caseless gun!

Apparent recoil in both the semiauto and full-auto modes was nil. The only visual movement was that of the magazine as it slammed back and forth, feeding rounds into the system. The magazine, which holds 50 in-line caseless cartridges, is as bizarre in location as the remainder of the weapon is in appearance. It rests as a horizontal bar atop the barrel and extends rearward from the muzzle all the way back to the receiver. The magazine is inserted and detached up forward, the contention being that it is safer to insert it in the direction of fire. The magazine can be loaded one cartridge at a time, or by use of 25-round loading clips. An indicator at the top of the magazine identifies the quantity of ammunition which remains.

Four separate rifles were used in the demonstration conducted by H&K reps. Two were dark gray in color and two were olive drab. The significance of this was not revealed. However, they were probably steps along the path toward lighter versions. The latest variant weighs in at 9.9 pounds with 100 rounds of ammunition. No malfunctions of any type were observed during the entire firing sequence, which consisted of approximately 300 rounds. The firing sequence ranged from single rounds, through three-shot bursts, and finally, in one instance, to a rather well-controlled firing of an entire magazine of 50 rounds. The man-sized silhouette targets were set up from 50 meters out to about 150 meters. Well-placed hits were consistently made at all ranges in all the firing modes.

At the briefing given by Michael W. Iten, H&K's military sales rep, we learned that the weapon, which is caliber 4.7x21mm, utilizes a rotating breech (somewhat reminiscent of the principle used in the 16th-century Cookson repeating flintlock). A cylinder, approximately 2 to 2.5 inches in diameter and about one inch in width, contains the partially rectangular-shaped chamber, and rotates 90 degrees out of and in to alignment with the barrel. Gas pressure is used to tilt this cylinder upward into the loading position. A round is then stripped from the magazine, bullet down and primer up, into the chamber. The cylinder, which rotates perpendicular to the barrel axis, is then turned forward and clockwise into alignment with the barrel and the firing pin, which is to the rear of the cylinder. During the firing sequence, the magazine, cylinder and barrel recoil together, which partially explains the lack of observed recoil. The hand guard encloses the recoiling barrel assembly.

The rifle fires multiple three-shot bursts while actually in recoil. The full-auto mode is in reality a series of three-shot bursts. The cyclic rate is quite high, possibly in excess of 1,000 rpm, and is made possible by the extremely short bolt travel. The polygonally rifled barrel has a gain twist, which means that the twist increases toward the muzzle end (1:6 inch at the muzzle). This is somewhat peculiar, as gain-twist barrels have not been successfully used much since the advent of conical bullets. The ammunition has a muzzle velocity of about 3,100 fps and of course the primer and propellant are totally consumed. The latest

Soviet PK medium machine gun, caliber 7.62x54mm Russian Rimmed. Designed features derived from AK-47, Goryunov Sg 43, Czech VZ52 LMG, and Degtyarev RPD. Weapon weighs about 19.8 pounds. Gas-operated with cyclic rate of about 650 rpm. Excellent, sturdy weapon.

FN factory representative fires FNC carbine, caliber 5.56mm NATO in full-auto mode. Weapon has just been dug up after having been buried in ground for four hours: flawless operation.

ammunition uses a propellant with a cook-off point that is 100 degrees higher than that of standard nitrocellulose powders (solving a problem which has plagued this weapon since its inception).

The entire system appears to be quite dirt-sensitive and must be sealed against all the elements—as evidenced by the rubber boot which surrounds the trigger mechanism. Debris entering the magazine opening under field conditions will undoubtedly induce functioning problems. The G-11 can be fitted with both a bipod and bayonet.

Imagine being able to see and examine just about every new and significant development in Western and ComBloc military small arms in one live-firing demonstration. The recent ADPA (American Defense Preparedness Association) Second International Symposium on Small Arms at Ft. Benning, Ga., gave just that.

Coordinated by Col. Paul H. Scordas (USA, Ret.), ADPA's director of weapons technology, the symposium brought together in one think tank the Western world's largest assemblage of military small-arms brain power. The 185 participants represented Australia, Austria, Belgium, Canada, England, Germany, Israel, Italy, Netherlands, Norway and Sweden, in addition to all five branches of the U.S. military, the FBI and the Secret Service.

One of the most interesting briefings we received was given by Angelo Mancini (Deputy Chief, JSSAP Project Office, ARRADCOM, Dover, N.J.) on the M16 Rifle Improvement Program. The proposed M16A2 rifle, intended to serve as an interim system, pending the arrival of the next generation of infantry rifles, incorporates a number of significant modifications: a heavier barrel with a 1:7-inch twist to accommodate the new 62-grain SS109 bullet, a more durable handguard (it resembles the M16A1 carbine [CAR 15] forearm, only it's longer), a redesigned slip ring which will make the handguard easier to disassemble, a sturdier buttstock and pistol grip, a new upper receiver with rear sights out to 800 meters, a flash suppressor without bottom cuts (the absence of which should aid in the reduction of muzzle climb) and three-shot burst control.

Hughes EX-34 Chain Gun, caliber 7.62mm NATO. Cyclic rate of 570 rpm, smallest member of Chain Gun family. Ejects fired cases forward (overboar). Long bolt-dwell time minimizes gas buildup for vehicular installations.

Karl Walter fires Steyr AUG assault rifle, caliber 5.56mm NATO.

Walther MPL submachine gun with Larand sound suppressor—extremely effective.

The Canadians, always models of terseness and efficiency, announced they are replacing the Sterling submachine gun and the FN C1A1 rifle with a single weapon—either the M16A2 just described or the Belgian FNC. The new Canadian squad automatic will be the XM 249 (MINIMI), with the only decision remaining whether to issue the weapon with one or two barrels.

The other symposium topics included presentations on Australia's future small arms; JSSAP updates on the pistol, submachine-gun, and advanced heavy-machine-gun programs; the FNC 5.56mm NATO carbine system; the Heckler & Koch G-11 caseless gun; helmet/body-armor developments; plastic training ammunition; naval line-throwing equipment; hypervelocity squeeze-bore, Raufoss multipurpose, and fin-forming cone ammunition; the Enfield semi-automatic anti-riot weapon; and European trends in light antiaircraft guns.

The weapons fired during the range demonstration included the Soviet AK74, AK74S, RPK-74 (all in 5.45mm ComBloc), the SVD sniper rifle and the PK medium machine gun (both in 7.62mm Russian Rimmed); Hughes Chain Gun (EX-34) in 7.62mm NATO; the new M60 Light Weight (LW); the amazing new Mini Uzi; Galils in all configurations and in both 5.56mm NATO and 7.62mm NATO; the FNC carbine; Steyr submachine guns and AUG assault rifles in all barrel lengths; the revolutionary H&K G-11 caseless ammo assault rifle; the Leatherwood submachine gun; and the Valmet M78 LMG in 5.56mm NATO and 7.62mm NATO. Larand sound suppressors were demonstrated, and both Steyr and Israeli Military Industries also demonstrated their respective rifle-grenade systems.

Without doubt, the objective of the symposium—which was to provide a common meeting ground for the exchange of information on small arms and crew-served weapons of current and future interest to NATO and Western nations—was more than well served.

Originally appeared in *Soldier Of Fortune*
March 1982.

ISRAEL'S DEADLY DESERT FIGHTER

SOF's Kokalis Evaluates Galil's AK

SAR—with optional 50-round magazine—on rock 'n' roll shows little muzzle climb. Note bolt moving forward to chamber another round.

Semiauto Galil ARM fired from bipod by Al Nordeen in Guat cammies and beret decked with Guatemalan Master Para/Pathfinder wings and paratroop battalion badge.

The Galil rifle is a phoenix, risen from the ashes; a result of lessons learned by Israeli desert fighters in the 1967 Six-Day War. Very much the progeny of my friend, Israel Galili, chief weapons designer for IMI (Israeli Military Industries), and Yaacov Lior, the Galil is a somewhat successful attempt at Candide's "best of all possible worlds."

Dissatisfied with the 7.62mm NATO FN FAL with which the Israeli Army was largely equipped, as it has always been a poor performer in high sand and dust environments, Galili went directly into the field to investigate the problem. He was told by everyone that the Kalashnikov was the "tiger of the desert."

Taking what he needed from the AK-47, Galili placed his rifle in competition with the M16A1, the Stoner 63, the AK-47, the HK 33 and a design by Uziel Gal. The test's greatest emphasis revolved around performance under arid-region conditions. The Galil emerged as the clear winner and won the Israeli Defense Award. It was officially adopted by the Israeli Defense Forces (IDF) in 1972. More than a decade later, it is now finally available through Magnum Research, Inc., its exclusive importer, in BATF approved semiautomatic versions. The selective-fire versions are available to law-enforcement agencies and qualified Class 3 dealers.

Although also produced in caliber 7.62mm NATO to increase its sales on the world market, the Galil rifle as issued to the IDF is chambered for 5.56mm NATO M193 ball ammunition.

The Galil's Kalashnikov heritage is apparent, even at first glance. Not so evident are its differences. It fires from the closed-bolt position and is gas-operated without an adjustable regulator. The change in caliber, from 7.62x39mm ComBloc to 5.56mm NATO, required numerous alterations. The AK-47's 4.2mm gas hole was reduced in diameter to 1.8mm. The Galil's most immediate predecessor was the Finnish Valmet M62 rifle and, in fact, early Galil prototypes were fabricated using M62 receivers made in Helsinki. However, as the 52,000 cup SAAMI (Sporting Arms and Ammunition Manufacturers' Institute) pressure limit specified for the 5.56mm NATO round is far greater than that developed by the 7.62x39mm ComBloc cartridge, Galili abandoned the pinned and riveted, stamped sheet-metal receiver of the Valmet M62/M76 series and went to a heavy milled forging.

In addition, the Galil does not utilize the usual Kalashnikov barrel-extension unit for lock-up of the bolt. The bolt lugs lock into recesses milled into the receiver body itself. Thus, heat dispersion occurs more rapidly, the cartridge remains cooler and the possibility of a cook-off, even under the most intensive full-auto conditions, is minimized.

While the method of operation is identical to the Kalashnikov, Soviet AK-47 parts most certainly cannot be used in the Galil, contrary to the statements of others. When the trigger is pulled, the hammer drives the firing pin forward to ignite the primer. Kalashnikovs have inertia firing pins without a spring. The initial lot of Galils brought into this country also had no firing pin springs. Military primers have hard cups, not easily touched off. American commercial ammunition, Winchester in particular, has relatively soft primer cups. The Winchester ammo caused several slamfires and all Galils offered for sale in the United States have now been fitted with strong firing pin springs. If yours does not, have it retrofitted before firing commercial ammunition.

After ignition of the primer, a portion of the propellent gases migrate into the 1.8mm vent, drilled at a 30-degree angle into the gas block which is pinned to the barrel. The gas enters the cylinder (to which a small spring has been attached to secure its retention during reassembly) and drives the piston rearward. The piston is hard-chrome-plated for ease of maintenance. It is also notched to provide a reduced bearing surface and permit excess gas blow-by. The bolt carrier is attached to the piston. After a short amount of free travel, during which time the gas pressure drops to a safe level, the cam slot engages the bolt's cam pin and the bolt is rotated and unlocked as the carrier moves rearward.

Primary extraction occurs as the bolt is rotated and thus the massive Kalashnikov-type extractor claw is not required. Empty-case ejection is typically violent. The cases are severely dented by the ejector and thrown to the right and front by as much as 40 feet (a defect with regard to position disclosure). At this time, the recoil spring is compressed and its return energy drives the carrier forward to strip a round from the magazine and chamber it.

The Galil's hammer spring is made of multistrand cable. The trigger and sear springs are conventional coil types. Like other Kalashnikov-system rifles, the trigger mechanism is that first used in the U.S. M1 Garand rifle.

The Galil's right-side selector lever is the same stamped, sheet-metal bar common to all Kalashnikovs. South African troops often wrap nylon line around this selector bar to quiet the sound of its manipulation. It can also be slightly bent to draw it away from the receiver notches.

The top position, marked "S," is safe, where the trigger is locked and the bolt can be retracted only far enough to inspect for a chambered round in this position.

The Galil also features a selector switch on the receiver's left side, intended to be manipulated by the thumb of the trigger hand. On the semiauto version, through use of a two-piece hinged bar inside the receiver, the rearmost position of this selector is safe and pushing forward with the thumb will place the weapon in the firing mode, marked "F." This is as it should be. However, on the selective-fire model the rearmost position is "R" (British terminology for Repetition, or semiauto), the middle position is "A" (full auto) and the forward. position is safe. Thus, to come off safe, using the left-side selector, one must pull rearward with the thumb, a most unnatural and awkward maneuver, especially under stress.

On the selective-fire Galil, two sears control the firing mechanism, the trigger sear and a safety sear. In full-auto fire the trigger sear is held back and only the first round of the burst is fired off this rear sear. Subsequently, the bolt carrier moves rearward and rolls the hammer over. The safety sear continues to hold the hammer back until the bolt carrier is fully forward again, at which time it trips the safety sear and the hammer rotates to fire another round. Thus, after the first round the trigger sear is deactivated entirely from control on the hammer. Releasing the trigger will catch the hammer on the trigger sear once more. In semiautomatic fire, no pressure is placed on the trigger sear, which is free to catch the hammer each time it is rolled back by the bolt carrier.

The entire safety sear assembly (sear, spring, cross pin and trip lever) is absent from the semiautomatic-only version of the Galil. In addition, certain receiver mill cuts have not been made, the hammer spring pin protrudes from the right side of the receiver to stop further downward travel of the selector lever and the bolt carrier has been altered to prevent full-auto fire. Unauthorized attempts to convert this rifle to selective fire would be most difficult and quite dangerous.

There are three basic configurations of the Galil, all available in calibers 5.56mm NATO or 7.62mm NATO: The ARM is equipped with a bipod, wooden handguard and carrying handle. It is intended for use as an assault rifle and

Galil Specifications

	ARM	AR	SAR
CALIBER	5.56mm NATO	5.56mm NATO	5.56mm NATO
OPERATION	All gas-operated, rotating bolt, magazine fed, fire from closed-bolt.		
CYCLIC RATE (selective fire models)	650 rpm also available semiauto only	650 rpm also available semiauto only	650 rpm available as selective fire only
WEIGHT, empty	9.6 pounds (with bipod and carrying handle)	8.7 pounds	8.2 pounds
LENGTH,			
butt folded	29.2 inches	29.2 inches	24.2 inches
butt extended	38.6 inches	38.6 inches	33.5 inches
BARRELS	All six-groove, right-hand twist, pitch of 1 turn in 12 inches.		
BARREL LENGTH (with flash suppressor)	18.5 inches	18.5 inches	13.5 inches
SIGHTS			
front	All adjustable post-type with protective hood.		
rear	All flip-type with apertures set for 300 and 500 meters.		
night	All folding, with tritium dots.		
PRICE	$1,499	$1,399	$1,399
ACCESSORIES	Magazines, 12-rd. ballistite, $47.95; 35-rd., $47.95; 50-rd., $54.95; adapter for M16 magazines, $81.95; cleaning kit, $23.95; sling, $28.95; wood stock, $124.95.		
MANUFACTURER	Israel Military Industries, Tel Aviv, Israel		
IMPORTER	Magnum Research, Inc., 2825 Anthony Lane South, Minneapolis, MN 55418.		

squad automatic weapon. The AR is equipped with a high-impact-plastic handguard without a bipod or carrying handle. The barrel length of both, in caliber 5.56mm NATO, is 18.5 inches with the flash suppressor (and 21.0 inches for the 7.62mm NATO models). Both are available in semiauto-only and selective-fire versions. The SAR is a short-barreled version of the AR model. It has a barrel length of only 13.5 inches in 5.56mm (15.8 inches in the 7.62mm version) and, as a consequence, is available in the United States as a selective-fire weapon only. Its gas tube and piston are 1 1/8 inches shorter than the other models. The 5.56mm NATO Galils all have six-groove barrels with a right-hand 1:12-inch twist for the M193 ball projectile. All three are normally issued with a folding stock, although a wooden buttstock is an available option.

At first glance, the folding stock appears to be that of the FN FAL. It is not. The FAL stock is constructed of tubular aluminum. The Galil folding stock is fabricated from tubular steel—stronger, but heavier. More important, the Galil stock has no button latch to confound the operator in opening or closing, no small consideration during high-stress situations.

The ARM's carrying handle is almost identical to the Belgian FAL's. Located to the rear of the wooden handguard, it is not positioned over the rifle's center of mass.

The wooden handguard remains somewhat cooler during sustained full-auto fire than the black plastic handguard. The squared-away shape of the wooden handguard is not entirely comfortable, but necessary to store the bipod.

The Galil bipod is a sturdy, rigid affair, certified so by my memory of Israel Galili jumping wildly and theatrically on top of the rifle with its two steel legs extended. When stored in the handguard, the bipod serves as a feed chute to speed insertion of the magazines. The bipod can be used as a wire cutter and to open beer bottles also.

The Galil's gray-plastic pistol grip is one of the very best ever put on an assault rifle and seems to be taken from the Hungarian AKM/AMD-65 series. Of more than adequate length, with a sharp bottom flare to prevent the hand from slipping, the grip has been mounted to the receiver at precisely the correct grip-to-frame angle. Somehow, it just feels right.

Galil offers tough, all-steel magazines in three capacities: The 12-round magazine, color-coded with white stripes, is blocked to accept only ballistite (blank) cartridges for launching rifle grenades. The standard magazine has a capacity of 35 rounds. A large capacity 50-round is also available. Difficult to load by hand, it is intended for use primarily in the squad automatic role. However, like all bottom-fed magazines of this length, it will "monopod" the weapon when fired with the bipod in the prone position.

An optional magazine adapter allows the use of 20- and 30-round M16 magazines. Unfortunately, the magazine wells of the semiautomatic and selective-fire Galils are of different dimensions and the adapter supplied by IMI can be fitted only to the semiautomatic version. Why this is so I do not know. However, the adapter is well-designed and the magazines can be inserted and released with no greater difficulty than in the M16. Valmet 5.56mm NATO magazines will likewise fit into the semiautomatic Galil, but cannot be used in the selective-fire rifle. South African R4 magazines are identical to their Israeli counterparts and can be inserted into all versions of the Galil. The magazine-release latch is of the flapper type, similar to the Kalashnikov.

The retracting handle is attached to the bolt carrier and bent upright to allow cocking with either hand, providing a useful feature.

The flash suppressor has six ports and is almost identical to the M16 "bird-cage" muzzle device. Those who still dream of charging up San Juan Hill will be pleased to note that the Galil accepts the readily available M7 bayonet issued for the M16.

The rear end of the Galil's recoil-spring guide rod, which serves as a retainer for the sheet-metal receiver cover, has been extended to ease disassembly and lock the cover more securely to the receiver body. This is especially important as the rear sight has been mounted on the receiver cover. While no less secure than its attachment to the gas cylinder on the Valmet M71, it does not provide the rigidity offered by the receiver-mounted rear sight of ComBloc Kalashnikovs. The trade-off is a longer sight radius.

Reassembly of the receiver cover on all Kalashnikov-type weapons is simplified if you first place the recoil-spring guide rod slightly below its notch in the receiver onto the rear interior wall of the receiver. Then set the receiver cover in place. Jack the retracting handle smartly to the rear and the guide rod will pop into its notch and the square-cut hole in the receiver cover.

Standard Kalashnikov disassembly and reassembly procedures apply to the Galil. But, a small, though important, correction to the preventive maintenance instructions given in the IMI operator's manual is required. After cleaning, we are instructed to lubricate the gas cylinder and piston. I say no to that. Keep lubricants of all types away from the piston and the interior of the gas system. The intense heat generated in this area of a gas-operated weapon will cause lubricants to bake and varnish these parts.

The rear sight is a flip-up peep type with 300- and 500-meter apertures adjustable for elevation only. The front-post sight is adjustable for elevation and windage zero. Elevation adjustments are by means of the UZI front-sight tool. Windage adjustment is achieved by loosening and tightening the two opposing screws which move the entire front-sight assembly in its dovetail on to the gas block. The diameter of the front-sight hood is such that it forms an additional aiming circle just within the rear aperture to further assist sight alignment and speed target acquisition.

Taking another cue from the Valmet, the Galil is equipped with tritium (betalight) night sights set for 100 meters. To use, at dusk or night, the front betalight is folded up to expose a vertical bar, which is aligned between the two rear luminous dots. When the rear tritium sight is flipped up for use, the rear peep sights must be placed in an offset position midway between the two apertures.

The left side of the receiver is dove-tailed for a scope side-mount. Mounting a scope on the receiver body usually results in maximum stability. But the IMI side-mount has exhibited a decided tendency to lose zero after take-down and remounting. As a consequence, Magnum Research, Inc. plans to market a Weaver-type base attached to the sheet-metal receiver cover (usually the worst place to mount a scope). The initial units will be equipped with the excellent Leatherwood ART II.

The Galil issue sling is admirable. Constructed of heavy, wide, black webbing with sturdy steel hooks at each end that rotate 360 degrees, it is easily the best assault rifle sling I have ever seen. Designers in the past have often neglected this piece of equipment, yet it is important to those in the field.

After phosphating (Parkerizing), all exterior metal surfaces on the rifle (except for the barrel, gas block and front sight) are finished with semi-gloss black enamel.

An interesting after-market accessory has already surfaced for the Galil. Produced by J.F.S., Inc., the Redi-Mag fast-action speed loader attaches in minutes to the left side of the receiver next to the magazine well. The Redi-Mag holds one spare magazine. By means of a connecting catch bar, its operation is synchronized with the rifle's magazine-release latch. To manipulate the Redi-Mag, drop the muzzle about 10 to 15 degrees and, with the left thumb, press the catch bar forward while grasping the spare magazine with the left hand. While rocking the loaded magazine out of the Redi-Mag, the empty magazine will fall to the ground. Insert the new magazine and you're back in business.

I have fired several thousand rounds through both the ARM and SAR in the off-hand, kneeling, hip-assault and prone positions, and can report no stoppages of any kind. Of course, I neither threw them in the mud nor rolled over them with a truck, as such tests have already been completed under controlled and repeatable laboratory conditions by IMI. And properly so, as such tawdry, unscientific displays demonstrate nothing but the vaudevillian inclinations of the popular gun press.

The five-inch differential in barrel lengths between the ARM and SAR did provide an excuse to chronograph their respective muzzle velocities. PMC (Pusan Arsenal, Korea) M193 ball ammunition was used throughout this portion of the test and evaluation. The 18.5-inch barrel of the ARM generated an average of 3,087 fps. The stubby 13.5-inch barrel of the SAR dropped the average velocity by only 183 fps. to 2,904 fps. The extreme spread and standard deviation were significantly lower for the SAR. But, the accuracy potential of both rifles was quite high, even with trigger pulls no better than the average Kalashnikov.

In addition to high marks for hit probability and target acquisition, the SAR exhibited phenomenal controllability in the full-auto mode. The cyclic rate is 650 rpm. Muzzle rise is barely perceptible with two- and three-round bursts. In fact, firing in the off-hand position, at 30 meters an entire and continuous 50-round burst can be contained within a standard military silhouette target!

Felt recoil was virtually nonexistent with both rifles. But, a *heavy* price must be paid for all these attributes.

All of the above operating characteristics are a function of the weapon's weight. At almost 9.5 pounds, empty, with bipod and carrying handle, the ARM is quite heavy in comparison with other state-of-the-art assault rifles. The M16 and AKM weigh only 7.0 pounds apiece. The Galil is only a quarter-pound shy of the U.S. M14. So what, you say? The South African troops who must constantly drag this beast through the bush have real cause for complaint. And reports to me indicate that their moaning and groaning have reached a discordant crescendo.

The Galil's weight is principally a consequence of the designer's attempt to create a weapon system which could serve the roles of submachine gun, infantry rifle and light machine gun. To date no other short-barreled assault rifle comes closer to stealing the submachine gun's final fading thunder than the SAR. With its stock folded, it measures only 24.5 inches in length. Most modern submachine guns fall somewhere between the 16.4-inch Beretta M125 and the 19.3-inch HK MP5A3. Presenting a package in size close to the SMG, the SAR chambers a cartridge far more potent at much greater ranges than the 9mm Parabellum.

As an assault rifle, the sturdy and reliable Galil is one of the very best. With a U.S. retail price of $1,499, whether or not it is worth the cost of almost three AR15s is a question only you can answer.

It is in the role of a squad automatic weapon that it falls short of the mark. By definition of its requirement for intensive sustained fire at the squad level, the ideal SAW should incorporate a quick-change barrel, adjustable gas regulator and belt-feed potential. The Galil has none of these features; the U.S. M249 has all of them. And, as I mentioned previously, the tendency of the 50-round magazine to "monopod" the weapon when fired off the bipod in the prone position seriously compromises the Galil's ability to effectively engage targets at any small degree of elevation above the operator.

In general, the Galil system is well-executed, and a fine example of the qualities one should look for in a modern assault rifle. It stands as testimony to the brilliance of Israel Galili as a military small-arms designer and is, without doubt, his crowning achievement to date. That it is not perfect is simply an axiom which has held since the invention of gunpowder and will lead to the continued evolution of military small arms. The Galil's important position in the history of such matters is secure.

Originally appeared in *Soldier Of Fortune*
July 1983.

SMALL ARMS OF THE PERSIAN GULF WAR

RPG-7s were part of small arms inventory for many of the Coalition forces and the bad guys. *(Photo: Robert K. Brown)*

Kuwaiti M-84 MBT, an improved version of the T-72 built by Yugoslavia, prepares for final thrust into Kuwait City. Note 12.7x108mm NSVT Soviet designed HMG that can be used in ground or antiaircraft role. *(Photo: Robert K. Brown)*

While media attention has focused almost exclusively on the amazing high-technology weapons systems arrayed against Iraq in the Persian Gulf, the fact remains that in the end it is always the common foot soldier who must meet with and destroy the enemy and occupy the ground he once held. His rifle and auxiliary small arms are still the principal tools he will employ to accomplish these primary missions of close combat, as they have been for the last hundred years.

Quite clearly the Patriot PAC-2 antimissile system stands as master over the 1950s-vintage Soviet Scud. Laser-guided bombs and Tomahawk cruise missiles were equally impressive. But, what about small arms? For years high technology's weak sister in the West, can it be that Iraq's largely Soviet-designed small arms systems were superior to those of the Coalition?

U.S. SMALL ARMS

A decision to adopt a service pistol chambered for the 9mm Parabellum cartridge—in order to conform with NATO standards—immediately enraged all those for whom only the .45 ACP round would do. This was followed by anger over the selection of a foreign design. Much to the glee of its detractors, the U.S. armed forces have experienced some slide failures with the M9. However, the problem is one of metallurgy, not design efficiency.

The M9 is more or less typical of the current generation of so-called "wonder-nine" pistols, which are all double-action and have large capacity magazines (15 rounds for the M9).

In general, the handling characteristics are excellent with acceptable hit probability and accuracy potential. The grip-to-frame angle and balance are adequate. Early reports from Saudi Arabia indicated the M9 was performing in a satisfactory manner.

Submachine guns—by definition selective-fire, shoulder-held weapons chambered for pistol ammunition—are moribund. The development of lightweight, short-barreled assault rifles has signaled their death knell. While 20 million submachine guns of one sort or another were fielded during World War II, their short effective range of rarely more than 100 meters, limited accuracy due to their open-bolt operation (with the exception of the H&K MP5 series), and relatively low power precluded their widespread employment in Desert Storm. There remains but one highly specialized application for these relics of the past. When their barrels are shrouded by a sound suppressor, burp guns were effectively employed by the elite units (U.S. Navy SEALs, U.S. Army Delta and British SAS) who were engaged in clandestine operations including ambush, assassination, prisoner recovery, EPW snatches and reconnaissance. The Heckler & Koch MP5 SD is one of the finest examples of the suppressed machine gun genre and was used with great success by U.S. Special Ops groups operating behind Iraqi lines.

No weapon ever adopted by the U.S. military has created as much hysterical controversy as the M16. It has been attacked at all levels, from its "pipsqueak" cartridge to its method of operation and "Mattel toy" construction. Yet it has survived its legion of detractors to become one of the finest assault rifles in the world. The M16A2 currently fielded by the U.S. Armed forces is not the same weapon that went to war in Vietnam a quarter of a century ago.

While the method of operation remains the same—direct gas action without a piston—a significant number of needed modifications were incorporated into the M16A2. Barrel twist was changed from one turn in 12 inches to one in 7 inches to stabilize the heavier M855 ball and M856 tracer ammunition. The 62-grain M855 bullet's performance in the human body essentially duplicates that of the older 55-grain M193 round with slightly increased fragmentation.

The 5.56x45mm NATO cartridge is usually effective out to 200 yards.

Images of Hollywood's *Beau Geste* aside, even in arid environments most infantry contacts will be under 100 yards. For that the M16A2 will more than suffice. Weakest link in the M16 series is the 30-round aluminum magazine. Most stoppages in the M16 can be attributed to damaged magazines, which, if suspect, should be pitched in the nearest ditch and replaced. In any event, never load more than 29 rounds.

A shortened version of the M16A2 was adopted by the U.S. Armed Forces as the M4. So far, only a modest quantity have been acquired. Reminiscent of the CAR-15s and XM177E2 fielded during the Vietnam War, these carbines have 14-inch barrels. Barrel configuration differs somewhat from the Colt Commando to permit attachment of the M203 grenade launcher and the hand guard has double heat shields. The M16A2's 800-meter sight system has been retained.

Also chambered for the M855 caliber 5.56x45mm NATO cartridge is the U.S. Army's M249 SAW. Although selected from a rigorous competitive evaluation of four candidate systems, the M249 has also seen its share of contentious commentary. Gas-operated with a short-stroke piston and two-lug rotary bolt, the M249 fires from the open-bolt position to inhibit "cook-offs," as do most belt-fed machine guns. It was designed to accept either disintegrating link belts or the M16 30-round box magazine without modification.

That has proved to be a major design error for, in my experience, the M249 will not function reliably with any magazine, loaded to any capacity. This weapon should be restricted to belt-fed operation only. Another area of concern is the M249's accuracy potential.

With those exceptions, the M249 is generally the very model of a modern SAW. When belt-fed, reliability exceeds all competing designs.

When employed for area-target fire support, as it most often should be, the accuracy potential is adequate. It has numerous desirable features, not the least of which is its weight, which is only 21.3 lbs. with a 200-round assault pack. It has a quick-change, chrome-lined barrel and exhibits excellent human engineering overall. It provides the fire power required at the squad level in a compact and lightweight envelope.

While it has been reported that U.S. Navy SEALs are employing some M14 rifles and both the M21 (M14 with ART scope) and M24 Sniper Weapon System are chambered for the 7.62x51mm NATO cartridge and were employed in Desert Storm, the most prevalent weapon utilizing this round is the M60E3 adopted by the USMC. M80 7.62x51mm ball is an excellent choice for the rare occasions when contacts occur beyond 200 meters.

The M60E3 machine gun chambered for this cartridge is a decided improvement over the despicable M60 GPMG (general purpose machine gun). Weighing only 18 lbs., a light bipod has been mounted to the receiver (where it stays when the barrel is changed, unlike the M60). The forearm has been redesigned and now ends in a pistol grip. The barrel has been lightened and incorporates an M16-style flash suppressor.

This lightened barrel has already caused serious problems. When high rates of sustained fire are employed without changing barrels within the prescribed intervals (and this is almost never done during the high anxiety of the battlefield), the barrel will slump with the possible consequence of damage to the gun and injury to the operator. A heavier barrel will probably eventually be adopted, but nothing as yet has been done about this deficiency. The M60E3's carrying handle has been moved to the barrel to assist barrel changing. The gas system has been revamped, which means the piston can no longer be installed backwards.

Far superior is the caliber 7.62x51mm M240 machine gun mounted on U.S. main battle tanks and armored fighting vehicles. This is the coaxial version of the famous FN MAG58 GPMG, which is in service with the armed forces of more than 80 countries. The "MAG" is belt-fed, gas-operated and fires from the open-bolt position. Reliability is a salient feature of this widely adopted weapon. It can be faulted in only one area. There are a lot of bits and pieces and it takes a long time to properly clean a gas-fouled "MAG."

John Browning's venerable .50 caliber M2 HB heavy machine gun, affectionately known as "Ma Deuce," is still very much alive and booming. It's attached to the turret of the M1A1 Abrams main battle tank and on the M113 armored personnel carrier (APC). In restricted roles it's still sitting on sand dunes fixed to the 44-lb. M3 tripod. Its 708-grain bullet with a hard steel core and muzzle velocity approaching 3,000 fps can defeat light vehicles and will give enemy personnel a grade-A migraine headache no matter where it strikes on the human body. Firing from the closed-bolt position, the M2 HB—when equipped with a sturdy MilSpec scope—can reach out to well beyond 2,000 meters and touch anyone.

Recoil-operated with a reasonable cyclic rate of about 550 rpm, the M2 HB is the most rugged and reliable machine gun in our inventory. Very little ever breaks and maintenance is relatively simple as its recoil operation throws very little fouling into the system. Ma Deuce is still the ultimate heavy machine gun—bar none—even after more than 65 years of use on all the world's battlefields.

FRENCH SMALL ARMS

While France adopted a modified version of the Beretta Model 92F 9mm pistol in 1987, known as the *Pistolet Automatique MAS 9mm G1*, the Model 1950 is still in service. A thoroughly reliable single-action 9mm Parabellum pistol with the swinging link and dual locking lugs of the Colt M1911A1, the M1950 is short-recoil operated. It cannot be fired with the magazine removed. It has a modular hammer/sear assembly which can be removed from the frame as one piece during disassembly.

French soldiers refer to it as *le Clairon*. But this bugle spits bullets. France's service rifle, the FA MAS *(Fusil Automatique, Manufacture d'Armes de St. Etienne*), has already demonstrated itself to be an effective and generally well-conceived piece of ordnance. Placed into production in 1979, the bullpup configured FA MAS is chambered for the 5.56x45mm NATO cartridge of the M193 type as the barrel twist is 1:12 inches. Weighing 8 lbs., empty, and firing from the closed-bolt position, the method of operation is by means of delayed blow-back.

To accommodate the bullpup configuration, the trigger mechanism and pistol grip have been mounted to the lower hand guard. At about 3 MOA at 100 meters, the accuracy potential is acceptable.

Reliability is of a high order. Because of its proximity to the operator, muzzle blast is more noticeable. There were buttstock failures during early series production, but this problem has been corrected. French troops participating in Desert Storm were issued plastic cases for the FA MAS. While it's downright ugly, principally because of the huge carrying handle on the upper handguard, the FA MAS is an excellent weapon. It's usually issued with a lightweight, tubular aluminum bipod and a cleverly designed, ambidextrous web sling.

Almost as bad as the M60, the French AA 52 *(Armee Automatique Model 52)* GPMG is chambered for the French 7.5x54mm rifle cartridge and is delayed-blowback in operation. Designed for ease of manufacture, the receiver body is made of semi cylindrical tubes welded together.

To prevent premature unlocking while chamber pressures are still too high, the AA 52 employs an unusual, troublesome, two-piece bolt. The bolt's two parts are joined by a fragile T-shaped connecting pin about 1 inch long. If this pin is broken or lost the weapon cannot be made to operate. French troops would have been better served to discard their AA 52 GPMGs and employ captured Soviet PK machine guns.

BRITISH SMALL ARMS

One of John Browning's greatest designs, the Browning High Power 9mm Parabellum pistol still serves in the British armed forces, where it is known as the L11A1. Locked-breech, short-recoil operated, this single-action classic has a magazine capacity of 13 rounds.

Sturdy and reliable, the High Power can only be faulted in the area of its frame-mounted safety lever which is difficult to manipulate—an important consideration for those who carry the arm in "condition one" (a round in the chamber, a full magazine seated in place, the hammer fully cocked and the thumb safety engaged).

U.S. Army Special Operations personnel providing security for U.S. Embassy in Kuwait City after its liberation. From left to right troops carry Heckler & Koch MP5SD and sound-suppressed MP5SDs. ***(Photo: Robert K. Brown)***

We can offer no compliments for the new caliber 5.56x45mm NATO British service rifle, civilian designation SA-80. Essentially an AR18 action in EM2 bullpup form, the L85A1 Individual Weapon (IW) represents a fiasco of major proportions. Its failure to perform adequately during Desert Storm is no more than the final straw in what will shortly become one of the biggest small arms scandals of this century.

Soon after fielding it was found that when dropped from

a height of several feet onto the muzzle, the L85A1 would fire. This was corrected, but the new trigger has such a small gap between its rear surface and the front of the pistol grip that small particles of sand immediately impede the operator's ability to pivot the trigger. This is literally a fatal flaw.

In the manner of the woeful U.S. M3 "Grease Gun," the L85A1's magazine catch/release button would dump magazines when inadvertently depressed by equipment or the operator's body. An interim solution, in the form of a shield attached with glue (!) was fielded. The newly redesigned magazine catch/release is reported to be no better than before.

The firing pin is fragile and breaks at the tip. Projecting from the bolt face, the jagged, broken tip results in primer ignition out of battery (called "a slam fire") with a very real potential for damage to the weapon and serious injury to the operator.

The barrel's gas port is drilled after the bore has been chrome-plated and this strips chrome from the port area. Both the crossbolt-type safety and bolt release button are made of plastic and break easily. The rear sling swivel is attached to the buttplate which is fixed to the receiver body by two screws. It doesn't take long before the buttplate tears away from the receiver. The captive upper and lower receiver retaining pin assemblies are of poor design and are easily damaged by repeated disassembly. The handguard's hinged top-cover opens spontaneously and breaks off. Receiver and trigger housing welds are of low quality.

The L86A1 Light Support Weapon (LSW) shares all of the L85A1's incredible defects and a few more besides. It has no quick-change barrel system and has limited potential for sustained fire due to its 30-round, bottom-fed magazine. The accuracy potential, even at close range, is dismal. After several hundred rounds, fouling clogs the gas cylinder preventing the bolt carrier from moving forward enough to trip the auto safety sear. When that happens, the gun will stop firing.

Gratefully, a substantial number of British troops are still armed with the caliber 7.62x51mm NATO L1A1 Commonwealth FAL. While too large (overall length approaches 45 inches) and too heavy (11 lbs., 3 oz. with a loaded magazine) by today's standards, it is vastly superior to the L85A1. Gas operated with an adjustable regulator, the FAL fires from the closed-bolt position and, at one time or another, was adopted by more than 90 nations.

However, it was never noted for reliable operation in high dust and sand environments. It did not perform well during the 1955 Suez Canal Zone crisis and as a result "sand cuts" were embodied in the bolt carrier and the hold-open device was omitted.

The Israelis experienced similar problems and were led, as a consequence, to develop and adopt the Galil—a Kalashnikov derivative.

L86A1 Light Support Weapons have as yet not completely replaced the FN MAG 58 GPMG in British service, where in modified form it is know as the L7A1/2 "Jimpy." Differing only cosmetically (i.e., a black plastic buttstock) and with a 10-position instead of three-position adjustable gas regulator, the L7A1/2 series demonstrated its value during the Falklands campaign, although the Royal Marines preferred the slightly lighter, magazine-fed L4 Bren. A thoroughly battle-proven machine gun, the MAG is not noted as particularly dust-sensitive.

IRAQI SMALL ARMS

With few exceptions, Iraq's military small arms are of ComBloc origin. In general, these weapons systems can be characterized as highly reliable with only mediocre accuracy potential and little design concentration on human engineering.

The principal Iraqi sidearm is the TT-33 Tokarev pistol. Chambered for the 7.62x25mm cartridge, the Tokarev is essentially a much modified Colt-Browning design. The method of operation is locked-breech, short recoil. All modifications to John Browning's original design were intended to simplify manufacture and enhance reliability. Although the hammer has a half-cock position, this pistol should not be carried with a chambered round. The 86-grain round-nose FMJ bullet has a muzzle velocity of about 1,450 fps and will most often over-penetrate in human targets.

The standard-issue Iraqi rifle is the Kalashnikov, called the "Tabuk," chambered for both the 7.62x39mm and 5.45x39mm cartridges. Most are of Soviet, East German or Romanian manufacture. Some were locally manufactured at a factory established for the Iraqi armed forces by technicians from the Izhmash Industrial Combine's Izhevsk weapons factory.

Fifty million Kalashnikovs can't be wrong and it is the world's most ubiquitous military small arm. Gas-operated and firing from the closed-bolt position, the *Avtomat Kalashnikov* (AK) has a much-copied rotary bolt that is piston-actuated. The pinned and riveted, sheet-metal receivers of the AKM series most often are seen with a five-component mechanical device that delays hammer drop until the complete cessation of all bolt-carrier bounce.

Probably the world's most reliable infantry rifle, the Kalashnikov falls short only in the areas of ergonomics and accuracy. The selector lever, on the right side of the receiver body, clatters excessively when manipulated.

AK magazines, although rugged and more reliable than those of the M16, must be rocked forward to remove and rearward to insert. As a consequence, tactical reloads are slow and awkward for all except highly trained operators. The open U-notch rear sight is far inferior to the M16's peep aperture and can only be adjusted for range.

Kalashnikov triggers are notorious for horrendous and variable creep with sudden, uncontrolled let-off. Except for Yugoslav variants, AK buttstocks are too short for most Westerners. All models of the Kalashnikov exhibit excessive flash signatures. Few AKs will shoot better than 5-6 MOA at 100 meters and Ivan apparently feels that's close enough for government work. The wound ballistics potential of the boat-tail 7.62x39mm bullet is mediocre.

Called the "Al-Kadisia," the Iraqis manufacture a Dragunov-type sniper rifle with technical assistance from Yugoslavia. Although its scope has an excellent range-finding reticle pattern, the Dragunov is overrated and will rarely shoot better than 3 MOA at 100 meters, even with match-grade ammunition. The skeletonized stock exhibits poor ergonomics and tends to exaggerate the perceived recoil of the full-size 7.62x54R cartridge.

Except for a few French AA 52 and Belgian FN MAG58 GPMGs, Iraqi machine guns are exclusively of ComBloc origin. The RPK is no more than a Kalashnikov rifle with a longer barrel, bipod and redesigned buttstock. Unless Soviet-type 75-round drums are employed, this weapon will "monopod" on its 40-round box-type magazine when fired from the prone position. Lacking a quick-change barrel, its sustained-fire potential is limited.

First introduced to the Soviet Army in 1961, the PK GPMG was eventually product-improved and lightened into the PKM *(Pulemet Kalashnikova Modernizirovanniy)* series. A quarter century of fighting from arid regions to tropical jungles has demonstrated it to be flawless, with the possible exception of an overly complex feed mechanism required to accommodate the 100-year-old 7.62x54R rimmed cartridge. Weighing less than 20 lbs., the PK's most distinctive characteristic is its skeletonized buttstock, fabricated from wood laminate material.

The cyclic rate is about 650 rpm and although it has no buffer system of any kind, the lack of perceived recoil when fired from the bipod is nothing short of amazing. If bursts are kept to three or four shots, the muzzle climb is negligible. The accuracy potential when fired from its lightweight aluminum tripod is more than adequate out to its effective range of approximately 800 meters. The handling characteristics are excellent, with a consequence of exceptionably high hit probability in the hands of experienced operators. This is truly an outstanding machine gun. It is gas-operated with an adjustable three-position regulator, has Kalashnikov-type rotary-bolt locking and fires from the open-bolt position.

Iraq's heavy machine gun is the caliber 12.7x108mm *Degtyarev Pekhotnyy* (DP) Model 38/46, a modification of the DShK 38 which featured a complicated rotary feed system, replaced by a more conventional shuttle system on the DP Model 38/46. With good reason, this weapon is usually encountered mounted on armored vehicles. Sans the barrel, the receiver group alone weighs 78.5 lbs.—18.5 lbs more than Ma Deuce. Add another 28 lbs. for the barrel which has totally superfluous radial cooling fins over almost its entire length.

The distinctive bulbous muzzle device is moderately effective in controlling muzzle jump at the expense of considerable side blast and muzzle flash. Great balls of fire spew out of the muzzle and burning embers of propellant all too frequently fly back into the operator's face. All in all, a reliable brute, but not quite in Ma Deuce's class. The 12.7x108mm cartridge matches the accuracy and effectiveness of the .50 cal. Browning (12.7x99mm) round.

SMALL ARMS VS. THE DESERT

Desert environments will play havoc with any small arms system. When they overran Iraqi defensive bunkers, U.S. Marines encountered Kalashnikov magazines so clogged with sand grains that the followers were jammed halfway up the magazine bodies.

Dust-sized particles can be every bit as abrasive as sand on a firearm's operating mechanism. The endless argument has always been to lubricate or not, and, if so, how much and what kind. Deserts represent an extremely adverse environment for military small arms, more so than even a tropical rain forest or the arctic. It's beyond the scope of current technology to design firearms that will operate reliably, if left unattended, in every possible environment. Increased maintenance and moderate lubrication are the only alternatives at this time. There are no magic potions that will preclude the drudge of repeated disassembly and cleaning.

This is not to say that some small arms systems are not more tolerant of harsh climatic regimens than others. Clearly, the Kalashnikov, with its built-in loose tolerances, is marginally superior in reliability to the M16. However, this is more than counter-balanced by the M16's greater accuracy potential, better human engineering and the more effective wound ballistics potential of its cartridge.

In general, the military small arms of the Coalition did not demonstrate themselves to be superior to the ComBloc weapons lined up against them on the other side of the trench. Iraq was quickly crushed in a 100-hour ground offensive because of high technology, air superiority, a massive 42-day air and artillery bombardment that scrambled their brains, and the vastly superior training of the non-conscripted U.S., British and French troops.

The Persian Gulf War ended so swiftly and was so one-sided that small arms development should not be greatly affected. Though hopefully the British will reexamine their abysmal L85A1.

Originally appeared in *Soldier Of Fortune*
June 1991.

Silent Six-Shooters

Knight's unusual Revolver Rifle, chambered for special .30-caliber telescoped ammunition, offers the potential for rapid follow-up shots and no empty cases to police up, along with sniper capability out to 200 meters.

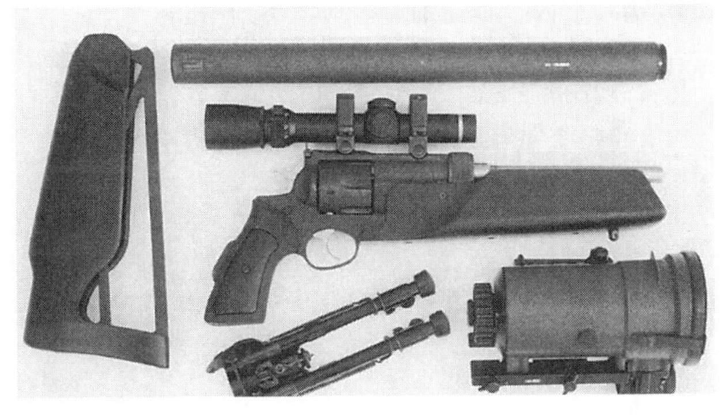

Disassembled for transport, the Knight Revolver Rifle's major components consist of a sound suppressor, detachable buttstock, Ruger .44 Magnum revolver with Leupold VARI-X III scope and forearm, Harris bipod and Simrad 1X KN250 image intensifier.

Nothing lifts self-styled weapons *cognoscenti* out of their armchairs faster than a villain on the boob tube waving about a revolver with a sound suppressor. They will quickly inform their novitiates that the gap between the cylinder and the rear end of the barrel precludes a significant reduction in the sound pressure level, no matter how efficient the suppressor attached to the barrel's muzzle may be.

They are theoretically correct. However, these experts appear to have had no influence upon either Irish Republican Army terrorists or Chicago gangsters, both occupational groups from whom have been seized countless examples over the years of all makes and calibers of wheelguns to which have been attached sound suppressors of one degree of sophistication or another.

In an effort to increase the efficiency of a sound-suppressed revolver, some years ago the designer Siegfried Hubner of West Germany developed a system that enveloped the barrel, cylinder and front portion of a Smith & Wesson Model-10 frame. The end result was incredibly clumsy, and most designers have restricted themselves to both locked-breech and blowback semiautomatic pistols.

While there is no front cylinder gap to contend with, semiauto pistols spit empty cases all over the scenery and there are some operational circumstances under which this might compromise the mission.

C. Reed Knight Jr., of Knight's Armament Co., designed the American 2000-model pistol, which Colt produces. Knight also reintroduced the Stoner 63A weapons system and recently commenced series production of the SR-25 countersniper system. He is known principally, however, for the sophisticated sound suppressors he supplies to the U.S. military. In my opinion, Knight easily ranks among the top three sound-suppressor designers in the world today (all three are located in the United States).

Faced with a Request For Proposal (RFP) for a relatively close-range (100 to 200 meters), sound-suppressed sniper weapon with the potential for rapid follow-up shots and no empty cases to police up, Knight responded with a tour de force totally unique in the annals of sound-suppressor history. Knight's successful response to the RFP combines special telescoped ammunition of two calibers with different modified Ruger revolvers (a GP-100 and a Super Redhawk) to which have been attached sound suppressors. It is, of course, all quite a bit more complicated than this.

The so-called "Revolver Pistol" starts out as a stainless steel Ruger GP-100. It's an excellent, and very rugged, starting point. Both the mainspring and trigger-guard latch spring are massive coils. In addition to the hammer and the trigger, most of the small internal parts are stainless steel. Mounted in the frame, the floating firing pin must be struck by a transfer bar, which in turn receives the hammer's hit. Should the finger relieve pressure on the trigger while the hammer is falling forward, the transfer-bar safety will drop down and the firing pin will not be driven forward.

The cylinder's capacity is six rounds, and the crane/cylinder assembly swings out of the frame in the usual manner after its spring-loaded latch button is depressed. In the firing position, the cylinder is securely locked to the frame in two places—by a strong pilot bearing at the rear of the cylinder, and by a robust locking bolt at the front of the crane, which engages a matching slot at the end of the frame. This mechanism was invented by Ruger and first used in the Redhawk revolver.

For increased strength, the cylinder locking notches are substantially offset, not cut into the weakest part of the cylinder walls over the centers of the chambers. Cylinder rotation is to the left; the ejector rod does not rotate with the cylinder and serves only to activate the ejector, not as a bearing point for front locking of the cylinder.

While the trigger system remains unaltered, the factory barrel is replaced with a caliber-5.56mm barrel, 3 inches in length. This stainless steel barrel has six grooves with a right-hand twist of one turn in 6 inches. The front of the cylinder face is cut back by a half-inch.

The fast-twist rifling is required to stabilize the 47-grain, flat-based projectile with its needle-sharp point that's guaranteed to penetrate most current body-armor types.

This screw-turned brass bullet leaves the muzzle at 1,025 feet per second (fps). The bullet is encased in a piston made from a space-age synthetic called Ultem. Both are loaded into a Winchester .38 Special case and powered by an unspecified propellant of undisclosed charge weight. Upon ignition, the piston moves forward a small amount and its beveled face interfaces with the rear end of the barrel to seal the

Knight's Revolver Rifle comes equipped with sophisticated optics, including a Leupold VARI-X III scope and a Simrad KN250 night-vision unit mounted directly over the Leupold. Emergency tritium sights complete the package.

front cylinder gap. A rubber O-ring on the piston seals the case from propellant gas blow-by, so that all of the propellant gas is driven into the sound suppressor attached to the barrel. Due to the proprietary nature of this unique ammunition, we were not permitted to publish close-up photographs of it, nor of the larger-caliber cartridge described further in this article.

The sound suppressor housing for the Revolver Pistol—6 inches in overall length—is made from 6061 T6 aluminum-alloy tubing, which has been hard-coat anodized. Internally, it contains both aluminum and steel baffles. In order to obtain the highest level of sound reduction, these baffles must be kept wet so that the main propellant gas jet can be cooled at an increased rate over conventional designs. For this purpose, a small amount of Shooter's Choice-brand All Weather High-Tech grease is applied to the first baffle at the rear end of the unit.

The method employed to attach the suppressor to the revolver is that of the Knight unit obtained by the U.S. Air Force for the Beretta 92F (M9) pistol. Between a top-piece at the rear of the suppressor housing and the knurled and rounded push-release on the underside are two sets of stainless steel pins. The two inner pins are notched so that the suppressor can be removed from the pistol when the push-release is pressed upward. When locked to the revolver, the un-notched surfaces of the two inner pins interface with a slot on each side of the muzzle. Simply apply thumb pressure to the push-release to instantly install or remove the suppressor.

The sound signature produced by this system is about 123 decibels (dB). This is outstanding performance, with a net sound reduction of 40 dB, as the revolver itself generates a sound-pressure level of about 163 dB without the suppressor. The sound of the hammer falling on an empty chamber is about 112 dB, so the muzzle blast is reduced to only 10 dB more than the sound of the action. The firearm's flash signature is also totally eliminated.

These sound signatures were measured by means of a Bruel and Kjaer type 2209 Impulse Precision Sound Level Meter, with a No. 4136 microphone placed 1 meter away from the front, and to the left, of the muzzle at an angle normal (90 degrees) to the barrel's axis. The meter was calibrated both before and after these tests: No instrument drift was noted. All of this is as per U.S. milspec.

21ST CENTURY SIGHTS

The sights provided for Knight's Revolver Pistol are every bit as elaborate and startling as the rest of the system. To the standard-type, open-notch rear sight (retained for emergency use) has been added two horizontal tritium bars. A single tritium dot has been installed on the serrated blade-type front sight, which is mounted in a dovetail on the barrel shroud.

A dovetail has been milled into the pistol's top-strap for installation of a battery-operated AIMPOINT 5000 red-dot sight with polarizing filter. A rotary switch with 10 positions adjusts intensity of the red dot to accommodate all light levels, from bright sunlight to pitch dark.

A dovetail milled into the frame accepts an Applied Laser Systems Terminator II laser sight, which has been modified by Knight's Armament Co. with an new switch and more rugged body and mount. This unit is available with either a visible or infrared (IR) laser beam.

The word "laser" is an acronym; its letters stand for Light Amplification by Stimulated Emission of Radiation. There are several processes used to produce lasers, and in each instance they generate an intense beam of light. This beam of light is both very pure—i.e., all of the light rays in the beam are nearly the same color (monochromatic)—and well collimated, i.e., all the rays are headed in almost exactly the same direction ("highly coherent").

The Terminator II is a state-of-the-art diodide laser unit, also known as an injection laser. It is a semiconductor diode having a "pn" junction and emitting laser radiation when the diode is forward-biased with a current above the threshold of the pn-junction material.

There are several types of laser diodes, the most commonly used being the gallium arsenide (GaAs) diode, which emits in the infrared spectrum in the range of 8,200 to 9,050 angstroms. The direction of forward current flow is from the n-type layer to the p-type layer in the pn junction, when a current of 100 milliamperes (threshold current) or more is applied across the junction operating analogous to a resonant cavity. The diode behaves like an ordinary light-emitting diode when biased with a current below the threshold level.

Visible laser sights (such as a gallium-aluminum-arsenide diode unit, which emits in the visible spectrum in the range of 6,300 to 6,700 angstroms) mounted on small arms have, in my opinion, extremely limited applications. However, when coupled with passive night-vision goggles, an infrared laser beam projects as a bright dot to the operator. It cannot be detected by others unless they are also equipped with night-vision viewing devices. In this latter mode we have something that can actually be useful to those who sneak about in the shadows.

REVOLVER RIFLE

All well and good for handgun ranges, which are most often under 7 or 8 meters and rarely more than 50 meters, but to pop human targets at ranges of 100 meters and more, we need a more stable firing platform. For this purpose, Knight designed the "Revolver Rifle," which although conceptually similar to the system described above, differs in several significant areas.

Starting envelope for the revolver rifle is a Ruger caliber .44 Magnum Super Redhawk revolver. The crane/cylinder assembly is essentially the same as that of the Ruger GP-100. All stainless steel components of both the Revolver Pistol and the Revolver Rifle have been finished by a black-oxide process.

The factory barrel of the Revolver Rifle has been replaced by a four-groove, .30-caliber barrel, 10 inches in overall length, with a right-hand rifling twist of one turn in 9 inches. The front cylinder gap is adjusted to a maximum of 0.005-inch and a minimum of 0.003-inch (as is that of the Revolver Pistol).

This is an exceptionally fast twist for a bullet weighing only 145 grains. However, the muzzle velocity is only 1,025 fps (muzzle velocity of the .30-06 150-grain military ball round is 2,700 fps). A rifling twist of 1:9 inches is required to stabilize the flight path of a lightweight projectile moving at such a reduced velocity, and to provide the necessary accuracy potential. To further minimize the tendency of the slow-moving projectile to yaw in flight, "driving" or "rotating" bands have been milled into the flat-based, solid brass bullet in the manner of many artillery shells. Driving bands transmit rotation from the rifling to the shell.

Screw-turned with a needle-sharp point, the bullet is encased in an aluminum piston with a black-plastic front face seal. Both are loaded into a Federal .44 Magnum case and powered by an unspecified propellant of undisclosed charge weight. As with the Revolver Pistol, upon ignition the piston moves forward a small amount until its front face impinges against the rear end of the barrel to seal the front cylinder gap. Once again, a rubber O-ring on the piston seals the case from propellant gas blow-by so that all of the propellant gas is driven into the sound suppressor attached to the barrel. Early prototypes of this concept featured 7mm bullets.

An unusually long sound suppressor (18.75 inches) fits over the barrel and is attached to the revolver by means of a knurled, threaded coupler on the front end of the frame. The suppressor's housing is made from black-anodized, 6061 T6 aluminum tubing. The interior consists of a series of steel and aluminum baffles, together with rolled metal screen material. This suppressor has an exceptional life span of 5,000 rounds or more.

When the suppressor is installed, a small amount of Shooter's Choice All Weather High-Tech grease should first be applied to the front of the barrel. There is a small index pin on the suppressor housing. During installation, this pin should be inserted into a keyway slot on the rear of the barrel before attempting to tighten the knurled coupling nut.

The sound signature produced by this system is about 119 dB. This sound-pressure level is about 4 dB lower than that generated by the suppressed Revolver Pistol, with a net sound reduction of 44 dB, as the Revolver Rifle itself generates a sound-pressure level of about 163dB without the suppressor. Once again, the sound of the hammer falling on an empty chamber is about 112 dB, so the muzzle blast is reduced to only 7 dB more than the sound of the action. And, once more, the firearm's flash signature is completely eliminated.

The black polyurethane forearm under the barrel has a quick-release stud to which can be attached a Harris bipod. A detachable polyurethane and steel buttstock is also provided. The overall length of the Revolver Rifle is 36.5 inches with the buttstock and sound suppressor attached. Without night vision equipment, the weight is 8.5 pounds. Add another 1 pound for the Simrad night vision unit.

Once again, to the standard-type, open-notch rear sight (which has been retained for emergency use) have been added two horizontal tritium bars. A single tritium dot has been installed on the serrated blade-type front sight mounted on the end of the suppressor housing.

Knight's Revolver Rifle comes with a Leupold VARI-X III scope. This compact, variable-power scope can be adjusted from 1X to 5X magnification. Its adjustment system consists of a 15-minute dial with clicks and markings at 1/4-minute intervals (a 1/4-minute equals a 1/4-inch at 100 yards).

A Simrad 1X KN250 image intensifier can be mounted directly over the Leupold VARI-X III scope. It can use both second- and third-generation 18mm wafer tubes and does not magnify the image.

Second- and third-generation passive night-vision equipment employs a so-called "micro-channel amplifier," which consists of a bundle of extremely minute fibers—several million of them in a tube no more than 3/4-inch (18mm) in diameter. A screen in front of the fibers receives the optical image and generates electrons which are fed into the micro-channels.

DARK ILLUMINATION

The amplifier unit is supplied by an electrical field. As the electrons travel down the micro-channels, they ricochet off the thin peripheral layer (known as "cladding," it is made of an optical material of lower index refraction than the core material) and by means of their zigzag path through the core, generate additional electrons so that the signal coming out the end of

With an overall length of only 36.5 inches, Knight's Revolver Rifle is equipped with a detachable buttstock, polyurethane forearm and Harris bipod, all of which provide the stable firing platform required for accurate head-shots out to 200 meters.

the micro-channels is much greater than that which went in. These outward-bound electrons strike a second screen and provide a visual image. This image can be then either viewed by an ocular or given further amplification by another micro-channel amplifier.

Micro-channel amplifiers are much more compact than first-generation "cascade tubes," and a single unit provides a much greater degree of amplification. The total amplification (often referred to as "gain") is about 64,000 times with the more sophisticated units. Thus, a very murky star- or moon-illuminated scene becomes almost as clear and distinct as if viewed under the noonday sun.

What's the accuracy potential of these esoteric snuffers? *Soldier Of Fortune* tested Knight's Revolver Rifle at 100 meters from a benchrest. We fired 2-inch groups until it became monotonous. That's more than sufficient to meet any Mission Essential Need Statement (MENS) likely to have been developed for this weapon system: likewise with the Revolver Pistol. At the close ranges handguns are employed in (the vast majority of all gunfights with a pistol take place at ranges under 7 meters), it will more than do. Furthermore, the AIMPOINT 5000 red-dot sight extends the Revolver Pistol's useful range out to a distance where it overlaps with the Revolver Rifle's range (at the low end of the latter's potential).

Without doubt, C. Reed Knight Jr. has created a most impressive and totally unique weapons system. However, this equipment is most definitely not available to the public under any circumstances. Unless you can demonstrate that you represent a U.S. government agency qualified, and with a substantial need, to possess such equipment, do not bother to contact Knight's Armament Co. Inquiries are accepted only from legitimate governmental and military users.

Originally appeared in *Soldier Of Fortune*
September 1992.

Return of the BAR

Old Soldiers Never Die . . .

France, 1918: U.S. Army 1st Lieutenant Val Browning demonstrates his father's new M1918 "machine rifle" to the American troops. The weapon is actually firing in this rare photo.
(U.S. Army Signal Corps/National Archives/Robert Bruce Photo)

It was the stuff of legends. Designed by the most famous firearms inventor that ever lived, John Moses Browning, the Browning Automatic Rifle (BAR) fought its way through World War II, the Korean War and even saw action during the early years of the Vietnam War. Serving as the squad's base of fire in the U.S. Army and as the centerpieces of the USMC's fire team concept, Marine Corps veterans of the Pacific campaign and Korea have literally canonized this weapon.

On 6 April 1917 when war was declared between the United States and the Imperial German Government, the U.S. Armed Forces had a total of 1,100 assorted Benet-Mercie, Maxim and Colt "Potato Digger" Model 1895 machine guns. Germany, on the other hand, had entered the war three years previously with 12,500 Maxims and another 50,000 under construction. The first machine gun issued in quantity to the American Expeditionary Force was the incredibly dreadful French Chauchat. One of the very first true automatic rifles, it was intended to provide "marching fire" to suppress enemy fire from the trenches while the infantry moved forward on the assault across open ground.

John Browning first demonstrated his automatic rifle, which was based upon the above tactical concept, on 27 February 1917. The BAR was originally designed to be carried by an advancing infantryman with the sling over his shoulder and the butt against the hip. In fact, early BAR web gear featured a steel box on the right side on the belt into which the soldier would place the butt as he walked forward, firing the weapon with each step of the right foot. The hit probability must have been unacceptably low by today's standards, but the idea was to keep the enemy hunkered down in their trenches as the assaulting units advanced.

Initially, Colt held the exclusive rights to manufacture Browning's designs in the United States, but as the need for machine guns became urgent after the United States entered the war in April 1917, contracts were also awarded to the Winchester Repeating Arms Company and the Marlin-Rockwell Corporation. The inventor's son, 2nd Lieutenant Val A. Browning, was the first to use the BAR in combat, firing on German positions on 13 September 1918. However, World War I ended before the BAR was available in quantity and its actual combat use was quite limited in that conflict. By the end of the war, 85,000 BARs were delivered before all the contracts were canceled. During the so-called gangster era of the 1930s, the M1918 BAR gained prominence almost approaching that of the Thompson submachine gun with both desperadoes—such as John Dillinger, Pretty Boy Floyd, Bonnie and Clyde—and J. Edgar Hoover's fledgling FBI.

The original Model 1918 (so called to avoid confusion with the M1917 Browning water-cooled machine gun although the BAR also was adopted the previous year) fired from the open-bolt position, either full auto or semiautomatic. The cyclic rate was about 480 rpm. The bottom-fed, 20-round magazine could be emptied in 2.5 seconds. Gas-operated with an adjustable regulator and a conventional piston below the barrel, it was chambered for the standard .30-06 U.S. service round of that period. Although reliable and widely acclaimed, the M1918 was difficult to control in full-auto fire, as it had no bipod and at 16 pounds, empty, was too heavy to fire effectively from a shoulder mount. A cavalry version was adopted as the model 1922, but only a few hundred were made. In 1937, a small number were converted to the M1918A1 configuration, which included a buttplate with a hinged shoulder strap and a bipod attached to the gas cylinder. This increased the weight to 18.5 pounds. With the bipod at this location, the gun could easily be pivoted to engage fast moving targets on the flanks. Colt manufactured a number of commercial versions, including the famous Colt Monitor with a short barrel, Cutts muzzle compensator and pistol grip which was marketed in the 1930s as a law-enforcement and bank-guard weapon. *Fabrique Nationale,* licensed to manufacture John Browning's designs in Europe, sold several versions of the BAR in various calibers throughout the world including a caliber 7.92x57mm variant to Poland known as the Model Wz28. After World War II,

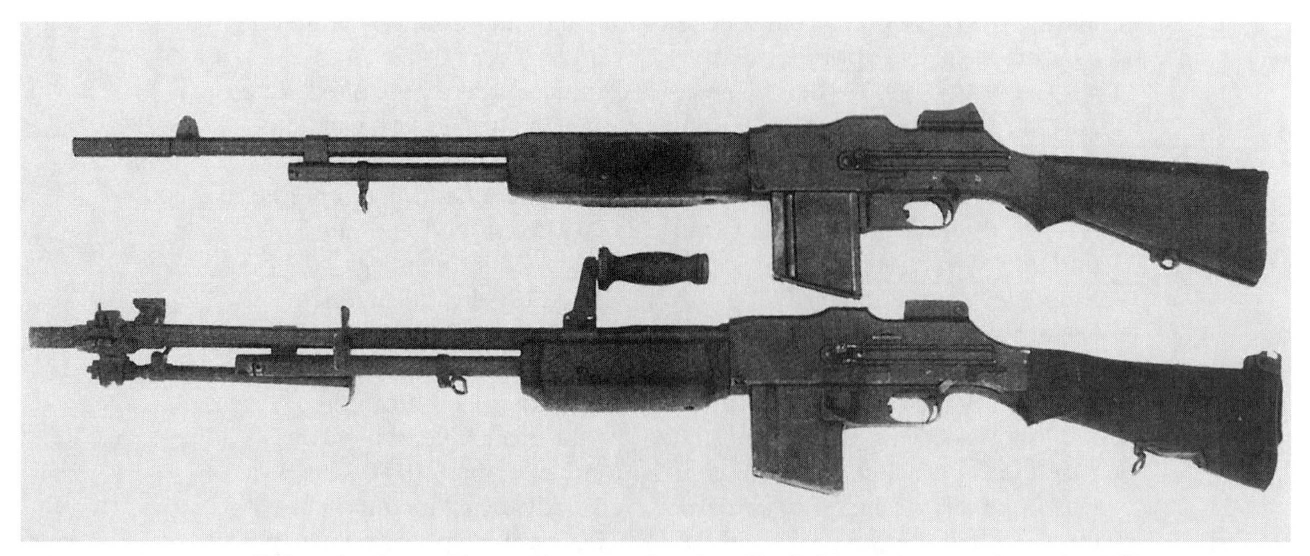

M1918 BAR (top) saw little action in World War I, but was a favorite of both desperadoes—such as John Dillinger, Pretty Boy Floyd, Bonnie and Clyde—and J. Edgar Hoover's fledgling FBI. The 1918A3 SLR (bottom) is a faithful semiautomatic reproduction of the M1918A2 BAR.

FN introduced the Model D version with a quick-change barrel and simplified disassembly. In 1937, Sweden developed a model of the BAR with a quick-change barrel and pistol grip.

The final U.S. military version of the BAR was adopted and issued in 1940 as the M1918A2. Provision to attach a monopod was added to the buttstock, but the monopods, of dubious value anyway, were never issued. The checkered wooden forearm was reduced in height to enhance cooling of the barrel and then eventually replaced by a shorter, uncheckered forend. The M1917 Enfield pattern rear sight was replaced by a unit similar to that of the M1919A4 .30 caliber BMG (Browning Machine Gun). Guide ribs were added to the front of the machined trigger housing to assist insertion of magazines. The M1918A1 buttplate with hinged shoulder strap was retained. In 1942 a plastic butt-stock replaced the walnut buttstock. Toward the end of World War II, a carrying handle was added to the barrel, but it did not see widespread use until the Korean War (at which time an odd-looking pronged flash hider was also adopted). All of this brought the weight up to 19.4 pounds, empty. There was no longer any provision for semiautomatic fire (it was felt that the adoption of the M1 Garand made this unnecessary), only two rates of full auto: 300-450 rpm and 500-650 rpm, respectively. The Marines preferred the original semiautomatic feature and frequently retained that capability. Initially, M1918A2 BARs were converted from existing M1918s and M1918A1s. Most of the M1918 configuration BARs in collectors' hands today were sent to England under the Lend-Lease program and then imported back to the United States by Interarms prior to 1968.

By the end of World War II, a total of 208,380 M1918A2 BARs were manufactured by IBM and (mostly) New England Small Arms. During the Korean War approximately 61,000 M1918A2 BARs were manufactured by Royal McBee Typewriter Company. Prior to 19 May 1986 a small number of M1918A2 type BARs were assembled from surplus parts and newly made receivers by Tony Maples of RAMO. An additional 250 BAR receivers were manufactured by Group Industries. The total number of BARs manufactured in the United States throughout its production life span was somewhat less than 355,000.

1918A3 SLR Specifications

CALIBER 30.06

OPERATION ....Locked-breech, gas-operated with three-position regulator, semiautomatic, closed-bolt, hammer-fired. Two-position selector lever.

FEED MECHANISM ....20-round, staggered-column, detachable box-type magazine.

WEIGHT, empty ....19.4 pounds.

LENGTH, overall ....47.8 inches with flash hider.

BARREL ....Four grooves with a right-hand twist of one turn in 10 inches.

BARREL LENGTH ....24 inches.

SIGHTS ....Leaf-type rear sight with open U-notch battle sight and peep aperture adjustable for windage and elevation from 100 to 1,500 yards; reverse ramp-type front sight blade with protective hood.

FINISH ....Phosphate ("Parkerized").

PRICE ....$2,650 including bipod, flash hider, carrying handle, web sling and two 20-round magazines.

MANUFACTURER .....Ohio Ordnance Works, Inc., 310 Park Drive, P.O. Box 687, Chardon, OH.

T&E SUMMARY: Faithful semiautomatic reproduction of the Browning Automatic Rifle that served as the squad's base of fire during World War II, the Korean War and the early years of the Vietnam War. Robust and reliable. Extremely accurate.

BATTLE-PROVEN BAR

Even in its final M1918A2 configuration, the highly regarded BAR was less than perfect. The location of its bottom-fed magazine limited capacity and 20 rounds is just not enough for a squad automatic. None of the models fielded by the U.S. Armed Forces had quick-change barrels. Melvin M. Johnson, Jr., designer of the Johnson M1941 semiautomatic rifle and Models 1941 and 1944 Light Machine Guns, stated that a BAR barrel could withstand 75 to 100 rounds per minute for up to 300-400 rounds without serious damage to the barrel. In one test the barrel was ruined after 1,000 rounds of firing at this rate.

Semiautomatic accuracy was degraded by the open-bolt firing mechanism which caused the reciprocating group to slam against the rear of the barrel to jar the weapon at the moment of primer ignition. In juxtaposition, full-auto fire off the bipod was almost too accurate. At the standard 1,000-inch machine-gun target, the M1918A2 BAR could place one full 20-round burst into a 2x3-inch rectangle. This is a direct consequence of the bipod's location at the muzzle. Group dispersion of this size is superior to that of the M1917A1 water-cooled Browning machine gun mounted on its tripod. As a result, at combat ranges of 200 to 600 yards the M1918A2 BAR almost comes close to lacking a useful distribution of fire. In addition, with the bipod attached to the muzzle, it becomes more difficult for the operator to engage flanking targets. At almost 20 pounds the M1918A2 BAR is too heavy for anything other than snap shooting from a standing shoulder mount.

Regardless of its idiosyncrasies, the BAR is a genuine, battle-proven piece of American military history. It ranks in desirability with collectors and shooters right up there with the Thompson submachine gun, .30 M1 Garand, Colt M1911 Government Model .45 pistol and the Colt Single-Action Army Revolver. Unfortunately, in private hands, BARs are quite rare. There are probably no more than 10 M1918A2 BARs, manufactured by either IBM, New England Small Arms or Royal McBee Typewriter Company, registered to individuals in the United States. When offered for sale they will bring up to $7,500. About 1,500 M1918 BARs, manufactured by either Colt, Winchester or Marlin-Rockwell, were imported by Interarms from Great Britain. In excellent condition they will sell for $5,000. All of these are Title II firearms and require a $200 tax stamp when acquired by individuals living in a state that permits their possession, as well as the fingerprint cards, photos and three- to six-month delays involved in BATF approval of ATF Form 4 transfer applications. A substantial number of collectors and shooters don't want the hassle and can't afford the price tag of a full-auto BAR.

BEST ALTERNATIVE BAR-NONE

There is now an intriguing alternative: Ohio Ordnance Works Inc. has recently introduced the 1918A3 Self-Loading Rifle (SLR) which is an exact external duplicate of the M1918A2 BAR. Complete with bipod, flash hider, carrying handle, web sling and two 20-round magazines, the price is $2,650 plus shipping. This caliber .30-06,

Korea, 1951: From a protected position behind a tank, this infantryman fires his Browning Automatic Rifle at Chinese Communist positions. Note the absence of a bipod, removed in order to lighten this heavy but effective one-man machine gun.
(U.S. Army National Archives/Robert Bruce Photo)

closed-bolt, hammer-fired semiautomatic-only rifle has been approved by the BATF as it has been designed and manufactured to permit only semiautomatic fire. The receiver, not modified from an existing BAR receiver, is made from an 86/20 steel investment casting, fully machined on a state-of-the-art, computer-controlled, machining center. Both M14 and M1 Garands were also made from 86/20 ordnance steel. In addition, since this rifle does not have either a sinister folding or telescoping buttstock, pistol grip that protrudes conspicuously beneath the action, a bayonet lug or grenade launcher, the gun gestapo has ruled that it is not a banned and extraordinarily dangerous "semiautomatic assault weapon."

The 1918A3 SLR receiver precludes the installation of the following BAR components: slide, firing pin, trigger group, hammer and bolt link. A second retaining pin hole was drilled through the receiver to support the rear end of the redesigned trigger housing. This retaining pin is attached to an additional takedown lever on the left side of the receiver. The slide was modified by machining its lower surface to completely remove the original sear surface. The rail channels in the receiver have been machined to accept only the modified slide. The firing pin has been altered and is designed to work only in conjunction with a secondary striker mounted in a redesigned bolt link. The new firing pin is designed for firing from the closed-bolt position when the secondary striker is impinged against by a spring-loaded hammer. The trigger mechanism was redesigned by the inclusion of a spring-loaded hammer and a .30 M1 Carbine-type sliding sear that also serves as a disconnector. The standard BAR bolt lock, gas cylinder and handguard have been utilized.

Just about every thing else remains pretty much the same. When the bolt group travels forward and is about 2 inches from battery, a circular cam surface on the bottom of the bolt lock starts to ride over the bolt support's rear shoulders, camming the rear end of the bolt lock upward. As the link pin rises above the line between the hammer and bolt pins, the bolt lock is aligned with its locking recesses in the receiver and pivots around the bolt-lock pin. The hammer pin revolves on its link and forces the bolt lock upward. A rounded surface on the lock slips over the locking shoulder in the "hump" of the receiver and provides additional momentum to force the bolt all the way up into battery. On the recoil stroke, the piston and slide, driven rearward by expanding propellant gases forced into the gas cylinder through the barrel's gas port, draw the bolt group down out of its locking recess and then rearward. This method of operation, turned upside down (locking onto shoulders below the action), was used 40 years later on the famous FN MAG 58 GPMG.

July 1950, Taejon, Korea: U.S. Army infantry soldier fires a Browning Automatic Rifle (M1918A2) while heavy street fighting continues during the evacuation of Taejon by U.S. troops.

(U.S. Army Signal Corps/National Archives/Robert Bruce photo)

The overall length, with flash hider, is 47.8 inches. The 24-inch barrel, which by itself weighs 3.65 pounds, has four grooves with a right-hand twist of one turn in 10 inches. The non-reciprocating cocking handle is on the left side of the receiver with the ejection port on the right. The selector lever, located on the left side of the trigger housing, has two positions: S to the rear indicating *safe,* and *F* in the forward position for *fire.* The top of the receiver is marked "SELF LOADING RIFLE MOD. 1918A3, U.S. CAL. 30 COLLECTORS CORNER, CHARDON, OHIO, PAT. PEND." Followed by the serial number. It carries the same phosphate ("Parkerized") finish as the M1918A2.

The gas cylinder assembly is composed of a body, body lock key and regulator assembly. There are three circles of different sizes on the face of the regulator assembly, which correspond to similar size gas ports in the body of the regulator. These gas ports control the amount of gas required to operate the weapon as fouling increases or other parameters such as ammunition, lubrication or atmospheric conditions alter the operating cycle. The BAR is normally operated with the regulator set to the smallest port and then varied only if the weapon exhibits signs of insufficient gas. To adjust the gas setting to the smallest port, turn the knurled regulator knob to the extreme clockwise position. All three regulator positions have audible click stops provided by a spring-loaded plunger indenting on one of three notches on the body lock key. This also prevents the regulator from rotating during firing.

A complex buffer and rate-reducing group is located in the M1918A2 BAR's buttstock. It serves two purposes: Most important in the case of the 1918A3 SLR, the buffer mechanism absorbs a great deal of the recoil of the reciprocating parts as they travel rearward. However, the rate reducing mechanism obviously has no effect on the 1918A3 SLR's performance, as it was intended only to permit the M1918A2 BAR to be fired at the slow cyclic rate of 350 to 450 rpm.

A FULL SUPPLY ROOM

BAR magazines hold 20 rounds, and are of the staggered-column, detachable box type. They are still available in quantity at very reasonable prices. Ohio Ordnance Works Inc. sells new USGI (United States Government Issue) BAR magazines for only $8.50 each. An original, but used, chest containing 12 new BAR magazines will cost you $150. BAR magazines represent 80-year-old technology. They should never be kept loaded for any length of time. Load them only prior to use or their accordion-shaped, flat follower springs will lose their compression strength.

Other USGI BAR parts and accessories available from Ohio Ordnance Works Inc. include an original World War II-era cradle for dash board and pedestal mounts at $250; winter trigger assembly for $35; complete gas regulator assembly for $25; blank firing adapter for $8.50; gas cylinder reaming tool for $7.50; and small parts pouch for only $3.50.

The M1918A2 BAR bipod uses wing nuts to both retain the bipod in its open and closed (against the barrel) positions and to adjust the command height (the distance from the ground to the bore's center line), which can be altered from 9 to 14.5 inches. The gun will rotate 360 degrees in the bipod body, which surrounds the muzzle end of the barrel.

The rear sight is a modified M1917A1 BMG leaf-type. The windage scale, located at the rear of the leaf, is attached to the base. The rear sight can be moved to the right or left by rotating the knurled windage knob located on the right side of the base. One click of windage moves the mean point of impact (mpi) 4 inches to the right or left at 100 yards (turn the knob backwards to move the aperture to the left). There are two rear sights. An open U-notch is used when the leaf is down and corresponds to a battle sight setting of 300 yards. The peep aperture is used when the leaf is in the upright position. It can be adjusted for elevation from 100 to 1,500 yards by means of the knurled knob on top of the leaf (turn the knob to the right to move the aperture up). One click of elevation adjustment will also move the mpi 4 inches at 100 yards. The front sight blade, a reverse ramp type, is dovetailed to its base on the barrel and can be drifted right or left to establish initial windage zero. It is protected by a robust hood. This is a classic example of the type of sophistication almost always found on the sights of U.S. military small arms. In juxtaposition, contemporaneous European small arms often featured sights that were crude in comparison and frequently had no provision for windage adjustment.

Detailed disassembly of the BAR series is somewhat complicated and is best accomplished through the use of a Department of the Army Field Manual. In my opinion, the best manual ever written about the BAR was *FM 23-15, BROWNING AUTOMATIC RIFLE* CAL. *.30, M1918A2*. The last edition of this field manual was published in May 1961 and it can still be found at gun shows or from the numerous book dealers who specialize in military manuals.

The 1918A3 SLR specimen sent to *Soldier Of Fortune* for test and evaluation has had a total of more than 10,000 rounds fired through it. Reliability is up to M1918A2 BAR standards. I would plan on stocking the following spare parts for any BAR: recoil spring and guide rod, extractor and extractor spring, firing pin, hammer, bolt link and complete gas cylinder assembly.

Weighing close to 20 pounds and firing from the closed-bolt position, the 1918A3 SLR's accuracy potential, when shot from the prone position off the bipod or rested on sandbags, is outstanding. This rifle will shoot close to 1-inch groups at 100 yards with USGI Armor Piercing (AP) ammunition and no worse than 2 inches with standard ball—all this with the issue iron sights. With its substantial mass and highly efficient buffer system, the 1918A3 SLR generates about as much perceived recoil as a pop gun.

Ohio Ordnance Works Inc. has just developed a National Match version of the 1918A3 SLR. Furnished with a Krieger heavy barrel, but without bipod, flash hider or threaded muzzle, this rifle produced 7/8-inch 10-shot groups at 100 yards with match-grade ammunition at the most recent Camp Perry high power competition.

The 1918A3 Self-Loading Rifle provides both military buffs and shooters the opportunity to own a piece of American firearms history without the red tape associated with acquiring a machine gun and at less than half the price of a selective-fire BAR. It also shoots like a match rifle.

Originally appeared in *Soldier Of Fortune*
January 1997.

Best of the Bullpups?

SOF T&Es Vektor's CR21

New South African CR21 caliber 5.56x45mm NATO infantry rifle, which may eventually prove to be the best of the bullpups.

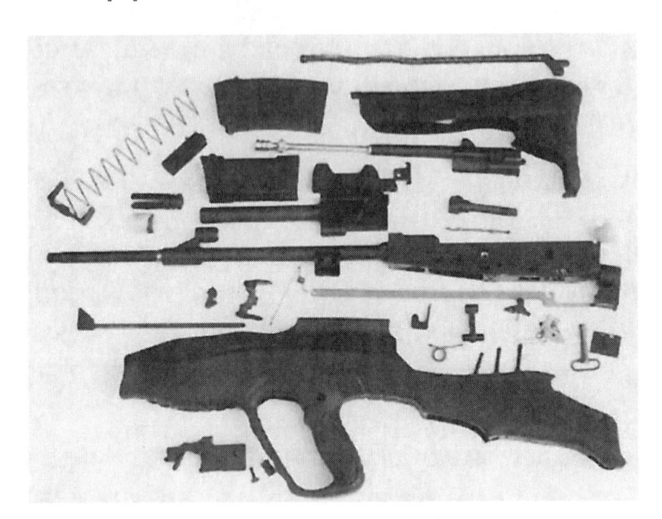

CR21 bullpup, completely disassembled.

A considerable amount of controversy has always swirled about bullpup infantry rifles. Although the Steyr AUG bullpup has recently found favor in the law-enforcement community as a tactical carbine for dynamic entry teams in lieu of the more traditional pistol-caliber submachine gun, this configuration has never been popular with the U.S. Armed Forces.

It is in fact, a British invention. In August 1902, British engineer J.B. Thorneycroft presented a prototype bolt-action rifle to the British War Office for consideration by the Small Arms Committee. In trials it was not impressive and all official interest in the Thorneycroft design ceased by 1903.

During the early part of 1944, work started at Enfield Lock on the design of a new bullpup sniper rifle. Called the Sniper Rifle Experimental Model I (SREM I), the weapon featured a radical design. The bolt traveled in a metal housing inclined 12 degrees below the bore's axis. The bolt was operated by a pistol grip which carried an arm engaging a cammed slot on the right side of the bolt. Rearward movement of the pistol grip first rotated the bolt to unlock it and then retracted it. But military weapons technology was firmly dedicated to self-loading rifles and the bullpup sniper rifle project was abandoned in 1945 at the war's end. Despite the slide into post-war obscurity, the early research had sparked some continuing interest.

Early efforts culminated in the controversial, ill-fated British EM-2 bullpup design, which was effectively torpedoed by the Americans in the 1952 light rifle trials held at Aberdeen Proving Ground. As a consequence, the British ended up by adopting an inch-equivalent of the FN FAL. However, eventually the British went back to the bullpup concept and launched the incredibly dreadful SA80 caliber 5.56x45mm NATO bullpup in 1985 at a cost of $800 million.

During the early 1950s, when the so-called light rifle concept was being developed by the NATO countries, *Fabrique Nationale* in Belgium produced a prototype carbine (serial no. 3) which was a bullpup design by Dieudonne Saive. But the project was subverted by the dictatorial Director General of FN, Rene Laloux. French weapons engineers also took a long look at bullpup technology and produced their own prototypes. One of these—the FAMAS F1—became a general-issue weapon in the French army more than two decades ago.

Using the Johnson/Stoner rotary bolt, the Enfield EM-2 layout and a host of carefully considered innovations, the Austrian firm of Steyr Daimler Puch, AG, fielded a bullpup called the AUG (Army Universal Gewehr-rifle). Adopted by the Austrian army in 1977 where it is called the StG 77 (SturmGewehr—assault rifle-1977) it has become one of the most familiar bullpup weapons in the world.

COMPACT FOR COMBAT

The bullpup's salient feature is compactness. By definition, a bullpup's barrel group is moved well back into the stock and both action and magazine are placed behind the trigger assembly. Ejection of spent cartridge casings occurs close to the shooter's face and some critics claim bullpups are inappropriate for left-handers.

Other critics—particularly those with combat experience in confined areas—claim the bullpup design forces a soldier to expose too much of his body when firing around corners. But, bullpup advocates have their own reasoned response to all that.

Only 20% of the world's population is left-handed, they argue, and most of them can be trained to fire effectively from the opposite shoulder. Even southpaws who can't manage the switch can be equipped with bullpups modified to accommodate them. The Steyr AUG features a left-side ejection option and the French FAMAS F1 extractor can be moved to either the right or left sides, as can the cheek rest, to provide for ejection on either side.

To commence fieldstripping the CR21, the sling swivel pin must be removed. The ambidextrous selector lever is somewhat longer on this pre-production prototype than current production-series specimens. Although the grip area is somewhat reminiscent of the Steyr AUG, the CR21's overall appearance is both startling and innovative.

In addition, those who favor bullpups say shooting around corners is not healthy, so soldiers armed with shoulder-mounted weapons should not be taught to do so. I have personally employed left-side barricade positions with the Steyr AUG that provide as much cover as you can expect firing from the right side. The bullpup debate will continue unabated well into the foreseeable future.

In the meantime, the South African firm, Vektor, formerly a division of LIW, has just introduced a new caliber 5.56x45mm bullpup infantry rifle called the CR21 (Compact Rifle, 21st Century). A company highly respected throughout the world, with a proven track record in defense products ranging from handguns to automatic cannons, Vektor's CR21 demands serious consideration.

21ST CENTURY BULLPUP

The CR21's unique stock configuration is the result of a complex design process between Vektor and Pentagraph, an internationally respected and well-known industrial design firm that was responsible for the exterior envelope of Vektor's racy-looking CP1 pistol. These graphic, product and environmental design consultants were commissioned to design a package that would propel Vektor's bullpup into the 21st century. They have, in my opinion, succeeded. The stock, a polymer injection molding, consists of two major components: the stock itself and a top cover. The top cover can be removed by a latch at the bottom of the butt-stock. This provides access to the working parts and also a storage compartment. A cleaning rod and other maintenance equipment are stored inside the stock housing along the top right side. A more elastic polymer (except for the scope housing which must remain rigid) than that used in the CP1 pistol's frame has been employed, because the rifle is heavier and must pass more severe drop testing. There is provision for the attachment of a bayonet.

There is also a polymer accessory attachment bracket at the rear which interfaces the rear end of the receiver body to the stock housing. A number of features have been incorporated into the design to allay the user's potential fear of a bullpup-type rifle "blowing up" in his face. The sight base is attached by a pin on each side to add rigidity to the front of the receiver. The entire selector system, including its axis pin, is made of polymer. In the highly unlikely case of a blowup when the system is in battery, the scope mount would force the direction of the blast downward.

Overall weight of the CR21, empty, is 8.4 pounds (3.8 kilograms). Overall length of the

VEKTOR CR21 SPECIFICATIONS

CALIBER .5.56x45mm NATO (and potentially 7.62x39mm and 5.45x39mm).

OPERATION . . . . .Locked-breech with rotary bolt; gas-operated; selective-fire capability; fires from the closed-bolt position.

FEED . . . . .Standard: 20-round detachable staggered-column, two-position-feed, box-type magazine; optional capacities: 35-, 30-, 25-, 15-, 10- and 5-rounds.

WEIGHT, empty . . . . .8.4 pounds (3.8 kilograms).

LENGTH, overall . . . . .29.9 inches (760 mm).

BARREL LENGTH . . . . .18.1 inches (460 mm).

SIGHTS . . . . .Reflex-type with fiber optics, 1X magnification, triangle aiming point with tritium illumination; mounted to the barrel and action.

FINISH . . . . .Molybdenum paint of the dry-film lubricant-type—extremely tough and both corrosion-resistant and high-temperature-resistant—baked over the exterior of steel components.

MANUFACTURER . . . . .Vektor, P.O. Box 5445, Pretoria 0001, South Africa.

T&E SUMMARY: Reliable Kalashnikov-type. High-accuracy potential. Low-recoil impulse. Excellent hit probability and rapid target acquisition. Extremely well-executed. 21st-century envelope. May eventually prove to be the best of the bullpups.

rifle is 29.9 inches (760 mm). The barrel length is 18.1 inches (460 mm). The barrel has been manufactured from cold-forged steel and features a hard-chrome-lined bore and chamber. The six-groove rifling is provided with a right-hand twist of one turn in 9 inches. This has become a popular rate of twist for SS109-type ball ammunition, as the 1:7-inch twist adopted by the United States is too fast and was dictated by requirement to stabilize L110-type tracer ammunition which features a long projectile. A Vortex-type flash hider, with a spring for holding rifle grenades, has been attached to the end of the barrel. The most effective flash hider ever designed, there is no detectable flash signature. A molybdenum paint of the dry-film lubricant-type—extremely tough and both corrosion-resistant and high-temperature-resistant—has been baked over the exterior of the steel components.

The safety is in front of the trigger. Of the so-called cross-bolt-type, it must be pushed from right to left to fire the weapon. The ambidextrous selector lever is located at the rear of the buttstock on each side (and is now shorter than the one shown in *SOF's* photos of early prototypes). International picture symbols are used to indicate semiautomatic or full-auto in production-series versions. They are raised for tactile recognition. The selector system is in the opposite sense as that of the Kalashnikov. The upper position will provide semiautomatic fire. The bottom position will produce full-auto fire.

MODERN MAGAZINES

The magazine catch/release is directly in back of the magazine and similar to that of the

CR21 Method of Operation

The CR21's method of operation remains identical to the Kalashnikov's. It fires from the closed-bolt position and is gas-operated without an adjustable regulator.

When the trigger is pulled, the hammer drives the firing pin forward to ignite the primer. After ignition of the primer, a portion of the propellant gases migrates into the 1.8mm-diameter gas port, tapped at a 24-degree angle to the gas block which is pinned to the barrel. The gas enters the cylinder (to which a small spring has been attached to secure its retention during reassembly) and drives the piston rearward.

The piston is hard-chrome-plated for ease of maintenance. It is also notched to provide a reduced bearing surface and permit excess gas blow-by. The bolt carrier is permanently attached to the piston. After a short amount of free travel, during which time the gas pressure drops to a safe level, the cam slot engages the bolt's cam pin and the bolt is rotated and unlocked as the carrier moves rearward.

Primary extraction occurs as the bolt is rotated, and thus a massive Kalashnikov-type extractor claw is not required. Empty case ejection is typically violent, in the manner of all Kalashnikovs. The cases are severely dented by the ejector and thrown to the right and downward. At this time, the recoil spring is compressed and its return energy drives the carrier forward to strip a round from the magazine and chamber it.

Two sears control the firing mechanism, the trigger sear and an auto-safety sear. In full-auto fire, the trigger sear is held back and only the first round of the burst is fired off this rear sear. Subsequently, the bolt carrier moves rearward and rolls the hammer over. The auto-safety sear continues to hold the hammer back until the bolt carrier is fully forward again, at which time it trips the auto-safety sear and the hammer rotates to fire another round. Thus, after the first round, the trigger sear is deactivated entirely from control on the hammer. Releasing the trigger will catch the hammer on the trigger sear once more. In semiautomatic fire, no pressure is placed upon the trigger sear, which is free to catch the hammer each time it is rolled back by the bolt carrier.

R4. Like other Kalashnikov types, the magazine must be rocked forward for release from the magazine-well.

Similar to those of the South African R4 rifle, CR21 detachable staggered-column, two-position-feed, box-type magazines contain only four components (a steel follower spring, nylon follower, floor-plate and body). The body is a one-piece injection molding. The follower ends are closed to inhibit the accumulation of debris. More durable than steel, and just as reliable, these nylon magazines can absorb tremendous shock without damage. Their strength and drop tests surpass steel and, of course, they're not susceptible to corrosion. Impervious to all commonly encountered cleaning solutions, Vektor's magazines have successfully passed accelerated ultraviolet tests (five years of exposure lying on a steel plate on Vektor's roof) with only slight discoloration. Cold weather tests were conducted without failure of any kind. Tests indicated an endurance life of more than 20,000 rounds. A 20-round magazine will be standard with the CR21, however, 35-, 30-, 25-, 15-, 10- and 5-round capacities are also available.

The cocking handle is located on the left side and is of the non-reciprocating type. It is a polymer injection molding and does not fold down. It is, therefore, considerably more robust than the hinged Heckler & Koch type. The rifle's bolt carrier has a stud welded to its top which engages a rod to which the cocking handle is attached.

There is a molded socket that serves as a sling mounting point at the front of the rifle on top of the stock. At the rear, the sling swivel can be changed over from the right or left sides. This permits a number of different carrying options and allows the rifle to be fired from a slinged position.

A gas relief hole in the top of the stock is located directly over three ports in the gas cylinder. There is one ventilating port on top of the buttstock and three on the bottom. A polymer heat shield inside the stock protects the forearm area. At the front of the stock, a glass-filled phenolic bushing, surrounding the barrel and touching it at only three points, also permits heat transference to the atmosphere. All of this is quite effective, as the CR21 has passed all cook-off tests performed on it, including firing 350 rounds as quickly as possible.

A considerable amount of design effort was directed toward the CR21's optical sight. For

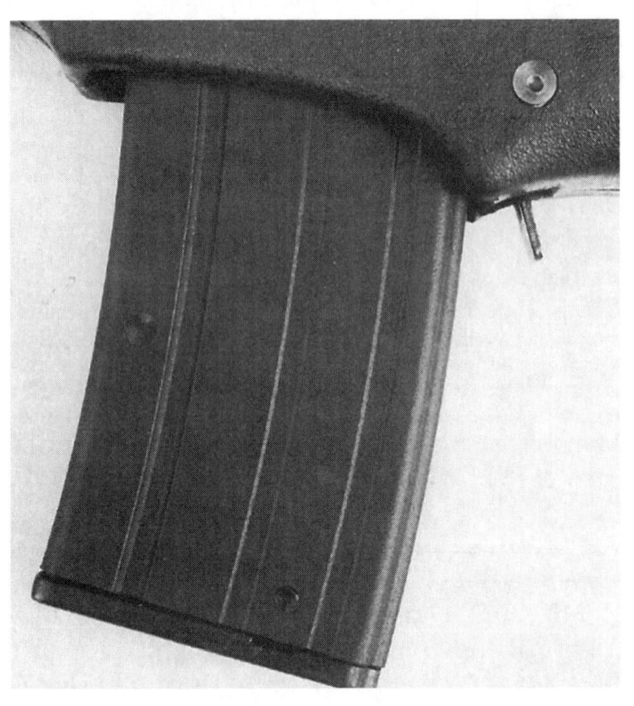

The magazine catch/release is directly in back of the magazine and similar to that of the R4. Like other Kalashnikov types, the magazine must be rocked forward for release from the magazine-well. A 20-round magazine is standard, but 35-, 30-, 25-, 15-, 10- and 5-round capacities are also available.

CR 21 bullpup, fieldstripped.

more rapid target acquisition it features only 1X magnification. This is a reflex-type sight using fiber optics at the objective end to gather light. The intensity of the aiming point—a solid triangle—is self-adjusting. A triangle presents a more precise aiming point for long range shooting than a round dot. The triangle fits into a 5-6 MOA circle. An amber-colored tritium capsule inside the housing illuminates the triangular aiming point in the dark. Amber is easier to see than red. The optical sight uses no batteries and thus has no switches—important advantages over red-dot-type sights.

The optical sight can be removed from the rifle by drifting out two roll pins. This presents the opportunity to install other sighting devices or to easily replace a damaged unit. Two pockets in the scope housing interface with two pillars on the stock. This optical sight is compatible with monocular or binocular night vision devices. This is an important advantage for the modern soldier, as replacing the scope with night vision equipment requires the operator to scan the terrain with the weapon itself. Adjustment of elevation and windage zero is by means of two screws, one on the front and the other on the right side. These screws can be turned with the rim of a cartridge case. The scope base is milled from a solid block of aluminum and attached to the barrel and receiver. Thus when the barreled action is removed from the stock, the optical sight remains with these components.

While the ejection port is located on the right side of the CR21's stock, the rifle can be fired from either the left or right side without changing any components. This is the consequence of a polymer deflector plate at the top of the ejection port which directs empty cases downward and away from the operator.

The long trigger bar required of bullpup designs runs along the right side of the stock's interior. It engages a hole in the rear trigger. The rear trigger itself has only a right-side hook unlike most Kalashnikovs which have a hook on each side. This makes the trigger-pull weight a bit lighter and involves less machine work.

While it is free floating, the firing pin does not duplicate that of the R4. R4 firing pin assemblies include a synthetic bushing and steel flat washer retained by a pin through a small hole at the rear end of the striker. This was done to prevent so-called "ghost shots." Lighter than the R4's striker and made of stainless steel, the CR21's firing pin has an AK configuration, with no bushing. Its large diameter head precludes ignition of more sensitive primers during cocking.

To remove the main operating components from the stock housing, first remove the magazine and clear the weapon by retracting the bolt, then press the trigger to release the hammer. Push up on the sling swivel's axis-pin lock and remove the sling swivel pin. Then separate the barreled action and scope from the stock. Further disassembly duplicates that associated with all Kalashnikov-type rifles. The gas cylinder can be removed by merely pushing out its retaining pin. The gas cylinder can then be withdrawn from the rear.

KALASHNIKOV UPGRADES

Although the grip area is somewhat reminiscent of the Steyr AUG, the CR21's overall appearance is both startling and innovative. South African R4 rifles and almost all Kalashnikov types can be converted by Vektor to the CR21 configuration. The South African Defense Force is seriously considering an upgrade of all its R4 rifles to the CR21 format. With today's shrinking defense budgets this should prove to be an attractive alternative to countries equipped with aging R4s, Galils and Kalashnikovs.

My test firing of the CR21 demonstrated the reliability found with all Kalashnikov-type rifles. The accuracy potential is quite high—certainly far beyond the capabilities of other Kalashnikov-type rifles. It also scores high marks in both hit probability and target acquisition, when employed by an experienced and properly trained operator.

The recoil impulse is lower even than that associated with the R4, quite noticeably so in burst-fire sequences. This is undoubtedly a consequence of the in-line stock design and the wider and longer butt area which distributes the felt recoil over a larger area.

Bullpups such as the popular Steyr AUG can expect fierce competition from Vektor's CR21. In addition to more than matching any existing bullpup's performance capabilities, the CR21 provides the alluring prospect of the economical conversion of Kalashnikov rifles already in inventory and approaching replacement age.

Originally appeared in *Soldier Of Fortune*
March 1998.

KALASHNIKOV'S KOMBAT KLASSIC

SOF Celebrates the 50th Anniversary of the AK47

Victor Kalashnikov, son of Mikhail T. Kalashnikov who, together with Alexi Dragunov, youngest son of Evgeni F. Dragunov, designed the Bizon submachine gun with its unique helical-type magazine, shown here with a sound-suppressed version of his design.

SOF's Contributing Editor, Valery N. Shilin, fires a sound-suppressed version of the Bizon submachine gun. Note propellant gases escaping from the front of the sound suppressor, a common phenomenon.

It has been conservatively estimated that more than 70 million Kalashnikov-type small arms have been manufactured to date throughout the world. The AK47 rifle and its many variants are the most ubiquitous infantry weapons ever fielded. Nothing else even comes close to their dominance on the battlefield during the second half of the 20th century. Today it is still used by the armed forces of 55 nations. Six countries carry its image on their national seals.

Russian AK rifles are principally manufactured in the formerly closed city of Izhevsk at the IZHMASH factory. At the beginning of the Great Patriotic War (World War II), the Izhevsk plant was the only one in the Soviet Union producing rifles. By December 1941, a daily output of 10,000 Mosin Nagant rifles was achieved. No other factory in the world has ever approached this level of production. Located 1,130 kilometers due east of Moscow in the Udmurt Republic, Izhevsk has a population of 730,000. Its small-arms factory was established in 1807 making it the third such facility established in Russia.

During the first week of November 1997 Izhmash celebrated the 50th anniversary of the AK47. *Soldier Of Fortune* was the only U.S. publication invited to participate in this historic event. The week's events

SOF's Technical Editor, Peter G. Kokalis, and Mikhail T. Kalashnikov toast in the Russian manner—Kalashnikov with vodka (of course) and Kokalis with Coke (alas).

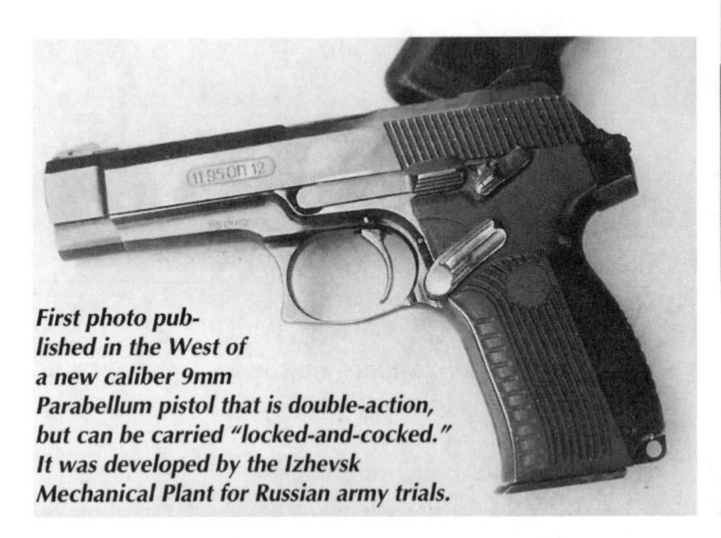

First photo published in the West of a new caliber 9mm Parabellum pistol that is double-action, but can be carried "locked-and-cocked." It was developed by the Izhevsk Mechanical Plant for Russian army trials.

Russian arms designer Gennadiy Nikolayevich Nikonov, of IZHMASH, holding one of the many prototypes of his AN-94 "Abakan" assault rifle.

Left: General Mikhail T. Kalashnikov, whose AK rifle is in service with the armed forces of 55 nations. More than 70 million Kalashnikov-type small arms have been manufactured.

This would have been an impossible occurrence during Russia's 50 years of communism. But, in November 1997, the Russian Orthodox Archbishop of the Udmurt Republic honored the 50th anniversary of the AK47 rifle on the steps of Alexander Nevsky Cathedral in the presence of Vladimir R. Grodetsky, Director General of IZHMASH and Mikhail T. Kalashnikov.

Designer Mikhail E. Dragunov, close personal friend of the author and eldest son of Evgeni F. Dragunov, holding a specimen of the KLIN/KEDR submachine gun family which was designed by his late father but brought to production-series development by him.

A .410-gauge semiautomatic shotgun of the SAIGA family. Based upon the Kalashnikov method of operation and sharing some parts interchangeability with the AK, this particular version has been configured as an AK and is already in service with border-guard units of several states of the CIS.

included two formal celebrations held in the city's principal theaters and a banquet honoring the AK47's designer Mikhail Kalashnikov, an arms fair and several small dinner parties attended only by General Kalashnikov, his immediate family and a handful of his personal friends from the United States, Japan and Germany. What follows is a photographic odyssey of some of the events and people I saw during that memorable occasion.

I would never have believed for a moment when we were in Afghanistan in 1983 that someday the world would so change that I would sit down to dinner with a Hero of the Soviet Union who designed the rifle that symbolized everything I was fighting against with all the passion and energy within me. How incredible. On the 18-hour train ride from Moscow to Izhevsk I had ample opportunity to gaze out the window at the snow-covered immenseness that is Russia and contemplate this strange turn of events.

Originally appeared in *Soldier Of Fortune*
April 1998.

KALASHNIKLONES

A Consumer's Guide to AKs

Introduced to the American public in 1982 by the U.S. division of Steyr-Daimler-Puch in Steyr, Austria, the Egyptian ARM is an almost exact semiauto-only duplicate of the Russian AKM.

Valmet M62/S, although a faithful copy of the Kalashnikov-type issued to the Finnish armed forces, did not fit the AK stereotype and few were sold in the United States.

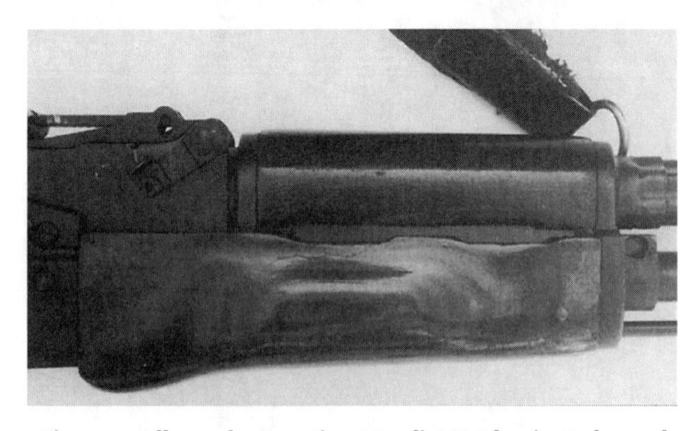

Finger swells on the Egyptian Maadi ARM laminated-wood handguard exactly duplicate those of the Russian AKM.

There is no more ubiquitous assault rifle than the Kalashnikov. It has been estimated that more than 50 million have been manufactured to date. Although principally associated with the present and former ComBloc nations, it will be encountered, to one degree or another, in nearly every country of the world. During the 1980s, tens of thousands of Kalashnikovs were imported to the United States in a semiautomatic-only format.

In that form, they were not *assault rifles*, which, by definition, have selective-fire capability. No matter, the word "assault" became a sinister buzzword to anti-gun leftists in this country and their myrmidons in the media. No matter that they represented less than 3% of the firearms associated with crimes. They became touted by liberals with an agenda that included disarming the populace of "the criminals' weapons of choice." With an unrelenting focus placed on this outright lie by the mainstream media, the outcome was only a matter of time. Liberals are forever mesmerized by "feel good" legislation that has no real effect on a problem, as long as it doesn't impact upon their own personal rights and chattel property. First, President Bush banned the importation of a number of so-called "assault guns," which included all of the semiautomatic-only Kalashnikov types. President Clinton and Congress finalized the purging process by passing a "crime bill" that prevented not only the future importation, but also the domestic manufacture of firearms with such incredibly frightening features as pistol grips, flash hiders, folding stocks, bayonet lugs, and magazines with a capacity of greater than 10 rounds.

But, those which were imported legally can be possessed and transferred freely, except in the People's Democratic Republics of California, Connecticut, and New Jersey, and some Marxist municipalities here and there throughout the country. Various Kalashnikov models were imported from no fewer than five nations. How do they differ? Are some more desirable than others? And, more importantly, what are they worth today? While banning them had absolutely no effect on crime, it certainly increased their value, in some cases by a great deal. Let's take a close look at them all and try to separate the wheat from the chaff. No need to concern ourselves with the method of operation or disassembly procedures, as they are almost identical in every instance, and I've covered that ground many times before. I will arbitrarily omit semiauto versions of the RPK-type since, without the full-auto option, these squad automatics, in my opinion, serve no useful function. I'll also limit my discussion to those Kalashnikovs chambered for the original M43 7.62x39mm cartridge, leaving out the Israeli Galil and the 5.56x45mm NATO models of the Finnish Valmet and Chinese Kalashnikovs. No thumb-hole-stocked mutants will be discussed either.

FINNISH VALMET

The very first semiauto Kalashnikov to reach these shores was the Valmet (*Valtion Metallitehtaat* or State Metal Works) M62/S, imported from Finland by Interarms in Alexandria, Va. Based upon the third model AK-47, the M62/S was a decided failure. Although internally a Kalashnikov in every detail, except for its selector lever, magazine and mill-finished forged receiver, externally it bears little resemblance to the Kalashnikov stereotype.

Its odd but effective three-pronged flash hider is pinned to the barrel and carries an integral bayonet lug that accepts only the indigenous Finnish knife/bayonet with its green leather, Lapp-style sheath. The gas tube is unprotected; there is a black plastic bottom handguard only. The oversize, flapper-type magazine release lever has a large protective housing which serves as the front of the trigger guard. Ribbed and oval-shaped, the black plastic pistol grip is totally unorthodox in appearance. Even more grotesque is the buttstock, which consists of plastic-coated steel tubing with a sheet-metal buttplate. M62/S rear sights are mounted at the rear of the sheet-metal receiver cover. They are adjustable for elevation zero only and can be set in 100-meter increments from 100 to 600 meters with a 150-meter battle-sight position. As the receiver cover is of somewhat thicker gauge than that found on most Kalashnikovs, the rear sight rests securely and in this location provides a longer sight radius. Enclosed within a protective hood, the round post front sight can be adjusted for windage zero more easily than on other Kalashnikovs, by merely turning opposing screws on the sight base. It came with a single-dot, flip-up luminous night sight on the front sight's housing. They have by now lost all of their luminosity.

In the mid-1970s, Finland introduced an AKM-type rifle with a pinned and riveted sheet-metal receiver. Designated the M62-76, it was also imported in semiauto-only form as the M-76. Early specimens were identical in appearance to the M62/S except for the receiver, a more appealing pistol grip and the tubular buttstock, which was now hinged and folded to the left. It didn't help.

This well-made ugly duckling just didn't look like a Kalashnikov. Eventually a wooden-stock version was introduced, and the forearm was redesigned to include a heat shield over the gas tube. In this configuration, the M-76 was imported by Valmet Sporting Arms Division in Elmsford, N.Y. If you want the version used by the Finnish army, it will have to be the M62/S as imported by Interarms. Since few were sold, the M62/S or the AKM-type with folding tubular buttstock will now bring $1,800 to $2,100 (this and all subsequent prices cited are for specimens in new condition) on the rare occasions they are encountered for sale. Wooden-stock M-76 rifles are worth approximately $100 less.

EGYPTIAN MAADI ARM

In contrast, the next semiauto Kalashnikov imported to the United States was, in almost every regard, an exact duplicate of the Russian AKM. During the 1950s, the Soviet Union, as part of its military-aid program, established the production of Soviet-pattern small arms in the Arab Republic of Egypt. AKM rifles were manufactured at "Factory 54," the Maadi Company for Engineering Industries in Cairo. Key Egyptian personnel were trained in the Soviet Union, and the plant was supervised by Russians prior to their abrupt expulsion from Egypt.

Designated as the "ARM" (Automatic Rifle Misr), a semiauto version of the Egyptian AKM was introduced to the American public in 1982 by the U.S. division of Steyr-Daimler-Puch in Steyr, Austria. For all intents and purposes this is as close as most of us will ever come to a Russian AKM. The laminated-wood hand-guards and buttstock are correct in every detail. The lower handguard has the proper hand swells, and the buttstock has been reinforced with steel pins in all of the right places. All of the wood furniture was imported from Finland but finished in Egypt. The plastic pistol grip was injection-molded to the correct Soviet configuration.

The metal components are exactly those of the Russian AKM. Small recesses on each side of the receiver, directly over the magazine well, serve as magazine guides and instantly distinguish this as an AKM. The four gas-escape holes found on each side of the AK-47's gas cylinder have been omitted, and there are two gas-relief holes on each side of the gas block where it mates with the gas tube. The bayonet lug is directly under the gas block. Rear sights are graduated to 1,000 meters (with a 300-meter battle-sight setting marked "P") instead of the AK-47's 800-meter maximum. An AKM muzzle brake was installed. The sheet-metal receiver cover has transverse ribs and a rolled edge over the ejection port. Chambers and bores are hard-chromed. All of the exterior metal surfaces have been finished with a baked-on black enamel over phosphate. Only the green web sling was of Egyptian origin.

No doubt about it, except for the selective-fire option and markings on the receiver, this is a Soviet AKM. Yet no more than a few thousand were brought in before importation ceased. In 1986 about two dozen semiauto Egyptian ARMs with a side-folding buttstock, similar in appearance to that of the East German MPiKMS-72, entered the United States via a small importer. Today Steyr/Maadi ARM rifles are worth little more than their original price, $750 to $900. The folding stock variant is worth considerably more, and when one can be located it will fetch $1,500 to $1,800.

Why did the Steyr/Maadi ARM fail? There were two principal reasons: First, at the time it was being imported a supply of reasonably priced ammunition was not available. Berdan-primed, noncorrosive, Finnish Lapua 7.62x39mm ammunition (usually headstamped "VPT" with two digits indicating the year of production) cost 45 to 55 cents per round. Norma's Boxer-primed ammo in this caliber was selling for 85 cents per round. Yugoslav ammunition was just starting to trickle into this country. Americans don't buy firearms unless they can afford to shoot them.

CHINESE AKS

Second, within a year after the introduction of the Steyr/Maadi ARM, the People's Republic of China (PRC) gained Most Favored Nation status with the United States, and Clayco Sports Ltd., of Clay Center, Kan., commenced importation of the first PRC semiauto Kalashnikovs at a price substantially lower than the Egyptian version. Both rigid- and folding-stock variants were imported under the designation "AKS." While these rifles were of the AKM-type—that is, the receiver body had been fabricated from a 1mm-thick U-section of stamped sheet-metal, extensively supported by pins and rivets—numerous features of the AK-47 were retained.

The rear sight had a maximum elevation of 800 meters, and the 300-meter battle-sight setting was marked "D." The receiver cover was without transverse ribs or a rolled edge over the ejection port. There were four gas-escape holes on each side of the gas cylinder. There were no hand swells on the lower handguard. Handguards, pistol grip, and buttstock on early Clayco AKS rifles were of reddish-brown plastic. This was quickly changed to black plastic in an attempt to broaden its appeal to American shooters. All exterior metal surfaces were salt-blued. Bolts and bolt carriers were left "in-the-white" (unfinished) or phosphated. Clayco provided these rifles with only one magazine (blocked to accept only five rounds), plastic oil bottle, sling and buttstock cleaning kit. Spare magazines and bayonets were also available. Today, any of the Clayco Kalashnikovs are worth $650 to $750 each.

The semiauto Kalashnikov was off and running. Clayco dropped along the wayside, but several other importers picked up the slack, and rifles started to pour in from the PRC.

PRC TYPE 56-2

One of the more interesting was a side-folder (PRC Type 56-2) with tubular buttstock struts that was standard issue with the People's Liberation Army. First imported by Pacific International Merchandising Corporation in Sacramento, Calif., and manufactured by Arsenal 626 (abbreviated as "66" in a triangle on the receiver) in Hei Long Jing Province, it was fitted with a pistol grip resembling those of the Browning M1917A1 and M1919A4/A6 machine guns (apparently an export-only feature, as I observed these grips on PRC Type 56 rifles in Afghanistan). This rifle will presently bring between $750 to $850.

These rifles, fitted with a PLA-issue AKM-type pistol grip, were eventually imported by PTK International, Inc., in Atlanta, Ga., which represented the small-arms products of Poly Technologies, Inc. All Poly Tech Kalashnikovs were manufactured at Arsenal 386 in Fu Jian Province, although a few early specimens were produced at Arsenal 416 (actually Arsenal 976) in Shan Dong Province.

The original AK-47/AKM folding stock was in need of a PIP (Product Improvement Program) since its inception. A double-strut type, controlled by a large press-button release located above the pistol grip on the left side of the receiver, it folds under the receiver, and the magazine passes through it. Patterned directly after the German World War II MP38/40 submachine gun's stock, it's adequate for firing the 9mm Parabellum pistol round, but will not take the sustained pounding of a rifle cartridge. I once examined an early Soviet folding-stock AK-47 in South West Africa (now called Namibia) near the Angolan border. The stock latch was so worn that the rifle could be rotated 15 degrees in either direction after the stock was securely braced against the shoulder—not very beneficial to the Kalashnikov's already limited accuracy potential.

Soviet designers addressed this problem in the caliber 5.45x39mm AKS-74 by designing a side-folding stock. PRC designers arrived at an even more robust solution. Their stock's skeletonized frame (as imported by PTK) was fabricated from a single piece of heavy-gauge sheet metal, bent

Semiauto Chinese Type 56-2 sidefolder was fitted with a peculiar pistol grip resembling those of the Browning M1917A1 and M1919A4/A6 machine guns.

and folded into the proper shape. Five horizontal ribs have been stamped into the buttplate portion. There is a 4.75-inch-long, reddish-brown plastic panel on each side of the stock at the forward end. These panels have ribbed gripping surfaces and are retained by a threaded screw and pin. They serve as a compartment to hold the cleaning kit. Depress a spring-loaded pin on top of the stock and the cleaning kit will pop out to the rear. The latch mechanism is operated by a checkered, spring-loaded release button on top of the stock. Press down on the button, and the stock can be swung to the right, where it locks in place. The button must be pressed again to open the stock.

The latch mechanism is quite substantial, and this stock is every bit as stable as a non-folding wooden stock. When folded, access to the trigger is maintained, although manipulation of the selector lever is somewhat more difficult. Both hand-guards and the pistol grip were injection-molded reddish-brown plastic. All other features of the Type 56-2, as imported by PTK International, Inc., were those of the other PRC AKM-type rifles, except that there was no provision for the attachment of a bayonet. If you want the compactness of a folding-stock AK, then this is the one to acquire. These PTK side-folders now sell for $850 to $1,000.

PTK International also distributed rigid, wooden-stock (PRC Type 56) and down-folder (PRC Type 56-1) versions of the AKM-style (sheet-metal receivers) Kalashnikov with cruciform-shaped folding spike bayonets. Either variant is now worth between $650 to $800.

AK-47/S LEGEND SERIES

A faithful semiauto copy of the third model AK-47 (*Avtomat Kalashnikova obrazets 1947g* or Kalashnikov assault rifle model 1947) was produced for PTK and imported under the designation AK-47/S Legend Series. Its cold-hammer-forged receiver required a total of 105 individual machining operations. Barrels were selected for uniformity and accuracy potential. Chambers, bores, and pistons were hard-chromed. Polish and bluing of the external metal surfaces were exceptional. Both the bolt and bolt carrier were lightly phosphate-finished. The wooden butt-stock, pistol grip, and handguards were stained and varnished to an exact image of a Vietnam-era AK-47. Cross-checkering on the wooden pistol grip was crude and correct. An AK-47 muzzle nut was attached to the barrel. The front sight post had open protective ears instead of a hood with a hole for the adjustment tool (either is proper for the AK-47).

A Russian-style AK-47 bayonet was supplied with the AK-47/S rifle. It has an 8-inch satin blade with a spear-point shape and centrally placed fullers. The blued-steel hilt has mahogany-stained wooden grips (usually wood-fiber-impregnated plastic on Soviet specimens) held in place by two bolts and slot-head nuts at each end. A web hanger is attached to two metal loops at the rear of the blued-steel scabbard. Except for the selective-fire option, this is a genuine AK-47. Nothing else like it was ever imported and today they are worth $1,000 to $1,200 each, either down-folder or rigid-stock versions.

Rifles imported by PTK International, Inc., differed in one very important regard from all of the other caliber 7.62x39mm semiauto

A faithful semiauto copy of the third model AK-47 (Avtomat Kalashnikova obrazets 1947g or Kalashnikov assault rifle model 1947) was produced for PTK and imported under the designation AK-47/S Legend Series.

Kalashnikovs imported to the United States: Military-issue Kalashnikovs and their semiauto equivalents do not have spring-loaded firing pins. If commercial or reloaded ammunition—usually with primers more sensitive than milspec because of a thinner cup and sometimes a difference in the primer mixture—is fired in rifles of this type, the free-floating firing pin can, and eventually will, result in a slam-fire with ignition out of battery. The resulting detonation can lead to self-destruction of the firearm and anatomical damage of varying severity to the shooter. At my suggestion, all Kalashnikov-type rifles distributed by PTK International, Inc., were fitted with *spring-loaded* firing pins.

If you are not going to fire Bulgarian, Chinese, Czech, East German, Finnish, Romanian, Russian, or Yugoslav ammunition with hard military primers, then I suggest you use the new 7.62x39mm ammunition introduced by Black Hills. Aware of this primer-sensitivity problem, Black Hills is using the CCI No. 34 milspec primer, together with the excellent Hornady 123-grain Soft Point bullet.

HUNGARIAN SA 85M

Two other countries exported semiauto Kalashnikovs to the United States. In 1986, Kassnar Imports, Inc., in Harrisburg, Pa., obtained a waiver from the State Department's restricted-munitions list to import 7,000 AKM-type semiauto Kalashnikovs from Hungary. Along with the model designation (SA 85M), caliber, importer and serial number, the left side of the receiver was marked with the state arsenal's logo, "FEG" *(Femaru Fegyvar es Gepgyar* in Budapest).

Available with either a downfolding or rigid wooden stock, the furniture was cut from native blond beechwood (twice steamed, disinfected, dried to a humidity level of 10-12% and then oil-varnished with a waxed paste). The wooden pistol grip's shape was quite distinctive and taken from the plastic grip on the Hungarian AMD short-barreled assault rifle.

While the four gas-escape holes on each side of the gas cylinder have been retained from the AK-47, the pinned and riveted sheet-metal receiver, the two gas-relief holes on each side of the gas block where it mates with the gas cylinder, the muzzle compensator, sheet-metal receiver cover with transverse ribs and rolled edge over the ejection port, bayonet lug under the gas block, rear sight graduated to 1,000 meters, and all other features are exactly as found on the Soviet AKM. All exterior metal surfaces, including the bolt group and magazines, are finished with a matte black enamel baked over phosphate. Like all other semiauto AKMs, there is no auto safety sear or anti-bounce device.

Complete with three 30-round magazines, wire-cutter bayonet and scabbard, leather sling, two plastic lubricant bottles in a tray, and the usual buttstock cleaning kit with jag tip, nylon brush and front-sight adjusting tool, the Hungarian SA 85M could originally be purchased for approximately $400. They were all sold in less than a year. Now you can expect to pay from $1,000 to $1,200 for either variant.

Hungarian SA 85M features native blond-beechwood furniture. Only 7,000 were imported and they were all sold in less than a year.

YUGOSLAV AKMs

Last, but most assuredly not least, were the Yugoslav AKMs imported first by American Arms, Inc., in North Kansas City, Mo.

(in small quantity only), and then subsequently by Mitchell Arms. Inc., in Santa Ana, Calif. Marked as either the "Model AK" (American Arms) or "Model AK-47" (Mitchell Arms), these rifles are actually the semiauto equivalents (without the grenade-launching sights and gas cut-off) of the wooden-stock M70B1 and folding-stock M70AB2, both of which are essentially AKMs with sheet-metal receivers.

The receivers of all other AKM series rifles are fabricated from a 1mm-thick U-section of sheet metal. M70B1 and folding-stock M70AB2 receiver bodies are constructed using a U-section of sheet-metal that is 1.5mm in thickness. This 50% increase in wall thickness substantially decreases the accuracy-reducing twisting and flexing of the AKM receiver, which is associated with the recoil and counter-recoil cycles. In addition, the barrel extension, which contains the bolt's locking recesses, is considerably more substantial than that of all other AKMs. There's no free lunch, however: The M70B1 weighs 8.2 pounds, empty.

There are some other interesting features on these Yugoslav AKMs. They are equipped with beta-light night sights: a flip-up at the rear with a horizontal tritium bar on each side of the open U-notch, and a flip-up with a single vertical tritium bar that blocks the round-post front sight.

A spring-loaded cross-pin at the rear of the receiver, just below the receiver cover, must be depressed from the left side before the recoil-spring guide rod/cover latch can be pressed forward to remove the receiver cover. This feature is of dubious value.

All of the wooden furniture is teak. Dense and attractive, it was never properly oiled, so I suggest application of a mixture composed of equal parts of linseed oil, turpentine, and household white vinegar. Although standard-issue M70B1 and M70AB2 rifles come with ribbed, black plastic pistol grips, the semiauto version imported to the United States was fitted with teakwood grips from the M76 caliber 7.92x57mm sniper rifle (a highly modified version of the Russian Dragunov SVD). They may not look like Kalashnikov grips, but they exhibit excellent human engineering. M70B1 rifles have no butt traps for a cleaning kit, but instead are equipped with a 1/2-inch rubber recoil pad which increases the length-of-pull by 0.8 inches. That's a significant difference as, in general, the Kalashnikov's buttstock is too short for most Westerners.

All of the steel components, except for the bolt group (left in-the-white), have been salt-blued. With the exception of the receiver cover, which is that of the AK-47, all of the other features on the Yugoslav rifles are those of the AKM series. Well-made and exhibiting above-average wood-to-metal fit, the Yugoslav AKMs were especially attractive because of their somewhat enhanced accuracy potential. Complete with one magazine, buttstock cleaning kit, twine pull-through and web sling, the M70B1 retailed for $675, while the folding-stock version sold for $698 when originally imported. They remain highly desirable, and today you will have to pay $1,300 to $1,450 and $1,650 to $1,850, respectively.

AK MAGAZINES

Kalashnikov magazines come in several different configurations and, in my opinion, the

Yugoslav AK receiver bodies are constructed using a U-section of sheet-metal that is 1.5mm in thickness. This 50% increase in wall thickness substantially decreases the accuracy-reducing twisting and flexing of the receiver during the recoil and counter-recoil strokes.

Yugoslav variants are the worst. In outward appearance they are the standard 30-round, blued, all-steel magazine with reinforcing ribs on the sides of the body. However, the magazine follower is raised on the left side to block the bolt group in the rear position after the last round has been fired. Since there is no hold-open mechanism in the receiver, as soon as the empty magazine is withdrawn the bolt group will jump forward into battery. That's dumb. These magazines are more difficult to remove, because the full force of the compressed recoil spring presses the bolt group against the magazine follower and, after insertion of a loaded magazine, the bolt group must still be retracted manually. Just use standard Kalashnikov magazines in the Yugo AKM.

These ribbed magazines will be encountered either blued or painted black. With one exception (other than the Yugoslav), the country of origin cannot be determined. Finnish magazines have a rectangular steel ring on the floorplate to secure them to the soldier's LBE (Load-Bearing Equipment). Early Soviet AK-47 steel magazines had slab-sided bodies without reinforcing ribs. They are uncommon.

Even more rare is a Russian ribbed magazine made of aluminum. It was fielded briefly when the AKM was introduced in an effort to reduce the overall weight of the system. However, they proved to be insubstantial and were recalled and turned to scrap.

More recent Soviet 7.62x39mm magazines were fabricated from a glass-reinforced, rust-colored synthetic. Molded in two parts, the magazine body was assembled using a viscous, two-part epoxy resin adhesive. The adhesive residue was removed by hand-grinding. They were marked with a factory code and either the Russian five-pointed star, arrow-in-triangle or, more rarely, star-in-shield.

Most of the Soviet "red" plastic magazines in this country came from either Afghanistan or South Africa. They were followed by a magazine with a body of dark-brown buterate plastic (also called ABS) of the type commonly used in appliance manufacture in the United States. The two body components, probably vacuum-formed, were heat-molded together rather than glued. There are very few in the United States.

The new AK100 series magazines are made from black, fiberglass-reinforced, thermosetting (this indicates that heat is used to cure the resin) polyamide (epoxy-based resin). Injection-molded polyamides are super industrial-strength synthetics well known for their resistance to high temperatures, corrosion, wear, chemicals and radiation. Lighter than steel, they have a higher tensile strength than aluminum.

Chinese Type 63 20-round, ribbed steel magazines can also be used in the Kalashnikov. However, like the Yugoslav magazine, the follower has been raised to operate a hold-open mechanism, and the bolt group will be held rearward until the empty magazine is removed. A small quantity of these were imported by Keng's Firearms Specialty, Inc.

RELIABILITY AND ACCURACY

I have fired tens of thousands of rounds through several hundred Kalashnikovs of every make and configuration in at least four different calibers (7.62x39mm, 5.45x39mm, 5.56x45mm NATO, and 7.62x51mm NATO) without a single stoppage of any kind. There is not a more reliable assault rifle. Easily maintained and almost idiot-proof, Mikhail Timofeyevich Kalashnikov's creation will always remain one of the great classics in the history of modern military small arms.

To be sure, it's far from perfect. Its selector lever is far too noisy (although this can be partially corrected by wrapping monofilament fishing line around the lever). Ejection is far too violent, and cases will be thrown up to 30 feet to the right and front of the weapon. While this may irritate reloaders searching for their empties, it's a far more serious consideration for those firing from cover and concealment. There's no hold-open mechanism, but that's a debatable feature. Trigger-pull weights will vary between 3 to 6 pounds, with an average of about 4 pounds. But, you travel a long, creepy road to get there.

However, the Kalashnikov rifle needs no test and evaluation by me to demonstrate its merits and idiosyncrasies. That would be the kind of pompous joke that only the popular gun press would perpetrate on its readers. However, I must make note of the fact that the 5.56x45mm NATO versions I fired in Izhevsk, Russia, displayed noticeably greater accuracy than any of those chambered for the 7.62x39mm and 5.45x39mm rounds. They were the accuracy equal of the M16 and any other assault rifle I have fired that was chambered for this cartridge. It has long been suggested that the AK's incredible reliability (I once fired 400 rounds through a North Korean AK-47 that later proved to have the rear portion of a broken case resting in the bottom of the receiver to the rear of the trigger mechanism) was partially at the expense of its accuracy potential. It now appears to me that this may have been mostly a consequence of the mediocre ammunition produced by Eastern Bloc factories. The 5.56x45mm NATO ammunition I fired at the Izhmash test range was manufactured by Fabrique Nationale. All the more reason to use Black Hills' new 7.62x39mm ammunition, which has demonstrated exceptional accuracy potential.

No weapon system I have ever examined has been without flaws, but the Kalashnikov shines through its peculiarities. More than "good enough," it has proved itself on almost every battlefield in every conflict since WWII.

Which of the semiauto Kalashnikovs is best? Those who want the version closest to Russian original must select a Steyr/Maadi. If you want to maximize the system's accuracy potential, then you should opt for the Yugoslav M70. For me, the PTK Legend Series AK-47/S is especially attractive, as it is an authentic reproduction of the Vietnam-era AK-47. All of them will continue to escalate in value. It is unlikely that any will be imported again anytime in the foreseeable future.

My thanks to Gene Guilaroff (ArcLight Unlimited, P.O. Box 173, Alvaton, KY 42122) for accurate information on the current value of all of the semiauto AKs. If you're looking to acquire one, he is the single best source for any semiauto assault rifle.

Originally appeared in *Fighting Firearms*
Summer 1997.

Section 8
Handguns

Beretta's 93R Machine Pistol

A Burst Controlled Blaster

A machine pistol in the truest sense: With fore-grip extended, Beretta's 93R provides accurate close-range shooting in three-shot mode.

Beretta 93R machine pistol (left), with folding fore-grip extended and folding buttstock attached, shows relationship to Model 92 SB, its immediate antecedent.

Machine pistols: ill-defined, misunderstood, misused, and most often poorly conceived, a very minor limb on the tree of military small arms—until now.

Exactly what is a machine pistol? Is it different enough from a submachine gun to justify a separate category? Both fire pistol cartridges. Both are capable of full-auto fire. The shoulder stocks of many submachine guns are detachable (some more quickly than others). Many machine pistols are also fitted with detachable shoulder stocks (which in many instances have also served as holsters). The essential differences are size and origin.

Submachine guns start in the design phase as just that (carbine-like selective-fire weapons chambered for pistol cartridges). Invariably, machine pistols have been more-or-less-extensive selective-fire modifications of pre-existing auto-pistol designs. They are then, by definition, smaller and more compact than most submachine guns. Weapons like the Mini-UZI and Sterling Mk7 Para Pistol tend to blur this distinction. But remember their origin. They are, in essence, scaled-down submachine guns.

Machine pistols were first successfully marketed by the Spanish. The Royal machine pistol, an external copy of the Mauser Model 1896 "Broomhandle" pistol, with cleverly simplified lockwork, was produced in Eibar by the Beistegui Hermanos and introduced in 1927. It was followed in 1928 by the Astra 901, another weapon which resembled the Mauser Model 1896 in external appearance only. In 1930 the Star factory introduced its first selective-fire machine pistol. It was based upon an improvement of the locked-breech Browning system.

Threatened by the success of the Spanish-made machine pistols, in 1930 Mauser responded with a selective-fire version of the Model 1896 called the *Schnellfeuerpistole* (rapid-fire pistol). Early models were designed by Joseph Nickl, but their functioning proved unsatisfactory and a second version, designed by Karl Westinger, was brought out in 1932. Both are referred to as the Model 712, but the Westinger design was the successful one. Its production continued until 1938, by which time about 98,000 Schnellfeuers had been made. Far and away the greatest number of these were sold to the Chinese, who were greatly enamored with the machine-pistol concept.

Despite isolated instances of its use (such as by the Waffen SS) during WWII, the machine pistol was relegated to relative obscurity during the war years, and the submachine gun prevailed. Machine pistols were given scant attention in the years following the war. Large stocks of WWII-vintage weapons remained and there was little incentive for new small-arms-design projects, especially in so esoteric an area.

But the concept was not dead. During the surge of international rearmament which started in the 1950s, the Soviet Union introduced the Stechkin machine pistol (also known as the APS = Automatic Pistol Stechkin) in caliber 9x18mm Makarov. Issued with a shoulder-stock holster, the APS comes with a 20-round staggered box magazine. It is blowback-operated and is provided with a cyclic-rate reducer, which brings it down to 750 rpm. The Stechkin is a machine pistol in the classic sense.

With the exception of Heckler & Koch's interesting VP-70, NATO producers have paid little heed to the machine pistol. Enter the Beretta 93R (R = *raffica* = burst). While its direct antecedents are quite clearly the 92 series pistols, the 93R is the extension of a design process which commenced with the 951A machine pistol. The 951A is, in turn, derived from the Beretta Model 951 pistol (also known as the Model 1951 SL). This pistol's short recoil system makes use of a falling locking block which is driven downward to disengage the slide from the barrel and halt the rearward travel of the barrel. This operating principle, taken from the Walther P-38, was carried over to the 92 series pistols. All have inertia-type firing pins with coil springs.

The Model 951 pistol, used by the Italian, Israeli, Egyptian and Nigerian armed forces, has an eight-round magazine (a 10-round magazine, which protruded slightly below the frame, was also available), an awkward push-through pin-type safety located at the top rear of the frame and is single-action.

The Model 951A machine pistol was provided with a bulky folding-forward handgrip which helped to control muzzle climb during full-auto fire. No shoulder stock was designed or issued

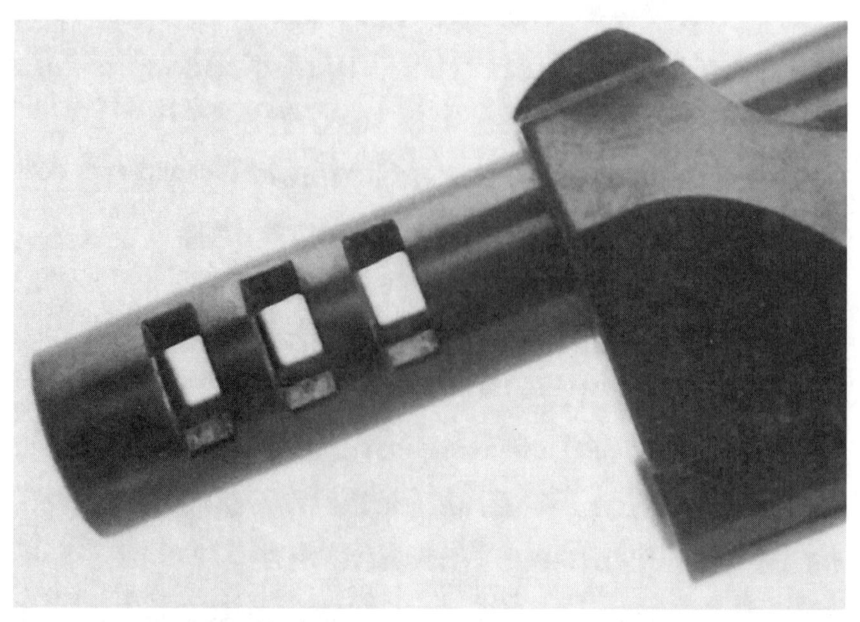

Barrel of 93R is 6.14 inches long. Three vertical cuts on each side of the barrel make an effective muzzle brake.

for this weapon. The selector switch was located on the frame's right side, just above the rear of the trigger guard. The cyclic rate was 750 rpm. A few were exported to Pakistan, and limited quantities, fitted with sound suppressors, went to Italian internal security forces.

Beretta has expended no small amount of time and funding on development of its 93R. The 300-year-old Pietro Beretta Company of Brescia, Italy, was among the very first to manufacture submachine guns. It comes as no surprise that their expertise has produced a state-of-the-art machine pistol. But it's refreshing to see a major design program in this long-neglected area of small-arms technology.

Beretta 93R Specifications

CALIBER .... 9mm Parabellum.

OPERATION .... Oscillating block locking system, short recoil, exposed single-action hammer, semiautomatic or 3-shot burst fire.

FULL-AUTO CYCLIC RATE .... 1,100 rpm.

WEIGHT, empty
With 20-round magazine .... 41.3 ounces.

OVERALL LENGTH .... 9.45 inches.

LENGTH OF STOCK
extended .... 14.5 inches,
folded .... 7.7 inches.

WIDTH .... 1.46 inches.

HEIGHT
With 20-round magazine .... 6.7 inches.

BARREL LENGTH .... 6.14 inches.

SIGHTS .... White-dot blade front, integral with slide; notched white-dot fixed rear, dovetailed to slide; sighting radius: 6.3 inches.

MAGAZINES .... Staggered box type, 20- and 15-rd. capacities.

PRICE .... Approximately $900.

EXCLUSIVE IMPORTER .... Beretta U.S.A. Corp., 17601 Indian Head Highway, Accokeek, MD 20607.

Starting with the proven and highly regarded Model 92 pistol, the Italians focused on the single greatest bedevilment of the machine pistol—unacceptably large burst groups (or "cones of fire") during full-auto fire.

The 93R's barrel is 6.14 inches long, about an inch and a quarter longer than that of the 92 SB. Three vertical slots have been cut on each side of the barrel near the muzzle end, toward the top surface. Propellent gases escaping from these slots exert a downward force on the barrel to assist in counteracting the sharp upward climb of the weapon during full-auto fire. This slotted barrel can also be used in the 92 SB pistol. In addition, its muzzle brake serves as a flash hider.

The trigger guard has been enlarged considerably and an ingenious spring-loaded folding fore-grip has been added to the beefed-up front end. When holstered or firing in the semiauto mode, the fore-grip remains folded under the frame. In burst-fire, the support hand grasps the extended fore-grip with the thumb looped through the trigger guard. This arrangement affords an extremely secure, comfortable two-hand hold.

When there is time to do so, a folding metal buttstock may be attached to the 93R. Two pins on the front part of the stock quickly mate with two holes in the frame. A spring-loaded latch is pushed into a notch at the rear of the frame, firmly fixing the stock to the weapon. When the stock is used in conjunction with the fore-grip, the viability of the machine pistol concept—engaging targets at longer-than-normal pistol range—begins to have substance. When not in use, the stock folds compactly, but only when removed from the weapon. Folding downward, the butt portion will hit the magazine before full closure. Bad design, that—but easily remedied by redesigning the butt itself to fold.

Cyclic rate of the 93R is 1,100 rpm, which is somewhat higher than that of the Mauser Schnellfeuer (900 rpm), but about average for a machine pistol. It is this very peculiarity—cyclic rates hovering around 1,000 rpm—combined with their light weight and usually unstable handling characteristics, that most often has crippled the machine pistol's potential. And most are thus useful for little more than assassination at close range or crowd control—bullet dispersion is so great that to fire accurately you must use the semiautomatic mode. On full-auto, if you jam the muzzle of a machine pistol into someone's groin, you will see the final round of the magazine enter the brain cap before you've had enough response time to release the trigger.

However, the Germans learned during WWII that if MG 42 gunners could be trained to fire short bursts only, the gun's high cyclic rate (often approaching 1,500 rpm) would produce smaller cones of fire downrange than machine guns with slower cyclic rates fired in longer shot bursts. And so, the 93R's cyclic rate of 1,100 rpm turns vice to virtue when coupled with its very positive three-shot burst control.

The 93R's massive slide is heavier than the 92 SB's. Its square-cut top portion shows in its weight (41.3 ounces with empty 20-round magazine for the 93R as opposed to only 34.5 ounces for the 92 SB with empty 15-round magazine). The added weight further inhibits muzzle climb during full-auto fire.

The 20-round 93R magazine extends about an inch and a quarter below the frame. The exposed portion is covered by a plastic sleeve which is grooved on the front side as is the frame itself. With one round in the chamber you have seven three-shot bursts to work with. (If that's not enough, you needed more than a machine pistol.) The 20-round 93R magazine and the 15-round 92 SB magazine are interchangeable in either weapon.

The 93R is single-action only, the three-shot burst mechanism having obviated the 92 SB's double-action feature. I suppose we just can't have everything.

The 93R's thumb safety is just plain dreadful. Located immediately behind the selector switch, its shape and location resemble that of the Mauser "Broomhandle" and its copies. If anything, it is far worse. In contrast to the Broomhandle's safety, pushing downward on the small, checkered button places the 93R on "safe." Pushing upward takes you off "safe"—an impossible task for anyone but Plastic Man while the hand is in the firing position, since the safety lever is to the very rear of the frame.

The magazine-release button is the same as the 92 SB's, located to the rear of the trigger guard where it can be pushed easily by the thumb of the shooting hand. It can be moved from the left to the right side of the frame for left-handed shooters. Magazines consistently fall freely and without hang-up. Walnut grips are standard, but they are not interchangeable with those of the 92 SB due to the inletting required to accommodate the three-shot burst mechanism.

To date, more than 1,500 rounds have shuffled through my 93R. I can report only two malfunctions Both were stovepipes. Both occurred after more than 500 rounds without any cleaning of the weapon. And both occurred with 1949 Yugoslav submachine-gun ammo which is undoubtedly starting to turn sour.

The 93R's trigger pull was clean and crisp right out of the box—not the slightest hint of drag or backlash. And very light compared to the average submachine gun.

Beretta has really put it all together in this compact little package of doom and destruction. Anyone who goes into a test and evalua-

Sensible selector switch points out three dots for burst fire, single white dot for semi-auto. Unfortunately, safety lever behind selector cannot be reached by hand in firing position without grip shift.

tion of a machine pistol without some very black reservations hasn't fired many. Surprise, surprise! The 93R's hit probability borders on the twilight zone.

My cohort, Mark Yuen (a graduate of Gunsite's 499 course), and I decided to first try the 93R's hand on the "Dozier Drill" developed at Gunsite—an ironic title, as Gen. Dozier was kidnapped by five Italian Red Brigade terrorists. The drill usually consists of a draw, pivot and the engagement of five silhouette targets. The distance can be anything from two to seven meters. We had to eliminate the draw as there is no holster for the 93R (Beretta has informed me that shoulder rigs in both leather and nylon will be available soon) A good time is four seconds and the record is 2.5 seconds.

At two meters, including a 180-degree pivot, but without a draw, Yuen was able to put three-shot bursts into all five targets' kill zones (a total of 15 rounds!) in 2.2 seconds. Without a pivot (precluded by my gimp leg), I was able to duplicate this effort in 2.4 seconds. I guess I just have to plan on never turning my back to the bastards. Mark's time at seven meters was a respectable 3.4 seconds. Some of the three-shot groups, I might add, were less than four inches.

All of this was accomplished using a somewhat modified Weaver hold. With the left arm bent and stabilizing the front portion of the weapon by grasping the fore-grip in the prescribed manner and exerting heavy downward pressure, the right arm must also be slightly bent. The sights were used, but not the buttstock.

When the buttstock is fitted and the 93R employed at longer ranges as a semiauto carbine, the results are equally amazing. In the kneeling position, kill-zone strikes at 150 meters are so common that they offer little challenge after the first magazine All this at a weight far less than half that of the despicable MAC 10.

Field-stripping procedures are identical to the 92 SB. Just press the disassembly latch's release button (after clearing the weapon) and rotate the latch counter-clockwise until it stops. Pull the entire slide group forward and separate it from the frame. Reassemble in the reverse order. The slide stop doesn't have to be aligned with a link upon reassembly since the barrel recoils along a straight path and there is no link or cam. Some designers feel this induces less vibration in the system, but the benefits, if any, are probably inconsequential.

Already in service with Italian security forces, the 93R deserves a close look by antiterrorist and special operations units, such as GSG9, the British SAS and U.S. operatives, who can use the firepower and hit potential it offers in a 2 1/2-pound package. Beretta has combined a three-shot burst control, high cyclic rate, an effective muzzle brake and a cleverly designed fore-grip to bring the machine pistol concept from moribund anachronism to glory.

Fit and finish of the 93R, as to be expected from Beretta, are outstanding. Costing approximately $900, the 93R is available to law-enforcement agencies and qualified Class 3 dealers from its exclusive importer, Beretta U.S.A. Corp.

Originally appeared in *Soldier Of Fortune* November 1983.

CZ 83

Czech Pistol Checks Out

Czechoslovakia's latest advance in the small-arms field is the CZ 83 pistol, shown here in caliber .380 ACP.

Communist Czechoslovakia's latest entry in the field of small arms is a so-called pocket pistol: the CZ 83. It was initially introduced in caliber 7.65mm (.32 ACP), but *SOF* obtained the first specimen imported to the U.S. chambered for the more potent .380 ACP (9mm Kurz) cartridge. Nothing revolutionary here—just your basic PPK with every feature Walther should have added years ago.

Modern Czech arms-making commenced with the formation of Ceskoslovenska Zbrojovka Akciova Spolecnost of Brno in 1922. Owned by the Czech government, Skoda Works and its employees (another Workers' Paradise), their first serious effort was the ZB Model 1926 light machine gun. An application of principles used in weapons such as the Berthier, BAR, Hotchkiss and Chatellerault, the ZB 26 moved forward to become the basis for the Bren light machine gun. Over the years the Czechs stayed on the front lines with the Vz 23, 24, 25 and 26 submachine guns, Vz 58 assault rifle, BESA (Vz 37) tank machine gun, Vz 52 LMG, Vz 59 GPMG and CZ 75 pistol.

Double-action in design, the CZ 83 fires by blowback without a locked breech. The barrel is permanently pinned to the frame, and with the exception of grip panels and several magazine components, construction is all steel. This said, we begin to depart from the Walther PP/PPK series.

Hard beryl (aluminum beryllium silicate) plating covers the ejector, extractor, magazine catch and some components in the trigger mechanism (disconnector and hammer spring plunger) for wear resistance. Exterior surfaces have been polished only moderately and tool marks abound in a manner reminiscent of small arms manufactured in Europe during World War II. All parts, except those beryl-plated, have been salt blued. The top of the slide has been glass beaded to a matte surface.

Beware: Future versions may feature a black enamel finish over phosphate. This painted finish now appears on all recently produced CZ 75 pistols—much to the consternation of collectors and shooters alike—and has increased the value of the older blued version by at least $300.

The immediately obvious difference between the Walther pistols and the CZ 83 is in the grip portion of the frame, which is enlarged on the CZ to hold a higher-capacity magazine. PPK/S magazines hold seven cartridges in either 7.65mm or 9mm Kurz chamberings. The CZ 83 magazine will carry fifteen 7.65mm rounds or thirteen 9mm Kurz cases. The price we pay is an increase in thickness of 0.125 inches over the PPK/S and a frame which is 0.7 inches longer. With its 3.8-inch six-groove barrel the CZ 83 has an overall length of 6.8 inches (6.1 inches for a PPK/S). Width is 1.4 inches and height five inches. Weight of the CZ 83 with an empty magazine is 26.25 ounces (23 ounces for a stainless-steel PPK/S). When size and weight balloon on pistols of this type, they approach the bulk of the new breed of compact 9mm Parabellum weapons—such as the 12-rd. Smith & Wesson Model 469. Are they really pocket pistols then? (That may depend on the size of your pockets.)

The single-position feed magazine has a steel body and spring accompanied by a plastic follower and an aluminum floorplate which is painted black. The floorplate is 0.175 inches thick and serves to extend the front grip area of the frame.

Located where it belongs—to the rear of the trigger guard—the magazine release is ambidextrous: There is a checkered release button on each side of the trigger guard. This is excellent, except that the grip panels' finger swells rise above the release button and require a grip shift to change magazines. Empty magazines fall cleanly when released.

A slide stop lever, missing on the PP/PPK series, is conveniently located on the left side of the frame where it can be easily manipulated with the right thumb without altering the grip position. Since the magazine follower is cut for a hold-open, this is a useful device. After a fresh magazine has been inserted into a Walther PP/PPK-series pistol, the slide must be pulled slightly rearward and then released.

Safety mechanisms on the CZ 83 are superb. Spur-shaped, the hammer rebounds to a half-

Chronograph Results: .380 ACP Ammunition

Instrumentation: Oehler Model 33 Chronotach with Skyscreen III detectors positioned five feet from muzzle. Ambient temperature, 80 degrees F. All readings in feet per second (fps). Ten shots per lot. Firearm: CZ 83 with 3.8-inch barrel.

.380 ACP ammunition	Low velocity	High velocity	Extreme spread	Average	Standard deviation
Philippine 95-gr. FMJ	951	1,061	110	1,000	30
Winchester 85-gr. Silvertip JHP	1,007	1,066	59	1,036	18
Super Vel 88-gr. JHP	1,064	1,119	55	1,088	16
Smith & Wesson 84-gr. JHP	1,003	1,042	39	1,024	17
Peters 95-gr. FMJ	845	911	66	866	23
Hirtenberg 95-gr. FMJ	971	1,019	48	995	14

cock position after striking the spring-loaded firing pin. A hammer-block safety has been incorporated; unless the trigger has been pulled completely rearward the hammer cannot move forward to touch the firing pin. A U-shaped, ambidextrous thumb safety wraps around the top of the frame at the rear and pivots on an axis pin through the grip tang. It can be engaged only when the hammer is fully cocked and does not drop the hammer. Thus the CZ 83 can be carried cocked and locked. The manual safety can be disengaged and re-engaged by the thumb without shifting the grip. If desired, the hammer can be carefully lowered to its rebound position and carried in this manner with the first shot fired double-action. Hammer-drop safeties are an abomination and all double-action auto pistols should offer such a cocked-and-locked option.

The trigger moves but a modest 0.5 inches between its double- and single-action positions. After removing the slack, the trigger must be drawn through some irritating drag before it lets off in the single-action mode with a final pull weight of four pounds. Double-action pull weight is a remarkably light and consistent 9.5 pounds. The trigger guard is large enough to accommodate the use of gloves.

Black plastic checkered grip panels with finger swells on either side are standard issue. Like the CZ 75, the grip portion of the frame has a slight reverse taper which provides a grip-to-frame angle that enhances the pistol's natural pointing characteristics.

A low-profile, ramped front sight has been roll-pinned to the slide. It contains a non-luminescent white stripe designed to mate with two white dots on the square-notch rear sight, which is fitted to a dovetail in the slide and staked in place after factory zeroing. While luminescent dots might be of some use in subdued light, I have never noticed colored dots while firing under stress in the daylight. In any event, these white marks, as well as the red dots on the frame above the manual safety levers, were applied by a shaky hand. A solid, slightly raised rib has been milled into the top of the slide with ten longitudinal grooves, presumably intended to break up heat mirage, but which certainly are no more than cosmetic value in a firearm of this type. The sight radius is five inches.

The CZ 83 comes from the Agrozet Uhersky Brod factory equipped with two magazines, a nylon bore brush, a steel cleaning rod with a screwdriver tip to disassemble the grip panels, a 25-meter test target and a manual written in quaint English.

"Knockdown" (disassembly) is straightforward and proceeds almost in the manner of the Walther PP/PPK series. First, remove the magazine and clear the pistol. Then pull straight out and down on the trigger guard. It

CZ 83 Specifications

CALIBER . . . . .9mm Kurz (.380 ACP) or 7.65mm (.32 ACP).

OPERATION . . . . .Unlocked blowback; semiautomatic only; exposed double-action hammer, can also be carried "cocked and locked" for first-round single-action fire; internal hammer block safety; ambidextrous manual safety and magazine release; slide stop lever, Walther PP/PPK-type disassembly.

WEIGHT, empty . . . . .26.25 ounces.

LENGTH, overall . . . . .6.8 inches.

BARREL LENGTH . . . . .3.8 inches.

HEIGHT . . . . .5 inches.

WIDTH . . . . .1.4 inches.

STOCKS . . . . .Checkered black plastic grip panels with finger swells.

SIGHTS . . . . .Low profile; fixed ramp front with white line; fixed square-notch rear with two white dots.

FINISH . . . . .Blued (future versions may be painted).

PRICE .Approximately $375 including transportation charges, F.O.B. Toronto, Canada, and U.S. duty.

MANUFACTURER . . . . .Agrozet Uhersky Brod, Czechoslovakia.

will stay down and does not have to be shifted to either side of the frame as with a Walther pistol. Note that you cannot pivot the trigger guard unless the magazine is removed since the magazine catch mechanism is mounted to, and rotates with, the trigger guard. Pull the slide back and lift the rear end off the guide rails on the frame. Move the slide forward to separate it from the barrel and frame. Remove the recoil spring from its position surrounding the barrel. Depress the firing pin with a small punch and slide down the firing-pin stop just as you would with a Colt M1911A1. Withdraw the firing pin and its spring. Using the screwdriver tip on the end of the "gunstick" (cleaning rod), remove the single retaining screw on each grip panel and lift the grip panels up and off. No further disassembly is required or should be attempted by anyone except a competent pistolsmith. Reassemble in the reverse order, remembering not to insert a magazine until the trigger guard is back in place. The magazine can be disassembled by depressing the bottom plate attached to the follower spring and sliding off the floorplate.

Designed by John Browning and introduced in 1912 by FN as the 9mm Browning Short, the .380 ACP (Automatic Colt Pistol) cartridge is considered by many to represent the absolute minimum in adequate stopping power for use on human targets. Known in Europe as the 9mm

CZ 83 disassembled. Knockdown is similar to that of the Walther PP/PPK series.

Kurz, "CAL. 9 Browning" is the designation chosen by Agrozet Uhersky Brod to mark the left side of the CZ 83's slide. While acknowledging the origin of the cartridge (unusual for communists, who claim to have invented everything), it's a bit imprecise. In 1903 FN introduced a pistol chambered for the 9mm Browning Long cartridge. It was adopted by Sweden as an official military sidearm. Approximately between the .380 ACP and .38 Colt ACP in power, it was also chambered for pistols manufactured by LeFrançais and Webley & Scott, but never adopted by American manufacturers.

One thousand rounds from six different lots of ammunition were punched through the CZ 83 to evaluate its idiosyncrasies, if any: Philippine Arsenal (SB 85 headstamp) 95-gr. FMJ (full metal jacket); Winchester 85-gr. Silvertip Hollow Point; Super Vel 88-gr. JHP jacketed hollowpoint); Smith & Wesson 84-gr. JHP; Peters 95-gr. FMJ; and Austrian Hirtenberg 95-gr. FMJ. All of these loadings were chronographed through the CZ 83's 3.8-inch barrel and the results are given in the table (accompanying this article).

The excellent Super Vel loading, unfortunately no longer available, gave the highest average velocity at 1,088 fps. The S&W JHP, nothing more than a full metal jacket projectile with a hole drilled in it (also no longer available), was included merely because this ammunition has caused stoppages in other caliber .380 ACP pistols. It did not expand, but fed smoothly through the CZ 83. As expected, the U.S.-manufactured Peters load at 866 fps was almost 130-135 fps slower than either the European Hirtenberg (lowest standard deviation) or Philippine (highest standard deviation) cartridges. Without doubt, the Winchester Silvertip is your best bet in this caliber. Its serrated aluminum jacket assures dependable expansion and it races out of the CZ 83 barrel at 1,036 fps.

Admirably reliable, the CZ 83 didn't have a single stoppage during the entire 1,000-round test. Balance and handling characteristics are superb. With a full load of 14 cartridges, the CZ 83 weighs more than 30 ounces. As a consequence, felt recoil is mild and the hit potential is high at the expected engagement distances for this type of firearm. With its fixed barrel, the CZ 83 delivered consistent two-inch 10-shot groups at 15 feet from the offhand position with all of the ammunition tested.

Now there are only two holsters available for this pistol and fine ones they are. Lou Alessi's custom concealment holsters are famous for quality and their use by Jerry Ahern's fictional character, John Rourke, in "The Survivalist" series. Lou makes an open top version of his well-known Belt Slide holster for the CZ 83 which sells for $39. Made from eight-ounce shoulder cowhide and sewn with 200-lb. drop-test nylon cord, all the stress points are double-stitched for strength and endurance. Alessi's holsters are wet-molded and boned by hand (a whalebone or ivory tool is used to highlight the pistol's contours to increase the holster's retention capabilities). Lou's Belt Slide concealment holster is made for all auto pistols, large and small, and is available in either cordovan or black. Belt loops are cut to fit a 1 3/4-inch heavy leather belt. Alessi believes a rig's ability to hold a pistol snug against the body is a function of a proper belt.

Galco Gun Leather (formerly Jackass Leather Co.) sells an excellent shoulder holster for the CZ 83. Using premium saddle leather split to eight ounces per square foot, this horizontal rig is lock-stitched with #207 polydacron thread. Also wet-molded and hand-boned, the Galco Defender model costs $44.95 and features a half-harness (no provision for opposing side magazine pouches or handcuff cases) for maximum concealment. A Dot fastener on the holster body will accept an optional belt tie-down. The thumb-break strap has an internal polypropylene reinforcement and its fastener is countersunk to prevent scuffing the pistol. After the leather safety strap has been split, its fastener is attached and the strap cemented so the fastener does not touch the firearm. Rivets holding the nylon hangers onto the holster are also countersunk. The shoulder harness itself is fabricated from 4 1/2-ounce chrome-tanned cowhide. The countersupport strap is made of

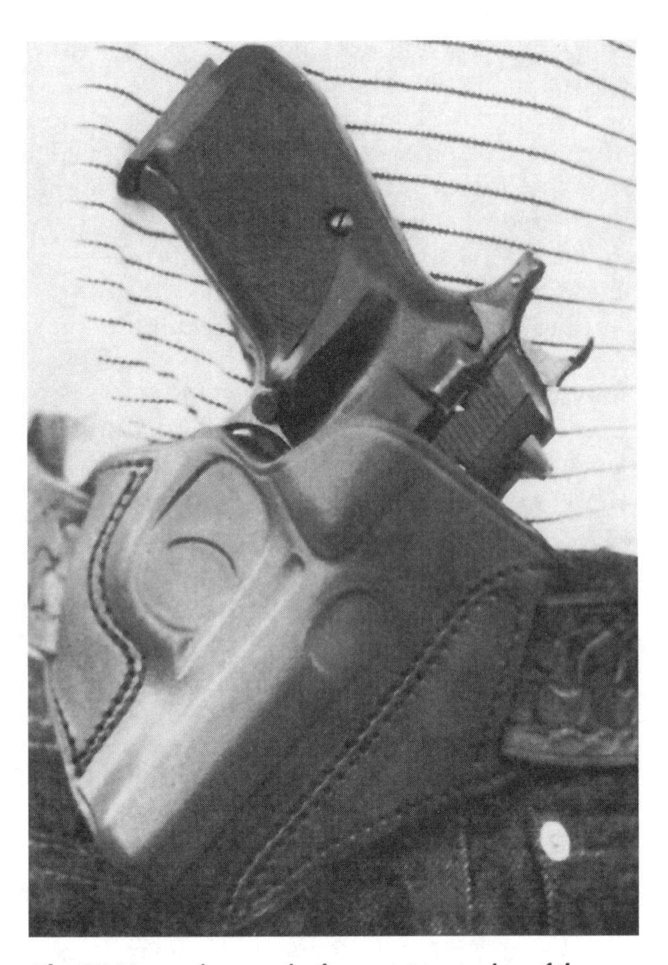

The CZ 83 remains snug in the open top version of the Alessi Belt Slide holster.

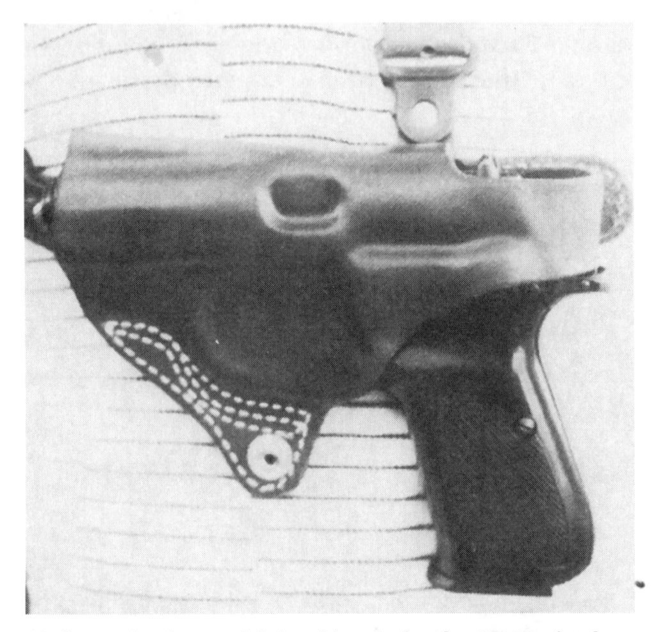

Galco Defender model shoulder rig for the CZ 83 pistol.

polyelastic. Further information on this hand-rubbed oil-finish shoulder holster can be obtained from Galco Gun Leather.

And what do you obtain with the CZ 83? A well-designed, reliable .380 ACP pistol that is entirely too big. Carrying a pistol in this caliber can be justified only if it is very small and thus very concealable. Either the S&W Model 469 or the Heckler & Koch P7 in 9mm Parabellum are better choices in this size class. Better yet, take a close look at the new .45 ACP Colt Officer's Model with an alloy frame.

Note: CZ firearms are imported by CZ-USA. The CZ 83 is no longer imported because of its high-capacity magazine.

Originally appeared in *Soldier Of Fortune* February 1986.

TITANIUM BULLS

Taurus' Ultra-Lite Guns Lead the Herd

The Model 450Ti Total Titanium caliber .45 Colt revolver features a titanium-alloy cylinder, frame, yoke, barrel shroud and side plate, along with many small parts.

The Taurus Model 85 UL/Ti MultiAlloy .38 Special revolver has a titanium-alloy forged barrel shroud and extruded cylinder, a high-tensile-strength stainless steel barrel, forged aluminum alloy frame and side plate, and carbon steel hammer and trigger.

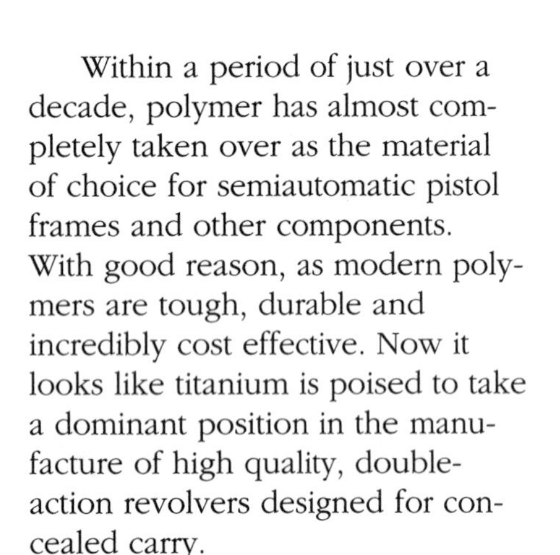

Within a period of just over a decade, polymer has almost completely taken over as the material of choice for semiautomatic pistol frames and other components. With good reason, as modern polymers are tough, durable and incredibly cost effective. Now it looks like titanium is poised to take a dominant position in the manufacture of high quality, double-action revolvers designed for concealed carry.

Smith & Wesson's AirLite Ti series of revolvers feature the use of titanium in the cylinder, firing pin and the center pin bushings and the frame studs. Taurus International has recently introduced the Taurus Total Titanium series of revolvers that moves significantly beyond S&W. Not only are the cylinders and many small parts made of titanium alloy (containing vanadium and other metals), but the frames, yokes, barrel shrouds and side plates as well. Why titanium? Simply because it's amazingly strong, corrosion-proof and ultra lightweight.

Soldier Of Fortune was sent a specimen of a Model 450Ti Taurus Total Titanium revolver for test and evaluation. The finish I requested and received is matte Stealth Gray. Two other finishes are available: Spectrum Blue and Gold. In my opinion, neither will appeal to *SOF* readers. This is a proprietary process in which variations in temperature and electrical charge result in different colors. Matte and high-gloss finishes are achieved by the degree of polishing employed prior to processing. Be advised that the color is a thin layer of specially treated tita-

Titanium: What is it and Why Use It?

Titanium is a lustrous, silver-white metallic element in Group IV B of the Periodic Table. Titanium (symbol: Ti) has an atomic number of 22 and an atomic weight of 47.90 with a melting point of 1,800 degrees C.

Its principal sources are the minerals Ilmenite and Rutile. Ilmenite is titanic iron ore found as beds and lenticular bodies enveloped in gneiss and other crystalline metamorphic rocks. Rutile, a titanium dioxide, is found in granite, granite pegmatites, gneiss, mica schist, metamorphic limestone, and dolomite.

Titanium's chemistry shows similarities to that of both zirconium and silicon. Its various compounds have a wide range of commercial importance. Titanium dioxide is used as a white paint pigment. Barium titanite is piezoelectric and is used as a sound and electrical transducer. Titanium tetrachloride is used for smoke screens and in skywriting. Titanium esters are employed as waterproofing chemicals for fabrics.

Because of its high strength-to-weight ratio titanium metal has been long-favored in the aerospace industry for the construction of both frames and engines. These properties of both light mass with correspondingly very high tensile strength, even at high temperatures, are what attracted the Taurus designers. Furthermore, titanium has a much higher "modulus of elasticity:" (the property of returning to an initial state without rupture after deformation) than steel—an important property because of the sudden and very high pressure peaks associated with the ignition of small arms ammunition. And, rightly or wrongly, those who carry concealed weaponry most often desire as light a handgun as possible.

However, although it's one of the earth's most common elements, titanium is difficult to refine. Titanium metal cannot be produced by the usual dioxide reduction method, as it reacts with both oxygen and nitrogen at high temperatures. Most often, titanium is extracted by passing chlorine over the ore. The titanium tetrachloride that results is condensed, purified by fractional distillation and then reduced with molten magnesium at high temperature in an atmosphere of argon. This is an expensive operation, with the result that titanium barstock costs 10 times as much as the stainless steel Taurus uses in many of its revolvers.

Titanium is also highly corrosion-resistant. It withstands saltwater spray and marine environments even better than stainless steel and most Teflon-based gun coatings. Amazingly, it is immune to a substantial number of both acids and alkalis. This is principally because of the tough oxide skin that forms on its surface. If scratched or ruptured in some manner, this surface oxide film regenerates itself almost immediately.

Using modern CNC equipment, Taurus machines its titanium revolver cylinders from titanium alloy extrusions. The Taurus frames, yokes, barrel shrouds and side plates are drop-forged from titanium alloys in their justifiably famous Brazilian forging facility. Forging is commonly recognized as the preferred process when high-precision, titanium components are desired.

nium that is subject to the same effects of wear and abrasion as bluing.

Cylinder-to-forcing-cone gap on our test specimen is 0.003-inch (0.003- to 0.008-inch is the generally accepted tolerance range). Fore-and-aft cylinder play is almost nonexistent and the lateral play is minimal. The cylinder holds five rounds, and rotation is to the left (counterclockwise) in the Smith & Wesson manner (Colt cylinders rotate to the right—clockwise). To insure that the cylinder remains in precise alignment within the frame at the moment of ignition, Taurus has installed a spring-loaded detent system to the top of the yoke on the Total Titanium series (and the Model 85Ti MultiAlloy .38 Special as well). This, in turn, has permitted them to extend the overall length of the ejector rod, providing more positive ejection of fired cases than most snubnose types. The cylinder latch together with its retaining screw are steel components, as are the hammer and trigger. Patterned after that of S&W, it must be pushed forward to swing out the cylinder.

The 2-inch stainless steel barrel (because titanium galls too much

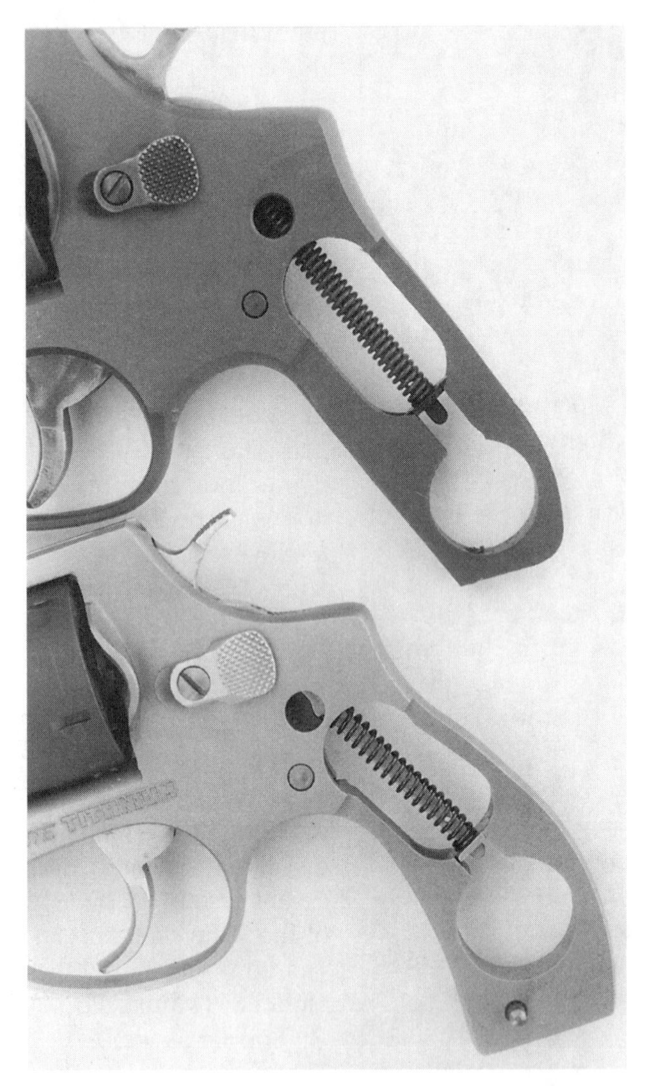

Taurus uses guide-in-coil-type mainsprings set into either round-butt or square-butt frames.

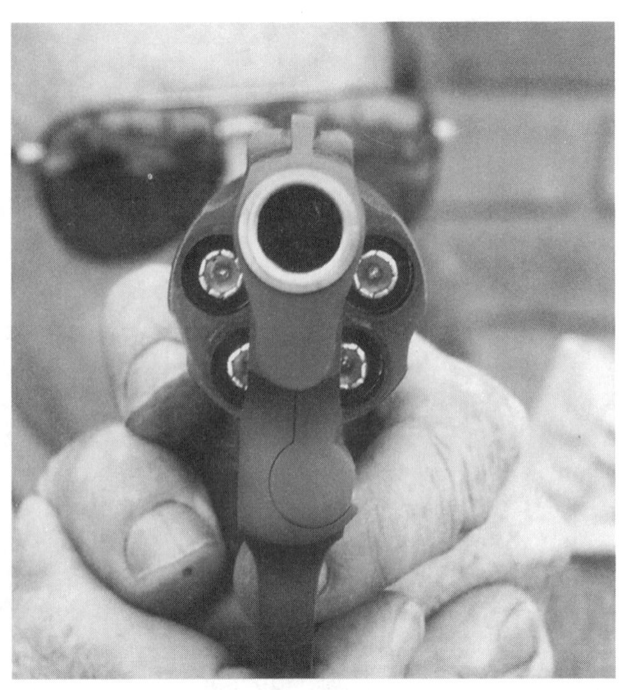

Taurus Total Titanium series sets the standard for the most comprehensive use of this exotic metal in a revolver.

Five beans in the wheel: Both the Taurus Model 450Ti .45 Colt (left) and Model 85UL/Ti .38 Special (right) revolvers have five-shot cylinders.

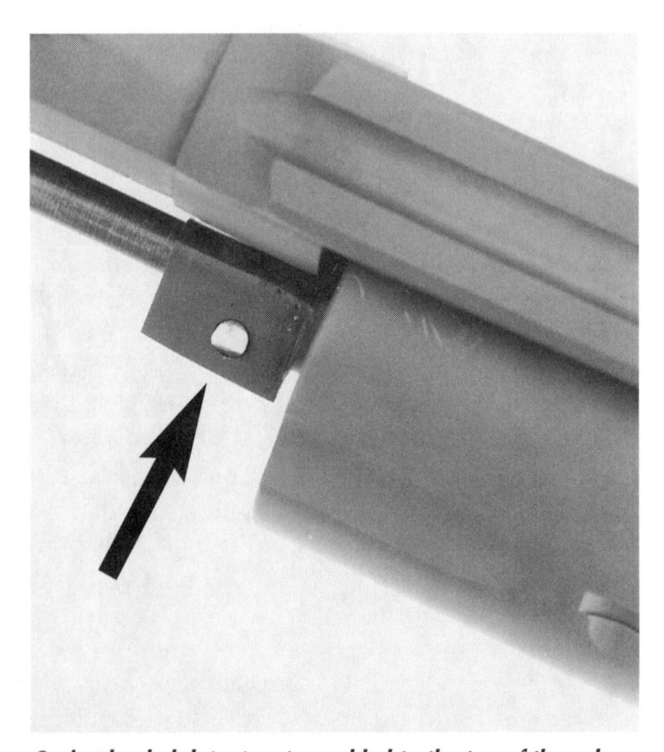

Spring-loaded detent system added to the top of the yoke on the Total Titanium series (and the Model 85Ti MultiAlloy .38 Special as well) permitted a longer ejector rod, providing more positive ejection of fired cases than most snubnose types.

TAURUS MODEL 450Ti TITANIUM REVOLVER

CALIBER . . . . .45 Colt (also available chambered for .38 Special, .32 H&R Magnum, .357 Magnum, .41 Magnum, and .44 Special.

OPERATION . . . . .Five-shot revolver. Exposed-hammer, single- and double-action. Coil-spring ignition with floating firing pin and transfer-bar safety. Unique key-lock safety system.

WEIGHT, empty . . . . .19.2 ounces.

LENGTH, overall . . . . .8.5 inches (from the muzzle to the tip of the grip).

HEIGHT . . . . .5.5 inches.

WIDTH . . . . .1.5 inches.

BARREL . . . . .Six-groove with a left-hand twist of one turn in 16 inches. Stainless steel barrel with titanium-alloy shroud. Three ports on each side of the front sight.

BARREL LENGTH . . . . 2 inches (2.5 inches, .41 Magnum only).

SIGHTS .Fixed; serrated-ramp front sight blade, 0.125-inch wide; open, square-notch rear sight, 0.130-inch in width, milled into the rear end of the top strap. Sight radius: 3.9 inches.

GRIPS . . . . .Ribber, one-piece synthetic rubber.

CONSTRUCTION AND FINISH .Titanium-alloy cylinders, frames, yokes, barrel shrouds and side plates and many small parts. Stainless steel barrel. Carbon steel hammer and trigger. Stealth Gray, Spectrum Blue or Gold titanium finish.

SUGGESTED RETAIL PRICE . . . .$599.

MANUFACTURER AND IMPORTER . . . . .Taurus International, 16175 N.W. 49th Avenue, Miami, FL 33014-6314, Web site: www.taurususa.com.

T&E SUMMARY: . . . . .Lightweight titanium technology taken to its maximum potential in a revolver. Recoil impulse substantially moderated by ported barrel and especially effective synthetic rubber grips. Test specimen chambered for a proven man-stopper.

TAURUS MODEL 85 UL/Ti MULTIALLOY REVOLVER

CALIBER . . . . .38 Special +P.

OPERATION . . . . .Five-shot revolver. Exposed-hammer, single- and double-action. Coil-spring ignition with floating firing pin and transfer-bar safety. Unique key-lock safety system.

WEIGHT, empty . . . . 13.5 ounces.

LENGTH, overall . . . . 7 7/8 inches.

HEIGHT . . . . 4 5/8 inches.

WIDTH . . . . 1.25 inches.

BARREL . . . . .Six-groove with a right-hand twist of one turn in 18.75 inches. Stainless steel barrel with titanium-alloy shroud. Three ports on each side of the front sight.

BARREL LENGTH . . . . 2 inches

SIGHTS .Fixed; serrated-ramp front sight blade, 0.125-inch wide; open, square-notch rear sight, 0.130-inch in width, milled into the rear end of the top strap. Sight radius: 3.7 inches.

GRIPS . . . . .Two-piece black Santoprene, a rubber-like synthetic.

CONSTRUCTION AND FINISH . . . . .Forged barrel shroud and extruded cylinder made from a titanium alloy frame and sideplate, and carbon steel hammer and trigger.

SUGGESTED RETAIL PRICE . . . .$515.

MANUFACTURER AND IMPORTER . . . . .Taurus International, 16175 N.W. 49th Avenue, Miami, FL 33014-6314, Web site: www.taurususa.com.

T&E SUMMARY: Innovative, multiple-alloy technology. Lightweight. Ported barrel, possible only because of titanium-alloy barrel shroud, moderates recoil and permits use of the highly effective 158-train, lead, hollow-point +P "FBI" load.

to be used for this purpose) is equipped with a titanium shroud. The importance of this must not be underestimated. Barrel shrouds on the S&W AirLite Ti series are made of aluminum. Thus, they cannot be ported, as propellant gases would eat them away in short order. By using a titanium barrel shroud, Taurus designers were able to place three barrel ports on each side of front sight blade. As a consequence of this, and the unique grip panels, no ammunition restrictions have been placed on the Taurus Total Titanium revolvers. S&W has stipulated that only

THE HISTORY AND PERFORMANCE OF THE .45 COLT CARTRIDGE

The history of the .45 Colt (often incorrectly referred to as the ".45 Long Colt") cartridge is inexorably interwoven with the development of the Model 1873 single action revolver for which it was designed. However, this round actually lasted only a brief time as an official U.S. Government type. It was manufactured at Frankford Arsenal from 1873-1874. The Smith & Wesson Schofield revolver was adopted in 1875. As the .45 Colt case was too long (approximately 1.26 inches) for the S&W Schofield cylinder, the shorter .45 S&W Schofield cartridge (approximately 1.10 inches) was adopted for use in both revolvers.

The .45 Colt cartridge also must not be confused with the Model 1909 .45 Colt round which it closely resembles. Six thousand Colt New Service revolvers were purchased by the government in 1909 and sent to the Philippines. The ammunition supplied with these revolvers was made only in U.S. Government arsenals and never available on the commercial market. It differed from the earlier .45 Colt case by virtue of an increased case rim diameter.

While commercial production of the .45 Schofield round was discontinued in about 1939, the .45 Colt cartridge, now 126 years old, continues to march straight into the millennium. It was, and remains, one of the most effective handgun cartridges ever developed. Throughout its long history, the most commonly encountered factory load was a 255-grain lead-alloy, round-nose bullet with a flat point (RNFP) that left the muzzle at approximately 860 fps. While that is not fast enough for today's ballistic speed freaks, the fact is that this round has done great execution and has never been found wanting as a man-stopper. Its large, heavy bullet makes a big hole and penetrates deeply enough to crush, cut and break through the human body's vital structures and organs. Remember, once we've obtained the required penetration, the bullet that makes the biggest hole will do the most damage.

The .45 caliber bullet certainly makes a big hole, but we can further enhance this characteristic if we employ a soft-point or hollow-point bullet that will reliably expand to no more than twice the original diameter. It's important to remember that if we drive a hollow-point at too high a velocity, it will inevitably overexpand and reduce the penetration to an unacceptable depth. It may also fragment to an undesirable extent. There's a correct velocity for every handgun bullet, and it's *never* the highest velocity possible.

The only factory hollow-point .45 Colt ammunition presently available provides excellent wound ballistics performance. Winchester's Personal Protection .45 Colt ammunition (#X45CSHP2) features a 225-grain Silvertip hollow-point bullet that leaves the muzzle at about 960 fps. In soft tissue it will usually expand to about .70 caliber with very little fragmentation and penetrate well beyond the minimum 12 inches desired by most authorities.

Today, bore diameters for .45 Colt barrels are usually 0.451-inch, which duplicates that of the .45 ACP. The SAAMI maximum average pressure for this cartridge is 14,000 psi. While the original factory load was 40 grains of black powder (granulation not specified) over a 250-grain lead projectile, this was eventually dropped to 35 grains (Frankford Arsenal used 30 grains) for safety reasons. Today, about 8 grains of Alliant Unique propellant will provide about 900 fps from a 5 1/2-inch barrel.

jacketed +P bullets can be fired from their AirLite Ti series, as lead projectiles, due to the severe recoil impulse, will jump forward and bind up the cylinder's rotation. This is a serious deficiency, in my opinion, since in several instances (such as the .38 Special "FBI" load with its 158-grain lead hollow-point bullet) unjacketed bullets furnish superior wound ballistics potential.

The barrel has six grooves with a left-hand twist of one turn in 16 inches, standard for this caliber. Overall length of the Model 450Ti from the muzzle to the tip of the grip is 8.5 inches, with a weight, empty, of only 19.2 ounces. The height is 5.5 inches and the width is 1.5 inches.

A guide-in-coil-type mainspring is set into the round-butt frame. The steel trigger is smooth, as it should be for double-action shooting. The single-action trigger-pull weight is an incredibly light 2.75 pounds. Double-action pull weight is also excellent at 9.0 pounds with absolutely no loading at the end of the stroke.

Mounted in the frame, the floating firing pin must be struck by a transfer bar which, in turn, receives the hammer's full impact. Should the finger slip off the trigger while the hammer is falling forward, the transfer bar will drop downward, out of the hammer's path, and the firing pin will not be driven forward.

"LOCKED" AND LOADED

All Taurus revolvers are now equipped with their excellent key-lock safety system. To deactivate the revolver, you must insert a key over a hexagonal-shaped pin at the rear of the hammer. Turning the key clockwise will raise the pin and its shroud to impinge against the frame if someone attempts to roll the hammer rearward. Turning the key counterclockwise will lower the pin and shroud and permit the hammer to be fully rotated to the rear. This is quite an innovative and effective system. However, no one in my loop will use it, as the most effective safety system of all remains Rule Three (Keep your finger off the trigger until your sights are on the target) and the intelligence and training to apply it.

The fixed sights on this revolver consist of a serrated ramp-type front blade, 0.125-inch wide, and an open, square-notch rear sight, 0.130-inch in width, milled into the rear end of the top strap. They are more than adequate for the applications this handgun was designed to cover. The sight radius is 3.9 inches. Winchester's Personal Protection .45 Colt ammunition with its 225-grain Silvertip hollow-point bullet shoots directly to the point of aim at 7 yards.

The so-called "Ribber" one-piece grips installed on the Taurus Total Titanium series are

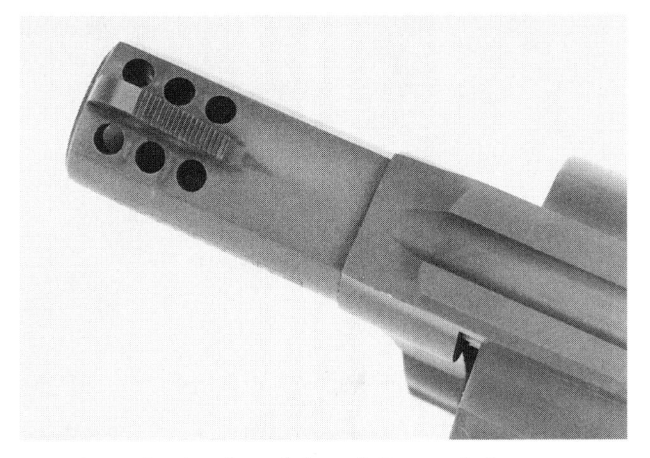

By using a titanium barrel shroud, Taurus designers were able to place three barrel ports on each side of front sight blade. As a consequence of this, and the unique grip panels, no ammunition restrictions have been placed on the Taurus Total Titanium revolvers.

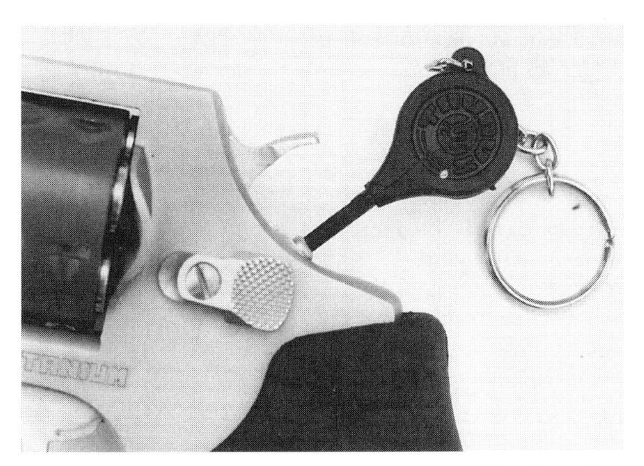

Taurus key-lock safety system raises a pin and its shroud to impinge against the frame if someone attempts to roll the hammer rearward.

another important factor in taming the not-inconsiderable recoil impulse generated by these lightweight revolvers. Full-length on each side of these black synthetic rubber grips are a series of ribs in horizontal alignment. On the front strap area these ribs protrude by as much as about 0.15-inch in the form of raised fins that collapse under recoil and then restitute. The grips come up under the trigger to protect the middle finger. All very innovative. The Ribber grips are retained by a 3mm allen-head screw threaded to the butt of the frame.

Like a young widow's grief, the recoil generated by the Taurus Model 450Ti is sharp, but of short duration. It should be of little consequence to anyone with a modicum of experience. The Ribber grips really absorb recoil. However, the flash signature generated by the six muzzle ports could be somewhat disconcerting in a low-light-level environment.

When fired offhand using a strong two-handed Weaver hold, slow-fire, double-action shooting will produce groups under 3 inches. Picking up the cadence will enlarge the group size—and this is as it should be. You should never attempt target accuracy in stress environments. Placing all your shots within the approximately 8x12-inch area of the target's upper torso (commonly referred to as the "center of mass") with as much speed as practical is far more important than obtaining precisely aimed 1-inch groups.

This is a superb revolver, chambered for a magnificent, gunfight-proven, man-stopper cartridge. It has all the attributes a wheelgun advocate could ever desire. The suggested retail price

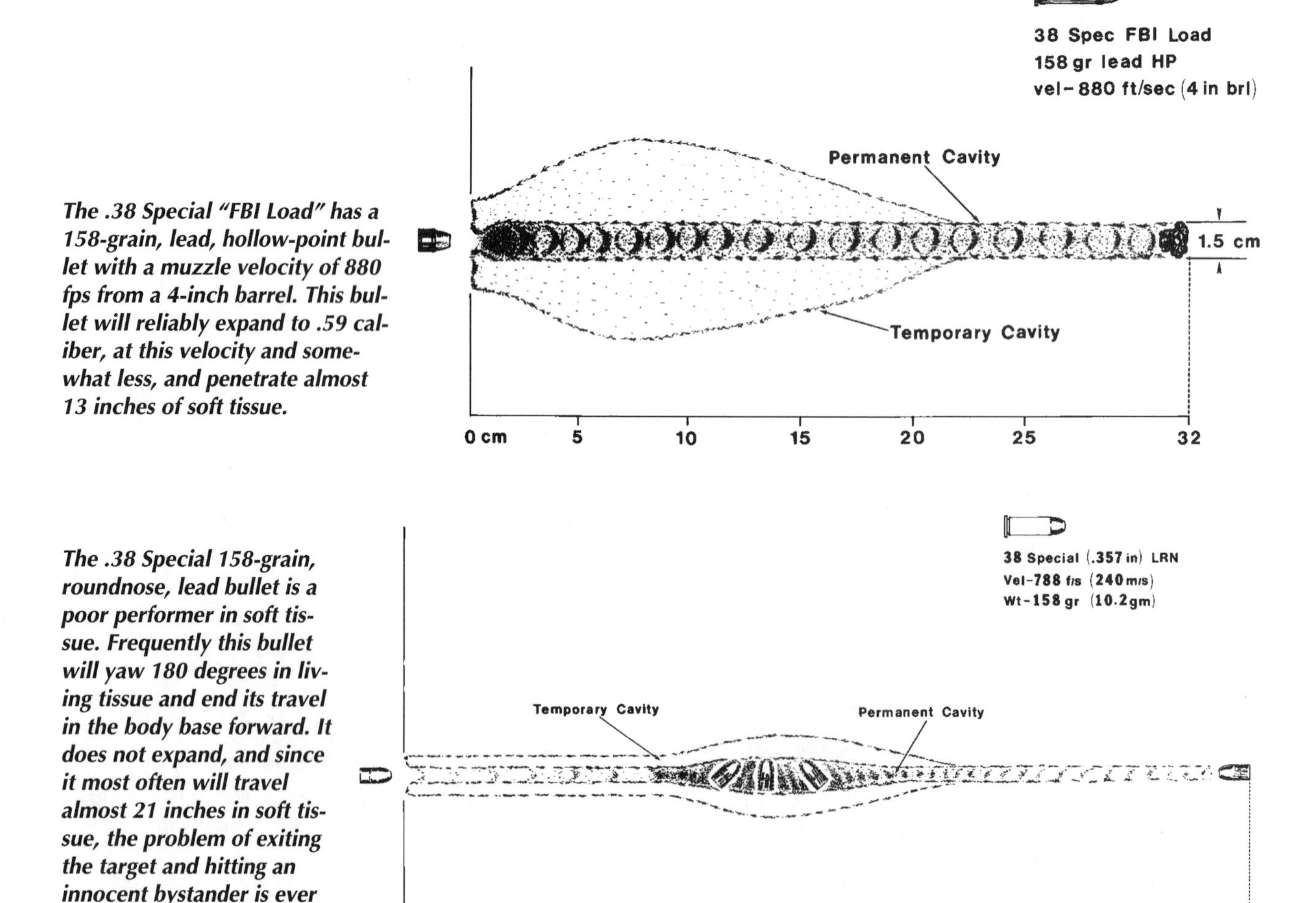

The .38 Special "FBI Load" has a 158-grain, lead, hollow-point bullet with a muzzle velocity of 880 fps from a 4-inch barrel. This bullet will reliably expand to .59 caliber, at this velocity and somewhat less, and penetrate almost 13 inches of soft tissue.

The .38 Special 158-grain, roundnose, lead bullet is a poor performer in soft tissue. Frequently this bullet will yaw 180 degrees in living tissue and end its travel in the body base forward. It does not expand, and since it most often will travel almost 21 inches in soft tissue, the problem of exiting the target and hitting an innocent bystander is ever present.

.38 Special: 100 Years and More Special Than Ever

Development of what was to become the .38 Special cartridge commenced a century ago. During the Philippines Campaign, which began in February 1899, the .38 Long Colt cartridge used by the U.S. Army failed to stop the Moros led by Filipino Emilio Aguinaldo. That the Moros were more often than not sky-high on drugs was of no consequence to those looking to lay blame for the guerrilla warfare that cost far more money and took far more lives than the Spanish-American War. The insurrection essentially ended in 1901 with the capture of Aguinaldo by General Frederick Funston. It's highly doubtful that any handgun round could have guaranteed consistent stopping power against "true believers" under the influence of strong drugs.

With the thought that he might convince the U.S. Army to adopt an improved .38 caliber round, Daniel B. Wesson designed the .38 Smith & Wesson Special cartridge. It contained 3.5 more grains of black powder (from 18 grains to 21.5 grains) than the .38 Long Colt and increased the projectile weight by 8 grains (from 150 grains to 158 grains) because of its flat base bullet. The increased bullet weight was made possible because the relation between the bullet and groove diameters was held to a much closer tolerance, making the previously required expansion of the bullet's skirt on firing unnecessary.

This was, however, to no avail as by this time the U.S. Army became locked into a .45 caliber round for its future handguns. Toward the end of 1899, the .38 Smith & Wesson Special cartridge, together with the new S&W Military & Police revolver, was introduced to the public. Subsequently, what became known as simply the .38 Special was adopted by almost every police department in the United States. It remained the premier law enforcement handgun cartridge until the 1970s when the .357 Magnum began to replace it as a service round. By the 1980s the revolver itself was replaced in department after department by large-capacity, large-frame, 9mm Parabellum service sidearms.

In 1930 Smith & Wesson introduced their .38/44 revolver—a heavy-duty .38 on their .44 frame. This was a result of experimentation by Elmer Keith and research by Major Earl Witsil, Chief Ballistician of Remington Arms Company. During its law enforcement heyday, the 158-grain roundnose lead bullet was the most commonly employed projectile. It was a poor performer in soft tissue. Frequently, this bullet will yaw 180 degrees in living tissue and end its travel in the body base forward. It does not expand, and since it most often will travel almost 21 inches in soft tissue the problem of exiting the target and hitting an innocent bystander is ever present. The 200-grain roundnose was not better.

The .38 Special's wound ballistics potential began to improve with the development of expanding hollowpoint ammunition. At first, these bullets were driven at such high velocities that they invariably overexpanded and thus underpenetrated. However, in recent years bullet designers have heeded the advice of Dr. Martin L. Fackler, who conducted his early research at the U.S. Army's Wound Ballistics Laboratory at the Presidio, and are now producing bullets that reliably expand to no more than twice their original diameters at moderate velocities up to +P velocities. The very best .38 Special ammunition you can stuff into the cylinder of your Taurus 85UL/Ti MultiAlloy is Federal's so-called "FBI Load." This features a 158-grain, lead, hollow-point bullet with a muzzle velocity of 880 fps from a 4-inch barrel. This bullet will reliably expand to .59 caliber, at this velocity and somewhat less, and penetrate almost 13 inches of soft tissue. this is just the performance we are looking for. And remember, you can shoot this lead +P load in the Taurus 85UL/Ti but you cannot in the S&W Model 337 AirLite Ti revolver.

is $599. The Total Titanium series is also available chambered for the .38 Special, .32 H&R Magnum, .357 Magnum, .41 Magnum (comes with 2.5-inch barrel), and .44 Special rounds.

MULTI METALS

SOF was also sent one of the new Taurus Model 85UL/Ti (Ultra Lite/Titanium MultiAlloy .38 Special snubnose revolvers for test and evaluation. This slick little 2-inch snubnose features a forged barrel shroud and extruded cylinder, both made from a titanium alloy; a high-tensile-strength stainless steel barrel; forges aluminum alloy frame and side plate; and carbon steel hammer and trigger. Can there be any question as to why Taurus refers to this as a "MultiAlloy" revolver?

The six-groove bore has a right-hand twist of one turn in 18.75 inches. The height of this revolver is 4 5/8 inches, with a width of 1.25 inches. Overall length is 7 7/8 inches. The weight empty is 13.5 ounces. The two-piece grip panels on the Taurus Model 85UL/Ti MultiAlloy revolver are made of black Santoprene, a rubber-like synthetic. They also come up under the trigger to protect the middle finger. The sights are those of the Taurus 450Ti Total Titanium revolver, except that the sight radius at 3.7 inches is slightly less.

The cylinder holds five rounds. The cylinder-to-forcing-cone gap on this revolver is 0.004-inch. Again, the fore-and-aft cylinder play is almost nonexistent and the lateral play is minimal.

The guide-in-coil-type mainspring and transfer-bar trigger system are those of the Total Titanium series and the other Taurus revolvers. The double-action trigger-pull weight was measured at 9.75 pounds using dead weights. The single-action pull weight is exactly 3.0 pounds.

As it also carries three barrel ports on each side of the front sight blade, no ammunition restrictions have been placed on the Model 85 UL/Ti MultiAlloy revolver either. While it weighs just 2 ounces heavier than its S&W equivalent, the M85Ti sells for $515, which is more than a hundred dollars less than the S&W M337 AirLite Ti. More important, the barrel ports do moderate the perceived recoil to some extent. Make no mistake, any extremely lightweight revolver firing +P ammunition is going to kick hard. The Taurus Model 85 UL/Ti MultiAlloy and Total Titanium series revolvers were not designed for plinking. They are concealed-carry wheelguns intended for handing from your belt on a daily basis. These are dedicated self-defense firearms and the recoil they generate shouldn't bother an experienced shooter anymore than a .505 Gibbs you fire at a charging Cape Buffalo 10 paces from you.

Accuracy duplicated that of the Model 450Ti. With 158-grain bullets, the Model 85 UL/Ti MultiAlloy .38 Special Revolver shoots directly to the point of aim. This is a fine piece and it has been a frequent companion of late. Taurus has made tremendous strides in the last decade. Originally little more than a less expensive copy of Smith & Wesson, in recent years they pulled up even and now, in my opinion, with their Total Titanium series they may have surged somewhat ahead.

Originally appeared in *Soldier Of Fortune*
February 2000

.45 ACP Glock Model 21 disassembled.

Plastic Perfection in .45 ACP

Glock's Model 21

Gaston Glock's Model 17 9mm Parabellum pistol was first introduced to the American public by *Soldier Of Fortune* magazine almost six years ago. Since that time more than 2,000 U.S. local and federal law enforcement agencies have adopted or authorized the Glock as duty weapons. In addition to Austria, the armed forces of both Norway and the Netherlands have adopted the Glock. Law enforcement agencies and military units in Belgium, Canada, Ecuador, Hong Kong, India, Jordan, the Philippines, Taiwan, Thailand, Venezuela and West Germany issue the Glock as their standard sidearm. Tens of thousands have been sold to the American public, and hundreds of thousands have been sold worldwide.

Now, at long last, this highly acclaimed handgun has been chambered for America's justifiably famous .45 ACP cartridge—mating 21st-century technology with an octogenarian of combat-proven effectiveness.

Dubbed the Glock Model 21, it is very similar in size to Glock's previously announced Model 20 10mm pistol. In fact, the frames appear to be identical. *SOF*'s test specimen of a prototype Model 21 has an overall length of 8.27 inches, a height, with sights and inserted magazine, of 4.85 inches and a width of 1.2 inches (at the grips). The barrel length is 4.6 inches. The weight is 29.5 ounces with an empty magazine (almost 10 ounces less than a Colt Double Eagle or M1911A1). Almost 85 percent of this mass is accounted for by the steel components.

While some of the smaller components are interchangeable with the 10mm Model 20, you cannot assemble a Model 20 slide group to a Model 21 frame, as the locking block has been altered to prevent this. As with the other models in the Glock series, there are only 35 parts including the magazine. Glock says there are 33, but I count the sights and trigger spring cups as two components each. Of small consequence, as in either case this is still fewer than half the number of bits and pieces found in competing designs.

The Glock's remarkable success in just six years is matched by its even more remarkable design—the salient features of which are all retained by the Model 21. Glock's only concession to conventionality is the pistol's method of operation. Short recoil operated, the barrel is locked to the slide by a single lug that recesses into the ejection port, in the manner of the SIG-Sauer series. During the recoil stroke the barrel moves rearward approximately 3mm until the bullet leaves the barrel and pressures drop to a safe level. The barrel then drops downward, separating from the slide and terminating any further motion. The slide's continued rearward movement and counter-recoil cycle are those of the Browning system.

Hammerless and striker-fired, the Glock's trigger and firing pin mechanisms are innovative and mostly unique. There is no manually operated thumb safety or de-cocking lever. A so-called "Safe Action" trigger system, patterned after that encountered on the Sauer Behorden ("Authority") Model 1930 caliber 7.65mm pocket pistol, constitutes the first fail-safe. A wide, serrated, outer trigger encompasses a small, spring-loaded inner trigger, both fabricated from polymer. The outer trigger cannot be actuated, such as by contact with a holster, unless the inner trigger is depressed first. Thus the trigger can be pulled only from the center, not the edges.

A spring-loaded firing pin safety in the slide blocks forward movement of the striker, and is raised and deactivated by a projection on the sheet metal trigger bar as the trigger is pulled to its final rearward position.

When the trigger is in the forward position, the firing pin's spring remains lightly compressed. As the trigger is pulled about 10mm through its first stage (with a pull weight of approximately 2.2 pounds), its full compression is almost complete. Removal of the finger from the trigger at this time will return the firing pin spring to its partially compressed, "relaxed" and completely safe state. Continued pressure at this point will 1) draw the firing pin fully rearward and its spring into complete compression; then 2) draw the T-shaped end of the trigger bar to its final rearward position in the trigger housing's stepped safety notch; so that 3) it is free to drop downward away from both the "connector" (sear) and a projection at the end of the striker to release the firing pin and fire the round.

The firing pin is rectangular in cross-section with a chisel-shaped tip. Although primers are left with an instantly identifiable indentation, the striker's unorthodox configuration produces less drag on the primer (eliminating the possibility of firing pin breakage) and concentrates its momentum onto a smaller area to insure positive ignition. Fluted firing-pin spring cups, which permit the Glock pistol to be fired underwater, are available to legitimate government agencies only. A stamped sheet-metal ejector, with an odd-looking inward cant, is permanently attached to the polymer trigger housing.

Glock's new .45 ACP Model 21 pistol mates 21st-century technology with a venerable big boomer cartridge of combat-proven effectiveness.

Further explanation of the connector is required. This sheet-metal component also serves as a disconnector. When the slide moves forward in counter-recoil, a thumb above the rail on the right side pushes the connector away from the trigger bar to prevent another round from being fired until the trigger is released and the trigger bar moves forward. The angle between the connector's upper face and its bottom face determines the trigger pull weight of the second stage. An angle of 90 degrees will produce the standard pull weight of 5 pounds. A pull weight of 8 pounds is achieved by increasing the angle to 105 degrees (this connector is stamped with a "+"). A pull weight of 3.5 pounds, available only with the Long Slide Target Model 17, is obtained when the angle is reduced to 75 degrees (stamped with a "–"). At the request of the New York State Police, a small polymer and steel component has recently been designed that increases the trigger pull weight to approximately 12 pounds when it is inserted into the trigger housing. That's too heavy for me, but should prove ideal for law enforcement agencies in transition from double-action revolvers. If the pistol is to be stored for any length of time, the trigger should remain in the retracted position to remove all tension of the firing pin spring.

This triple safe trigger mechanism is housed in a high-impact polymer frame that initiated the pistol's unjustified controversy (all the more strange as Heckler & Koch's VP70z and P9S pistols, both introduced more than a decade ago, were fabricated with largely polycarbonate frames). Four steel guide rails (about 0.4 inches in length) for the slide have been integrated into the injection molded frame—in pairs at the rear of the frame, and above and in front of the trigger guard. To meet BATF regulations, a steel plate carrying the serial number has been embedded into the frame in front of the trigger guard. The trigger guard has been squared off, recurved and checkered, but those who fire from the correct Weaver position will not employ this useless feature.

The grip-to-frame angle of the Model 21 remains that of the Glock 17/19, which is somewhat steeper than competing designs. There is a non-slip, stippled surface on the sides of the grip and both the front and rear straps are grooved and checkered. As there are no separate grip panels, the grip portion of the pistol, while larger in circumference than that of the Glock 17/19, accommodates normal-sized hands despite its large magazine capacity.

The locking block, which engages a 45-degree camming surface on the barrel's lower lug, appears to be the Glock's only investment casting. It is retained in the frame by the same steel axis pin that holds the trigger and slide stop. The trigger housing is attached to the frame by means of a polymer pin. A spring-loaded, sheet-metal pressing serves as the slide stop, which is protected from accidental manipulation by a raised guard molded into the frame. The slide lock, operated by a single bent flat spring, engages a step on the front of the barrel's locking lug to prevent the slide and frame groups from parting company during the counter-recoil stroke. The magazine catch-release, another polymer component—located where it belongs, on the left side of the frame, directly to the rear of the trigger guard—is held in place by an uncoiled piece of spring steel. Both interior surfaces of the magazine-well's mouth have a beveled contour to assist in the insertion of magazines.

Rectangular in shape, the slide is milled from bar stock using CNC (Computer Numerical Control) machinery. Three hardening processes are employed on both the slide and barrel. The final tenifer finish, .04 to .05 millimeter in thickness, produces a patented 69 Rockwell Cone hardness just below a diamond) by means of a nitride bath at 500 degrees Centigrade. Scratches, which are in this instance no more than deposits from the other object, can usually be removed with a cloth and solvent. This matte, non-glare finish is 99 percent salt water corrosion resistant and meets or exceeds stainless steel specifications. It's also 80 percent more corrosion-resistant than any hard-chrome finish.

Milled into both the top and right side of the slide, the Glock's large ejection port enhances functional reliability. A large claw extractor, fitted to the slide at the rear of the ejection port on the right side, maintains its tension from a spring-loaded plunger, which, together with the firing pin assembly, are held in place by a polymer backing plate. Cocking serrations on the Model 21's slide are cut deeply and provide an excellent purchase when the slide is retracted.

High profile, combat-type, fixed sights are standard. Four rearsight heights are available: 6.1mm (lower impact), 6.5mm (standard issue), and the higher impact 6.9mm and 7.3mm. A rear sight mounting and adjusting device can be obtained by certified Glock armorers. The polymer front sight carries a white dot and the rear sight has a white outline. However, best of all, in my opinion, are the self-luminous, tritium, low-light-level sights with which the Model 21 can be fitted directly from Glock, Inc. These tritium crystals will last more than 10 years. Sight radius of the Model 21 is 6.75 inches.

Glock's hammer-forged barrels are also innovative. Called "hexagonal," the rifling lies somewhere between conventional land and groove and H&K's "polygonal" bores. The rifling's hexagonal profile (in cross-section a series of six small arcs connected by flat surfaces) provides a better gas seal, more consistent velocities, superior accuracy and ease of maintenance. The direction of twist is right-hand. Although not yet specified at the time of our test and evaluation, I expect the rate of twist will be close to the standard 1:16 inches for this cartridge. A single-coil spring under the barrel rides on a polymer guide rod, which is hollow to serve as a cooling air pump.

The Model 21 magazine is of the single-position-feed, staggered-column type with a capacity of 13 rounds. With one up the snout, that gives you 14 rounds of .45 caliber medicine. Magazine bodies, followers and floor plates are fabricated from polymer. The magazine bodies have steel liners and indicator holes starting with round No. 4 up to the capacity of the magazine. When new, Glock magazines will drop freely from the magazine-well. After use, however, the magazine walls will set with an outward bulge that requires their removal by hand. In my opinion, this is a matter of small consequence. If you haven't solved your problem with 14 rounds, a pistol was an inappropriate choice for the confrontation. Each Model 21 is issued with two magazines, a polymer magazine loader and cleaning rod and a nylon bristle bore brash. The polymer storage box has been designed for armory stacking and retention with a steel rod or chain. Suggested retail price is $598.

While somewhat different from the norm, there is nothing complex about the Model 21's disassembly procedures and, unlike the Colt Double Eagle, no component will part company from the slide or frame unless you intend it to do so. First, remove the magazine and remove any round in the chamber. Then, and only then, pull the trigger. Wrap the four fingers of the right hand over the slide from the right side with the thumb wrapped around the rear of the frame and retract the slide about an 1/8-inch (any more than that and the trigger will move forward to prevent separation of the slide and frame). Pull the slide lock downward with the thumb and index finger of the left hand. While the slide lock is down, push the slide forward and off the frame. Push the guide rod forward and remove the rod and recoil spring. Push the barrel forward, lift up and pull it back out of the slide. No further disassembly is recommended. Do not attempt to manipulate the trigger system after the slide has been removed or you may damage the inner trigger's spring. Reassemble in the reverse order. To disassemble the magazine, merely squeeze the side walls at the base and slide off the floor plate.

There can be no question about the Glock design's levels of reliability or durability. In its 9mm Parabellum version, it has successfully passed tests every bit as rigorous as the U.S. XM9 trials, involving hundreds of thousands of rounds. That it was excluded from the most recent XM9 trials is a commentary on the U.S. Army's conventional mind-set, not the Glock design.

SOF's test and evaluation of the Model 21 did no more than confirm impressions already formed from tens of thousands of rounds fired through our Glock 17 and 19 pistols. There were no stoppages of any kind during the

Glock Model 21 Specifications

CALIBER .... .45 ACP.

OPERATION .... Short recoil, barrel locks with single lug into ejection port, semiautomatic, hammerless, Safe Action double-trigger system with two additional internal automatic safeties.

FEED MECHANISM .... 13-round, staggered column, detachable box-type magazine with removable floor plate; polymer construction with steel liners and indicator holes.

WEIGHT, empty .... 27.5 ounces; magazine: 2 ounces.

LENGTH, overall .... 8.27 inches.

HEIGHT .... 4.85 inches (including SIGHTS and inserted magazine).

WIDTH .... 1.2 inches.

BARREL .... Hexagonal rifling with a right-hand twist.

BARREL LENGTH .... 4.6 inches.

NUMBER OF PARTS, including magazine .... 35.

SIGHTS .... Fixed polymer front sight with white dot, 0.12 inches thick; fixed, polymer, white-outside (four heights available) or fully adjustable rear sights. Optional self-luminous tritium front and rear sights. Sight radius: 6.75 inches.

METAL FINISH .... Tenifer process—non-glare matte; 99 percent salt water corrosion resistant; meets or exceeds stainless steel specifications.

PRICE .... $598, complete with two magazines, magazine loader, cleaning rod, nylon bristle bore and storage box.

MANUFACTURER .... Glock GmbH, Produkte aus Kunststoffe, Metallwaren und Holz, A2232 Deutsch-Wagram, Hausfeldstrasse 17, Austria.

IMPORTER .... Glock, Inc., 6000 Highlands Parkway, Smyrna, GA 30082.

T&E SUMMARY: High impact polymer frame reduces perceived recoil; three safety systems eliminate the need for a manual thumb safety; hammerless design eliminates need for hammer-drop lever; highest possible level of reliability; hit probability and accuracy potential at maximum levels; recommended without reservations of any kind for all military, law enforcement and self-defense applications.

course of the 500 rounds fired through our test specimen. The frame's inherent elasticity dampens felt recoil considerably. As the barrel's axis lies close to the hand, the recoil momentum is perceived as an almost straight rearward thrust with much less muzzle climb that of either the Colt Double Eagle or a standard Government Model. Target reacquisition times between shots are minimal as the front sight barely leaves the point of aim if a strong Weaver hold is employed. Quite muzzle heavy, the Model 21 points instinctively and comes on target with great speed. With its clean and constant trigger system, the hit probability is high. There is, of course, no hammer bite to distract the shooter. The frame's grip ergonomics are excellent.

What about the accuracy potential? Most engagements with a handgun will take place at 21 feet or less. Firing a pistol from 50 yards off a Ransom rest will provide information concerning its theoretical accuracy potential, but nothing about its practical accuracy in a stress scenario. We fired the Model 21 at camouflaged combat targets from 21 feet in the Weaver position. The ammunition used in *SOF*'s test and evaluation included Black Hills 185-grain Jacketed Hollow Point (JHP) and 230-grain Full Metal Jackets (FMJ), 230-grain U.S. military ball and 230-grain hard-cast round-nose reloads with a powder charge weight of 6.3 grains of Hercules Unique. Best results were obtained with the Black Hills 185-grain JHP, which consistently dumped a magazine of double-taps into a ragged half-inch group. The ejection path was consistently two to three feet to the right and rear.

I predict that the recently completed ammunition evaluation and wound ballistics penetration analysis by the FBI's Firearms Training Unit will deservedly draw attention away from the 9mm Parabellum cartridge and focus it once more on the venerable .45. The best 9mm load, Winchester's 147-

grain subsonic JHP, was 30 percent less successful than the FBI/Sierra 10mm 180-grain JHP. That's a significant difference in performance. Remington's .45 ACP 185-grain JHP was only 2.5 percent less successful than the FBI/Sierra 10mm load. That's an inconsequential difference. Some law enforcement agencies, in copy-cat fashion, will jump on board the 10mm bandwagon. In my opinion, far more—both departments and individuals—will re-examine the .45. Most of us already have access to .45 ACP reloading dies and the components are plentiful. Long-established and battle-proven, the .45 has an aura steeped in the folklore of American history. The FBI tests should result in a revival of interest in the .45, only moderate acceptance of the 10mm and a waning of popularity in the U.S. for the 9mm Parabellum.

Glock's Model 21 has arrived at the right place, at the right time. With its large capacity magazine, brilliant design and superb reliability, we can expect its surge to the forefront in the wave of new popularity anticipated at both law enforcement and civilian levels for an ancient cartridge that makes a big hole and penetrates deep enough. Let's hope a compact version of the Model 21 with a single-column, eight-round magazine in a truly reduced envelope is shortly forthcoming.

Originally appeared in *Soldier Of Fortune*
June 1990.

BLACK AVENGER

Kokalis/Novak Custom Colt Commander

As it has the potential for making the biggest hole, the .45 ACP round is still our best choice when engaging human targets with a handgun. No better concealment envelope for this cartridge was ever devised than the Colt Light Weight Commander.

Kokalis/Novak Fighting Commander, fieldstripped—note that Birdsong's Black T Treatment has been applied to all metal components, including the recoil spring.

Anyone anticipating an imminent deadly confrontation would be well advised to arm himself with a fighting shotgun, submachine gun or battle rifle. Unfortunately, social taboos and mores proscribe the display of such apparatus in public places such as banks, post offices and shopping malls—the very places they might prove most useful these days. Had just one worthy been so armed in Luby's Cafeteria, it's doubtful that anyone other than cowboys would have ever heard of Killeen, Texas.

Although a distant last choice, most of us must be reconciled to packing a pistol most of the time. This said, its limited potential for defense must be maximized to the fullest extent possible. For professional gun handlers this means a handgun that is as utterly reliable as it is possible to devise, regardless of expense, with acceptable combat accuracy and chambered for a cartridge of proven wound-ballistics potential. It goes without saying that another key factor in this equation is the operator's ability to transform extensive training into high-hit probability under stress.

For many of us this translates into a customized M1911A1-type semiautomatic pistol. As it has the potential for making the biggest hole, the .45 ACP round is still our best choice, especially so with the recent availability of Winchester's 230-grain Jacketed Hollow Point

(JHP) load. Properly trained aficionados, when not inhibited by law-enforcement administrators, carry this single-action classic in "condition one" (a round in the chamber, a full magazine seated in place, the hammer fully cocked and the thumb safety engaged).

Handguns of this type are shot frequently in practice, carried constantly and seldom fired in anger. No better envelope has ever been devised for this set of parameters than Colt's Light Weight Commander. The concept dates back almost half a century.

The U.S. Army became interested in a lightweight pistol right after World War II, in 1946. A year later, the Army Ordnance Committee established a requirement for a semi-automatic pistol with a weight limit of 25 ounces and a maximum overall length of 7 inches. It's possible that Colt had already experimented with shortened slides prior to World War II.

Colt approached the Aluminum Company of America (ALCOA) in 1947, and by the following year sample forgings had been prepared. Several specimens were assembled, including some chambered for the 9mm Parabellum cartridge, as this was part of the Ordnance Committee's specifications. The slides were made of steel in the conventional manner, but the length was reduced from 7.38 inches to 6.63 inches. The barrel length was reduced from 5 inches down to 4.25 inches. A new barrel bushing, recoil spring plug, and guide rod were also designed.

FIRST LIGHT WEIGHT COMMANDER

By 1949, further modifications had been made. To achieve greater weight reduction, the mainspring housing was made of aluminum. The grip safety was shortened, the lanyard loop on the bottom of the mainspring housing was omitted, a rounded hammer with a serrated edge and pierced head (reminiscent of the so-called "ring" hammer found on later Mauser "Broomhandle" pistols) was introduced, the extractor and firing pin were modified from the commercial-type, and the sights were altered to compensate for the reduced sight radius.

Early in 1950, Colt introduced this short, lightweight pistol in 9mm Parabellum, .38 Super and .45 ACP as the "Commander Model." The serial numbers carry a "CLW" (Commander, Light Weight) prefix. Forty years later, this pistol is still available, chambered now for only the .45 ACP round, as a Series 80-type with a firing-pin safety. ALCOA still supplies the frame forgings, and the material used is an aluminum alloy, appropriately called "COLTALLOY"

The overall length of the Light Weight Commander is 7.75 inches. Width at the grip panels is 1.24 inches and the height is approximately 5.3 inches. The 4.25-inch barrel has six grooves with a left-hand twist of one turn in 16 inches. The weight, empty, is 27.5 ounces. An appealing package, but out of the box it just won't do.

Don't get me wrong. It will shoot straight and reliably, but devotees of the M1911A1 are never satisfied. For more than a half-century, pistolsmiths have been filing, stoning, polishing, grinding, soldering, welding, hammering and changing parts on the .45 auto pistol. Why? Is the design defective? Is it poorly executed in production series? Not at all. This constant tinkering is no more or less than an expression of devotion for what is arguably the best combat handgun that ever slapped leather.

Many still carry the M1911A1 into the deepest shadows of danger. Professionals whose lives may well depend on this tool of the trade will spare no expense to obtain the ultimate degree of reliability and performance. Is this obsession for perfection any different from that of the professional auto racer's?

Everyone who packs a .45 has his own ideas concerning the modifications required to bring John Browning's design up to his personal specifications and, most important, make him feel good. Feeling "good," i.e., confident, about the handgun you're toting is sometimes an important, albeit unquantifiable, ingredient in predicting success in a gunfight.

I STICK WITH WINNERS

After more than a quarter-century carrying the Browning-designed .45 in one variant or another, I too have some specific ideas about what should or should not be done to the M1911A1 series to make it street-ready. None of my .45s have muzzle weights or compensators. None have recoil springs designed for cream-puff loads. There are no ambidextrous, extended safety levers, as I have seen shooters' jackets rotate the right side lever of holstered .45s to the fire position. None of my sights have white dots or bars. The trigger guards are not squared-off and serrated.

When correctly adjusted to the ammunition employed, I prefer the robustness of a fixed rear sight. Enlarged magazine catch/release buttons will bump against the equipment or the body and dump the magazine at inappropriate times. Extended slide stop levers interfere with a proper Weaver hold, and thumb pressure is likely to inactivate the slide's hold-open after the last shot has been fired.

Beveling the magazine-well is pretty much standard procedure on a custom .45. It will marginally improve the time required for a speed reload. In the field, beveled cuts on the bottom of the magazine-well can attract debris. I usually choose to omit this modification, as I spend a considerable amount of time out in the bush. However, I recommend this option for those whose feet pound only pavement.

There are an infinite number of other doodads that serve no function but to provide unnecessary frosting. The M1911A1 series is no piece of cake. Important modifications are needed, however—most intended to enhance reliability and handling characteristics.

Unfortunately, custom pistolsmiths have largely focused on International Practical Shooting Conference (IPSC) competition, with little more than lip service paid to the needs of professional gun handlers. Those who have directed their efforts almost exclusively to fighting handguns are few and far between.

Among this small group, those who are truly competent are even smaller in number. Wayne Novak, who runs Novak's .45 Shop, is one of them.

Wayne began his career working for the famed Armand Swenson. After his return to the Ohio Valley region, Novak attended a short course taught by the highly respected former MTU armorer, CWO John M. Miller. While you will encounter Browning High Powers, Czech CZ-75s, plus SIG-Sauer, H&K, Glock and S&W pistols in various stages of modification on Novak's workbench at any given time, the bulk of his work is devoted to Colt Government Models and Light Weight Commanders.

MODIFYING THE COMMANDER

About a year ago, I obtained an early Colt Series 70 LW Commander in fine condition. It was the ideal starting envelope for a custom carry gun. Together, Wayne and I decided upon a number of alterations intended to enhance practical performance. The result is "Black Avenger," the "Kokalis/Novak Fighting Commander."

The modifications were extensive (and expensive) and all street-proven options. Let's start at the top of the slide, and more or less work downward.

Novak's LoMount Carry rear sight—without doubt the finest combat-type rear sight ever attached to a handgun—was installed ($57.95) with a Trijicon Self Luminous Dot on each side of its 0.125-inch open square-notch. Rounded and radiused in all the right places, Wayne's LoMount Carry rear sight will not impede the important "tap, rack, bang" drill. It can be drifted in its dovetail on the slide to adjust windage zero.

Colt M1911A1 front sight blades are too small and, worse yet, have a disastrous tendency to fly off into the sunset at inappropriate times. A dovetail was milled into the slide and a high-profile, blade-type front side was installed ($50). It can also be drifted in its dovetail to adjust windage zero, but more important, it is secure and will never part company from the slide. Of the same thickness as the rear sight's open square-notch, the front sight has a single Trijicon Self-Luminous Dot.

Most of my carry pistols (and all of my MP5 SMGs) are equipped with self-luminous sights. Tritium (an isotope of hydrogen) provides the energy source for self-luminous sights of this type. Tritium gas and a phosphor particle are pressurized within a tiny glass capsule. Tritium creates soft beta rays which are converted to visible light when they strike the phosphor particle. The capsules are resistant to oil, water, corrosion and temperature changes.

While white dots or outlines are never noticed under stress by those of us trained by Jeff Cooper to concentrate solely on "Front sight—Press," self-luminous tritium sights are useful adjuncts to firing at night or under subdued-light conditions. Three-dot Trijicon tritium inserts cost $135, installed.

As a previous owner with more guts than skill had attempted to "throat" (polish around the chamber mouth) the barrel of my LW Commander with a Dremel tool and had botched the job badly, Novak installed a stainless steel, match-grade BarSto barrel, to which was fitted a King's National Match bushing, machined out of solid 416 stainless-steel stock ($245 for both).

NOVAK'S NOVEL ENHANCEMENT

As part of his "reliability" package, Wayne throated this barrel correctly, checked the chamber specifications with a .45 ACP chamber reamer, polished the barrel's and frame's feed ramps, and "crowned" the barrel's muzzle. Crowning, which is done with a special reamer, is a beveling process that protects the critical part of a barrel's rifling from damage. In addition, if the muzzle is not squared, propellant gases will escape prematurely, tipping the bullet's base as it leaves the muzzle and degrading the accuracy potential.

The reliability package also includes adjusting, and if required, reshaping the extractor; inspecting and refining (removing sharp edges and polishing) the breech face; adjusting the magazine catch/release; and proving (by firing at the range) all magazines shipped with the pistol. Total cost of the reliability package is $50.

I also requested Wayne's famous "carry bevel" package ($55), in which all sharp edges on the slide and frame are rounded off. At this time, Novak also put a light scallop on the rear lip of the ejection port to avoid dents on the empty cases (important for those of us who reload). In addition, almost everyone can benefit from an extended (left side only) thumb safety on the M1911A1, and one was installed for $45.

If you fire from the Weaver position, hammer bite will cut the web of your firing hand, between the thumb and the trigger finger, after just several magazines have been fired. This can be eliminated by installation of the so-called "beavertail"-type grip safety ($70)—a mandated option for those who practice frequently.

No combat M1911A1-type .45 would be complete without a trigger job. Wayne recommends no lighter than 4- to 4.5-pound trigger pull weights, and mine measures a crisp, consistent 4.25 pounds. It may seem contradictory, but the more you shoot, the more trigger-sensitive you become.

After all this hammering and sawing, my LW Commander needed to be refinished. Nickel-plated .45s are for fighter pilots (who are also usually encountered wearing cowboy boots, sheepskin-lined leather jackets, "mirror"-type sunglasses, silk scarves, shoulder holsters and carrying extended pistol magazines that stick a foot below the magazine well—all accompanied with incredible lies about their sexual exploits). Seriously dangerous people prefer black.

BASIC BLACK PROTECTION

As a consequence, Wayne sent my pistol to W.E. Birdsong & Associates for the "Black T Treatment" ($125). Black T is a resin-bonded lubricant coating, and a co-dispersion of fluoropolymer resin, Teflon and graphite, in a thermosetting binder. It produces a low-friction coating, free of so-called "stick-slip" (smear) characteristics, combined with excellent resistance to corrosion. Available colors are non-reflective black and NATO (olive) green.

This finish coats the firearms and ancillary equipment of numerous, albeit unnamed, federal agencies and certain unspecified military units.

All metal (steel and aluminum) parts of the firearm are treated. After vapor-degreasing, they are low-pressure blasted with very fine, 150-grit aluminum oxide. Then the parts are both phosphate- and chromate-finished. Black T is then applied by hand and heat-cured. The complete finish is no more than 2 to 3 ten-thousandths (0.0002-0.0003) of an inch in thickness. This compares favorably to salt bluing (black or blue oxide) which is usually about 4 ten-thousandths (0.0004-inch) of an inch thick. The final result has a high salt spray and humidity resistance (greater than 500 hours), a very low coefficient of friction (0.08) and a service temperature-high of 300 degrees Fahrenheit continuous (325 F for intermittent service).

This is an extremely low-maintenance finish—very little, if any, lubrication is required. In most cases, a lightly oiled rag is sufficient to wipe away accumulated debris and carbon fouling. I know, someone told you that about your M16 rifle, 25 years ago in Vietnam, and look what happened. That was then, this is now; protective coatings have come a long way in the past quarter-century. Believe me, Birdsong's Black T is a superb finish for combat weapons.

At the author's request, Milt Sparks Holsters Inc. altered their sophisticated new Executive Companion inside-the-waistband holster to a crossdraw design, which is both comfortable and provides maximum concealment. A valuable accessory for this holster is the new inside-the-waistband #1PS single magazine pouch.

A fitting complement to the Black T finish was the installation of a set of African Blackwood grip panels ($65) handcrafted by Craig Spegel. All of Spegel's grips are hand checkered. I chose his M1911 diamond pattern, but they are also available smooth. I have a Novak-customized Browning High Power that sports a set of fully checkered Spegel grips. If specified, the right grip panel can be cut for an ambidextrous thumb-safety lever. Other exotic woods are available and include Cocobolo, Kingwood, Tulipwood, Madagascar Rosewood and Bocate. Spegel is well known among professional *pistoleros,* and his work is highly regarded.

This pistol was begging for a concealment holster rig to match the master touches of Novak, Birdsong and Spegel. Tony Kanaley of Milt Sparks Holsters Inc. was up to the challenge.

PERSONALIZED HOLSTER RIG

Tony recently introduced an inside-the-waistband holster called the #EX Executives Companion, which has several features setting it apart from their famous Summer Special. The metal-lined top band is positioned directly underneath the gun belt, assuring that the holster will not collapse at this point. This allows the pistol to sit deeper in the pants for added concealment and security, while still permitting a full firing grip. The holster is smooth-side-out and molded to the gun it is meant to carry—both to reduce bulk, add to retention and improve appearance.

All Milt Sparks holsters are made of top-quality cowhide, and their finish helps repel perspiration. One of the most important features of the Executive Companion is the interchangeable belt-loop system, which permits the holster to be fitted to belts varying in width from 1-inch to 1.75 inches. Finally, it is fabricated of somewhat lighter leather than the Summer Special (5.5-ounce leather instead of 6.5-ounce, which translates into about a 0.25mm difference in thickness).

Great, but this holster has a rearward rake (angled to point the muzzle of the handgun to the rear—also called the FBI rake), and thus can only be worn strong side. I prefer crossdraw exclusively for belt holsters. A crossdraw position permits the support hand to sweep aside the concealing jacket or coat without fumbling and, equally important, the shooter does not telegraph his draw stroke to potential threats on the flanks or to the rear.

At my request, Tony removed the rearward rake and provided a straight up-and-down cant. He then took the pattern and reversed it, with the metal reinforcement moved to the rear—in back of the trigger guard. The result is the most comfortable inside-the-waistband holster I have ever carried, while also providing maximum concealment.

Most crossdraw rigs are more difficult to conceal than a strong side holster, as the pistol's butt faces forward, and a forward rake will cause the weapon to fall slightly away from the body. Kanaley's re-design of the Executive Companion to my specifications minimized this latter tendency to the fullest extent possible. The price is $62 and the only options are colors, which include natural, black or cordovan. Extra belt loops for the Executive Companion Crossdraw holster are $4 each. A valuable accessory for this rig is the new inside-the-waistband #1PS single magazine pouch ($22), which features a rear flap to protect the body from the front edge of the magazine's floorplate.

At the present time, the ammunition of choice in caliber .45 ACP is Winchester's 230-grain Jacketed Hollow Point (JHP) load—product number Q4243. The velocity of this ammunition averages about 820 feet per second (fps) from the Commander's 4.25-inch barrel. The depth of penetration in soft tissue is about 15.5 inches, with consistent, concentric expansion of the bullet to approximately .65 caliber.

That's precisely the performance we want. Penetration is without doubt, the most important single parameter in the wound ballistics of handgun ammunition.

The bullet must penetrate deeply enough to crush, cut and break through the human body's vital structures and organs. Any handgun bullet not capable of penetrating at least 12 inches is not acceptable; the capacity to penetrate up to 20 inches of soft tissue is desirable. Once we've obtained the required penetration, the bullet that makes the biggest hole will do the most damage.

Black Hills Ammunition will soon introduce a .45 ACP cartridge with either Winchester's or Hornady's 230-grain JHP bullet loaded to velocities equivalent to the Winchester round. I envision a bandwagon effect, as the popularity of this type of heavy, expanding .45 bullet accelerates.

The Kokalis/Novak Fighting Commander will deliver match-grade accuracy, but this was not a primary goal. When we practice for combat, as our group dispersion gets smaller, we need to increase the speed of the draw stroke. Remember, our motto is "DVC" *(Deligentia—Vis—Celeritis)*—accuracy, power and speed. All three elements are equal factors in the equation.

Before we meet the requirements of DVC, our weapon must be reliable—to the maximum degree obtainable. All of Wayne Novak's customized handguns meet that parameter at the highest level. This one is no exception. There were no stoppages during *SOF*'s extensive test and evaluation and there will be none.

Those accustomed to the M1911A1 series and its cartridge will experience no discomfort firing the alloy-framed Commander. The perceived recoil with self-defense loads is very little greater than that of the heavier steel-framed models.

Black Avenger provides professional gun handlers and serious students of the gunfighting discipline with the best handgun combat cartridge, in the most sophisticated rendition of the unparalleled LW Commander ever available.

Originally appeared in *Soldier Of Fortune*
May 1992.

Ruger's New 22/45

Pop Gun in Body of a Big Bore Boomer

Ruger's new 22/45 pistol is the progeny of a popular and successful series of .22 LR Rimfire handguns, of which more than 1 million have been made.

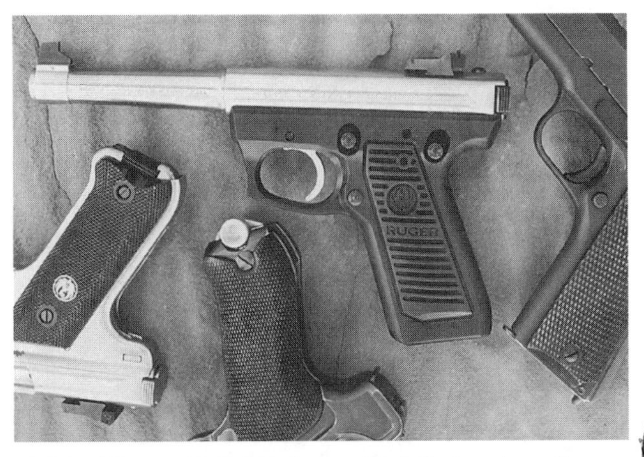

Original Ruger .22 LR semiauto pistols featured configuration and grip angle (35 degrees) of the German Luger. New 22/45 pistol takes the grip angle (17 degrees) of the M1911A1 .45 ACP pistol.

Literally billions of .22 LR (Long Rifle) rimfire cartridges per year are sent downrange by target shooters, hunters and plinkers (Warning: Be advised that the wizards in the Bureau of Alcohol, Tobacco & Firearms aviary have determined this latter activity serves no "legitimate sporting purpose."). A substantial number are fired through Ruger handguns and rifles.

Caliber .22 rimfire cartridges had their origin with the Flobert BB Cap, developed in 1845. The .22 Long Rifle (LR) cartridge was introduced by the J. Stevens Arms & Tool Company in 1887.

In August 1949, William B. Ruger and Alexander M. Sturm introduced the .22 Ruger pistol, their first product. Its price then was $37.50. The method of operation was unlocked blowback, with the barrel fixed to a tubular receiver in which the bolt reciprocated.

Thirty years after this, in 1979, the 1-millionth Ruger Standard Automatic Pistol was manufactured.

In 1982, the so-called Mark I—by this time produced in either blued or stainless steel versions with 4 3/4-inch or 6-inch tapered barrels, or a 5-inch bull barrel—was re-designed.

The new Mark II pistol had a new trigger pivot retainer and a reformed trigger. With the safety engaged and the sear locked, it was now possible to manually operate the bolt to clear the chamber. The magazine was redesigned to increase its capacity from nine to a total of 10 rounds. Scallops were milled in each side of the receiver tube at its rear to permit cocking ears in the bolt to be grasped more securely.

Finally, a bolt hold-open device with a release button above the left grip panel was added.

The pistol's basic configuration remained as before—that of the German Luger (9mm Parabellum *Pistole 08).* The grip angle of both pistols was 35 degrees. This is the angle the front-leading edge of the grip frame makes with an imaginary line drawn perpendicular and downward from the bore's axis. The Ruger pistol has retained this form for more than 40 years in far more than a million copies.

This has never really satisfied competition shooters who prefer a grip angle similar to that of the M1911A1 .45 ACP pistol. Since the demise of High Standard, no high-quality, target-grade .22-rimfire pistol has been available that duplicates the M1911A1's grip angle.

Ruger's recently introduced 22/45 Pistol provides shooters with both the M1911A1's 17-degree grip angle and its left-side controls, as well as a frame fabricated from a space-age synthetic material.

The Ruger 22/45 pistol's grip frame is made of matte-black, glass-filled, injection-molded Zytel nylon. Zytel is unaffected by salt water, sweat, oils or bore-cleaning solvents. It will not crack, chip or peel, and never needs refinishing. Depending upon the model, the Zytel frame reduces weight by 7 to 8 ounces (17% to 23%) compared to standard Ruger Mark II pistols. The frame's center panels are ribbed and carry the Ruger logo. The front of the trigger guard is recurved to comply with the current fetish for a completely useless feature. Shooters firing from the correct Weaver position will just ignore the front of the trigger guard.

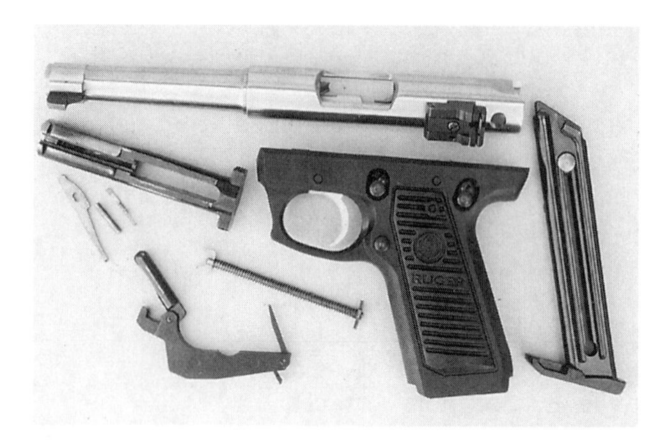

Ruger 22/45 Specifications

CALIBER . . . . .22 LR (Long Rifle) Rimfire.

OPERATION . . . . Unlocked blowback. Bolt reciprocates in a receiver tube, to which is attached the fixed barrel. Manual safety, bolt hold-open release and magazine catch/release located on the left side of the frame in the M1911A1 manner.

FEED . . . . 10-round, detachable, single-column, box-type magazine.

WEIGHT, empty
KP-4 . . . . 29 ounces;
KP-514 . . . . 32 ounces;
KP-512 . . . . 36 ounces.

LENGTH, overall
KP-4 . . . . 8 13/16 inches;
KP-514 . . . . 9 5/16;
KP-512 . . . . 9 9/16 inches.

HEIGHT . . . . Approximately 5 1/2 inches.

THICKNESS . . . . 1.1 inches (grip frame area).

BARREL . . . . Six-groove with a right-hand twist of one turn in 16 inches.

BARREL LENGTH
KP-4 . . . . 4 3/4 inches, tapered;
KP-514 . . . . . 5 1/4 inches, tapered heavyweight;
KP-512 . . . . 5 1/2 inches, bull barrel.

SIGHTS . . . . .Black oxide, blade-type front, 0.125 thick; rear sight eight fixed (KP-4 only) or black-oxide, open square-notched type, adjustable for both windage and elevation zero.

FINISH . . . . .Brushed satin stainless-steel barrel and receiver; matte-black, glass-reinforced Zytel frame.

SUGGESTED RETAIL PRICE AND ACCESSORIES
KP-4 . . . .$329.25;
KP-514 and KP-512 . . . .$388.
Prices include a lockable, fitted gray-plastic box; lock with keys; two magazines and an instruction manual.

MANUFACTURER . . . Sturm, Ruger & Company, Inc., 200 Ruger Road, Prescott, AZ 86301.

T&E SUMMARY: Progeny of a successful series; reliable; provides grip angle and controls of the M1911A1 in an economical .22 LR Rimfire envelope; stainless steel and Zytel construction exhibit exceptional durability. Highly recommended.

The magazine catch/release has been moved from the heel of the frame, where it is found on the standard Ruger .22 pistols in the European manner, to the left side of the frame directly in back of the trigger, where it is located in the M1911A1 series.

The manual thumb safety is also located in the left side of the frame, in the same location as that of the M1911A1's thumb safety and where it also is in the Ruger Mark II series: Push down for firing. In the safe position, the mechanism blocks the sear and prevents it from pivoting forward to release the hammer when the trigger is pulled. As the safety button barely protrudes above the Zytel frame, it is somewhat difficult to re-engage.

The bolt's hold-open release button will be found the left side in approximately the same place as the M1911A1's slide-stop lever. The hold-open is activated after the last shot has been fired, when a steel button protruding from the left side of the magazine (and also used for manually lowering the magazine follower when loading the magazine) pivots a sheet-metal lever in the frame upward to hold the bolt to the rear. This lever is spring-loaded on Mark II pistols and, after inserting a loaded magazine, it will drop downward when the bolt is pulled slightly rearward, permitting the bolt to travel forward after it is released to strip a round from the magazine and chamber it.

With the 22/45, the hold-open release button must be manually pressed downward to release the bolt, strip and chamber a round after a new magazine has been inserted with the bolt held to the rear.

Ruger 22/45 receivers are made from heavy-walled, seamless 400-series stainless-steel tubing that has been machine-finished. The stress-relieved barrel forgings are also made of heat-treated, 400-series stainless steel. All Ruger .22 LR semiauto-pistol barrels have six-groove riflings with a right-hand twist of one turn in 16 inches. The finish is a natural brushed satin.

The bolt is made of polished, black oxide-finished, chrome-moly steel. Ruger 22/45 bolts have the Ruger logo stamped onto their rear faces. The fixed ejector, a stainless steel stamping, is permanently attached to the receiver.

The two-piece extractor is spring-loaded. Also spring-loaded, the firing pin's protrusion is restricted by a cross-pin that passes through the bolt body. The recoil spring, held captive in its guide rod, rides on top of the bolt body.

All triggers for Ruger .22 LR semiauto pistols are serrated and made of aluminum. Trigger pull weight on our test specimen was a crisp 3 1/4 pounds with no perceptible overtravel. Of three Mark II pistols in my possession, two have identical pull weights to the 22/45 pistol we tested and one, a "U.S." target bull-barrel model, has a 3 1/2-pound pull weight.

Manual safety, magazine catch/release and bolt hold-open release controls on the Ruger 22/45 approximate the location of those on the M1911A1.

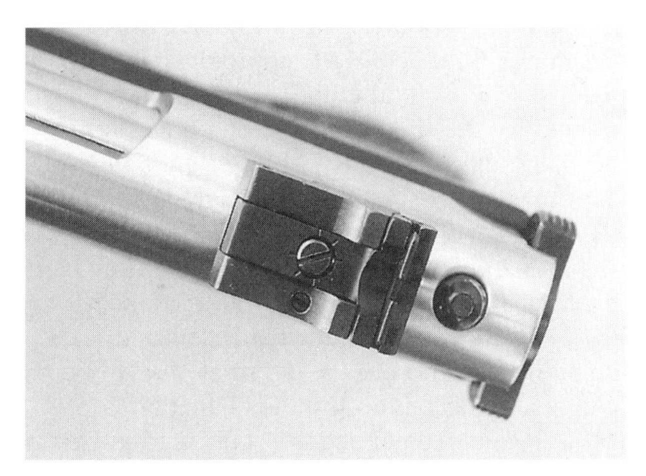

Click-type rear sight on KP-514 and KP-512 models of Ruger 22/45 provides a 0.125-inch-wide open square-notch that can be adjusted for both windage and elevation zero.

TRIGGER TALK

That's good enough for me, but those striving for trigger perfection usually opt for Jim Clark's 3/8-inch-wide serrated-steel Ruger trigger, which is longer and has precision-drilled and reamed pinholes. Clark's trigger costs only $15 and can be obtained from Brownells Inc.

There are three different models of the 22/45 pistol; these differ only with regard to barrels and rear sights. The KP-4 model is equipped with a 4 3/4-inch tapered barrel and fixed rear sight. *Soldier Of Fortune* was provided with a KP-514 model for test and evaluation. It has a 5 1/4-inch tapered heavyweight barrel and a click-adjustable rear sight. The KP-512 model is equipped with the same rear sight, but has a 5 1/2-inch barrel.

Our test specimen has a black-oxide, blade-type front sight 0.125-inch thick and roll-pinned to the barrel. The black-oxide, open square-notch (also 0.125-inch wide) rear sight can be adjusted for both windage and elevation.

Each click of either the windage or elevation screws will move the point of impact of the bullet approximately 3/4-inch at a range of 25 yards.

No adjustment of the sight was required, as Ruger .22 LR semiauto pistols are factory-zeroed by means of a unique, and quite accurate, laser system.

Overall length of these pistols varies from 8 13/16 inches to 9 9/16 inches. Width at the grip-frame area is about 1.1 inches, just slightly less than that of the M1911A1 (approximately 1.26 inches). The weight, empty, varies from 29 ounces for the KP-4 model to 36 ounces for the heavier, bull-barreled KP-512. Our KP-514 test specimen weighs 32 ounces. Height is about 5.5 inches for each model.

Although the single-column, detachable box-type magazine retains the 10-round capacity of the Mark II magazines, the magazine body's angle with relation to the floorplate had to be changed to accommodate the change in grip angle, while maintaining the same feed angle with regard to the receiver body. Ruger 22/45 and Mark II magazines are therefore not interchangeable. The Ruger 22/45 magazine falls freely away from the pistol when the magazine catch/release button is depressed.

All three models of the 22/45 pistol are priced to compare with similar models of stainless-steel Mark II pistols. Suggested retail prices include a lockable, fitted gray-plastic box; a lock with keys; two magazines and an instruction manual. Suggested retail price of the KP-4 with fixed sights is $329.25. Both of the adjustable rear-sight models—the KP-514 with 5 1/4-inch heavy tapered barrel, and the KP-512 with 5 1/2-inch bull barrel cost $388 each. Contact Sturm, Ruger & Company Inc. for further information, or for ordering parts and/or servicing.

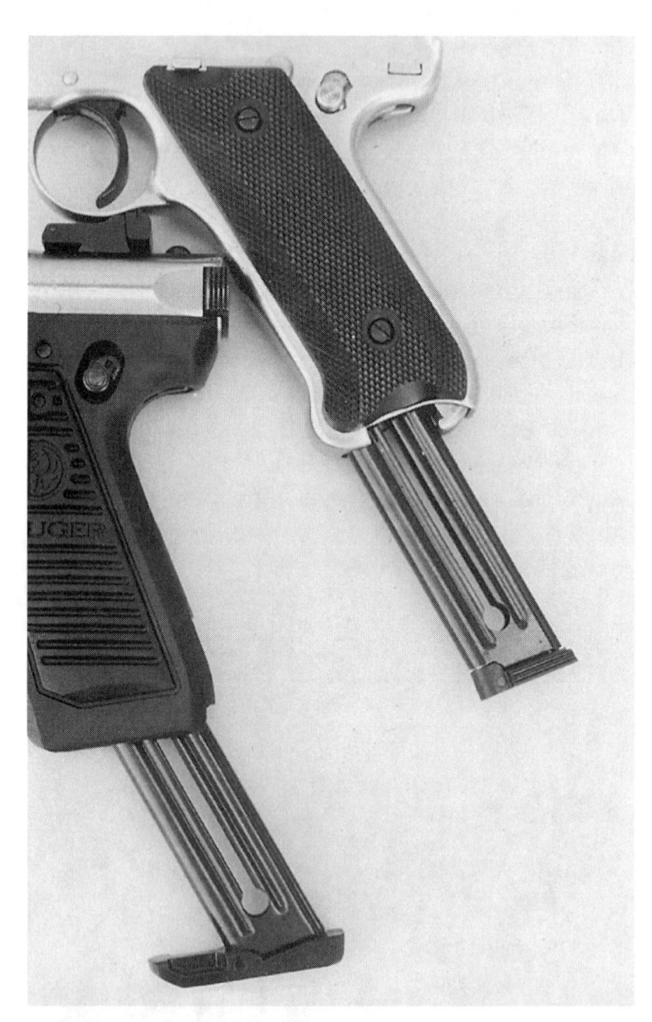

Ruger 22/45 (left) and Mark II (right) magazines are not interchangeable, as the new 22/45 magazine body's angle with relation to the floorplate had to be changed to accommodate the change in grip angle, while maintaining the same feed angle with regard to the receiver body.

If you don't follow the instructions explicitly—and a few hints are not found in the manual—reassembly of any Ruger .22 LR semiauto pistol can prove to be a nightmare. Let's cut through the manual's chaff, focus on the essentials and add a few tips from our own experience.

First, remove the magazine and clear the pistol while visually inspecting the chamber. The safety must be placed in the "off" ("F") position. Point the pistol in a safe direction and pull the trigger. The hammer must be uncocked before proceeding any further.

Use a plastic-coated paper clip or a similar non-marring tool to swing the housing latch on the back strap open, like the blade of a pocketknife. Swing the entire housing outward on its pivot and pull downward, removing the mainspring housing and bolt stop pin. Pull the bolt assembly out of the rear of the receiver. Push the barreled receiver forward and separate it from the frame. This latter step is easier to perform with the 22/45 than the Mark II pistols, which sometimes require a smart blow with a plastic or leather mallet to either separate and/or reinstall their barreled receivers onto their frame.

The mainspring and guide rod can be lifted off the top of the bolt body. To remove the firing pin assembly, drift out the cross-pin (the firing pin stop). The magazine should be disassembled by inserting a drift in the hole in the floorplate and depressing the magazine-block retaining plunger. Then slide the floorplate forward. Keep the plunger and follower-spring under control. Withdraw these components from the magazine body, along with the follower.

No further disassembly is required for normal maintenance and cleaning. After cleaning and lubrication, the magazine and bolt assembly can easily be reassembled in the reverse manner. Before attempting to reassemble the pistol itself, pray for divine guidance and then proceed as follows:

With the safety still on "F," place the hammer in the horizontal (cocked) position. Place the barreled receiver on top of the frame and slide it rearward until the steel lug on front of the frame is engaged in the recess on the receiver's underside. The rear end of the receiver will slightly protrude past the frame.

Point the muzzle up and pull the trigger to make sure the hammer is still cocked. Slide the bolt group into the receiver.

Point the muzzle down and pull the trigger. Gravity will cause the hammer to fall to its uncocked position. Insert the bolt stop pin through the holes in the receiver and bolt until the tip of the pin protrudes beyond the top of the receiver. The bolt stop pin must be fully seated, or you will not be able to swing the mainspring housing into the frame.

Rotate the mainspring housing on its pivot pin upward against the frame to drive the bolt stop pin fully home. Make sure the hammer strut is free to move and swing the mainspring housing back down into the frame. Close the housing latch.

Test reassembly by drawing the bolt to the rear. If the bolt cannot be drawn fully rearward, then the hammer strut is misaligned and you must remove the mainspring housing and make sure the strut is positioned correctly to seat in its recess in the housing.

None of this is either as simple as fieldstripping a Walther PPK or as complicated as disassembling a 20mm M39 revolver cannon. In general, the 22/45 seems to be somewhat easier to disassemble and reassemble than the steel-framed Mark II pistols.

A large assortment of .22 LR rimfire ammunition totaling close to 1,000 rounds was fired during *SOF's* test and evaluation of the Ruger 22/45. Included were CCI Stinger, Eley Tenex and Match (Black Box), Federal Hi-Power 22s and American Eagle 22s, Remington Pistol Match, High Velocity and Hyper Velocity Vipers, South African Swartklip, Winchester Super-X and Winchester Western T22.

As both the rear and front sights are mounted on non-reciprocating components—the receiver and barrel, respectively—the Ruger 22/45 pistol's accuracy potential is quite high. When fired from a rest at 25 yards, all of the ammunition tested punched 10-round groups of between 1 1/2 and 2 inches. Remington and Eley match-grade ammunition, not unexpectedly, performed best. If you want to improve on this, you'll have to change barrels for a heavy match type with a slightly faster 1:14-inch twist, tight chamber and 12-degree forcing cone. However, for most of us—intent only on watching beer cans spin into the air—the Ruger 22/45 as issued will more than do, as its accuracy potential exceeds our ability to hold the weapon steady.

Reliability was excellent and there were no stoppages of any kind. I wouldn't expect anything less from a series of which more than a million have been made. Without exception, all of *SOF's* test personnel preferred the 22/45's 17-degree grip angle over that of the Mark II series.

I concur. I don't make my living pointing German Lugers at either people or paper. Armed professionals, who carry and practice with some version or another of the M1911A1 series on a daily basis, will find the new Ruger 22/45 pistol to be a more than satisfactory complement to their battery.

Bill Ruger's uncanny ability to predict what U.S. gun owners want and need, and his superstar batting average at successfully doing so, continues to astound everyone in the firearms industry.

Although we have occasionally disagreed on matters of tactics with regard to the Second Amendment, I remain proud to count him among my friends. Ruger's new 22/45 pistol appears destined for success, and rightfully so.

Originally appeared in *Soldier Of Fortune*
September 1992.

H&K's USP was designed for law enforcement and civilian users to exceed the strictest NATO milspec standards.
(Photo courtesy H&K)

H&K's Universal Self-Loading Pistol

One Platform, 10 Variants Equals *Volks*handgun for the '90s

In spite of its innovative technology and superb quality, Heckler & Koch's P7 "squeeze-cocker" pistol was never an overwhelming success in the United States. Introduced in the late 1970s as the PSP (*Polizei-Selbstladepistole,* or Self-Loading Police Pistol), the P7 series was just too expensive for most U.S. law enforcement agencies and individual consumers. A pity, as I often carry a .40 S&W P7M10 and it is an excellent handgun.

Design work commenced in September 1989 on a new family of H&K pistols principally focused on the U.S. market. In addition to the qualities usually associated with H&K firearms—reliability, durability, safety, accuracy, advanced materials and innovative features—the new design was to stress a competitive price structure and, consequently, a conventional method of operation. In November 1989 I was informed the design would include a polymer frame. An attempt to copy Glock?—Hardly.

Back in the late 1960s, H&K introduced the P9S pistol, which featured, like most of its shoulder-mounted weapons, a delayed blowback, roller-locking system of operation. This pistol's "frame" was composed of two components: a plastic trigger-guard/front-strap and a sheet-metal receiver. A one-piece plastic grip was held to the rear of the receiver by two screws positioned through the backstrap.

Furthermore, the H&K VP70 caliber 9mm Parabellum machine pistol and its semiauto equivalent, the VP70Z— both designed during the same time frame as the P9S—had frames made entirely of plastic, although they were reinforced with steel at appropriate points. Both the P9S and

VP70 appeared more than a decade before Gaston Glock's pistol was ever even conceived. H&K's G11 rifle, its CAWS (12-gauge Close Assault Weapon System) and both the MP5/10 and MP5/40 submachine guns also make extensive use of polymer components.

In an unusual reversal of the normal course of military small arms development, H&K's entry in the OHWS (Offensive Handgun Weapon System) requested by the U.S. Special Operations Command (USSOCOM) was an outgrowth of the commercial project that led to the Universal Self-Loading Pistol (USP), formally introduced in January 1993.

ECLIPSING MILSPEC TESTS

As H&K was awarded a Phase I development contract for the special operations forces' OHWS on 28 August 1991, its USP prototypes were fortunate enough to be included in the rigorous milspec testing that H&K performed on its OHWS prototypes.

The USP method of operation is short recoil, locked breech. The quite conventional locking system is a modification of the Browning dropping-barrel method as encountered on the SIG-Sauer P226, the Ruger P85 and on some others. A large, rectangular lug over the barrel's chamber engages the slide's ejection port in the locked position. When a cartridge is fired, chamber pressure drives the cartridge case back against the breech face.

Locked together, the slide and barrel recoil rearward for about 3mm, after which a shaped lug on the underside of the chamber engages the hooked locking block at the end of the steel recoil-spring guide rod (which is held in place by the slide-stop lever's axis pin) to lower the rear end of the barrel and disengage it from the slide.

The locking surface on the front top of the barrel's locking lug is tapered with a forward slope. This tapered surface produces a camming action, which enables the USP to enhance its reliability in high-dust environments or after extensive firing has introduced severe fouling into the system.

USP40 & USP9 SPECIFICATIONS

CALIBER
USP40 . . . . .40 S&W;
USP9 . . . . 9mm Parabellum.

OPERATION . . . . .Locked breech, short recoil, semiautomatic, frame-mounted fire modes and control functions with options that permit combinations of "cocked and locked" carry, manual safety, double-action/single-action, de-cocking and double-action-only with right- or left-hand controls.

FEED MECHANISM . . . . .Staggered column, single-position feed, detachable box-type magazine with 13-round capacity (.40 S&W) and 15-round capacity (9mm).

WEIGHT, empty, both with magazine;
USP40 . . . . 29.62 ounces;
USP9 . . . . 28.46 ounces.

LENGTH, overall . . . . 7.64 inches (194mm).

BARREL . . . . .Six-groove with a constant right-hand twist. Rate of twist: USP40, 1:14.96 inches (380mm); USP9, 1:9.84 inches (250mm); cold-hammer-forged from high-grade chromium steel.

BARREL LENGTH . . . . .4.13 inches (105mm).

SIGHTS . . . . .Fixed, high profile; front: blade-type, 0.135-inch thick with single white plastic plug; rear: open square-notch, 0.135-inch wide with a white plug on each side of the notch. Self-luminous tritium night sights also available ($59).

FINISH .Black polymer frame; slide finished with a nitro-gas-carburized and matte-black oxidized process; internal components treated with low-friction Dow Corning "Molykote" process.

PRICE . . . .$624, complete with two magazines and lockable plastic storage box. $20 extra for left-hand controls; extra magazines $27 each.

MANUFACTURER . . . . .Heckler & Koch GmbH, Postfach 1329, D-7238 Oberndorf/N., Federal Republic of Germany.

DISTRIBUTOR . . . . .Heckler & Koch Inc., 21480 Pacific Blvd,. Sterling, VA 20166-8903.

T&E SUMMARY: Conventional method of operation. Competitively priced. Passed extensive milspec testing prior to introduction. Many desirable features. Fire mode and control-function options provide alternatives to satisfy all conceivable user specifications.

From the beginning, the envelope was designed to accommodate the .40 S&W cartridge (USP40) and is presently available chambered for this round and also for the popular 9mm Parabellum cartridge (USP9). Both models are approved for high-pressure +P and +P+ ammunition, which should surely satisfy those who believe in magic bullets.

Eventually a larger USP package will be fielded to handle both .45 ACP and 10mm loadings. A compact variant in this series is also anticipated. The specimen provided to *Soldier Of Fortune* for test and evaluation was chambered for the .40 S&W cartridge, which is rapidly overtaking the 9mm Parabellum round in popularity within the U.S. law enforcement community.

Dimensions are essentially identical for both the USP40 and the USP9. Overall length is 7.64 inches (194mm). At its widest point on the frame, the width is 1.26 inches (32mm). The height is 5.35 inches (136mm). The weight, empty but with a magazine, is 29.62 ounces for the USP40 and 28.46 ounces for the USP9.

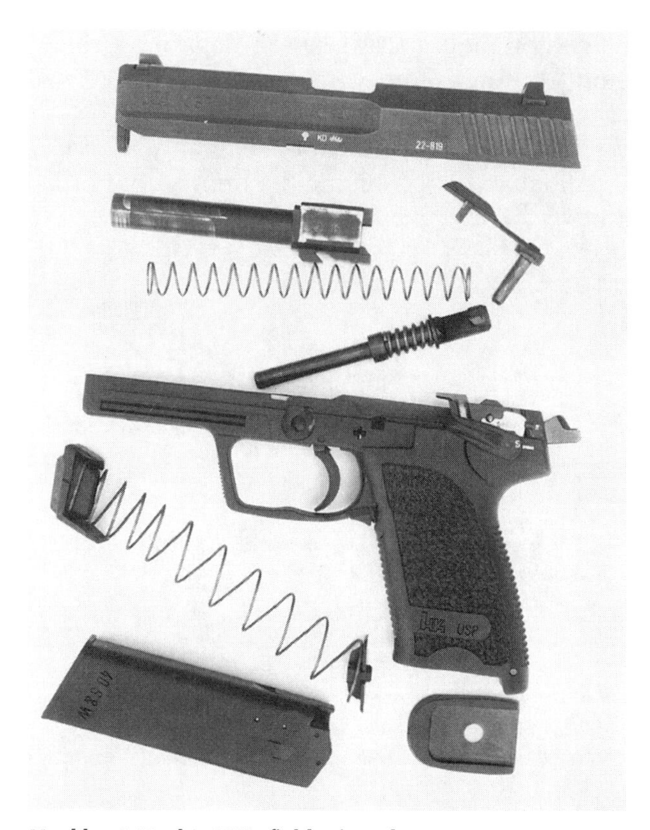

Heckler & Koch's USP, fieldstripped.

FIRST-CLASS ENVELOPE

Length of the USP barrels—cold-hammer-forged from high-grade chromium steel—is 4.13 inches (105mm). They have six lands and grooves with a constant right-hand twist. Rate of twist for the .40 S&W-caliber barrel is one turn in 14.96 inches (380mm) and one turn in 9.84 inches (250mm) for the 9mm barrel. The rifling is not of the polygonal type, as provided on a number of the other H&K firearms. While not excessively large, the USP as currently offered is by no means a compressed envelope. In dimensions, it generally compares favorably with other service-size semiauto pistols.

The injection-molded, two-piece polymer frame has been reinforced with a 15% microscopic glass-fiber content. There are four steel guide rails on the frame for the slide, in a manner reminiscent of the Glock.

The grip's front and rear straps are deeply cross-checkered with pointed tips, while the grip side panels have a rough, stippled texture and feature the H&K escutcheon.

The grip angle (defined as the angle between the front leading edge of the grips and a line drawn normal to the bore's axis) is 17 degrees, exactly matching that of the Colt M1911A1 Government Model. A step at the heel of the frame holds a polymer pin to attach a lanyard loop.

Injection-molded polyamides are super industrial-strength synthetics well-known for their resistance to high temperatures, corrosion, wear, chemicals and radiation. Lighter than steel, they have a higher tensile strength than aluminum.

The spring-loaded steel slide-stop lever is mounted on the left side of the frame and can be easily manipulated with the thumb of the firing hand by a right-handed shooter. When the last round has been fired, a shelf on the magazine follower forces a projection on the slide-stop lever upward to hold the slide in the rearward position. The slide-stop lever also serves as a dismounting lever to separate the frame and slide groups.

LASER DRAWBACKS

Grooves cut parallel to the bore's axis on each side of the frame in front of the trigger guard provide mounting points for a Laser Aiming Module (LAM); this was probably taken from H&K's OHWS. As I have stated many times in the past, I am opposed to the employment of visible laser-aiming devices on small arms. Operators invariably seek out the dot and instinctively react to place it on the target with no regard for the proper shooting stance or training, which has mandated that they bring the handgun up to eye level in the Weaver position.

Under stress, we do what we have trained to do. If we shoot correctly in daylight, but become accustomed to sloppy techniques that will nevertheless result in hits when a laser is employed in subdued light, then under the intense stress of contact with the enemy we will be forced to decide between two alternative firing techniques. There simply isn't time for decision-making like this in a gunfight. Furthermore, it has been demonstrated that laser sights actually impede response time, as the shooter must slow down and take whatever time is required to locate the red dot on his target.

However, infrared (IR) laser beams and the night-vision goggles required to employ them do have some valid applications for clandestine operations. H&K can provide you with a laser toy and wireless switch that attaches to the USP for $229. A small flashlight mounted on these grooves would make more sense. H&K has a tactical light under development that will be available soon. This is a more secure system for attaching accessories than the usual trigger-guard mounting.

The trigger guard is large enough to accommodate gloved operation in cold weather. The front of the trigger guard is squared-off with horizontal serrations, even though few armed professionals place the index finger of the support hand in this area. Furthermore, in this instance, the trigger guard is so long as to make the use of this useless fetish even more awkward. The bottom of the trigger guard is flared to protect the ambidextrous magazine catch/release and prevent accidentally dumping a magazine on the ground at an inappropriate time.

To release a magazine, the catch/release must be pressed down instead of inward; this takes a little getting used to for those of us trained on the M1911A1 series. Magazines, either empty or full, fall freely away when the catch/release is depressed. The magazine catch/release is somewhat difficult to reach with the thumb without shifting the firing grip. I find

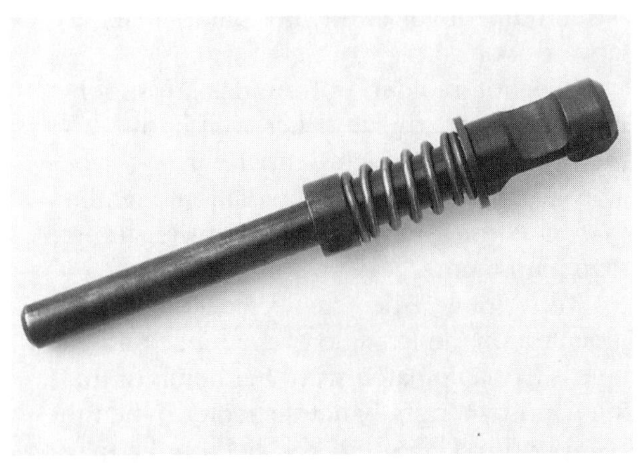

There is a hooked locking block at the end of the USP's recoil-spring guide rod that stops the barrel's rearward travel during the latter part of the recoil stroke. Forward of this is a mechanical recoil-reduction system that buffers unlocking of the barrel and consists of a heavy, captive coil spring around the guide rod.

All of the various fire modes and control functions available for the USP series are determined by a notched detent plate inside the frame, to which the safety lever, if present, is attached.

that the trigger finger is better positioned for this purpose. Half-moon-shaped, "tear away" notches have been molded into the grip on each side at the bottom to assist removal of a jammed magazine. There is no magazine safety and the USP can be fired without a magazine in place.

The staggered-column detachable box-type USP magazines have a capacity of 13 rounds in .40 S&W and 15 rounds in 9mm Parabellum. Magazine bodies are made of polymer with their top thirds internally reinforced with stainless steel. There are numbered indicator holes at the rear of a magazine body, while the magazine lips are rounded for easier loading.

The magazine floorplate, its locking plate and the follower are also made of polymer. Unlike those of the MP5, USP magazines are easy to disassemble.

The trigger and frame-mounted safety mechanisms provide one of the USP's most unusual and interesting features. Ten variations of fire modes and control functions are available.

FIRE MODE VARIANTS

Variant 1 has a safety lever mounted on the left side of the frame for right-handed operators. Rotating the safety lever upward prevents firing, whether the hammer is cocked (as in "condition 1," with a round in the chamber, the hammer cocked and the safety on) or uncocked. This variant permits "cocked and locked" carry. When the safety lever is in the horizontal position, the pistol can be fired whether the hammer is cocked (single-action mode) or uncocked (double-action). Rotating the lever below the horizontal position will drop the hammer (if it was cocked). The lever is spring-loaded; when released it will spring back to the horizontal position, permitting the first round to be fired double-action.

Variant 2 is identical in operation as the above, except that a mirror-image safety lever is mounted on the right side of the frame for left-handed shooters. Note that all odd-numbered variants are for right-handed operators while all even-numbered variants are for southpaws.

Variants 3 and 4 have no "safe" position and the lever is employed only to de-cock the hammer. The pistol can be fired in either the single-action or double-action mode, but cannot be carried "cocked and locked."

Variants 5 and 6 permit double-action fire only, but when rotated upward, the lever locks the trigger mechanism in the "safe" position. Variant 7 has a "bobbed" hammer, no safety lever and can be fired double-action only. Variant 8 is identical to Variant 7, but comes equipped with self-luminous tritium sights and is currently being evaluated by the U.S. Immigration & Naturalization Service (INS).

Variants 9 and 10 are equipped with a safety lever permitting "cocked and locked" carry, but no de-cocking function—*SOF* was provided with a Variant 9 USP for test and evaluation. These last two USP variants are the ones I would personally select. It's important to emphasize that, unlike variants 1 and 2, if the operator pushes down on the safety lever as he comes off the target in a high-stress environment, there is no possibility that he will go past the "safe" position and activate a de-cocker mechanism to inadvertently drop the hammer. Also, let's not forget that the variant 9 or 10 USP is not just another M1911A1 clone. Although most of us will never use it, the double-action mechanism is there if ever required, i.e., for a second strike on a hard primer.

All of these various fire modes and control functions are determined by a notched detent plate inside the frame, to which the safety lever, if present, is attached. These detent plates and their corresponding safety levers can be removed and installed in a matter of minutes by Heckler & Koch-certified armorers only. While the user can specify the variant he wants, it is not intended that individuals endlessly experiment with the different variants, so parts will be not be provided for this purpose.

TRIGGER HAPPY

Trigger pull weights on our test specimen were 9.75 pounds in double-action and a surprisingly light 3.75 pounds in single-action. The smooth, ungrooved trigger is also an injection-molded polymer component. It is connected to a sheet-metal trigger bar—inside the frame on the right side—that governs rotation of the hammer.

The fixed sheet-metal ejector is mounted in front of the hammer mechanism on the left side. The hammer mechanism also contains the lever that rotates upward at the moment of firing to depress the spring-loaded firing-pin safety in the slide and permits the striker to travel forward. There is also a passive disconnector safety. All of the metal surfaces are treated with "HE" (Hostile Environment) finishes. The exterior of the slide, a milled forging from 42 Cr Mo 4 steel, has been subjected to a proprietary nitro-gas-carburized and matte-black-oxidized process that is extremely hard (rated 732 HV1 Vickers) and highly corrosion-resistant. This finish is also applied to the H&K G3SG1 sniper rifle.

Internal components are treated with a unique Dow Corning "Molykote" process that has low-friction qualities. Heckler & Koch weapons issued to German navy *Kampfschwimmer* (combat swimmers) are also provided with this finish.

There is no barrel bushing; the barrel passes through a hole bored into the front of the slide. The spring-loaded extractor, installed in a slot on the right side of the slide, pivots around its retaining roll pin in the conventional manner.

Both the front and rear sights are installed by friction-fit only into dovetails milled into the slide. The sight radius is 6.2 inches. The unprotected front sight blade is 0.135-inch in width and has a cylindrical white-plastic plug inserted into a hole bored through its entire length. The open square-notch rear sight is also 0.135-inch in width and has two white-plastic plugs, one on each side of the open notch. Under stress, however, these colored dots will not be noticed.

EFFECTIVE BY NIGHT

A far better alternative is to order tritium night sights installed on the USP either directly from Heckler & Koch Inc. at a cost of $59, or have them installed by Innovative Weaponry Inc., who provides them to H&K.

Innovative Weaponry has designed a multi-color night-sight system that focuses attention on the front sight, speeding target acquisition by a considerable margin. Most tritium self-luminous sights glow green simply because this color is highest on the night-visibility spectrum; however, operator alignment of three green dots in subdued light can be confusing, as this detracts from emphasis on the front sight and retards target acquisition.

The largest IWI self-luminous dot is installed in the front sight and it is green. The two smaller (by .008-inch) dots in the rear sight are yellow or orange (or subdued green). In use, I have found that the eyes instinctively focus on the green front dot with a strong 3-dimensional triangulation effect that helps to funnel the shooter's vision directly onto the target. This feature has been installed on my H&K P7M10 and I can personally recommend IWI's multicolor sight system highly and without reservations of any kind.

The conventional coil-type recoil spring rides on a somewhat unusual heavy-steel, full-length guide rod. We retrofitted our USP with a new improved guide-rod assembly provided to us, which can be distinguished from the original only by an incipient groove ringing the rod toward the front end. As previously mentioned, there is a hooked steel locking block at the end of the recoil-spring guide rod that stops the barrel's rearward travel during the latter part of the recoil stroke.

Forward of this is a mechanical recoil-reduction system that buffers unlocking of the barrel and consists of a heavy, captive coil spring around the guide rod. It is insensitive to ammunition variances, requires no special maintenance and supposedly has an indefinite service life. Heckler & Koch claims a 30% reduction in recoil forces through using this device. The company states that this unit reduces the peak force acting upon the USP grip to less than 300 Newtons (66 pounds) and that peak-force shock on competing .40 S&W polymer and steel-framed pistols is more than 5,000 Newtons (1,102 pounds). This sounds, and is supposed to be, very impressive. However, neither I nor any member of the *SOF* staff who fired this handgun could detect any qualitative difference in perceived recoil or muzzle jump between the USP and any other caliber .40 S&W pistol of similar size and weight.

Personally, I believe that the shooter himself can control the effects of handgun recoil—by using the proper stance and grip tension—better than any of these buffering devices, of which there seem to be an endless number. Still, H&K also claims longer service life and reduced stress on components as a consequence of the mechanical recoil reduction system—this is probably true.

Testing conducted on the USP exceeded the strict NATO AC-225 Military Specification Standards. During these tests, a bullet was deliberately lodged in the barrel. Another bullet was then fired into the obstructing projectile. Both bullets cleared the barrel, resulting only in a barely noticeable bulge. Subsequent accuracy tests on this barrel produced group dispersion of less than 2.5 inches at 25 meters.

One USP test gun fired more than 10,000 rounds without a single stoppage. Endurance firing of test samples has exceeded 20,000 rounds of .40 S&W ammunition without any parts failures. Milspec environmental tests were conducted under low temperature (minus 44 degrees Fahrenheit), at high temperature (153 F) and in mud, rain, under water immersion and in salt spray.

Accuracy testing with several brands of ammunition demonstrated that the USP is superior to competing models manufactured in the United States. Safety testing exceeded the ANSI/SAAMI requirements, including dropping a USP with a primed cartridge in the chamber and de-cocked hammer onto a variety of surfaces without its discharging. German police tests were also successfully performed, consisting of repeated drop tests from 6 feet—hammer first—onto a steel-backed concrete slab.

Furthermore, the USP9 is a finalist in Germany's P90 trials, which will result in the adoption of a new handgun for their rapid reaction forces.

In light of this, the "results" of any "tests" conducted by *SOF* or by any gun magazine take on diminished value, to say the least. At this time, I have personally fired more than 1,000 rounds through the USP and have not experienced any stoppages. It is more than accurate enough for its intended purposes. Perceived recoil is moderate, slightly less than comparable envelopes chambered for the .45 ACP cartridge and slightly more than 9mm Parabellum pistols of similar size—exactly what we would expect from the .40 S&W cartridge.

A high level of human engineering has been applied to the USP design. Only those with short, stubby fingers might find the trigger somewhat difficult to reach. If you really slam a loaded magazine into the magazine-well with exceptional force, the slide will override the hold-open projection on the slide-stop lever and jump forward into battery. But this is trivia.

Right-side view of the H&K USP shows the large, rectangular lug over the barrel's chamber that engages the slide's ejection port in the locked position. Note groove on frame forward of trigger guard used for attaching accessories such as a laser sight. "Tear away" notches at the bottom of the grip assist removal of a jammed magazine. *(Photo courtesy H&K)*

FOREMOST LOADS FOR USP

Most .40 S&W ammunition we have tested is effective and it appears that major ammunition manufacturers are starting to produce handgun bullets of a proper design.

Newest and most exciting is Winchester's Black Talon. The lead core of the bullet is locked into place with a reverse-tapered jacket. The Black Talon's jacket wraps completely around the bullet's nose and into the cavity, providing the most reliable functioning of any hollow point to date. Retaining virtually 100% of its original weight, this design allows excellent penetration and expansion, even through barriers such as heavy cloth, wallboard, plywood or glass. Typically, caliber .40 S&W Black Talon bullets will penetrate 14 to 15 inches of soft tissue while expanding up to about .68 caliber.

Black Hills .40 S&W ammunition, when loaded with Hornady's 180-grain Jacketed Hollow Point (JHP) XTP bullet, is also an excellent performer. Muzzle velocity is about 925 feet per second (fps). Expansion is stellate-shaped with the points averaging .68 caliber and the flats at about .64 caliber. There is no fragmentation and this bullet will penetrate more than 14 inches of soft tissue.

There are a total of 56 components in the USP; fieldstripping procedures parallel those for handguns with this method of operation—Remove the magazine and clear the weapon by jacking the slide and inspecting the chamber to make sure it is empty. Retract the slide rearward until the slide stop's axis pin is aligned with the disassembly notch on the left side of the slide. Push the axis pin to the left from the right side and withdraw it completely. Slide the slide group forward and separate it from the frame. Remove the recoil spring, guide rod assembly and barrel from the slide.

USP magazines are disassembled in the conventional manner. Reassemble in the reverse manner, making sure that the angled locking surfaces on both the underside of the barrel and rear end of the recoil-spring guide rod are properly aligned.

COMPETITORS TAKE COVER

With the introduction of the USP, Heckler & Koch has signaled its intention to take a significant share of the substantial U.S. law enforcement and commercial defensive-handgun markets. By providing the option of five different fire-mode and control-function variants, with left-hand operation available on four of them, H&K has neatly sidestepped the unending arguments of "cocked and locked" vs. double-action only, manual safeties vs. de-cockers, and double-action vs. single-action, as the user can specify whatever he wants. Two calibers and the further option of tritium self-luminous sights expand the USP's alternatives to an even greater extent.

Already tested to tough milspec standards, the USP arrives fully developed as a complete system. In light of the variety of envelopes and calibers to follow from a maker whose name literally permeates with a justified reputation for high quality, H&K's competition have sufficient reason for major consternation.

Suggested retail price of the USP is $624 (add $20 for left-hand operation) in either caliber, complete with two magazines and a sturdy, lockable plastic storage box. Additional magazines are $27 each.

Originally appeared in *Soldier Of Fortune*
December 1993.

Belly Gun Ins & Outs

Half-Pint Wheelguns for Backup and Personal Defense

Belly guns, in this instance the S&W Model 640 stainless steel Centennial, remain popular in law enforcement circles and with armed professionals, primarily as concealed "backups" to more serious ordnance.

This round-butt Colt Detective Special with hammer shroud was manufactured in 1954 and rejuvenated 40 years later with Robar's NP3 process.

Revolvers are a long sight from moribund. They still have their aficionados and lethal applications—but no one can deny that they will rarely be found strapped onto the equipment belt of a police officer. Semiautomatic pistols, first in 9mm Parabellum and more recently in .40 S&W (Glock now sells two .40 S&W pistols for every 9mm), have pushed them off the belt and into ankle holsters and other concealment locations on the body.

Small-framed revolvers, usually chambered for the .38 Special cartridge, and with barrel lengths of mostly 2 inches (up to a maximum of 3 inches) are commonly known as "snub noses" or "belly guns." They remain popular in law enforcement circles and with armed professionals, primarily as concealed "backups" to more serious ordnance. However, never forget that a belly gun you've slipped into your pocket is far more potent that the M1911A1 you left on top of the nightstand.

Let's examine the top choices in this genre and the holsters, modifications, grips and ammunition that most professional gun handlers add to these half-pint wheelguns to enhance their hit probability and lethality. There are a number of excellent alternatives.

The epitome of the belly-gun type is the Smith & Wesson J-frame series. Designed as a five-shot .38 Special with the greatest emphasis on reduced weight and size, the first of this series, the so-called Chief's Special Model 36, was introduced in 1950. While it still remains popular in blued-steel, stainless-steel and alloy-frame models, in my opinion, the concealed-hammer Centennial and Bodyguard models are the very best of the J-frames.

In 1952, during Smith & Wesson's centennial year, they announced two new J-frame revolvers, each with a fully concealed hammer—thus combining the salient feature of the old Safety Model, introduced in 1887, with the two-year-old J-frame. Called the "Centennial" Model, the aluminum alloy-frame "Airweight" version (later called the Model 42) was the first to be produced. After the first 37 specimens were fabricated with aluminum-alloy cylinders (with a total weight of only 11.25 ounces), subsequent Centennial Airweights were equipped with steel cylinders and barrels for safety reasons, which increased the weight to 13 ounces. The all-steel version, known as the Model 40, followed shortly thereafter and weighed about 24 ounces. The Centennial revolvers were available with either a blued or nickeled finish and were equipped with 2-inch barrels.

These revolvers were also fitted with a grip-safety on the frame's backstrap. Many users contravened this dubious feature by pinning the safety lever in the depressed position.

Because the hammer was concealed within the frame, the Centennial revolvers could be fired double-action only—an acceptable tradeoff for loss of an exposed hammer which all too often served only to snag on clothing during the drawstroke. Designed for close-range contact, those employing snub-nosed revolvers should, in theory, rarely require the enhanced accuracy potential provided by single-action operation.

SANS HAMMER CULT

During the 24 years of the series production, sales of the Centennial revolvers were no more than mediocre. As a consequence, they were deleted from Smith & Wesson's catalog in 1974. Immediately after production ceased, a cult developed around the "hammerless Centennials" and prices for mint specimens began to escalate in an asymptotic manner—what people can't have, they want. In 1990, Smith & Wesson reintroduced this so-called hammerless revolver as the Model 640, which blended the features of the previous Centennials, sans the grip-safety, with stainless steel construction.

The Model 640 weighs 20 ounces, empty, with an overall length of 6.25 inches. The frame is a mill-finished drop-forging. The barrel, frame and cylinder surfaces carry a moderately high-gloss polish. As with all J-frame revolvers, the cylinder holds five rounds and rotation is to the left (as it is with all S&W revolvers). The short ejector rod, which is not covered with a shroud, does not permit complete ejection of the empty cases. The cylinder latch is that of the entire S&W series and must be pushed forward to swing out the cylinder. The 2-inch barrel has five grooves with a right-hand twist of one turn in 18.75 inches.

The sight system is simple and appropriate for the projected use of a snub-nosed revolver. A large square notch cut into the top strap complements a long, serrated ramp-type front sight blade. Both are fixed and neither have been covered with any type of white or red dots or squares—an apparently neverending and meaningless fetish. The front sight blade on the Model 640 is one-tenth-inch wide. The Model 640 is equipped with a smooth and narrow trigger. This is the preferred type for double-action shooting because it maximizes the trigger finger's sensory perception during the firing stroke. Wide target-type and even narrow-grooved triggers should be avoided on combat revolvers.

This model was shortly followed by the Model 642, a version with a lightweight, clear-anodized frame and a stainless-steel barrel and cylinder. Within two years this model was dropped. S&W claims that the steel and aluminum alloy finishes did not match closely enough to meet their standards.

There followed an almost confusing array of variants: the Model 940 chambered for the 9mm Parabellum cartridge, the Model 632 for the .32 H&R Magnum round and 3-inch types in both .38 Special and 9mm—with some available in blued, stainless steel or with factory satin nickel finish and aluminum alloy frames. My favorite remains the Model 640 with a 2-inch barrel.

In 1955, three years after introduction of the Centennial Airweight, the Model 38 Bodyguard

Airweight was placed in S&W's catalog. It was developed for law enforcement personnel who perceived a requirement for a concealable revolver with a protected hammer that would not snag on clothing, but could still be cocked for single-action firing. In no small measure this was also undoubtedly a response to the hammer shroud Colt introduced in 1950 for their Detective Special model.

In 1959 the Bodyguard was made available with a steel frame and catalogued as the Model 49. Both the Model 38 and 49 are still available with either blued or factory nickel finishes. A stainless steel version, called the Model 649, is also available. I frequently carry a nickeled Model 38 or a stainless steel Model 649. Although I have never had occasion to fire these revolvers single-action, the thought that they are capable of being fired in this more deliberate manner is a mildly comforting placebo.

MAIDEN BELLY SPECIAL

Introduced in 1927, Colt's Detective Special is the first of the modern belly guns. Through most of its production history, the Detective Special could be instantly distinguished by its completely exposed ejector rod. Its one advantage over the S&W J-frame series was a six-shot cylinder—an increase of 20% with very little increase in size or weight.

When chambered for the .38 Special cartridge, the six-groove barrel has a left-hand twist of one turn in 16 inches. Cylinder rotation is to the right (the direction opposite that of S&W revolvers—important to remember when reloading under stress with the Bianchi Speed Strip). The Colt cylinder latch must be pulled back to swing out the cylinder.

Over the years the Detective Special was offered in full range of .32 and .38 rimmed cartridges. There were both square and round-butt versions. The sights were almost always a half-moon blade with a square notch in the frame just forward of the hammer as a rear sight. A serrated ramp was added to the front sight blade in 1948.

The lightweight version was the Cobra, which was introduced in 1950. The Aircrewman Model is rarer than a Walker Dragoon. It had an alloy cylinder and the entire lot of 1,189 revolvers were returned by the Air Force to Colt and destroyed. The Agent Model was similar to the Cobra, but featured a shortened and squared-off butt.

In the early 1970s, a completely shrouded ejector rod was offered as an option on the Detective Special. Within a short time it became a standard feature on the entire series. In 1984 the Commando Special was added to the line. It was nothing more than a Detective Special with fully-shrouded ejector rod, rubber grips and a phosphate ("Parkerized") finish. By the end of the 1980s, the entire Detective Special series was dropped from production. However, Colt has recently announced the revival of the Detective Special. How many, if any, will be produced, is a function of Colt's shaky financial status. Be that as it may, there were close to half a million of all variants produced and they're out there if you want one.

I have a round-butt Detective Special that was manufactured in 1954. It has an incredibly smooth double-action trigger pull. I had the trigger grooves removed and a Colt hammer shroud installed. The Robar Companies Inc. then applied their excellent NP3 process to the entire revolver.

NP3 is a surface treatment for metals and metal alloys that provides the appearance of satin electroless nickel by combining submicron particles of PTFE (polytetrafluoroethylene, i.e., Teflon) with autocatalytically applied nickel/phosphorus. The result is a very accurate, dry-lubricated, low-friction surface that is extremely resistant to wear. The PTFE is evenly distributed and locked into the nickel phosphorus matrix, so when wear occurs, fresh particles of PTFE are exposed to keep the surface lubricated. No lubricants of any kind are required, and powder residue and carbon fouling can be easily removed with a dry cloth.

The Colt Detective Special round-butt frame and factory grip panels fit my large hand perfectly. Unfortunately, the grips issued with S&W J-frame revolvers, until recently, exactly matched the contour of the bantam-sized round-butt frame and just would not do. Here also, there are a number of excellent alternatives.

CUSTOM STOCKYARD

Keep in mind, however, a set of grips that direct the bore's axis into alignment with the arm, fill the void between the trigger guard and the frame's front strap, and drop down below the frame to adequately conform to the dimensions of the hand, will, by definition, increase the overall package. The increase in envelope can be minimized if the grips are custom-made to the specific shooter's hand and are, thus, no larger than necessary.

Herrett's Stocks Inc. has been producing made-to-measure handgun stocks for over 40 years. I purchased my first set of stocks (for a Clark-customized Ruger bull-barrel .22 LR target pistol) from Steve Herrett more than 30 years ago.

Almost all (98%) of Herrett's stocks are made from American black walnut, noted for its strength, beauty and resiliency. A precise hand diagram is required to order stocks from Herrett's. I use their Detective stocks with skip checkering on my S&W Model 640, and without checkering on my Airweight Model 642. Designed to assist in the concealment of small-frame revolvers, Herrett's even requires information about the manner in which the weapon will be carried when handcrafting a set of Detective stocks for your revolver.

However, the really hot grips for J-frames are those made by Craig Spegel. His handmade cocobolo "Boot Grips" for round-butt J-, K-, L-, and N-frame S&Ws are an outstanding example of equipment designed by an end user. Spegel, an IPSC competitor and Oregon combat master, specializes in custom grips that enhance control and natural aiming in as compact an envelope as possible.

The revolver's back strap is left exposed, so no extra trigger reach has been added. A slight, ambidextrous palm swell and comfortable finger grooves give a much firmer support than the usual factory grips. Unlike Herrett's Detective stocks, these grips do not extend below the frame's butt, so in most cases the little finger of the firing hand will extend below the grips. That's the tradeoff for increasing concealability. However, I have fired thousands of rounds with J-frame Smiths wrapped in Spegel's Boot Grips and have never experienced any problem with control or loss of accuracy, and they are light years ahead of the standard factory grips. They cost $55 (plus $3 shipping and $5 extra for old model Centennials with a grip safety). More professional gun handlers carry J-frames with Spegel's Butt Grips than any other type. They are, at this time, so dominant in these circles as to almost constitute a fetish.

In fact, Smith & Wesson now issues synthetic Boot Grips with all of their Centennial-series revolvers. Injection-molded from Craig Spegel's original Boot Grip design, they are made by,

Charter Arms .44 Special Bulldog with bobbed hammer, barrel cut back to 1.75 inches and Tyler T-Grip installed.

and also available from, Michaels of Oregon Co. Steel-reinforced and molded from a tough, durable, specially formulated elastomer called Santoprene, they are lightweight and much firmer than ordinary rubber grips. This latter characteristic is important. As normally encountered, so-called neoprene rubber grip panels will absorb recoil better than any other material you can attach to the frame of a handgun. But, if you should grip the weapon incorrectly during commencement of the drawstroke, their adhesive qualities do not permit rapid readjustment, with potentially fatal consequences. I have tried Michaels' Boot Grips and they are not subject to this criticism.

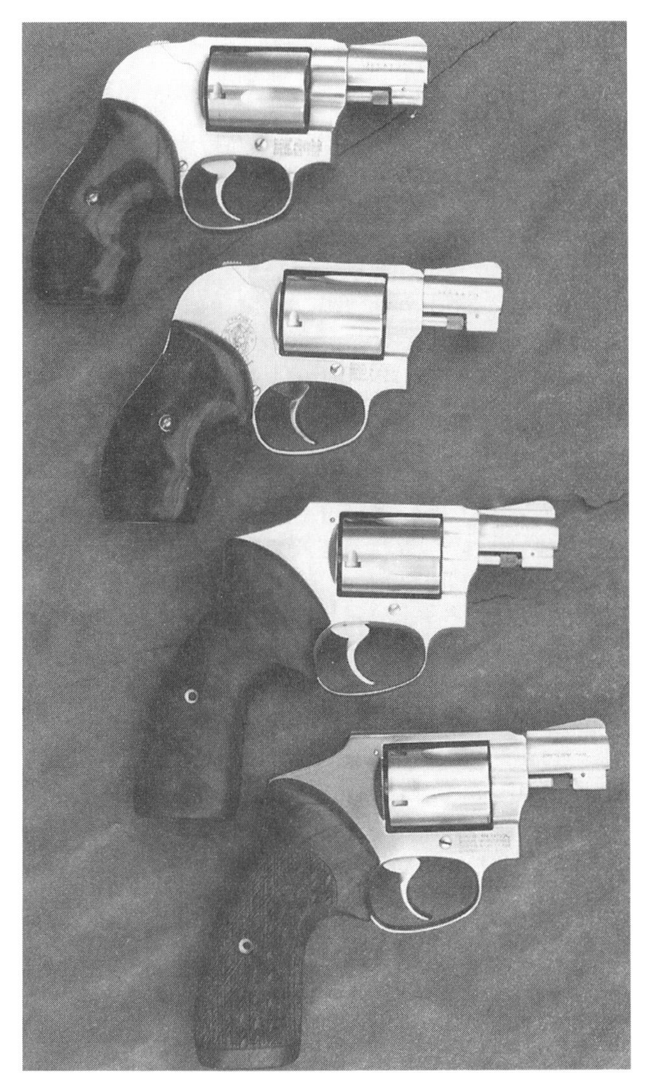

Smith & Wesson's best belly guns, from top to bottom: M649 stainless steel Bodyguard and M38 Airweight factory-nickeled Bodyguard, both with Craig Spegel "Boot Grips;" M642 Airweight Centennial and M640 stainless steel Centennial, both with Herrett's Detective stocks.

I can also recommend one other type of custom stock for J-frame revolvers. Bear Hug Grips Inc. has designed their Coyote grip panels for all S&W round-butt J-frame revolvers. Made from StaminaWood—a resin-impregnated, hardwood laminate—they are available in walnut, rosewood, ebony, mesquite and cocobolo and cost $49.95 (plus $4 shipping). I think the mesquite grips are especially attractive; however, more importantly they redirect the bore's axis into alignment with the firing arm and fill the void between the trigger guard and the frame's front strap to reduce the tendency of these small frame revolvers to roll violently upward in recoil.

BELT VS. BELLY

When equipped with round butts and 3-inch barrels, N-frame giants, such as the models 629 (.44 Magnum), 657 (.41 Magnum) and 624 (.44 Special) are significantly reduced in size. However, I own all of the above and they are still belt guns, not belly guns by any stretch of the imagination. Just imagine one of these brutes in an ankle holster strapped to your leg.

There is, however, one .44 Special revolver that qualifies as a legitimate belly gun. Eight years ago I obtained a Charter Arms .44 Special Bulldog. At that time it was available with a 2.5-inch bull barrel in a blued finish. Today it is manufactured by Charco Inc. and available only in a factory nickel finish with a 3-inch barrel.

Its five-shot cylinder rotates to the right (in the Colt manner) in a frame of high strength chrome-moly steel. There is no side-plate. Only the trigger guard is fabricated from an aluminum alloy. The coil-type mainspring provides maximum strength and longevity. An almost unbreakable beryllium copper firing pin can be dry-fired into infinity. When the hammer is cocked, firing either single or double-action, a small steel bar moves up between the firing pin and the hammer. Continued pressure on the trigger holds this steel bar in its raised position.

The falling hammer then strikes the steel bar, which in turn strikes the firing pin, discharging a cartridge. The bar is not raised unless the hammer is cocked and the trigger is pulled completely to the rear. If the finger is removed from the trigger while the hammer is falling forward, the bar will drop downward and ignition will not take place.

I had the 2.5-inch barrel cut back (from the rear) to 1.75 inches. The front sight's ramp angle was altered and reserrated. The forcing cone angle was recut to maintain a cylinder-to-forcing cone gap of .005-inch (.003-to .008-inch is the usual range). These alterations dropped the weight from 19 ounces, empty, to only 17 ounces.

The double-action pull weight was reduced to a smooth 9 pounds with a crisp single-action pull weight of 3.25 pounds. The case-hardened hammer was bobbed with serrations cut to permit cocking for single-action firing, if required.

To all of this I added Charter Arms A17 uncheckered walnut grip panels and a cast aluminum, black-anodized Tyler T-Grip (size No. 1) for $7.50. Melvin Tyler Manufacturing has been making this fine little accessory ever since 1940, and it has improved the qualification scores of many thousands of police officers during the years when service sidearms were still mostly wheelguns.

Chic Gaylord, a legendary figure in gunfighting circles—whose holsters are still highly prized and often copied—was convinced that in a holster, a 3-inch belly gun hugs the body far more effectively than a 2-inch snub nose. A coterie of pistoleros still adhere to this concept, preferring also the longer sight radius and superior balance of a 3-inch barrel on a small-frame revolver.

In my opinion, this has been best executed in the Ruger SP101 revolver chambered for the .38 Special cartridge. Made of 400 series stainless steel, the SP101 has an attractive brushed stain finish. Its 3-inch barrel has five-groove rifling with a right-hand twist of one turn in 18.75 inches. The sights are conventional for this type of revolver: a serrated ramped front blade and an open, square-notch rear sight on the frame directly in front of the hammer. My specimen (Model KSP-183) is equipped with one-piece black Monsanto Santoprene with walnut side panels that permit the grip to be shifted during the draw-stroke, if necessary.

Unbreakable coil springs are used throughout the SP101. The transfer bar trigger mechanism is coupled with a floating firing pin. When the five-shot cylinder (which rotates to left) is in the firing position, it's securely locked to the frame in two places—at the rear by the traditional cylinder pin, and at the front of the crane by a large, spring-loaded latch. Developed by Ruger, this mechanism ensures proper cylinder and barrel alignment.

It, together with the SP101's massive frame, certifies the Ruger revolver as the strongest ever built, with a tradeoff of slightly increased bulk.

I sent this revolver to Weigand Inc. to be transformed by their "Tame the Beast" package (most often performed on .357 Magnum Ruger revolvers). The most salient feature of the custom modifications performed on the revolver was porting the top of the barrel with five nozzles. These tapered holes are calibrated using a formula that calculates propulsion thrust by means of a ratio between the available gas pressure and the size of the nozzles. The smallest nozzles are closest to the frame. While muzzle velocity is decreased only 50 to 65 fps, muzzle flip is reduced about 75%. There is a significant flash signature generated by barrel porting of the Hybra-port type. In my experience, however, the flash initiates well above the front sight and does not disturb the shooter's sight picture.

Because of the reduction in muzzle rise, Weigand lowered the front sight about .03-inch to raise the point of impact. Next, all of the action's moving components were carefully matched, altered and polished. Factory springs were not altered or replaced. Thus both the double-and single-action pull weights remain as before. The trigger was polished and rounded to provide a smoother surface for the finger during double-action firing. The hammer was bobbed, but can be manually cocked by carefully rotating it rearward by means of the trigger until it

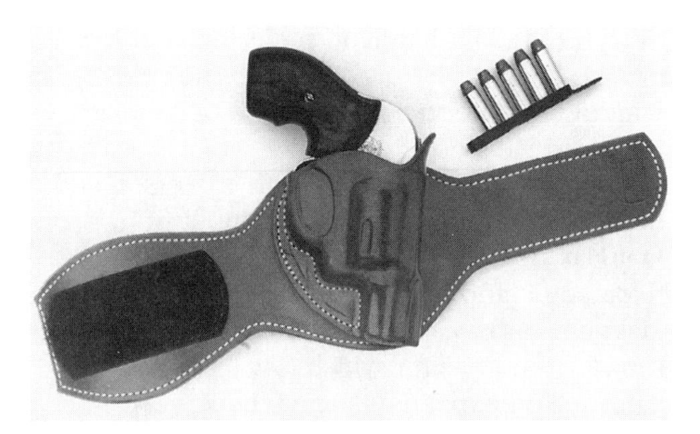

Milt Sparks' excellent Ankle Concealment Rig consists of a contoured leg strap, lined with sheepskin wool, and a holster hand-molded to the weapon it is meant to carry.

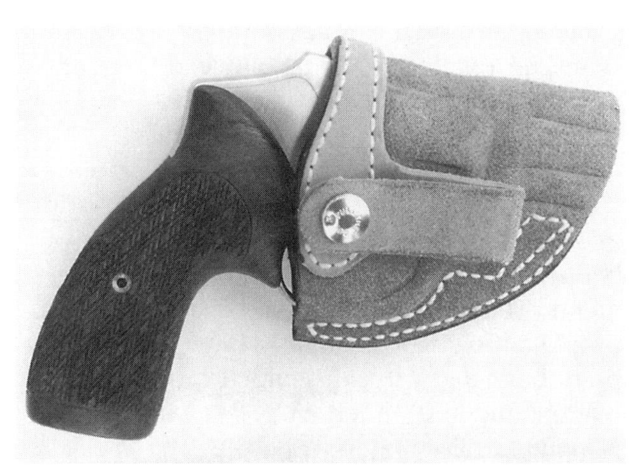

Bruce Nelson's well-known No. 10 Summer Special remains the hallmark among inside-the-pants holsters.

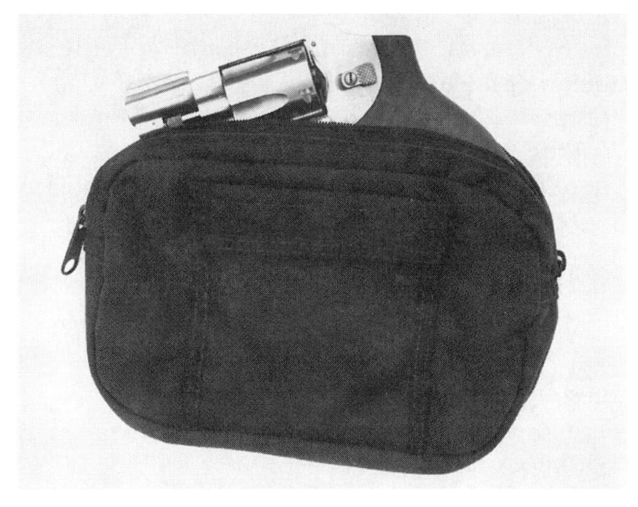

Gun Mate belt pack holster made by Michaels of Oregon is much smaller than most fanny-pack types and attaches directly to the belt.

can be grasped by the thumb of the firing hand. A double-action-only option is available.

The entire revolver was refinished by extra-fine glass-beading at a low pressure to produce a frosty, no-glare surface. Some of the pin heads were polished to break up the dull finish. Price of the package is $225, and S&W revolvers can be modified in the same manner. The package is not set in concrete and modifications can be added or deleted to suit the customer's specifications. I am very satisfied with this Weigand-modified SP101 and carry it frequently.

None of these revolvers, as issued or modified, respond to the problem of acquiring the target in subdued-light environments. No one installs better self-luminous sights on revolvers than Innovative Weaponry Inc. Their "BAR-DOT" night sights consist of a single green dot installed on the front sight blade and a horizontal green bar installed on the frame just under the rear sight notch. In the case of S&W Models 38 and 49 Bodyguard, IWI notches the hammer spur so the rear tritium sight bar is not blocked from view.

These tritium-phosphormiscrospheres are lightweight, insensitive to temperature variations and are always on. Decay of tritium H3 within the radioluminescent capsules creates soft beta rays that are converted to visible light when they strike the phosphor particles coating the inside of the glass miscrosphere. Tritium's half-life (the time required to reduce the illumination by 50%) is about 12.5 years, with a functional life exceeding 15 years. I know of no better method for delivering accurate fire at night and in low-light conditions than self-luminous sights, with the exception of cost-prohibitive passive night vision equipment—and those made by IWI are the very best, bar none.

BULLET ABUSE

The most important factor in this equation has yet to be mentioned. Nothing is more vital than the projectile your belly gun sends downrange. For all of its popularity, no cartridge has been subjected to greater abuse or more derision than the .38 Special. All of this tongue lashing would have been more justified had it been directed at the only bullet that was for so long the standard .38 Special loading with law enforcement agencies in the country.

The 158-grain .38 Special Lead Round Nose (LRN) bullet differs very little from that used in the .38 Smith & Wesson cartridge designed by Daniel Baird Wesson for the Model No. 2 Smith & Wesson single-action revolver, which was introduced in March of 1876.

Traveling at about 790 fps out of a 4-inch barrel, it does not expand, but will usually yaw 180 degrees in soft tissue and end its travel base-forward. Where the bullet yaw is at its maximum, 60 to 120 degrees, the height of the permanent cavity will increase to about 0.70 inch (the length of the projectile). Otherwise the wound track is no larger than the bullet's diameter (.357-8 inch). This load has a reputation for overpenetration and it will usually pass through about 21 inches of soft tissue.

That's not effective performance, but there is an excellent and well-proven alternative. Federal Cartridge Co. is one of several manufacturers who produce a version of the so-called "FBI load." Federal's load No. 38G is a 158-grain SemiWadCutter (SWC) Lead Hollow Point (LHP) loaded to +P velocity.

From a 4-inch barrel this ammunition will average approximately 880 fps. Average velocity from 2-inch barrels is 830 fps—only 50 fps less, still moving out and not enough of a drop to alter the round's wound ballistics potential. This bullet will expand to .59 caliber and penetrate about 12.5 inches of soft tissue. Both expansion and penetration are adequate and this load, which recoils sharply, can be recommended without serious reservations for those packing revolvers in this caliber.

We could load this bullet into a .357 Magnum case, which is one-tenth-inch longer than the .38 Special case, and drive it at considerably higher velocities. However, it's important to remember that if we drive a hollow-point bullet at too high a velocity, it will inevitably overexpand and reduce the penetration to an unacceptable depth. It may also fragment to an undesirable extent. There's a correct velocity for every expanding handgun projectile, and it is not the highest velocity possible. The velocity window for effective hollowpoint bullet performance is actually very narrow.

Introduced in 1935 by Smith & Wesson and developed by Winchester (with input from Douglas B. Wesson of S&W and renowned reloading authority, Philip B. Sharpe), the .357 Magnum cartridge was conceived as a hunting round, a purpose for which it will sometimes suffice. Non-expanding, heavy bullets driven at high velocity provide the deep penetration required to drop large and often dangerous game animals. In fact, all three of the original bullets introduced with this cartridge were 158-grain solids driven at an astounding 1,450 fps: a round nose, metal point; a solid lead SWC; and a conical, metal-piercing bullet. It was undoubtedly this latter projectile that Dick Tracy employed to crack engine blocks.

With the possible exception of comic strips, .357 Magnum revolvers should never have been applied to law enforcement environments. They are unnecessarily heavy and their propellant charges invariably drive hollow point bullets at velocities which cause them to overexpand and underpenetrate. With properly designed hollow point projectiles, .38 Special velocities are adequate. If you want to increase handgun performance against human targets, then you need to increase the diameter of the bullet, not its velocity.

Federal's bullet seems to be a bit softer than the other makes of 158-grain SWC LHPs. While this assures expansion, and is thus preferred loading, it will cause moderate leading after no more than a box of 50. Bore leading in handgun barrels is best removed by means of a Lewis Lead Remover kit which can be obtained from Brownells Inc.

I prefer the flat Bianchi Speed Strip for carrying an extra five rounds in the pocket, but these lead-alloy bullets deform easily and should be stowed, if possible, where they will not knock about.

Introduced in 1907, the .44 Special is an accurate and potentially powerful revolver cartridge which still remains quite popular, especially in the Smith & Wesson N-frame Models 24 and 624 series. The Bulldog can handle more bite than you can. But, remember you have to hit what you aim at; no one that I know of was ever killed by a loud noise.

I usually carry .44 Special loads developed by Black Hills Ammunition. They use a 240-grain SWC bullet cast from a hard lead alloy (92% lead, 6% antimony and 2% tin). Driven by a moderate charge of a fast-burning propellant (appropriate for short handgun barrels), this projectile will reach almost 700 fps in the 1.75-inch barrel of my modified Bulldog. This bullet does not expand, but it generates only mild recoil and penetrates deeply while making a .44-caliber hole.

Skeeter Skelton's favorite handloads for this caliber were the Lyman-Keith No. 429421 SWC bullet over either 7.5 grains of Hercules Unique or 17.5 grains of Hercules 2400. Both are stout loads, but the Bulldog can handle 7.5 grains of Unique, if the shooter can.

Safariland's synthetic holsters to a large extent look like leather, but wear like iron. Their inside-the-pants types include the Inside The Belt II (top) and the J Hook (bottom).

STUFF NOT INTO LOUSY LEATHER

Nothing will compromise a small revolver's concealment potential or presentation into the firing position more quickly than stuffing it into a poorly designed holster. Or worse yet—as some gunwriters have suggested—no holster at all. For example, some of these armchair commandos have stated that because the Centennial's hammer mechanism is concealed within the frame, it is ideally suited to—and was in fact designed for—carrying loose in a coat pocket and, if the situation required, firing from that location. The fact that some brain-dead couch potato may have actually pulled off such a stunt without managing to set himself on fire doesn't make it any less ludicrous. In the real world, combat handguns should be fired from the Weaver position with emphasis on the flash front-sight picture. For those who have been properly trained for such confrontations no further elaboration is necessary. Others should seek advice and training from instructors with the right credentials.

Small revolvers are best concealed by inside-the-pants or ankle holsters. In 1967, Bruce Nelson designed his justifiably well-known No. 10 Summer Special. By now, a much copied design (both acknowledged and otherwise), it corrected problems inherent at the time in inside-the-pants types. A metal-lined top band permits reholstering with ease. The belt retaining strap is secured by a one-way "pull the dot" snap. Fabricated from thin leather and precision hand molded (as are all Nelson's holsters), the Summer Special minimizes bulk inside the pants while its "rough out" texture allows the leather to adhere to clothing for position stability. This is normally a strong-side holster, but it can be fabricated with the retaining strap in the vertical position and thus the holster can, in this instance, be worn cross-draw also.

Safariland's synthetic holsters to a large extent look like leather, but wear like iron. I have experience with two of their inside-the-pants type holsters that I can recommend highly, the J Hook (catalog No. 027) and the Inside The Belt II (catalog No. 21).

My favorite ankle holster is made by Milt Sparks Holsters Inc., a small custom shop that has been highly regarded by such notables as Jeff Cooper, Ken Hackathorn and the late Elmer Keith and Mel Tappan. Milt Sparks' Ankle Concealment Rig consists of a firm, contoured leg strap to prevent shifting, which is lined with dense sheepskin wool that conforms to your ankle with use. The holster itself is meticulously hand-molded from top-quality cowhide to the weapon it is meant to carry to provide excellent retention without the use of safety straps. After 23 years, and now under the direction of Tony Kanaley, they are still doing it right. I use their ankle holster several times a week and it is the most comfortable I have ever worn.

Fanny pack holsters are now so ubiquitous that they actually focus attention on that fact that you are packing a rod. I have recently been experimenting with the Gun Mate belt pack holster made by Michaels of Oregon. Much smaller than a fanny pack, it attaches to your belt and can be worn either strongside or in the cross-draw position. The handgun is held in a rear compartment sealed by a Velcro-type flap. The front compartment can be used for a wallet, handcuffs or other items. Both compartments are sealed by a double-pull zipper. A small front pocket with a Velcro-type flap will hold keys and other small items. Don't expect a "par" time drawstroke from this rig, but it provides excellent concealment.

These are the belly guns and the associated equipment most often carried by armed professionals. Don't expect me to select one of each for you. That's up to you. To one degree or another I have used everything mentioned. I've discussed nothing that is without merit. The losers have been left out of this article. You can read about how great they are in the popular gun press.

Originally appeared in *Soldier Of Fortune*
May 1994.

Section 9
Shotguns

Benelli's Blaster

Modification Improves Revolutionary Shotgun

Bob Hall—wearing Swedish Fjallraven jacket—test-fires the formidable Benelli Model 121 M1 Military/Police 12-gauge shotgun.

Though extremely reliable, Benelli is most famed for speed: Spent shell in foreground is three inches in front of half-chambered next cartridge.

Allied in a losing cause during World War II, the Italians and Germans have united once more to market an absolute winner in the firearms field. Manufactured by Benelli Armi of Urbino, Italy, and imported into the United States exclusively by Heckler & Koch, Inc., the Benelli Model 121 M1, 12-gauge Military/Police autoloading shotgun deserves serious consideration by all those contemplating the purchase of a fighting shotgun.

Semiautomatic shotguns first appeared in quantity when Fabrique Nationale introduced John Browning's design in 1903. In method of operation most autoloading shotguns have taken one of two forms, either recoil or gas.

Browning's shotgun functioned using the principle of long recoil. When the gun was fired, the barrel recoiled rearward a distance slightly greater than the length of the shotshell. During this movement, the barrel and bolt remained locked together. At the end of the rearward travel, the bolt unlocked and was held back while the barrel returned to battery.

Extraction and ejection occurred during this time frame. The bolt then moved forward to chamber a new round.

Other guns of this type have utilized short recoil. Ammunition-sensitive, but fairly reliable, recoil-operated shotguns are noted for whaling the hell out of the shooter. Felt recoil is severe and they have given way to gas-operated shotguns.

In fact, America's best-selling autoloading shotgun is the gas-actuated Remington Model 1100. Pleasant to shoot, the 1100 is a perennial favorite at IPSC matches. In this system, gas is tapped from ports in the barrel and used to operate the weapon's ejection and feeding mechanism. However, this method of operation is not only ammunition-sensitive, but subject to fouling which can seriously compromise reliability under adverse conditions.

Thus, none of these autoloading methods has ever met military standards for reliability. As a consequence, law enforcement agencies and the military services have, for good reason, clung to the slide-action shotgun with the same stubborn intransigence American police have shown for the ancient, but reliable, wheelgun.

Enter the Benelli. It is most certainly not gas-operated. Nor does it make use of the usual recoil arrangement since the barrel is fixed. Instead, the bolt head is separated from the bolt body by a heavy six-coil accumulator spring. When the gun is fired, the bolt's inertia gives it forward movement, relative to the recoiling bolt-head and receiver, which compresses the spring by about four millimeters. The spring's return throws the bolt rearward, camming a locking bar out of engagement with a hole in the left side of the receiver. The spring is designed to delay opening of the action until the shot charges have left the barrel and pressures have dropped to a safe level, and also to assure functioning with shotshells of varying power.

After unlocking, the bolt assembly travels to the rear, extracting and ejecting the spent case. The recoil spring, housed in the buttstock, then drives the actuator rod (permanently hinged to the rear of the bolt body) forward and the weapon is reloaded in the conventional manner. All of the above is best described as an "inertia locking system."

Using the forward energy of a firearm's moving parts to pre-absorb a portion of the recoil energy and power the extraction/self-loading cycle is not new. The inertia locking system was the heart of the Robinson Constant Reaction machine pistol, designed and developed in Australia by Russel S. Robinson during WWII.

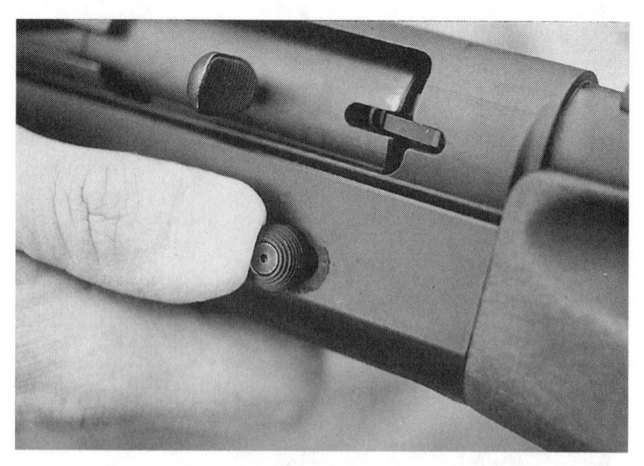

Newly-designed and installed push-button cartridge-release overcomes only military/police objection to Model 121.

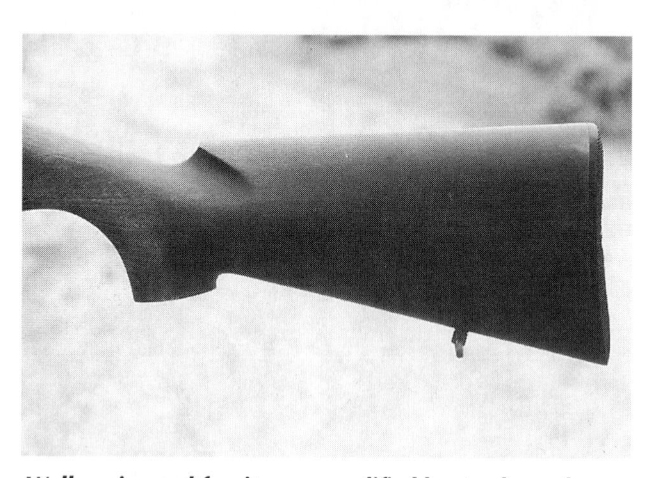

Well-engineered furniture exemplified by sturdy, sculptured pistol-grip.

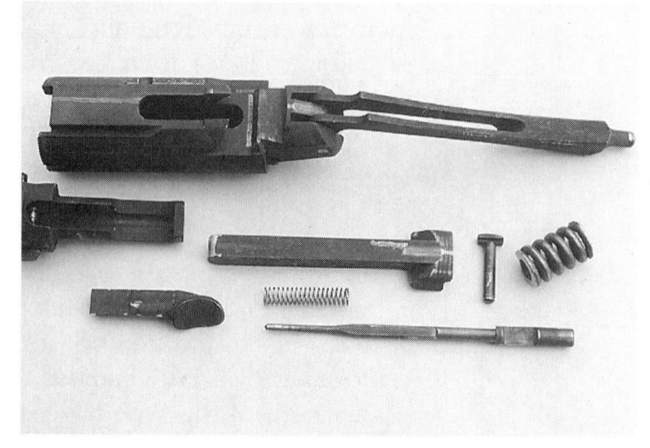

With field-use in mind, Benelli designed for quick, easy disassembly. Rugged eight-piece bolt group ready for cleaning.

Now this same inertia locking principle is also applied to the feed mechanism of the Benelli. As the gun recoils, the cartridge retaining lever, which is mounted in the receiver so that it can slide, tends to maintain its position due to inertia, so that it moves forward in relation to the rest of the weapon. At the same time a spring between the lever and the receiver rotates the lever clockwise on its pivot pin, releasing a round from the tubular magazine onto the lifter, which then moves upward into position for chambering the round. As the cartridge emerges from the magazine it causes the retaining lever to rotate counterclockwise, preventing a second cartridge from leaving the magazine.

Slick? Yes. Simple? Yes. Perfect? No. As the recoil energy alone is used to operate the cartridge retaining lever, we cannot charge an empty chamber by manually stroking the bolt's retracting handle as with the Remington 1100. A round must be inserted into the chamber by hand upon initial loading. Of no consequence throughout the rest of the world, where men of good sense carry the Benelli with a round in the chamber and the safety on, it is a monumental obstacle to American police who must commonly carry shotguns with the chamber empty.

Realizing this characteristic was seriously jeopardizing law enforcement sales in the United States, Heckler & Koch's solution, in contraposition to their ethnic inclinations, is simple and inexpensive.

Only $7.75 will get you H&K part No. 80950. Called the "Quick Release Carrier Latch Extension Button," this oversize, ribbed knob is easily installed by drilling and tapping the gun's carrier-release button to accept a 4-40 screw. After installation, the procedure for use is as follows:

With the magazine loaded and the bolt closed on an empty chamber, slap the button sharply forward and outboard with the thumb of the right hand. This will momentarily trip the cartridge retaining lever and a round will pop out of the magazine onto the lifter. Then retract the bolt and let it move forward, chambering the round.

It works. Maybe not as fast as simply jacking the retracting handle on the Remington 1100, but damn close. And it effectively counters the only major objection to the Benelli's use by U.S. law enforcement agencies. Eventually, all Benellis imported into the States will feature this device as standard-issue. Until then you must make the retro-fit yourself.

The Model 121 M1 Military/Police Benelli is finished for serious social purposes, not for display in a gun cabinet. The lower receiver, fore-end cap and trigger guard are fabricated from black anodized aluminum alloy. The lower receiver is black-enameled. After it starts to chip, paint stripper will remove the remainder, leaving only the matte-black anodizing. The beechwood handguard and buttstock have been walnut-stained to look about as elegant as a bookshelf made from pine orange crates.

But the modified-beavertail forearm and pistol-grip buttstock show thoughtful human engineering. The thin, checkered butt-plate is black plastic. The barrel, magazine tube, upper receiver and all other steel parts have been phosphate-finished (Parkerized).

The 20-inch barrel is choked cylinder bore, which means no constriction at the muzzle end whatsoever. The bore and chamber are chrome-plated. The magazine holds seven shells. Barrel and magazine are held together at the muzzle end by a detachable clamp intended to provide support for the long, extended magazine tube.

After it has been removed during disassembly several times, the retaining screw will strip the threads of the sheet-metal clamp. Just pitch it in a ditch and order one of Garth Choate's sling swivel base-clamp kits for $5. It not only solves this minor irritation but comes with a set of heavy-duty non-detachable 1.25-inch swivels to replace the puny .75-inch European-type swivels the Benelli already has. While you're at it, Choate's butt swivel can be mounted on top of the butt-stock, if you prefer the increasingly popular submachine-gun-style carry for the scattergun.

The common push-button safety is located in the usual place, at the rear of the trigger guard, but it offers small purchase for the fingers.

The front sight is a hefty Patridge type, silver-soldered to the barrel. Peculiar and offensive-looking at first glance, the rear sight is really quite effective. Its strange, interrupted-half-moon shape is snag-proof and significantly speeds sight alignment and target acquisition.

With no gas system and only eight components in the bolt unit, cleaning the Benelli is simplicity itself. To disassemble unscrew the fore-end cap and remove the handguard. Pull the barrel group forward and away from the lower receiver and magazine. Tug the bolt assembly back as far as it will go and lift out the firing pin stop. The firing pin and spring can now be freely withdrawn. Lift out the retracting handle and pull the entire breech block out the rear of the upper receiver. Take out the bolt's locking bar and separate the bolt head and accumulator spring from the bolt body.

To reassemble just repeat the above in the reverse order. The trigger group can be removed by drifting out its retaining roll-pin. I do not advise this since the trigger mechanism is sufficiently exposed for cleaning and roll-pins are usually difficult to remove without breakage.

The Model 121 M1 weighs seven pounds, three ounces. Felt recoil is moderate. Those accustomed to gas-operated guns, such as the Remington 1100, may find it to be severe. Those used to firing barrel-recoiling shotguns like the Remington Model 11 or Browning A-5 will consider felt recoil to be mild.

No other shotgun comes close to the Benelli in speed of mechanical action. I have watched John Satterwhite fire this gun at cyclic rates approaching that of the M3 submachine gun (450 rpm). Recovery time between shots is equally rapid and multiple targets can be engaged with ease at ranges out to 75 yards.

Accuracy with the buckshot and rifled slug loads, for which cylinder bore is intended, is excellent. However, the accuracy potential of both factory-loaded and cast slugs from Lyman molds of recent vintage is not what it was several decades ago. Stung by industry-wide product-liability litigation, manufacturers have reduced slug diameters, without notice, for fear they might be fired through full-choke barrels.

Arguments still rage between proponents of Nos. 00, 0, 1 and 4 buckshot (9, 12, 16 and 27 pellets, respectively, in 2.75-inch shells). There are convincing arguments for the use of each, but the U.S. military forces have generally stayed with No. 00. For indoor defensive purposes, I prefer the BB-size pellets. All buckshot sizes can penetrate two or more layers of sheet rock and result in injury to the very family members you are attempting to protect.

I've fired several thousand rounds through my Benelli and have experienced no failures of any kind with good quality factory and reloaded ammunition. The Model 121 M1 should come very close to passing the most rigid military specifications for reliability.

Shotguns reached their first prominence in the U.S. military during WW I. Gen. John J. Pershing, commander of the American Expeditionary Forces, had experience with the Winchester Model 1897 shotgun on the Mexican border just before the war. He anticipated its application to trench warfare and almost 30,000 shotguns were obtained for American use in France. The official trench gun, Model 1917, was

Benelli Model 121 MI Military/Police Shotgun Specifications

GAUGE . . . . .12 gauge, 2 3/4-inch chamber.

OPERATION . . . . .Inertia locking; semiautomatic.

WEIGHT, empty . . . . .7 pounds, 3 ounces.

BARREL LENGTH . . . . .20 inches.

CHOKE . . . . .Cylinder bore.

LENGTH OF PULL . . . . . . .Approximately 14 inches.

MAGAZINE . . . . .Tubular, 7-shell capacity.

SIGHTS . . . . .Front: Partridge-type; rear: fixed, interrupted half-moon.

FINISH . . . . .Phosphate and black anodized.

EXCLUSIVE IMPORTER . . . . . . .Heckler & Koch, Inc., 21480 Pacific Blvd,. Sterling, VA 20166-8903.

a 12-gauge Winchester Model 1897 riot gun with a 20-inch cylinder bore barrel equipped with the Springfield Armory-designed ventilated-sleeve bayonet adapter.

It should be made clear that Article 23 of the annex to the Hague Convention No. IV of 18 October 1907 most certainly does not prohibit the use of shotguns or shotgun ammunition in warfare.

During WWII, the U.S. Marine Corps made extensive and effective use of shotguns in the jungle campaigns of the Pacific islands. The terrain and tactics peculiar to the European and North African campaigns de-emphasized the shotgun's role and there were few instances of their use in these theaters.

After WWII, the British brought the shotgun to its zenith in the fierce jungle fighting of the war in Malaya. Detailed studies conducted during this conflict indicated that out to 75 yards, shotgun hit probability was superior to that of all other small arms. The studies further demonstrated that, although at that time less reliable, the autoloading shotgun produced more hits during brief engagements than slide-action guns by virtue of its higher rate of fire.

Both the U.S. Army and Marine Corps fought with shotguns under all tactical environments during the Vietnam War. Wherever used consistently, higher body counts were reported. The chances of a point man's survival were usually enhanced when he could respond to enemy targets with the dense pattern of projectiles the shotgun offered.

In an effort to define the shotgun's future role with the U.S. military services, a Mission Essential Need Statement has been formulated and assigned to the small-arms engineering team at the Naval Weapons Support Center as the JSSAP (Joint Services Small Arms Program) Multipurpose Individual Weapon (MPIW) program. Burst-fire weapons functioning with box or drum magazines are under evaluation.

Heckler & Koch itself is committed to its own combat-shotgun program and a prototype externally resembling their G11 caseless cartridge weapon has recently been displayed.

But the Benelli Model 121 M1 is with us, here and now. Until reliable weapons designed from the ground up as combat shotguns become a reality and can pass rigorous military test procedures (a prospect still years in the future), the Benelli is the very best fighting shotgun available. It merits your close examination.

Originally appeared in *Soldier Of Fortune*
August 1983.

Ithaca MAG-10

Big-Bore Blasts Birds But Not Bad Guys

Ithaca's big 10-gauge delivers more than enough lead, but weight, bulk and magazine limitations leave it stuck in the sporter category.

Roadblocker: Visions of state police manning highway barricades, waiting to blow speeding desperadoes off the road with a single shell from Ithaca's mighty 10-gauge combat shotgun. Is it the ultimate alley cleaner? Is twice the charge twice as effective? Awesome power or needless overkill?

Ten-gauge ammo is the only one of the old, big shotgun gauges still produced in the United States. Both four- and eight-gauge shells can be found now only on cartridge collectors' tables at local gun shows. (Gauge, by the way, is defined as the number of lead balls of bore diameter required to make one pound. Thus, in 10-gauge each ball would weigh 1/10 of a pound.) Ten-gauge magnum shotguns, chambered for the 3 1/2-inch shell, are widely touted as 100-yard waterfowl guns. But, in reality, the absolute maximum effective range is no more than 75 yards. This is not much more than the capability of the 12-gauge three-inch magnum load.

It was as a goose gun that Ithaca Gun Company, Inc. introduced the MAG 10 shotgun in 1975 with a 32-inch barrel. It remains the world's only 10-gauge magnum autoloading shotgun. In 1980 a police version of this gun, called the Roadblocker, appeared with little fanfare. Five years later there are still few who know of its existence.

The Roadblocker sports a 22-inch barrel which is cylinder bored (in theory, no constriction). Barrel and receiver have a dull phosphate finish and the machined 0.050-inch-high, 0.375-inch-wide flat rib machined into the receiver is tipped by a gold (actually brass) front head sight. This version carries a suggested retail price of $707. Another model, sporting a ventilated rib, will set you back $736. The walnut buttstock and forearm are uncheckered and feature a practical non-reflective oil finish. Gratefully, a 3/4-inch rubber recoil pad is standard. The metal to wood fit on the gun submitted to *SOF* for test and evaluation was only fair. The forearm would not stop rattling about no matter how snugly we cinched down the forearm retaining nut. Overall length is 43.5 inches: That's four to eight inches longer than most 12-gauge military/police riot guns. Even worse, Ithaca's 10-bore burdens the scales at 10 3/4 pounds.

Gas operated, the MAG 10 employs a system vaguely reminiscent of the Czech Vz52 Rifle. Upon firing, some of the propellant gases move through a barrel vent just 7 1/2 inches from the chamber into a stainless-steel stationary "piston" attached to the barrel's underside in front of the tubular magazine. A stainless-steel sleeve surrounding the piston serves as a cup to trap gas escaping from a port in the piston. Expanding gases force the sleeve rearward 3/4-inch to drive the slide extension tube, surrounding the magazine, against the bolt carrier by means of two steel rods brazed to the end of the steel tube. As the bolt carrier travels to the rear it pulls the bolt down, drawing the single locking lug on top of the bolt out of the slot in the barrel extension.

As you might expect from a weapon subjected to big-bore stresses, the barrel extension is machined into the barrel billet. And the barrel itself is hammered to shape by Ithaca's Roto-Forge process. If nothing else, the Roadblocker is *strong*. On top of that, triple reaming and careful hand-lapping polish the barrel's smooth interior surface to a mirror finish.

A five-inch-long link rod connects the bolt carrier to the recoil spring's guide rod. Spring and guide rod ride in a steel tube within the buttstock.

Solid stainless-steel fabrication helps protect the gas system and other parts prone to corrosion and wear. Roadblocker's bolt body, shell lifter, trigger, retracting handle and bolt release are also stainless. All are investment cast and show little machine finishing.

An aluminum casting forms the trigger housing, which is finished with black paint. This would have been acceptable if the paint hadn't dissolved in perchlorate ethylene during cleaning, leaving a bright, shiny, alloy trigger guard. If we can't expect a steel trigger guard on a $700 shotgun, the least we demand is proper anodizing or a baked enamel finish.

Trigger pull weight was only 3 3/4 pounds, but scratchy. Burrs on the trigger mechanism were removed by gunsmith Burke C. Hill Jr. Again, why should we encounter this on such an expensive firearm? The trigger housing is held to the receiver by two pins, which are easy to remove, but infuriating to re-install. The receiver has been milled from solid bar stock.

Behind the trigger in the housing casting, the conventional cross-bolt safety sits in its usual position. It's big enough to hit under stress . . . no problem here. In the safe position the cross-bolt blocks trigger movement by engaging a large spur cast into the trigger's rear surface.

The Roadblocker's shell lifter has been slotted, a now common feature on fighting shotguns. This is supposed to assist clearing a stoppage involving a shell wedged above the lifter that has failed to chamber. Nevertheless, on three occasions during our test, when shells jammed above the lifter were pried back into the magazine, the shell stop would not retain them: a potentially fatal shortcoming.

The location of the gas block just 7 1/2 inches from the chamber limits the magazine capacity to just two 3 1/2-inch 10-gauge shells. Geese do not shoot back. Law enforcement targets aren't so docile. Here is another example of a weapon designed for sport, then chopped, stripped and improperly marketed as a police weapon.

Ithaca's big gun has another problem for police use. Many police departments will not permit shotguns to be carried in the vehicle with a shell in the chamber. So, what does this mean when you've got a two-shot magazine? You guessed it: a two-round shotgun that's half as fast as a double-barrel. We end up with an autoloading shotgun—and all the compromises in reliability inherent in this method of operation—with a quarter the shell capacity of many riot guns. I cannot sign off on this defect and it'll have to be addressed before the Roadblocker merits serious consideration for law enforcement applications.

In an effort to partly correct this deficiency, I turned to the grand master of custom leather, Milt Sparks. Milt has a lace-on emergency ammo carrier for shotguns, designed by the late survival expert and *SOF* staffer, Mel Tappan. Of course, you can buy cheap elastic shell holders that fit over the buttstock. They're made in Mexico and Hong Kong, they'll stretch to fit any stock and then they'll keep on stretching until they fall off. Milt's handcrafted leather shell holder, euphemistically called the "Cold Comfort," totes seven shells in 12-gauge. It's made to last and exudes the quality that professional pistoleros have come to expect from Sparks.

After some consultation, Milt decided to limit the cheek piece for the Roadblocker to only five shells. Once attached, it never moved under the Roadblocker's pounding recoil. There are no rivets to mar the stock and the lacing eyelets are padded by a soft-leather gusset. Attractive and durable, Sparks' shell holder provides a comfortable interface between the cheekbone and walnut stock. Priced at $32 in natural finish, cordovan or black will cost you only $5 more. Milt offers immediate delivery on Cold Comfort shell holders (#C/C) in 12-gauge. He also makes some of the best no-nonsense holsters for serious social purposes. I use Milt's cross-draw holster (#55BN) and 1 3/4-inch lined belt to carry my Randall Curtis LeMay on the streets of San Salvador.

Now, by the time we've stuffed three shells into the Roadblocker and attached Sparks' shell holder we're up to 12 pounds, five ounces. No problem for Arnold Schwarzenegger, but too much for me. I need a sling to lug it around the range. Thank God, the Roadblocker comes ready to accept Uncle Mike's quick-detachable sling swivels (#1061, Model QD 115 MAG 10). To this I added their 48-inch padded carrying sling with one-inch straps. Similar to the excellent black nylon padded sling on the M60 GPMG, this rig will ease the burden of those poor devils forced to drag this monster more than 10 feet from the squad car. Both these items are available from Michaels of Oregon.

The Roadblocker can be loaded in one of two ways. In either case, engage the safety first (push the button to the left). Pull back the retracting handle to open the action. Drop a shell in the chamber. Depress the bolt/carrier-release lever below the ejection port on the right side of the receiver to close the action. Hold down on the bolt/carrier-release lever. This will permit the shell carrier to swing upward. Insert two shells into the magazine past the shell stop. If you must carry the shotgun with an empty chamber, merely jacking the cocking handle will chamber a round. To unload you must work the shells through the action (with the safety on) manually, depressing the carrier-release lever with each cycle.

Disassembly procedures are forthright and simple. Unload the weapon. Pull back the retracting handle to lock open the action. Unscrew the forearm cap and slide out the walnut forearm. Pull out the barrel with its gas block. Remove the slide extension tube from

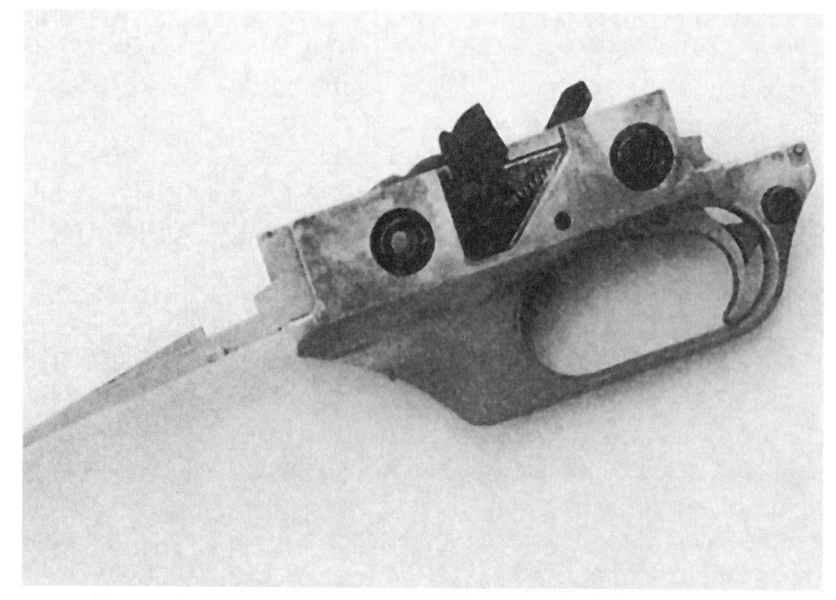

Roadblocker trigger housing group with black-painted finish removed after just one cleaning in solvent.

Ithaca MAG 10 Roadblocker Specifications

GAUGE . . . .10 gauge magnum, 3 1/2-inch chamber.

OPERATION . . . . .Gas. Stainless-steel stationary piston surrounded by a stainless-steel sleeve which drives a slide extension tube. Single lug which locks into a slot cut into the barrel extension. Slotted shell lifter.

FEED MECHANISM . . . .Two-shell tubular magazine.

WEIGHT, empty . . . . .10.75 pounds.

LENGTH, overall . . . . .43.5 inches.

BARREL LENGTH .22 inches.

SIGHTS . . . . .Bead front with short rib milled into receiver or ventilated rib entire length of barrel.

SAFETY . . . . .Cross-bolt on trigger guard.

FINISH . . . . .Phosphate and unfinished stainless steel.

PRICE . . . . .$707, standard. $736 with ventilated rib.

MANUFACTURER . . . . . .Ithaca Gun Company, Inc., 123 Lake Street, Ithaca, NY 14850.

around the magazine. To remove the trigger housing group, first allow the bolt to move completely forward, under control. Knock out the two retaining pins and remove the trigger housing group by pulling downward on the trigger guard. To disassemble the gas system, remove the circular retaining spring from the rear of the pistol and slide off the cupped sleeve. No further disassembly is suggested.

Ten-gauge ammunition can cause more than the usual amount of fouling. Use stainless-steel armorers' brushes and small scrapers (carefully) to remove the carbon residue. Shotgun bores are best cleaned by "Tornado"-type brushes—a spiraled design of looped non-corrosive wire. Unfortunately, no one seems to make one in 10-gauge so you'll have to use the bronze brush supplied by Ithaca. Do not lubricate the gas system. Re-assemble in the reverse order.

Four man-stopping loads are available off-the-shelf to those who wish to do their killing with a 10-gauge magnum. None of it is cheap. It ranges in retail price from $1.15 (00 Buck) to $1.42 (slug) per shell. Tapping the trigger on the Roadblocker is every bit as expensive as shooting a .50-caliber Browning machine gun. What kind of buck do we get for our buck?

Federal's Super Slug is a hollowpoint .78-cal. projectile weighing in at 765 grains (12-gauge slugs weigh about 440 grains). This is 57 grains heavier than the .50-cal. M2 Armor Piercing bullet. But the Federal soft lead-alloy slug comes out of a smoothbore barrel at an average muzzle velocity of only 1,240 fps. The M2 boat-tailed bullet has a hard-steel core, full metal jacket and roars out of Ma Deuce's mouth at 2,900 fps. They are hardly comparable, either in penetration or long-range potential. The best accuracy we obtained with the Federal slug was about 10 MOA, acceptable when the target is no greater than 50 yards away.

Ithaca Roadblocker disassembled.

As to penetration, certainly it's all you need on a human target, but questionable on a vehicle (it is called the "Roadblocker," isn't it?). Shooting at an abandoned car, one slug entered through the trunk door, passed through the rear and front seats and embedded itself in the dash. Another went through one door, but failed to make it through the other. And what does all this prove? Not as much as some gun-writing hacks would lead you to believe. An automobile body is the very worst kind of test medium for penetration tests. It's not homogeneous and thus the results are not repeatable. Who knows what support structures brace the body's skin at any given point? Based on their junkyard tests some sages will lead you to believe the M16's M193 ball projectile is worthless against a car body. Yet, I've seen cars and trucks in El Salvador with hundreds of .223 entry and exit holes, at every angle, through every portion of the body and windows.

Federal also offers a 00 Buck shell with 18 .33-inch pellets in a granulated filler (twice the number found in the usual 12-gauge military/police shell). At 15 yards 14 of these nasty pills will consistently strike your opponent in the head and chest. Instant mincemeat. At 25 yards we'll get 10 hits just about every time. Termination with extreme prejudice. But, by 35 yards three pellets is the best we can hope for. It's no longer certain he'll drop immediately. The 10-gauge shotgun is not a rifle.

Winchester provides a Super X loading of No. 4 Buck (.240-inch diameter) with 54 pellets. It's impressive. At 15 yards about 50 pellets will strike the cranial and chest area. Scrambled brains. When fired at a range of 25 yards almost 35 pellets will hit the lethal zone. At 35 yards about half the load will strike home to cause a massive migraine.

Those who wish to preserve the sanctity of their homes with an Ithaca Roadblocker are well-advised to employ Winchester's Super Double X load of BB shot. There are 125 .180-inch-diameter pellets in each shell. Buck shot will usually penetrate through two layers of sheet rock. BBs will not. There's little to be gained in killing your family while protecting them.

The Roadblocker's felt recoil falls somewhere between a Beeman R1 air rifle and firing a 100-round belt through an HK21 GPMG, depending on how many rounds you fire in succession, The weight of the gun and the well-designed recoil pad are mitigating factors, to be sure. Anyone used to firing a high-power rifle should be able to handle 10 or 15 rounds through the Roadblocker without problem. However, two of us fired more than 200 rounds through the MAG 10 during one session. After 30 rounds, shooter fatigue begins to slide upward in an asymptotic curve.

Recoil is only a minor consideration. The eight stoppages encountered were not. There were two failures to eject and six failures to feed. Three of the feed stoppages were cleared only with great difficulty. This just won't do

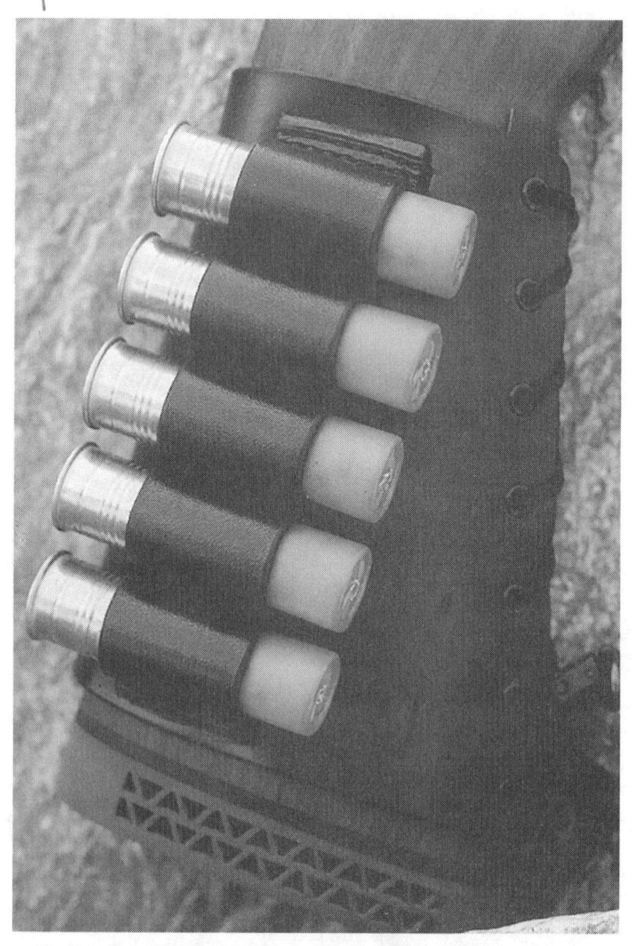

Milt Sparks' Cold Comfort leather shell holder isn't a luxury for a three-shot combat shotgun. It's a necessity.

when engaging targets that shoot back. The normal ejection pattern is six feet to the right.

In spite of its weight, the Rockblocker balances well when put to the shoulder. It's muzzle heavy and this assists in the acquisition of moving targets (unlike B21 silhouette targets, dirtballs usually refuse to stand still when you shoot at them). Out to 35 yards the Roadblocker always shot and patterned right to the point of aim.

Elmer Keith used to maintain that "A shotgun with buckshot is one of the most deadly gun fighting weapons up to 30 yards. Beyond that range it is problematical, and a good six-gun shot has the advantage." In my opinion, the Roadblocker extends that range by another 10 yards, at most. But, at what price? Is the tradeoff worthwhile?

Not to me. The gun and its ammunition are both too expensive. Even the most rabid power maniac would eventually groan under its 12-pound burden. It fails to meet the standards of reliability required of a combat weapon (as do almost all autoloading shotguns in 12-gauge as well). Its two-shell magazine is a most serious defect. At normal shotgun contact ranges 12-gauge is sufficient. There is only one degree of death and nine 00 Buck pellets in the body cavity will do nicely; eighteen just makes the county coroner's job more tedious.

Oh well, it's still a damn good goose gun.

Originally appeared in *Soldier Of Fortune*
November 1985.

LOVEABLE, YET LETHAL

Scattergun Technologies' "Politically Correct" Combat Shotgun

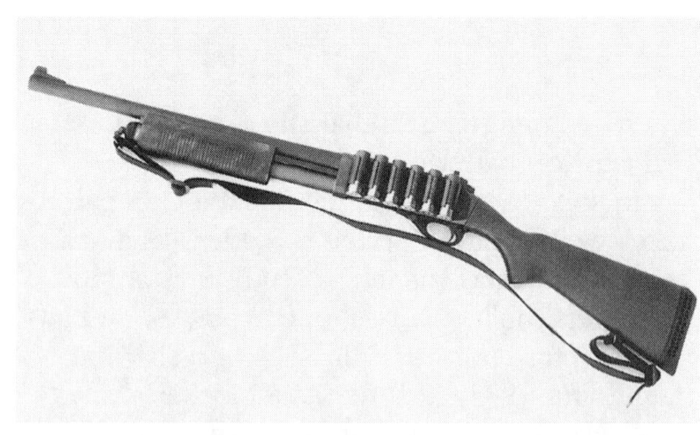

Reasonably priced, the Louis Awerbuck Signature Model starts out as a Remington Model 870 with a number of important added features, such as: "ghost ring" rear sight, front sight blade with vertical tritium bar, SideSaddle shotshell carrier, combat-style magazine follower and nylon sling with quick-detachable swivels.

Scattergun Technologies' Louis Awerbuck Signature Model shotgun is an assemblage of all the right components to provide an excellent self-defense smoothbore with legitimate applications for hunting as well.

Secretary of the Treasury, Lloyd Bentsen, at the bidding of his master, Slick Willie, has recently directed the ATF gun gestapo to declare three U.S.-produced smoothbore shotguns as "destructive devices" (i.e., the classification that includes mortars, M79 and M203 40mm grenade launchers, poison gas, bombs, grenades, mines and rockets): the Striker, Street Sweeper and USAS12. It seems they are not "Particularly suitable for sporting purposes"—a shibboleth placed into the Gun Control Act of 1968 not by our enemies, but at the insistence of major U.S. firearms manufacturers who feared competition from the importation of cheap, surplus military rifles.

"Political correctness" is an essential ingredient of the left-of-center, knee-jerk liberalism that currently controls our government. While most liberals hate most firearms, guns with a military appearance are especially terrifying to those who fear an armed citizenry. All liberals instinctively know, without recourse to reality, that drug lords decorate their weapons with bayonet lugs, large capacity magazines, pistol grips, flash hiders and black plastic furniture, apparently to intimidate rival drug lords. It is currently politically correct to acknowledge such ridiculous concepts as dogma and respond by attempting to disarm honest gun owners.

Possibly in response to this nonsense, Scattergun Technologies Inc., together with Louis Awerbuck, a *Soldier Of Fortune* and *Fighting Firearms* contributing editor and Yavapai Firearms Academy director, have designed what could be euphemistically referred to as a "politically correct" combat shotgun. Scattergun Technologies builds production series practical shotguns and provides a wide range of what they refer to as "Tactical Response" shotguns to armed professionals and to those civilians concerned with providing themselves with a self-defense potential at the highest possible level. Awerbuck's book, *The Defensive Shotgun: Techniques & Tactics*, is generally regarded as the definitive work on this subject.

The so-called "Louis Awerbuck Signature Model" (Item No. 90110) starts out in life as a Remington 870 Police Magnum, a rather plain Jane shotgun with a phosphate ("Parkerized") finish and a totally unnecessary 3-inch Magnum chamber. All Scattergun Technologies' shotguns with 18- and 14-inch barrels are choked cylinder bore, except the Awerbuck Model, which features a modified choke barrel to more effectively utilize the broader range of pellet sizes used by civilians for hunting and self-defense. (The degree of choke, or constriction of the muzzle of a shotgun barrel, is measured by the percentage of pellets delivered to a 30-inch circle at a range of 40 yards. A cylinder bore usually delivers 25% to 35% of the shot into the circle, while a modified choke will put 45% to 55% of the pellets into the 30-inch circle.) Overall length of this model is 38.25 inches with a weight, empty, of 8.25 pounds. It has an 18-inch barrel.

The Model 870 was introduced in 1950 to replace the Model 31 and shares a high percentage of component commonality with the Model 11-48 that appeared in 1949. Side-ejecting and without an external hammer, the Model 870 has a tubular magazine and takedown, interchangeable barrels.

Its method of locked-breech operation has been taken from the John Browning-designed Model 11 semiautomatic shotgun. When in battery, the bolt remains locked to the barrel by means of a locking block within the bolt body, which engages a recess in the barrel extension. Receiver strength is not critical in this system as the recoil forces are mostly absorbed by the bolt assembly and the barrel extension. A steel slide, to which has been attached a stepped lug (also housed within the bolt body), is mated to twin action bars attached to the forearm. When the slide is

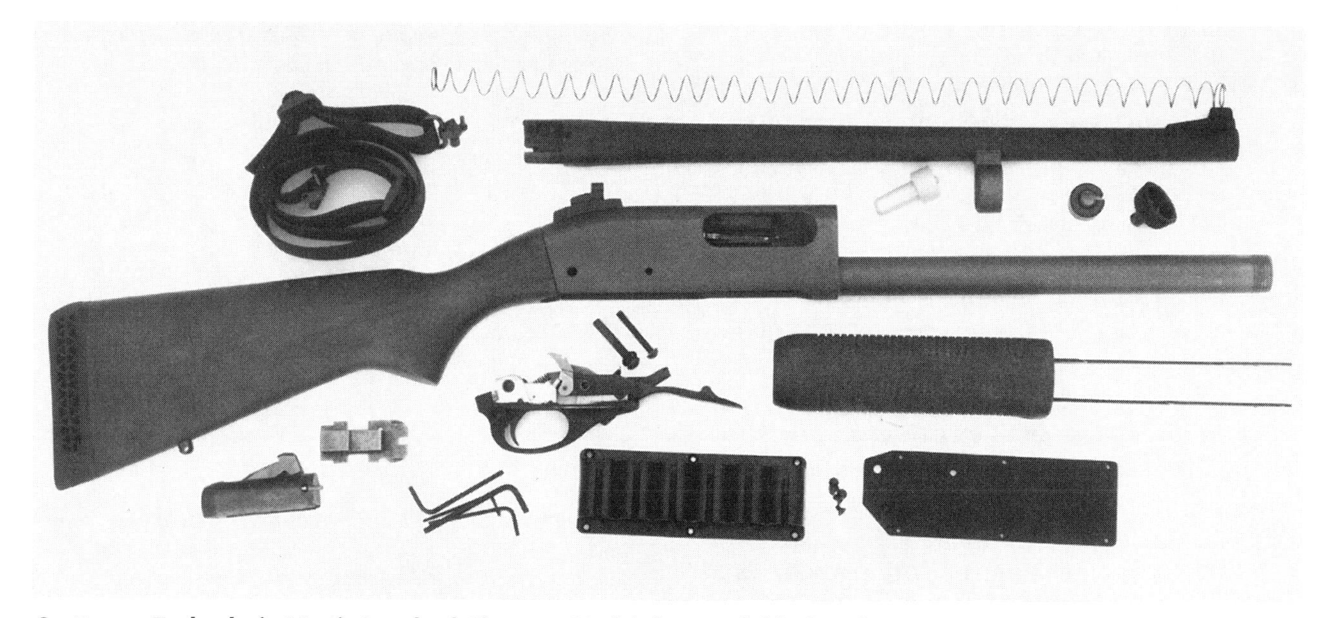

Scattergun Technologies' Louis Awerbuck Signature Model shotgun, fieldstripped.

racked forward, the front step on this lug moves under the front end of the locking block and pivots it upward to engage its forward projection in the recess on top of the barrel extension. After firing, when the forearm is manually driven rearward, the slide moves back through about a half-inch of free travel, after which the locking block drops downward to unlock the action. Completion of the rearward cycle by the operator will draw all of the reciprocating parts to the rear and eject the empty shell out the ejection port on the right side of the receiver.

The trigger mechanism's disconnector, which requires the trigger to be released between each shot, can be actuated both manually by depressing the action-bar lock button and by the fall of the hammer. There is a standard cross-bolt-type safety at the rear of the alloy trigger guard.

The factory wood (dense-grained birch) buttstock and forearm have been retained to lower the anxiety level of those frightened by "black guns." The recoil pad has a beveled top edge to facilitate speed mounts and inhibit snagging on clothing. To further minimize perceived recoil, a sealed aluminum tube filled with mercury and ball bearings has been installed in the buttstock. Length of pull, as measured from the center of the trigger to the end of the recoil pad, is about 14 inches.

One of the most important features of almost all Scattergun Technologies' fighting shotguns, including this one, are the "Trak-Lock" sights. The rear sight consists of a patented chrome-moly steel assembly with a counterbored "ghost ring" aperture, which provides a shadow-like effect with extremely fast sight alignment at close-range buckshot distances—without compromising the requirement for precision sighting with solid projectiles at longer ranges.

Adjustments for both elevation and windage are accomplished through the manipulation of a single screw (an allen wrench of the correct size is provided), which also locks the aperture to the sight base, mating and locking notches on the underside of the aperture body with a corresponding notched surface on the base. A

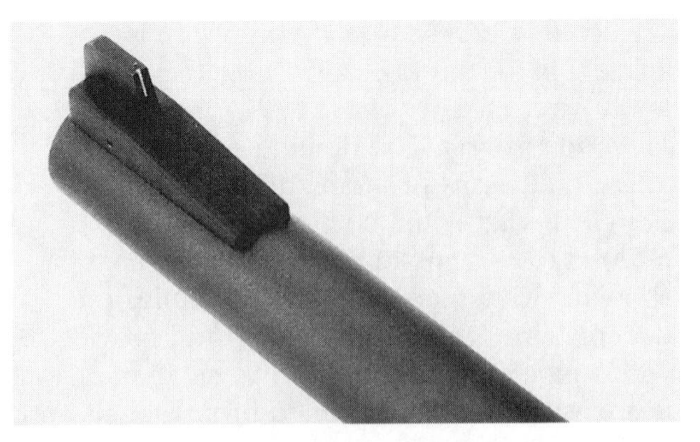

The exceptionally sturdy, high-profile front sight blade, which is cross-pinned in place, has a self-luminous, tritium vertical bar of Awerbuck's specifications.

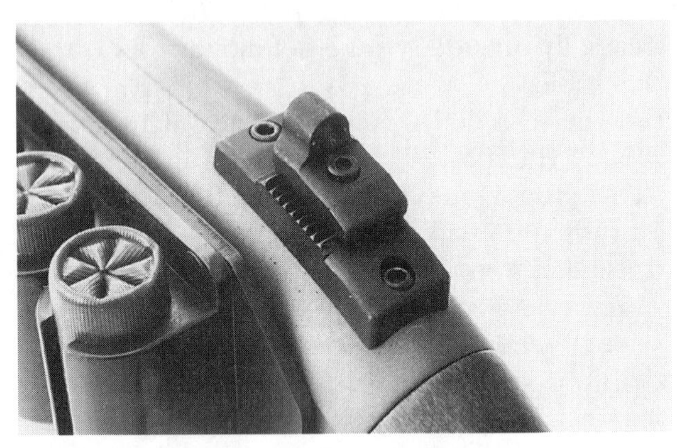

The rear sight consists of a patented chrome-moly steel assembly with a counterbored "ghost ring" aperture that provides a shadow-like effect with extremely fast sight alignment at close-range buckshot distances—without compromising the requirement for precision sighting with solid projectiles at longer distances.

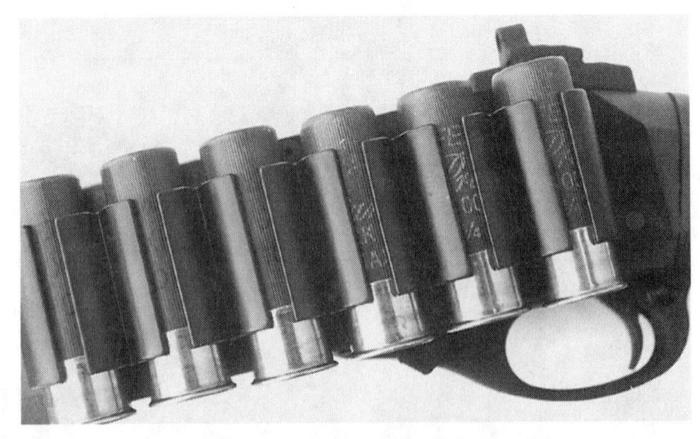

A SideSaddle shotshell carrier has been attached to the left side of the receiver. Manufactured from injection-molded black polymer, the SideSaddle holds six rounds and provides instant access to spare rounds or alternative types of ammunition, such as slugs.

tritium dot on each side of the aperture is standard on this model.

The exceptionally sturdy, high-profile front sight blade, which is cross-pinned in place, has a self-luminous, tritium vertical bar of Awerbuck's specifications and is installed (as are the rear-sight tritium dots) by IWI.

No extension has been added to the standard four-round magazine tube and thus with a round in the chamber, the total capacity is five rounds. This is sufficient for most civilian self-defense applications.

The magazine tube contains a fluorescent green combat-style follower with a projection on the head that instantly indicates by sight, or by touch, if the magazine tube is empty. Molded from non-rusting, high-impact ABS, the follower has a tapered tail to prevent the magazine spring from binding. The magazine spring, designed by W.C. Wolff Co., has been coated by electrostatic deposition with an epoxy-based, paint-like material that is only .050-inch in thickness. Used in the automotive industry as a base coating, this finish will purportedly withstand 600 continuous hours of salt spray.

As a supplement to the four-round magazine tube, a SideSaddle shotshell carrier has been attached to the left side of the receiver. Manufactured from injection-molded black polymer, the SideSaddle holds six rounds and provides instant access to spare rounds or alternative types of ammunition, such as slugs. Rounds should be loaded into this carrier from the bottom, bases down. To remove a shell from the carrier, simply push down on the top of the shell.

An excellent adjustable sling has also been installed. Fabricated from 1.25-inch milspec black nylon, it's designed for either standard over-the-shoulder carry, hands free over-the-chest or a backpack carry position. It attaches to steel, quick-detachable swivels at the buttstock end and on the magazine tube's front cap.

To facilitate manipulation of the cross-bolt-type safety mechanism, a jumbo-head safety designed by Scattergun Technologies has replaced the standard-issue safety. Manufactured from 4140 chrome-moly steel, it has undercut grooves to inhibit slipping. Trigger pull weight on *SOF*'s test specimen was a slightly creepy 3.75 pounds.

Louis Awerbuck Signature Model Shotgun Specifications

GAUGE . . . . .12 gauge.

OPERATION . . . . .Slide-action, locked-breech. Locking block engages a recess on top of the barrel extension. Twin action arms.

FEED MECHANISM . . . . .Four-round tubular magazine under the barrel.

WEIGHT, empty . . . . .8.25 pounds.

LENGTH, overall . . . . .38.25 inches.

BARREL . . . . .Smoothbore, modified choke.

BARREL LENGTH .18 inches.

FINISH . . . . .Phosphate ("Parkerized").

FURNITURE . . . . .Dense-grained birch buttstock and forearm. Sealed aluminum tube filled with mercury and ball bearings installed in buttstock. Rubber recoil pad with beveled top edge.

SIGHTS . . . ."Trak-Lock" rear sight with counterbored "ghost ring" aperture and tritium dots; adjustable for both elevation and windage. High-profile front sight blade with self-luminous, vertical tritium bar.

SPECIAL FEATURES . . . . .Six-round SideSaddle shotshell carrier attached to left side of receiver; combat-style magazine follower; magazine spring protected by electrostatic deposition of an epoxy-based, paint-like material.

MANUFACTURER . . . . .Scattergun Technologies Inc., P.O. Box 24517, Nashville, TN 37202.

PRICE . . . .$665.

T&E SUMMARY: An assembly of all the right components to provide an excellent self-defense smoothbore with legitimate applications for hunting as well. Reasonably priced. Highly recommended.

Workhorse loads for the 12-gauge combat shotgun are rifled slugs and No. 00 buckshot. Federal is my choice for both. Their so-called Hi-Shok\Hollow Point slug (product No. F127-RS), is a one ounce (437 grains, nominal), hollow-base, rifled lead-alloy slug with no more than an incipient dimple at the tip. Nevertheless, this .70-caliber projectile will penetrate up to 14 inches of soft tissue while expanding to 1.1 inches in diameter. The temporary cavity produced is significant and will add to tissue disruption. At 50 yards, this round produced incredible one-hole three-shot groups, measuring only 1 inch center-to-center, using the modified choke barrel of the Louis Awerbuck Signature Model shotgun sent to *SOF* for test and evaluation.

I also prefer Federal Premium No. 00 buckshot (product No. P154-00B), which features copper-plated shot granulated filler and a long-range shot cup. Select the nine-pellet load because the 12-pellet, 2.75-inch Magnum load just plain kicks too much. No. 00 pellets have a diameter of .33-inch. Federal Premium tower-dropped pellets are 97.5% pure lead with 2.5% antimony for added hardness. Two polishings guarantee sphericity. Copper-plating further increases resistance to deformation during firing. Shot is arranged in a spiral configuration within the long-range shot cup; granulated buffer is added to fill the gaps. The granulated buffer will eventually leak into the gun's action and chamber, so the mouth of these shells should be sealed with nail polish or clear lacquer.

Another excellent buckshot round is Federal's Tactical Load, a low-recoil police round with nine copper-plated pellets of No. 00 buckshot in granulated filler (product No. H132-00).

At 10 yards, Federal Premium nine-shot No. 00 buck impacted the target with all the pellets in an irregular 3.5-inch vertical oval. At 15 yards, all nine pellets would still impact into the chest area of the target. At 25 yards—in my opinion the outer limit of acceptable performance with buckshot—seven pellets would still impact into the torso area of a humanoid target most of the time.

Remember, every shotgun barrel, even those of the same make, choke, model and with sequential serial numbers, will throw a pattern different than any other and you must pattern your shotgun at varying distances with the exact loads you intend to employ.

The price of the Louis Awerbuck Signature Model is $665 and includes a copy of Awerbuck's book and the allen wrenches necessary for adjustment of the rear sight and disas-

FIELDSTRIPPING THE LA870

There are a few tricks involved in disassembly of the Model 870 series shotguns. First empty the magazine tube and clear the weapon. Set the cross-bolt to safe. Rack the action rearward slightly to disengage the locking block from the barrel extension. Remove the sling from the front swivel. Grasp the magazine tube's end cap and turn it counterclockwise to separate it from the magazine tube. Pry out the sheet-metal spring cap inside the magazine tube and remove the follower spring. Tilt the gun downward and the follower will drop out the end of the magazine tube.

Pull the barrel away from the receiver. Insert your finger through the underside of the receiver and depress the left shell stop. Move the bolt group and action arms out toward the front. Separate the slide and bolt group from the action arms after they are clear of the receiver. Remove the six small allen-head screws that hold the SideSaddle hotshell carrier to its mounting bracket. Then remove the two larger allen-head screws that hold the bracket to the receiver and replace the trigger mechanism's retaining pins.

Wrenches of the appropriate size are provided. This will permit you to separate the trigger housing assembly from the receiver body. No further disassembly is usually required.

After cleaning and lubrication, reassemble in the reverse order. The most difficult aspect of reassembly is installation of the bolt group and action arms since both the right and left shell stops must be depressed at the same time.

sembly of the gun. The receiver is marked under the loading gate with the "L. Awerbuck" signature and "LA 870" under the ejection port, instead of the usual "TR 870" (Tactical Response Model 870).

Scattergun Technologies, with insights from my friend, teacher and co-instructor Louis Awerbuck, has assembled all the right components to deliver an excellent self-defense smoothbore with legitimate applications for hunting as well.

Originally appeared in *Soldier Of Fortune*
July 1994.

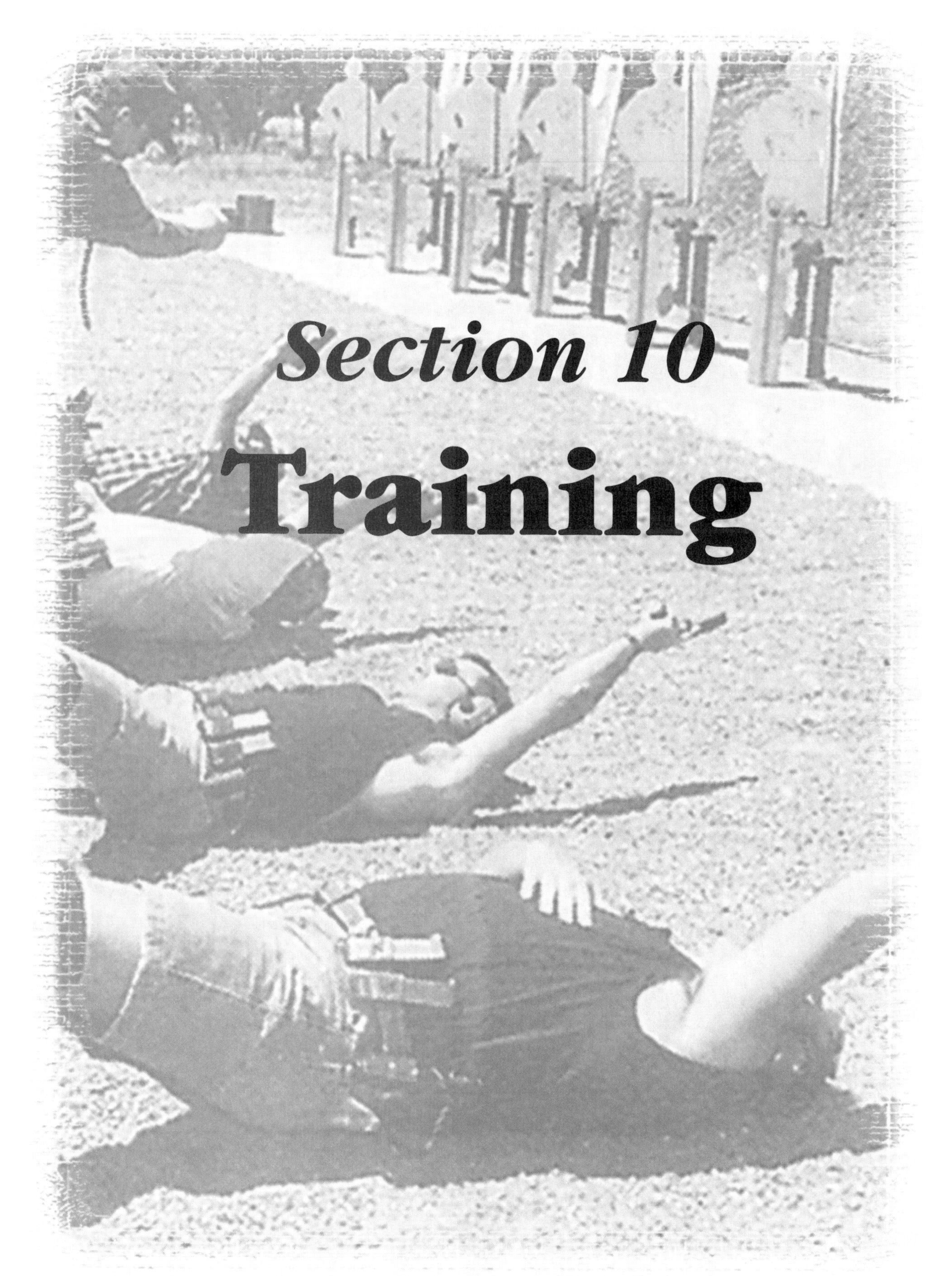

Section 10

Training

HITTS AND MYTHS

Awerbuck's Course on Cutting Edge of Sniping

Louis Awerbuck observes two-man sniper team moving through exercise involving stepped barrier during grueling HITT countersniper course.

Robar SR60 is just one step below the SR90 and is currently in use with the Los Angeles Police Department SWAT (Special Weapons And Tactics) Team.

Training is the key ingredient in a successful response to lethal aggression, whether that response is made by an individual, a law enforcement Special Reaction Team (SRT) or a military unit.

Unfortunately, most firearms training does not simulate reality. Generally, live-fire range training takes place at known distances with either static, two-dimensional paper targets or reactive steel plates facing straight on. On the streets or in the bush, live enemy targets bob and weave erratically and rarely present a frontal shot.

There is no more prominent exponent of realistic firearms training than Louis Awerbuck, currently owner and director of the Yavapai Firearms Academy (YFA), a mobile small arms training facility operating out of Prescott Valley, Ariz. Immediately prior to opening YFA, he served as chief rangemaster at Jeff Cooper's American Pistol Institute, where he was awarded the title of shooting master.

An adjunct firearms instructor for the Central Training Academy, Department of Energy, Awerbuck has trained civilians, law enforcement agencies and U.S. military personnel throughout the country. A native South African, he also served in the South African Defense Force.

Realistic training is a hallmark of all of Louis Awerbuck's courses, and he has also written a seminal book on practical range training titled *Hit or Myth*.

ROBAR SNIPER RIFLES

Ideally, sniper rifles should provide bench rest accuracy with hunting rifle portability. They usually come closer to the former than the latter. The tactical division of The Robar Companies Inc. designs, builds and markets a line of military and law enforcement sniper rifles with a guaranteed accuracy potential of half-minute of angle with Federal 308M ammunition. *SOF* was provided with two of these rifles for test and evaluation during the HITT course in Prescott, Ariz.

Robert K. Brown used Robar's top-of-the-line SR90 throughout the course. Although heavy (about 14 1/2 pounds complete with scope and bipod), it has the precision of a surgical instrument. Its Remington 700 BDL action is completely accurized, machined, ground and lapped. The bolt is modified with a military-style extractor. To this has been fitted a fluted Schneider stainless steel, match-grade barrel with Robar's exclusive contour.

With a bench-rest-quality chamber and crown, the muzzle diameter is approximately 0.880-inch. Exterior metal finish is black chrome sulfide per MIL-C-13924B. The bolt and all internal components have been finished with Robar's NP3. The modified Remington trigger mechanism had a crisp 2 1/2-pound pull weight.

The stock is built exclusively for Robar under license by McMillan Fiberglas Stocks Inc. This ambidextrous synthetic stock has an adjustable comb, three way adjustable butt assembly, accessory forend rail to which has been added the excellent Parker Hale bipod (a scaled-down version of the Bren LMG bipod), and a more vertically-oriented pistol grip with thumb rests and palm swells. The forend is slightly tapered in thickness from the front to the trigger guard to aid in gross elevation adjustments.

Internally, the action is pillar bedded with a completely free-floated barrel channel. The exterior features Robar's "non-slip" surface texturing. Quick-release sling swivel studs are standard. The stock is available in black, gray or various camouflage patterns.

Our test specimen was equipped with a Leupold & Stevens Ultra 10X-M1 military/police scope which has 20mm solid-tube construction and 1/4-minute click-adjustable windage and elevation, a separate focusing knob and a conventional duplex reticle pattern. Other scopes are available, including Phrobis International, Bausch & Lomb, Zeiss, Swarovski, Kahles, Redfield and Weaver.

The SR90 is priced at $2,475, about half the government's cost for the M24 Sniper Weapon System (SWS)—although price does not include an optical system or the bipod. Several SR90 rifles were used by special ops groups during Desert Storm with considerable success. As Brown commented on several occasions during the HITT course, nothing brings a shooter to a higher level of confidence than knowing his rifle is capable of the highest level of accuracy.

Our backup rifle was a Robar SR60. Just one step below the SR90, this rifle is currently in use with the Los Angeles Police Department Special Weapons And Tactics (SWAT) team and the Special Response Teams (SRT) of numerous other law enforcement agencies nationwide.

It is in most regards similar to the SR90, except for its stock. Also built by McMillan, the ambidextrous Monte Carlo is non-adjustable, but possesses most of the other features described. So equipped, the SR60 costs $1,750 without a scope or bipod. Our test specimen was finished with an urban camouflage pattern and equipped with a Harris bipod.

While I wouldn't want to tote this or any other sniper system of this weight category for any substantial distance in the bush, Robar's sniper rifle series, which also includes lightweight (touted as merely 8 1/4 pounds, sans scope) and plain Jane models, provides outstanding accuracy in an acceptable envelope for most urban law enforcement applications.

Courses at YFA include training with the handgun, submachine gun, combat shotgun, assault rifle and sniper rifle. They also address specific user requirements for these weapons, with customized training designed to meet the user's Mission Essential Need Statement (MENS).

In preparation for our return to El Salvador to provide Level II countersniper training for the Special Response Team of the *Policia Nacional*, I asked Awerbuck to provide High Intensity Tactical Training (HITT) focused on advanced urban countersniper techniques for myself and Chris Mayer, my co-instructor. *SOF*'s Editor/Publisher and an avid high-power shooter, Robert K. Brown decided to tag along. Two others also took part in the three-day course.

FIVE EASY SNIPERS

Four participants were armed with bolt-action rifles. Brown used a Robar SR90 (see sidebar) and I had my personal Steyr SSG with Parker Hale bipod and Swarovski 6-power ZFM scope with rangefinder reticle pattern. Both Brown and I employed Federal 308M match ammunition, which has the superb Sierra 168-grain MatchKing Hollow-Point Boattail (HPBT) bullet.

Mayer was equipped with a customized Remington Model 700 Varminter in an early McMillan stock with Harris bipod and a Leupold & Stevens 3.5-10x Law Enforcement scope. He used reloads with the Sierra 168-grain MatchKing bullet.

Another student had a caliber .30-06 pre-64 Winchester Model 70 fitted with a Leupold & Stevens 2.5-8x variable scope and Harris bipod. He used recently manufactured Yugoslav ball ammunition.

There was also an M14-type semiautomatic-only rifle with a Federal Ordnance receiver and Chinese parts. This rifle was equipped with a Harris bipod, Fabian Brothers muzzle brake, Leupold & Stevens 4-12x variable scope and Weaver rings. The scope mount failed on this latter rig

SOF's Editor/Publisher Robert K. Brown used Robar's top-of-the-line SR90 caliber .308 countersniper rifle throughout the HITT course. Completely accurized Remington 700 BDL action has been mated with a McMillan stock and fluted Schneider barrel.

during our second day and this shooter switched to an H&K 91.

Our first morning was spent zeroing the rifles at 25 yards and then firing at the torso from the prone, kneeling and sitting positions at 25 yards—two sets of three-shot groups from each position without a time limit. This was followed by 25-yard head shots from the prone position, then torso shots at 50 yards from the prone and finally head shots at 50 yards from the prone.

This culminated in a "10-minute" drill which consisted of a single head shot fired from the prone, no later than three seconds after the instructor's command (which could be given at any moment within the 10-minute time frame). We waited approximately six minutes for the command to fire. This is an excellent and realistic exercise in mental discipline.

After this we fired from the prone at a reactive steel plate target at 75 yards. The first day ended with a "leap-frog" drill. Working in two-man teams, participants shot from four stations with one man shooting while the other moved to the next station. All firing was carefully aimed at the steel plate. Wild, suppressive-type fire was not permitted. Precise, surgically placed shots were emphasized during this exercise and throughout the course.

The shooters were forced to use a code to indicate they were out of ammunition and needed to reload, moving only while their partner was shooting. In most instances, military or law-enforcement snipers operate as either two- or three-man teams. Their efforts must be coordinated and interdependent. Movement must be choreographed as precisely as a ballet and as carefully as the shot is placed on the target.

1,000-YARD KILLS EVERYDAY

Shooting at only 25, 50 and 75 yards in a sniper course? U.S. Marine Corps snipers routinely take out targets at 1,000 yards or more every day before they eat their Wheaties. But do they? I have personally never witnessed a sniper-kill even out in the bush at a distance greater than 300 yards, though I'm sure shots at far greater distances have been taken and, on occasion, made. But not in the city, and don't forget this was "urban" HITT countersniper training, not a long-range, open-field stalking course.

The average law-enforcement countersniper's engagement distance is well under 100 yards in city environments. His targets will be moving about, sometimes almost spastically, and he must connect with the utmost precision as hostages may often be covering a large portion of the target.

YFA's paper target is of the so-called "option" type, that is to say with a squared-off head and torso. Its camouflage pattern inhibits shooters from looking over the gun for their hits, and in so doing throwing the weapon downward after the trigger has been pulled, causing low impact. The torso kill zone measures approximately 12 inches long by 8 inches wide.

Challenging 3-dimensional moving targets used in HITT countersniper course stress realism and the erratic movements encountered on the street. Note unacceptable peripheral hits on left side of rear target's face.

An outline representing the eye-socket kill zone is about 2 inches by 4 1/2 inches. The target's rear side has a zero target with 1-inch squares, a 100-yard small-bore bull's-eye target and three black option-type targets scaled to represent 50, 100 and 200 yards at 25 yards. This excellent and versatile target is used for most of *SOF*'s tests and evaluations of combat firearms.

On our second day everything was quickly moved up several notches. After an initial exer-

The Back-Packing Sniper

Drag bags are not necessarily purses for transvestites. In fact, the "Drag Bag" as manufactured by Eagle Industries Unlimited Inc. was designed and built to meet needs of the military and law enforcement countersniper. It is a completely self-contained system for carrying a scoped countersniper rifle, bipod, spare ammunition, binoculars and/or spotting scope and other equipment up to the point of anticipated contact, and then moving quietly into position.

Constructed of heavy-duty 11-ounce abrasion-resistant Cordura nylon, milspec webbing and hardware are used throughout. Available colors include woodland camouflage pattern, olive drab, black, and chocolate brown. The case is fully padded with 3/8-inch high-density, shock resistant closed-cell foam. For easy access and use as a ground pad, the case opens fully with an extra heavy-duty, two-way foam-backed zipper. A security flap over the zipper snaps shut for additional protection.

Every feature on the Drag Bag is designed to meet requirements of professional users. The interior has two one-inch-wide nylon straps with Fastex quick-release buckles for securing the rifle. The interior accessory pouch holds two boxes of 7.62x51mm NATO ammunition, with room for a sectional cleaning rod and accessories.

The top of the bag has 11 loops for attaching camouflage. A nose cover protects it and eases movement through heavy brush. Drag loops on the front and top of the bag allow it to be pushed or pulled through the bush. Both cargo pouches open with heavy-duty, two-way zippers and have two full-length compression straps with Fastex quick-release buckles for securing gear.

The rear cargo pouch will accommodate a pair of binoculars and has a flat pocket on its flap. The front cargo pouch will hold a spotting scope and tripod. Three pieces of one-inch-wide abrasion-resistant Cordura webbing on the bottom of the bag serve to protect it when being slid into position.

Padded shoulder straps with quick-release Fastex buckles permit back-packing and can be switched for muzzle up-or-down carry. The shoulder straps stow quickly behind the bottom abrasion webbing to ease hand carrying or dragging. A Cordura nylon outer bag is available for low-profile transportation.

SOF's staff used the Eagle Industries Drag Bag throughout the HITT course in Prescott, Ariz. Except for some minor scuffing caused by dragging it through sagebrush, mesquite and creosote during one of the tactical exercises, it survived relatively unscathed and served admirably. Highly specialized, this is a superb piece of equipment for counter-sniper teams. Suggested retail price is $253.

Eagle Industries' Drag Bag is constructed of heavy-duty 11-ounce abrasion-resistant nylon and features padded shoulder straps for back-packing countersniper's rifle and equipment.

cise firing "cold" shots at the torso and then the head, Mayer and I were sent on a daytime tactical exercise with no time limit. Although we both carried our rifles, Mayer was selected as the shooter and I as the team observer.

I kept my Steyr SSG in an Eagle Industries Drag Bag (see sidebar) and used a pair of Yugoslav military 7x40 binoculars as we deployed under maximum cover and concealment across the scorching hot desert surface.

Crawling several hundred yards on your stomach across the thorn of every cacti and mesquite bush in the Upper Sonoran Life Zone at mid-day in August will demonstrate to anyone the unforgiving and brutal nature of arid region environments. We located our target in an arroyo, hidden behind dense sagebrush. Shooting from less than 20 feet, Mayer's bullet was deflected by a branch and I was forced to launch the coup de grace.

SNIPERS IN THE SONORAN DESERT

As cut and bruised as we were, this proved to be an easier go than what followed, when we were introduced to moving targets. Bobbing and weaving side-to-side and back and forth, we first practiced frontal torso and then frontal head shots on the YFA camouflage target. This was followed by a moving humanoid paper target which presented a 3/4 view. We were given two shots only at this target from 50 yards and could select either the torso or head. A torso shot had to be placed well back toward the spine to avoid shooting through nothing more than its coat.

We then moved up to two separate paper targets, with one serving as a hostage while a rear target moved back and forth. Finally we were presented with a clothed 3-dimensional head and torso. We each fired one shot at the body from an acute angle and then moved to a frontal position for a head shot.

When the target is wearing a cap and rolling his head back and forth, and you must concentrate with every fiber of your body to hit it in the brain cavity at 50 or 75 yards, then fantasies of routine "1,000-meter kills" begin to take on a different perspective. This scenario was engaged by two-man teams with only one shot permitted.

After sundown, we fired frontal-torso and then head shots, each five-shot groups, on the camouflage target from 50 yards while the instructor waved a flashlight back and forth across the line of targets. The evening culminated with a night tactical problem for two students who had not gone on the day exercise.

Our third and final day of the course began with a surprise. Awerbuck informed us that few plan ahead for unexpected emergencies. To demonstrate this, we were instructed on his command to retrieve our rifles from the vehicles. This exercise ended when the first shooter was able to set up and hit the reactive steel plate at 200 yards. In the real world, contact with the enemy never occurs on your timetable.

THEY SHOOT GONGS, DON'T THEY?

Using the YFA camouflage target, we then zeroed our rifles for 200 yards with five torso shots followed by five head shots. Every student then shot at the 200-yard steel plate. During the two-man, 200-yard steel plate exercise, the first man fired, then moved to the next station while his partner was shooting.

There were a total of five stations; the second exercise demanded each man in a team take two shots at the gong before moving to the next station. From a stepped barricade with a window and low opening, everyone fired at the 200-yard steel plate from low prone, squatting and other tactically appropriate positions.

The final exercise consisted of five shots fired at the 200-yard steel plate, starting from the prone at 200 yards and then running to four closer positions, several of which required kneeling to visually acquire the target.

While fun to shoot at, Awerbuck stressed that it's a bad practice to shoot exclusively at reactive steel gongs as the shooter will eventually invariably fail to follow through and look only for target movement as he pushes the weapon down and away.

Once training received in a course of this type has been absorbed, it becomes the responsibility of the individual or team to repeat it over and over again, with as many different variations as may be contrived, until proficiency is maximized.

When your very life, or others' lives, may rely upon consequences of your response to a deadly confrontation, it should be self-evident why you can never train too much—and why the world's elite special operations groups spend a majority of their time in the live-fire fun house.

Originally appeared in *Soldier Of Fortune*
January 1992.

WHO'S AFRAID OF THE DARK?

Shooting in the Shadows at Thunder Ranch

Working up the stairwells in the Tower at Thunder Ranch will test your tactical skills to a level just below an actual confrontation with an armed opponent.
(Photo: Chris Mayer)

Most gunfights take place in subdued light or in the dark. Most defensive firearms training takes place in the daylight. Thunder Ranch, recognizing this dangerous disparity, each year schedules a Low Light Level 2 handgun course which I and *SOF* staff photographer, Chris Mayer, recently completed. This was my third class at Thunder Ranch and, as usual, it was a high-speed, low-drag experience from start to finish.

Thunder Ranch's highly respected honcho, Clint Smith, is a Marine Corps veteran with two infantry tours in Vietnam. His experience includes seven years in law enforcement, during which time he served as head of his department's FTU (Firearms Training Unit), as well as the senior countersniper on the SWAT unit. He was Operations Officer for Jeff Cooper's API, started and served as director of Heckler & Koch's training services division and with his wife, Deborah, founded International Training Consultants, Inc., a highly regarded mobile training program.

Thunder Ranch has a guest staff of 28 instructors. The three who taught our Low Light course were representative of the impressive quality of Thunder Ranch's adjunct personnel. Bill McLennan retired from the San Antonio Police Department after more than 30 years of service, the last six and a half of which were spent as OIC of firearms training.

McLennan's unflappable demeanor, dry wit and encyclopedic knowledge of gunfighting tactics were tremendous assets to the course. Bill Black, a 16-year veteran of the Littleton, Colo., Police Department is a patrol supervisor and tactical team leader for his agency. His background complements that of McLennan and together they provided us with enough tactical information to fill several field manuals. They were more than ably assisted by Lori Hauserman, a police officer with 15 years' service with the Glendale, Ariz., Police Department. Lori is an armorer and firearms instructor responsible for training 275 law-enforcement personnel.

Thunder Ranch classes never contain more than 20 students, as that is the number of firing positions on each range. As a consequence, students get more than their money's worth, since there is none of the usual down time associated with firearms classes that must be run in relays. You can expect to fire about 2,000 rounds in one of Thunder Ranch's handgun courses.

ULTIMATE FIGHTING HANDGUN

The students in this class represented the usual cross-section of individuals that attend courses to hone their gunfighting skills. Eight of the 20 were law-enforcement personnel, two from as far away as Australia. The rest were men and women who realize that purchasing a firearm is only the smallest part of the self-defense equation.

Their mix of weapons provided revealing information. There was one SIGARMS P226, six Glocks and 13 M1911-type pistols. Three of the Glocks were either G17 or G19 9mm Parabellums and two were chambered for the .40 S&W round. Most unusual was a police officer from the Clay County, Fla., Sheriff's Department who was armed with a Glock Model 20 caliber 10mm Auto pistol. His department is one of only a handful who issue Glocks in that caliber. However, there can be no doubt that the pistol Jeff Cooper popularized as the ultimate fighting handgun still reigns supreme. For all of its crankiness and the often extensive custom gunsmithing many variants require, the caliber .45 ACP M1911 is still perceived by most armed professionals in this country to be the handgun most likely to win the fight when drawn from leather by an experienced pistolero.

No small part of that perception is based upon confidence in the .45 ACP cartridge. Penetration is, without doubt, the single most important parameter in evaluating the wound ballistics performance of handgun ammunition. The bullet must penetrate deeply enough to crush, cut and break through the human body's vital structures and organs. However, once we've obtained the required penetration, *the bullet that makes the biggest hole will do the most damage.* It is largely because of this that the .45 ACP round remains the cartridge of choice among a substantial majority of armed professionals after more than eight decades.

I went through the course with a Kimber Custom Classic .45 ACP—without doubt the best production series M1911-type pistol ever assembled. I used a holster and magazine pouches made by the Hanson Leather Co. Ernie Hanson has 18 years' service with the Lincoln City, Oregon, Police Department and knows what is required of gear designed for slapping leather on the mean streets. He is also a guest instructor at Thunder Ranch.

Ernie made his first holster in 1978. For a period of time in the 1980s the famous George Lawrence Company of Portland, Ore., and then of Lillicut, N.C., made his holsters. As he packs a piece for a living, a great deal of personal experience goes into the Hanson designs. All of Hanson's holsters are individually made to order by hand out of Herman oak leather, which is of exceptional quality. Stitching is done with harness needles using the two-needle back-and-forth method. After stitching, a handgun of the proper model is placed in the holster and it is once again wet-molded by hand for a perfect fit. Ernie applies Fiebings oil dyes and lacquer finishes to his completed holsters and pouches.

The holster I used was his H-1011, an off-duty vertical holster that can be used crossdraw or strongside and costs $61. This holster features an internal steel tension lining at the bottom that maximizes the holster's retention qualities. Oil, black or brown finishes are available. Hanson also makes single and double vertical magazine pouches, a horizontal magazine pouch, off-duty cuff cases and badge holders, an ankle rig with a magazine pouch, inside-the-waistband holster, cowboy holsters and belts. This is superb leatherwork, handcrafted in the most painstaking and fastidious manner. While as yet relatively unknown, Ernie Hanson must rank among the very top of the true custom holster makers.

PLAYING IN THE DARK

The morning of the first day of this five-day class was spent in the classroom being exposed to the important concepts we were to apply during the next four and a half days on the square ranges and in the Terminator and Tower. That afternoon found us on the Red Range, starting first with dry-fire practice and then with live-fire on the only stationary targets we would see for the rest of the week on that range.

At Thunder Ranch shooting a "Plan A" means firing at the torso, or the body's energy source, the number of shots determined by the threat. Firing a "Plan B" indicates a shot to the head—the body's computer, while a "Plan C" signifies the first two zones plus shooting at the pelvis, which should affect your opponent's mobility. Clint Smith says, "Shoot what is available as long as it is available and until something else is available." At Thunder Ranch you shoot until you win or the threat stops. When moving targets stop, you cease fire, as they have either gone down or you have achieved compliance.

Obtaining verbal compliance is an important aspect of training at Thunder Ranch and is stressed continually. This is, without doubt, a consequence of the law-enforcement background of most of the staff. As my background is entirely from the military, where this concept is a non sequitur, I have a great deal of trouble verbally communicating with targets. Bill Black informed us on several occasions that expressions such as, "Oh God, don't make me kill again!" were not acceptable examples of verbal compliance.

The important difference between *cover,* which stops or deflects direct and indirect enemy fire, and *concealment,* which only prevents you from being observed is also heavily emphasized at Thunder Ranch. This is because Clint Smith strongly advocates fighting your way out of the so-called "hole"—a distance of about two arm lengths or approximately 6 feet—back to cover. Throughout the week we fired most often at the Red Range while moving back to barricade cover. At Thunder Ranch, the barricades on the ranges are used to simulate cover and not as a brace for firing a handgun.

The third element in Thunder Ranch's fighting triad is an assertion of the importance of

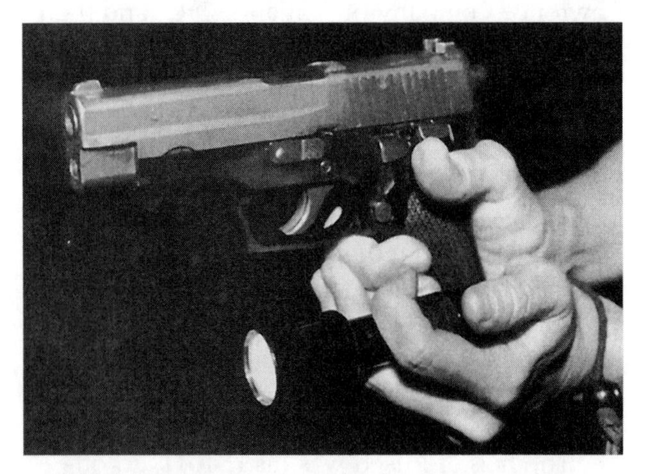

The SURE-FIRE Model 6Z COMBATLIGHT permits the unit to be used like a "syringe" that puts the support hand back in more or less the standard Weaver position.
(Photo: Chris Mayer)

The so-called "Chapman Technique" was developed for flashlights with a button switch near the head.
(Photo: Chris Mayer)

team tactics. For example, magazine changes are preferably executed in conjunction with your partner, whose responsibility becomes covering your target as well as his. Most often, the shooter's commands will be something like, "cover" and then "clear" after he has changed magazines and is back in the fight, but any verbal sequence would work, providing it was agreed upon beforehand.

Three types of magazine changes are taught at Thunder Ranch. The first, and *most* important, is the so-called "empty magazine change" which is unique to this facility, but in my opinion presents the most realistic scenario. Traditionally, those in my loop are supposedly programmed to keep track of our shots and perform a tactical reload before the slide locks rearward after the last shot in the magazine sails downrange. It never happens this way in the real world. I have never been in a fire fight where anyone knew how many rounds he had fired. When the M16's bolt locks to the rear or the AK's trigger is pressed and nothing happens, you reload. There are too many insane things happening in a gunfight to count the bullets going downrange. So when the slide locks rearward, you dump the empty magazine, slam in the new one and press down on the slide lock lever with the thumb of the firing hand.

Still, if possible, it's better to load when you can—during a lull in the action—and not when you have to. This so-called "tactical load" is performed by first reaching with the weak hand for the new magazine. Move it up to the magazine-well between the index and middle fingers. Use the thumb and index finger to remove the hold magazine on the down stroke and insert the new magazine on the up stroke. The depleted magazine goes in your pocket, *not* to the ground.

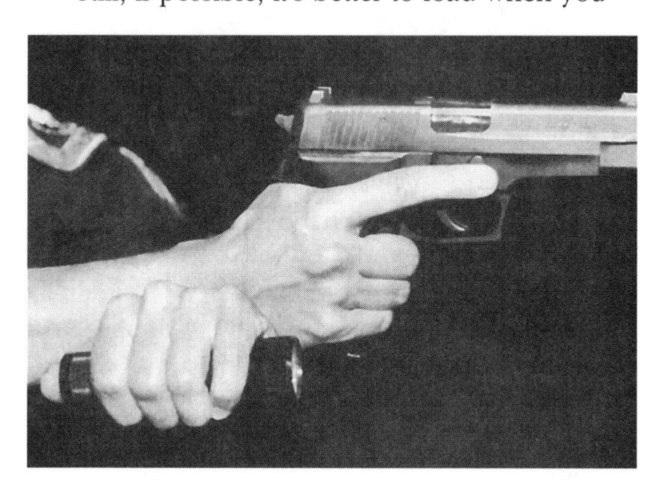

While it is the most commonly employed, the so-called "Harries Flashlight Technique" requires the support hand to control the light without providing the customary isometric pull associated with the standard Weaver position.
(Photo: Chris Mayer)

The "expedited" or "speed" load is based upon a four count. First, the opposite hand goes to new magazine just before releasing the magazine in the pistol's magazine-well. Second, the flat back of the new magazine goes to the flat back of the magazine-well. Third, seat the new magazine half way into the well. And finally, seat the magazine completely with one sharp stroke, reinforce the firing grip, then "front sight and press," if required.

Most commonly, when practicing these drills, beginning students will release the magazine in the pistol *before* they have the new magazine firmly in hand. This is a potentially fatal error.

On Tuesday, Wednesday and Thursday of the Low Light course, the class commences at 1500 and ends at 2300. Tuesday afternoon we began work on the Red Range with malfunction drills.

How important are they? What are the chances you will have a stoppage during a gunfight? I don't know. Statistics on this topic have never been tabulated, except in an anecdotal fashion. I have experienced several "bolt-over-base" stoppages in M16 rifles at inappropriate times in El Salvador. Modern handguns, even Glocks and highly tuned 1911-types, can go down at any time and that's why so many who walk in harm's way carry backups.

MURPHY'S MALFUNCTIONS

The most common stoppage and easiest to clear is a failure to fire. This can be a result of either an empty chamber due to an improperly seated magazine or a defective or hard-cupped primer. One prominent gun writer has advised his faithful myrmidons to pull the trigger again if they have a double-action pistol. This is typical of the bad advice that so often fills the pages of the popular gun press. If the primer is defective, pulling the trigger again will only provide your opponent with more time to send missiles downrange at you.

The correct response is the so-called "tap, rack, bang" sequence. This simply involves slapping the bottom of the magazine to certify that it is properly seated, racking the slide and pulling the trigger. All of this should be accomplished with your eyes and the pistol on the target.

A "stovepipe" stoppage can be caused by problems with ammunition, extractors, ejectors or recoil springs, or even "limpwristing." Clearing this kind of malfunction is almost identical to the procedure used to overcome a failure to fire. After the magazine has been tapped to assure its proper seating, a horseshoe-shaped swipe to the rear is used to wipe the empty case away while racking the slide. Then, fire if required.

The most severe stoppage you can encounter with a semiautomatic pistol is one variation or another of the so-called "double feed." In my experience, the most common stoppage of this type is not actually a double feed, but a round which stubs on the feed ramp, driving the bullet back into the case with

THUNDER RANCH: A WORLD-CLASS FACILITY

The Thunder Ranch Training Facility, which opened in 1993, is located 96 miles west of San Antonio, and 46 miles to the east of Kerrville in the rolling-hills karst topography of southern Texas on 2,400 acres. This region, once known principally for Texas Longhorn cattle, is now a center for exotic game ranches.

Thunder Ranch's physical plant is world-class throughout. There are four ranges. Both the Red Range and Orange Range are 50x50-yard square ranges. Red Range is the base range for all handgun-related classes. This 20-position range has turning, stationary, wobbler and interface target potential with "cover" structures, when required. While the Orange Range also has stationary and wobbler target potential, its most prominent facets include a unique "charger" system that represents threats closing quickly on the student. In addition, it also has a two-or four-target setup for high-speed lateral running targets with an 80-foot range of operation. The Yellow Range measures 50x100 yards and has paper, hostage, stationary, wobbler and movers. This 20-position range is the base range for preliminary work in shotgun and some pistol courses. Also holding 20 positions, the 50x100-yard White Range has stationary, wobbler and high-speed turning targets available. Its numerous barricades represent protective cover and/or concealment. The two-section Black Range has a 20-position conventional rifle range in 100-yard increments out to 300 yards. The other portion is a field range for target engagements from 400 to 1,200 yards for both known and unknown distance applications.

The most visually impressive facilities at Thunder Ranch are the Tower, Thunderville and the infamous Terminator. The Tower is a monolithic four-story structure with its own tactical simulator that permits live fire in all room and stairwells. Problems of both internal and external entry can be addressed. The stairwells vary in dimensions and geometry. All of the window ports can be plugged. Students in the Countersniper classes fire from the Tower. Thunderville is an elaborate, day or night, in-depth facade over 200-feet in length representing a two-sided street setting. It's used in all of the rifle programs that present dynamic scenarios. With over 60 target options, as well as movers, hostages, runners and wobblers, Thunderville features computer-controlled lighting, doors, movement, timing and hit documentation. This simulator has been designed to test decision-making and target identification under maximum stress. The Terminator, used in handgun, shotgun and rifle courses, can simulate home, motel or office environments with over 400 floor plan options and targets that emphasize identification and selection prior to engagement. This building is covered by a free-span roof for training under all lighting or weather conditions.

Thunder Ranch's unique "charger" system represents threats closing quickly on the student and can be adjusted to run toward students at varying speeds. *(Photo: Chris Mayer)*

Day or night, at Thunder Ranch you always fight your way back to cover. *(Photo: Chris Mayer)*

a consequent failure to chamber the round. If at all possible, I would transition to a backup handgun if this occurred, as this malfunction takes the longest time to clear. After first observing what has gone wrong, you must lock the slide rearward, remove the magazine, rack the slide at least, insert the magazine and then tap, rack, bang. Even a J-frame in an ankle holster is faster to access than performing a double feed drill.

Firing from the "retention" position is also covered in Level 2 handgun courses at Thunder Ranch. Used only at contact ranges, the handgun is held in line with the pectoral, with the wrist locked and the firing hand held tightly against the body. Special emphasis was placed on holding the support arm upright and tucked tightly into the body to plug the gap in your body armor under the arm pit and to present the "hard" or bony part of the support arm to your opponent's blows or an edged weapon. The fingers of the support arm should be kept open and spread apart so your view is not obstructed. As always, after firing from this position you must get out of hole quickly while continuing to throw lead into your opponent.

By nightfall, we were ready to begin working with our flashlights. Every student in the class was equipped with a SURE-FIRE unit manufactured by Laser Products. Since their introduction in 1987, they have come to totally dominate the field of combat flashlights.

LIGHT UP THE NIGHT

The switch used on the hand-held SURE-FIRE 6P series is quite simple. Push on the neoprene diaphragm in the middle of the tail cap, and the light comes on. Stop pushing and the light goes off. Rotate the tail cap until it's tight, and the light stays on. Unscrew it a bit and the light goes off. These flashlights are most often employed with the so-called "Harries Flashlight Technique." This technique requires the support hand to control the light without providing the customary isometric pull associated with the standard Weaver position. A variation of the Harries method is the "Chapman" technique, which was developed for flashlights with a button switch near the head. Using this technique, the light is held on the opposite side of the handgun from that used with the Harries method. Another technique that was demonstrated is the "FBI" method which involves holding the light above your head and forward of the body. It can be used for looking over fences or other obstacles or glimpsing into vehicles.

However most of us had the now prevalent SURE-FIRE Model 6Z COMBAT-LIGHT. It has a smaller diameter body just in back of the head and two neoprene rings around the rear of this portion of the body that in conjunction with the neoprene diaphragm in the tail cap permit the unit to be used like a "syringe" that puts the support hand back in more or less the standard Weaver position.

Each technique was practiced by shooting at wobbler targets. At first we turned on the light, fired, turned the light off and then moved: right, left or out of the hole. But, we also fired sequences with the light kept on to illuminate the target during movement. We also fired without use of the flashlight to demonstrate that muzzle flash will not totally blind you. However, Bill Black informed us at this time that firing a pistol with a muzzle compensator while dressed in a polyester leisure suit would probably turn you into a 6-foot road flare.

Several of the students had pistols fitted with self-luminous tritium night sights. They can be useful, but they have their limitations. Self-luminous tritium sights do not assist target discrimination in total darkness, Furthermore, when a flashlight is used, light reflecting off the target downrange is sufficient to back light the weapon's sights and permit proper sight alignment. However, when goblins are shooting at you in almost total darkness their muzzle flash identifies them and self-luminous night sights can be used in lieu of a flashlight with great effect.

I prefer the tritium night sights manufactured by Innovative Weaponry, Inc. IWI's optional multicolor night-sight system focuses attention on the front sight and speeds target acquisition by a considerable margin. Most tritium self-luminous sights glow green, simply because this color is the highest on the night visibility spectrum. However, operator alignment of three green dots in subdued light can be confusing. This detracts from emphasis on the front sight and retards target acquisition.

The largest self-luminous dot is installed in the front sight and it is green. The two smaller dots in the rear sight are either yellow or amber. In use, I have found that the eyes instinctively focus on the green front dot with a strong 3-

Used only at contact distances, Bill Black demonstrates the "retention" position while Lori Hauserman describes its tactical application. *(Photo: Chris Mayer)*

Students practice firing over their heads after they have been presumably knocked down during a gunfight. They must continue to fire as they get up and fight their way back to cover. *(Photo: Chris Mayer)*

dimensional triangulation effect that helps to funnel the shooter's vision directly onto the target. I can personally recommend IWI's multicolor system highly and without reservations of any kind.

By the third day, shirts had been placed on the wobbler targets to more accurately simulate reality. And, we began work in the Terminator. The application of the tactical techniques required to survive a gun fight are an essential ingredient of all training at Thunder Ranch. Moving through the unending labyrinth of rooms in the Terminator or working up the stairwells in the Tower will test your tactical skills to a level just below an actual confrontation with an armed opponent. Each step you take brings you into another potential threat area. Manmade funnels, such as doors, halls, corners and windows constrict movement and represent maximum hazard areas. They are bullet magnets and to survive you must stay on the outside of corners and visually "slice the pie" as you move forward. Shoot an innocent bystander in one of the simulators and the instructors will gleefully inform you that "you're going to share a cell with Bubba in the big house." There is a large sign in front of the Terminator which states, "Remember, you have the rest of your life to solve the problem. How long your life lasts depends on how well you do it."

During our night runs through the Terminator we learned not to set patterns of search with either movement or the flashlight. You should move away from the weapon's flash signature after firing. Crouching lower can help to back light the threat. In addition, you must stay away from windows or anything that would silhouette you.

Interspersed with our runs through the Terminator and Tower were a seemingly endless variety of drills on the Red and Orange Ranges. Most important to me were a series of stoppage drills, set up by each student's partner with both your pistol and his, that you had to clear in complete darkness from behind the barricade. Each of us was forced to clear about 80 malfunctions and then fire from cover.

During the daylight hours, we practiced both the drawstroke, shooting and reloading with one hand, first the normal firing hand and then the weak hand. Laying on our backs, first facing the target and then away from it, with the pistol on the ground, we fired laying on our backs, then sitting up and finally standing and firing while moving back to cover. Sound preposterous? Just picture yourself laying on the ground in a parking lot after being knocked down. No handgun course at Thunder Ranch would be complete without a run with the "chargers" on the Orange Range. This is especially challenging shooting, as the target advances almost faster that you can fire and get out of the hole.

All you need is money to acquire a pistol. Far more is required to insure that your opponent, and not you, goes down like "an asphalt snow angel" should you be forced into a deadly confrontation. You will need more than just a little true grit to attend a course at Thunder Ranch. Your physical endurance and emotional fortitude will be tested to their absolute limits. Personally, I wouldn't have it any other way. Only those who are really serious about survival end up on the firing line at Thunder Ranch. Their courses are booked far in advance with long waiting lists. Those looking beyond dry firing in their hallway mirror should contact them *now*—many of the classes in Thunder Ranch's 1998 schedule are already filled.

Originally appeared in *Soldier Of Fortune*
January 1998.

SOURCES

A

Aerotek, CSIR, P.O. Box 395, Pretoria 0001, South Africa;
e-mail: tgdent @csir.co.za

Alessi Custom Concealment Holsters, 2465 Niagara Falls Blvd., Tonawanda, NY 14150

Armor Metal Products, 2500 Phoenix Ave., Helena, MT 59604

B

Beretta U.S.A. Corp., 17601 Beretta Drive, Accokeek, MD 20607

W.E. Birdsong & Associates, P.O. Box 9549, Jackson, MS 39286

Black Hills Ammunition, P.O. Box 3090, Rapid City, SD 57709-3090

Brownells, 200 South Front Street, Montezuma, IA 50171

C

Colt Industries-Firearms Division, P.O. Box 1868, Hartford, CT 06101

Choate Machine and Tool Co., Inc., P.O. Box 218, Bald Knob, AR 72010

E

Eagle Industries Unlimited Inc., 400 Biltmore Drive, Suite 530, Fenton, MO 63026

F

Federal Cartridge Company – Blount, Inc., 900 Ehlen Drive, Anoka, MN 55303

G

Galco, 2019 W. Quail Ave., Phoenix, AZ 85027

Glock, Inc., 6000 Highlands Parkway, Smyrna, GA 30082

GSI, Inc., P.O. Box 129, Trussville, AL 35173-0129

H

Hanson Leather Co., 2724 N.W. Keel Ave., Lincoln City, OR 97367

Heckler & Koch Inc., 21480 Pacific Blvd., Sterling, VA 20166-8903

Herrett's Stocks Inc., P.O. Box 741, Twin Falls, Idaho 83303-0741

I

Innovative Weaponry Inc., 2513 E. Loop 820 North, Fort Worth, TX 76118

Iron Brigade Armory Ltd., 100 Radcliffe Circle, Jacksonville, NC 28546

IZHMASH, 3 Derjabin Street, 426006 Izhevsk, Russia

J

J.F.S., Inc, 515 Gordon, P.O. Box 1892, Klamath Falls, OR 97601

K

Knight's Armament Company, 7750 9th Street S.W., Vero Beach, FL 32968

L

Laser Products, 18300 Mount Baldy Circle, Fountain Valley, CA 92708-6122

Leupold & Stevens, Inc., P.O. Box 688, Beaverton, OR 97075

Lyttelton Engineering Works Limited, a division of Denel (Pty) Ltd., 368 Selborne Avenue, Lyttelton, P.O. Box 5445, Pretoria 0001, South Africa

M

McMillan Fiberglass Stocks, 21421 N. 14th Avenue, Suite B, Phoenix, AZ 85027

Michaels of Oregon Co., P.O. Box 1690, 1710 Red Soils Court, Oregon City, OR 97045

Magnum Research, Inc., 7110 University Avenue Northeast, Minneapolis, MN 55432

N

Novak's .45 Shop, P.O. Box 4045, 1206 1/2 30th St., Parkersburg, WV 26101

O

Ohio Ordnance Works Inc., 310 Park Drive, P.O. Box 687, Chardon, OH

R

The Robar Companies Inc., 21438 N. 7th Ave., Suite B, Phoenix, AZ 85027

S

Safariland Ltd., Inc., 3120 E. Mission Blvd., Ontario, CA 91761

Scattergun Technologies Inc., c/o Wilson's Gun Shop, Inc., 2234CR719, Berryville, AR 72616-4573

Milt Sparks Holsters Inc., 605 E. 44th, No. 2, Boise, ID 83714

Craig Spegel, P.O. Box 108, Bay City, OR 97107

Steiner Binoculars, 97 Foster Road, Suite 5, Moorestown, NJ 08057

Steyr-Mannlicher AG, Postfach 1000, Steyr, Austria

Sturm, Ruger & Company Inc., 200 Ruger Road, Prescott, AZ 86301

T

Taurus International Firearms, 16175 NW 49th Avenue, Miami, FL 33014-6314;
Web site: www.taurususa.com

Thunder Ranch Training Facility, HCR 1, Box 53, Mountain Home, TX 78058

Turner Saddlery, P.O. Box 120, Clay, AL 35048-0120

V

Vector Arms, Inc., 270 West 500 North, North Salt Lake UT 84054;
Web site: www.vectorarms.com

W

Weigand Inc., P.O. Box 239, Mountain Top, PA 18707

Y

Yavapai Firearms Academy Ltd., P.O. Box 27290, Prescott Valley, AZ 86312

ABOUT THE AUTHOR

Peter George Kokalis was born and raised in Chicago, and obtained his Bachelor of Arts degree from Northwestern University, Evanston, Illinois. After graduation he enlisted in the United States Army. He moved to Phoenix, Arizona, with his family in 1962, where he still resides. He obtained his Master of Science degree from Arizona State University.

A passionate student of military history and small arms, his technical reference library contains over 5,000 volumes. Attracted to firearms of all types, he has hunted extensively in Africa, is an active participant in Cowboy Action Shooting (with an alias of "Poison Pete"), and spends a great deal of time shooting weapons of all kinds from handguns to machine guns. To that end, he converted a 1957 2 1/2-ton military truck into a motor home, which he uses at "machine gun shoots" periodically held throughout the year in a remote area near Bagdad, Arizona. His impressive firearms collection places heavy emphasis on military small arms, and in particular, automatic weapons, on which subject he has written and instructed extensively. By nature eclectic, he also collects bayonets, fighting knives, helmets, inert ordnance and all of the other many accouterments of war.